Information Geometry

Information Geometry

Special Issue Editor

Geert Verdoolaege

MDPI • Basel • Beijing • Wuhan • Barcelona • Belgrade

Special Issue Editor
Geert Verdoolaege
Ghent University
Belgium

Editorial Office
MDPI
St. Alban-Anlage 66
4052 Basel, Switzerland

This is a reprint of articles from the Special Issue published online in the open access journal *Entropy* (ISSN 1099-4300) in 2014 (available at: https://www.mdpi.com/journal/entropy/special_issues/information-geometry)

For citation purposes, cite each article independently as indicated on the article page online and as indicated below:

LastName, A.A.; LastName, B.B.; LastName, C.C. Article Title. *Journal Name* **Year**, *Article Number*, Page Range.

ISBN 978-3-03897-632-5 (Pbk)
ISBN 978-3-03897-633-2 (PDF)

Cover image courtesy of Geert Verdoolaege.

Contents

About the Special Issue Editor

Geert Verdoolaege obtained an M.Sc. degree in Theoretical Physics in 1999 and the Ph.D. in Engineering Physics in 2006, both at Ghent University (UGent, Belgium). His Ph.D. work concerned applications of Bayesian probability theory to plasma spectroscopy in fusion devices. He was a postdoctoral researcher in the field of computer vision at the University of Antwerp (2007–2008), working on probabilistic modeling of image textures using information geometry. From 2008 to 2010, he was with the Department of Data Analysis at UGent, where he worked on modeling and estimation of brain activity, based on functional magnetic resonance imaging. In 2010, he returned to the Department of Applied Physics at UGent, first as a postdoctoral assistant and from 2014 onwards, as a part-time assistant professor. Since 2013, he has held a cross-appointment as a researcher at the Laboratory for Plasma Physics of the Royal Military Academy (LPP-ERM/KMS) in Brussels. His research activities comprise development of data analysis techniques using methods from probability theory, machine learning and information geometry, and their application to nuclear fusion experiments. He also teaches a Master course on Continuum Mechanics at Ghent University. He serves on the editorial board of the multidisciplinary journal Entropy and is a member of the scientific committees of several conferences (IAEA Technical Meeting on Fusion Data Processing, Validation and Analysis; International Workshop on Bayesian Inference and Maximum Entropy Methods in Science and Engineering; Conference on Geometric Science of Information). In addition, he is a consulting expert in the International Tokamak Physics Activity (ITPA) Transport and Confinement Topical Group and member of the General Assembly of the European Fusion Education Network (FuseNet).

Preface to "Information Geometry"

The mathematical field of information geometry originated from the observation the Fisher information can be used to define a Riemannian metric on manifolds of probability distributions. This led to a geometrical description of probability theory and statistics, which over the years has developed into a rich mathematical field with a broad range of applications in the data sciences. Moreover, similar to the concept of entropy, there are various connections to and applications of information geometry in statistical mechanics, quantum mechanics, and neuroscience.

It has been a pleasure to act as a guest editor for this first Special Issue on information geometry in the journal Entropy. For me, as a physicist working on the development and application of data science techniques in the context of nuclear fusion experiments, the interdisciplinary character of information geometry has always been one of the main reasons for its appeal. There are, of course, many other domains in physics where geometrical notions play a key role, including classical mechanics, continuum mechanics (which I have been teaching at Ghent University for several years now), general relativity, and much of modern physics. This interplay between the beautiful and elegant formalism of differential geometry on the one hand and physics and data science on the other hand is both fascinating and inspiring. The variety of topics covered by this Special Issue is a reflection of this cross-fertilization between disciplines.

"Information Geometry I" has been a great success, and although the papers were published already several years ago, it was decided that it was worthwhile to reprint the collection of papers in book form. Indeed, even though all papers present original research, many have a strong tutorial character, and we were honored to receive multiple contributions by authorities in the field. The papers have been structured according to their main subject area, or field of application, and we briefly discuss each of them in the following.

We start with two papers related to the foundations of information geometry. We were very pleased to receive a contribution by one of the founders of the field of information geometry, prof. Shun-ichi Amari. In his paper, the dually flat structure of the manifold of positive measures is discussed, derived from a class of Bregman divergences. These so-called (ρ,τ)-divergences, originally proposed by J. Zhang, are defined in terms of two monotone, scalar functions (ρ and τ) and form a unique class of dually flat, decomposable divergences. This is extended to the set of positive-definite matrices, additionally requiring invariance of the divergence under matrix transformations. It is well known that such dually flat manifolds have computationally desirable properties in applications to classification and information retrieval.

Harsha K. V. and Subrahamanian Moosath K. S. introduce F-geometry as a generalization of α-geometry, based on a representation of a probability density function through a function F. They then combine this with another function G to define a weighted expectation, from which an (F,G)-metric and connection are deduced. A condition for two of such structures to lead to dual connections is also derived. However, it was shown by Zhang (J. Zhang, Entropy 17, pp. 4485–4499, 2015) that this framework is equivalent to the (ρ,τ)-geometry introduced earlier by him. Although the present paper is slightly different in perspective, it should be read with this equivalence in mind.

The next four papers deal with applications of information geometry in statistics. The paper by Frank Critchley and Paul Marriott presents an important research program aimed at rendering some of the most useful results of information geometry more accessible to statisticians in practical applications. Indeed, the formalism of differential geometry and tensor algebra can appear daunting at first sight and may present an obstacle to adoption of many useful results by practitioners. The paper describes a computational framework that facilitates implementation of results from information geometry, based on an embedding of various important statistical models in a (sufficiently large) simplex. Challenges related to extension of the framework to the infinite–dimensional case are touched upon as well.

In the paper by Paul Vos and Karim Anaya-Izquierdo, the goal is to identify one-dimensional exponential families enjoying a number of properties that are convenient for statistical modeling, i.e., parametrization by a measure of central tendency, unimodality, and monotone likelihood ratio. The basis for the framework is the multinomial distribution, modeled geometrically by the simplex. The selection of exponential families with desirable properties is then based on a partitioning of the natural parameter space of the family of multinomial distributions by means of convex cones.

Guido Montúfar and co-workers consider various possibilities to define natural Riemannian metrics on polytopes of stochastic matrices, which describe the conditional probability distribution of two categorical random variables. Inspired by the classical result regarding the uniqueness of the Fisher metric by requiring invariance under Markov morphisms, they define metrics derived from a natural class of stochastic maps between such polytopes, or, alternatively, through embeddings in various possible model spaces. They provide recommendations as to which metric to use, depending on the application.

André Klein, in his article, provides a survey of several matrix algebraic properties of the Fisher information matrix corresponding to weakly stationary time series. The link with various structured matrices arising from a number of time series models is demonstrated. A statistical distance measure is built using the Fisher information matrix in the context of classical and quantum information. Finally, conditions are obtained for the Fisher information of a stationary process to obey certain forms of the Stein equation.

We continue with three papers concerning applications of information geometry in Bayesian inference and simulation. Keisuke Yano and Fumiyasu Komaki, in their paper, construct constant-risk Bayesian predictive densities using the Kullback-Leibler loss function when the distributions of the data and the target variable to be predicted are different but have a common unknown parameter. Specifically, the issue of prior selection is investigated, and several applications are given.

Samuel Livingstone and Mark Girolami provide an introduction to recent enhancements of Markov chain Monte Carlo simulation techniques inspired by information geometry. They apply this to the Metropolis–Hastings algorithm driven by Langevin diffusion, gradually transforming the ingredients to the setting of a Riemannian manifold equipped with a metric similar to the Fisher information metric. Pointers to various applications are given. The paper is written in a way that also makes it accessible to practitioners with little background in stochastic processes and geometry.

The paper by Hui Zhao and Paul Marriott concerns Bayesian inference making use of variational methods for approximating the posterior distribution. In the context of inference for time series models that switch between different regimes, variational Bayes is shown to be a computationally attractive alternative to Markov chain Monte Carlo simulations. The geometry related to the projection of the posterior onto a computationally tractable family of distributions is elucidated by means of a simple example. This is followed by an application wherein it is shown that variational Bayes is successful in estimating the regime-switching model, including the number of regimes.

Applications of information geometry in machine learning are represented by the following three papers. The article by Frank Nielsen and colleagues considers κ-means histogram clustering, with applications to, e.g., information retrieval. Based on the α-divergences as similarity measures, clustering is performed using either the sided (asymmetric) or symmetrized divergence, or by means of the interesting notion of a mixed divergence. An important computational advantage is that the centroids based on the sided and mixed divergences have a closed-form expression. Next, the scheme is extended to algorithms with optimized initialization of cluster centroids, as well as soft clustering.

Salem Said and co-workers present a class of distributions on the manifold of the univariate normal model equipped with the Fisher information metric. Expressed in terms of the Fisher-Rao distance, the distributions are used as priors for modeling the classes in Bayesian classification problems of normal distributions. Characteristics of this "Gaussian" distribution on the manifold are discussed, as well as estimation of its parameters and the posterior using the Laplace approximation. In an application to classification of image textures, the improved performance of these priors over conjugate priors is demonstrated.

Luigi Malagò and Giovanni Pistoni address optimization on manifolds of exponential distributions on a discrete state space using Newton's method, which is based on second-order calculus. In particular, the goal is to find maxima of the expectation of a function with respect to the distribution (stochastic relaxation). Details of the computation are provided, including calculation of the Riemannian Hessian. A nonparametric formalism is used, with a view to extension to the infinite–dimensional case.

The next three papers are related to the role of information geometry in complex systems research. Domenico Felice and colleagues consider the time-averaged volume explored by geodesics on a statistical manifold as an indicator of complexity of the entropic dynamics of a system. The parameters of the model play the role of macrovariables conveying information on the system's microstate. Examples are given for the case of univariate, bivariate, and trivariate normal distributions, providing interesting results depending on correlations between the microvariables.

Alexandre Levada investigates the role of entropy and Fisher information in pairwise isotropic Gaussian Markov random fields, acting as models for complex systems. Expressions for these quantities are derived and the evolution of the Fisher information, and entropy is studied as a function of the inverse temperature of the system. An interesting interpretation is given of asymmetries between these curves in terms of the arrow of time.

Masatoshi Funabashi presents a framework for measuring statistical dependence between subsystems of a stochastic model, based on the model's graph representation. A description in terms of the mixed coordinates of the system is used to quantify the complexity loss incurred by cutting an edge of the graph. In addition, a complexity measure is defined as a geometric mean of Kullback–Leibler divergences between decompositions of the system in terms of subsystems with fewer statistical dependencies. This quantifies the degree to which the system can be decomposed.

The following paper concerns an application to physics, specifically quantum mechanics. Roger Balian gives an overview of a geometrical framework for measuring information loss in quantum systems resulting from the mixing of states. A Riemannian metric is defined, based on the von Neumann entropy, generating a mapping between states and observables. The metric is compared to other quantum metrics, as well as the Fisher–Rao metric, and various geometrical properties are derived. Applications are given to quantum information, as well as equilibrium and non-equilibrium quantum statistical mechanics.

The final paper in the Special Issue is situated at the interface between physics and neuroscience.

Xiaozhao Zhao and colleagues consider the principle of extreme physical information based on the Fisher information, which has been used before in an attempt to establish an information-theoretical basis for physical laws. They extend the idea to cognitive systems and aim at narrowing the gap between the information bound and the data information for such complex systems, by transforming the model to a simpler one. This is done by means of a dimensionality reduction technique, also based on the Fisher information. The approach is applied to derive the model for single-layer Boltzmann machines and interpret their learning algorithms.

We are convinced that the varied collection of papers in this Special Issue will be useful for scientists who are new to the field, while providing an excellent reference for the more seasoned researcher. Furthermore, it is worth mentioning that the second *Entropy* Special Issue in this series, "Information Geometry II", will also be published as a book, and that a third Special Issue is being prepared. We hope that the reader will enjoy browsing and reading through this collection of papers as much as we enjoyed guest editing this Special Issue "Information Geometry I".

Finally, I would like to thank the Editor-in-Chief of *Entropy*, Prof. Dr. Kevin H. Knuth, for suggesting the opportunity to guest-edit a Special Issue on information geometry. Furthermore, I wish to thank the editorial staff at MDPI for their great help with contacting authors, organizing paper reviews, and editing the original Special Issue in *Entropy*, as well as the reprinted version in the present book.

Geert Verdoolaege
Special Issue Editor

entropy

MDPI

Article

Information Geometry of Positive Measures and Positive-Definite Matrices: Decomposable Dually Flat Structure

Shun-ichi Amari

RIKEN Brain Science Institute, Hirosawa 2-1, Wako-shi, Saitama 351-0198, Japan;
E-Mail: amari@brain.riken.jp; Tel.: +81-48-467-9669; Fax: +81-48-467-9687

Received: 14 February 2014; in revised form: 9 April 2014 / Accepted: 10 April 2014 /
Published: 14 April 2014

Abstract: Information geometry studies the dually flat structure of a manifold, highlighted by the generalized Pythagorean theorem. The present paper studies a class of Bregman divergences called the (ρ, τ)-divergence. A (ρ, τ)-divergence generates a dually flat structure in the manifold of positive measures, as well as in the manifold of positive-definite matrices. The class is composed of decomposable divergences, which are written as a sum of componentwise divergences. Conversely, a decomposable dually flat divergence is shown to be a (ρ, τ)-divergence. A (ρ, τ)-divergence is determined from two monotone scalar functions, ρ and τ. The class includes the KL-divergence, α-, β- and (α, β)-divergences as special cases. The transformation between an affine parameter and its dual is easily calculated in the case of a decomposable divergence. Therefore, such a divergence is useful for obtaining the center for a cluster of points, which will be applied to classification and information retrieval in vision. For the manifold of positive-definite matrices, in addition to the dually flatness and decomposability, we require the invariance under linear transformations, in particular under orthogonal transformations. This opens a way to define a new class of divergences, called the (ρ, τ)-structure in the manifold of positive-definite matrices.

Keywords: information geometry; dually flat structure; decomposable divergence; (ρ, τ)-structure

---

1. Introduction

Information geometry, originated from the invariant structure of a manifold of probability distributions, consists of a Riemannian metric and dually coupled affine connections with respect to the metric [1]. A manifold having a dually flat structure is particularly interesting and important. In such a manifold, there are two dually coupled affine coordinate systems and a canonical divergence, which is a Bregman divergence. The highlight is given by the generalized Pythagorean theorem and projection theorem. Information geometry is useful not only for statistical inference, but also for machine learning, pattern recognition, optimization and even for neural networks. It is also related to the statistical physics of Tsallis q-entropy [2–4].

The present paper studies a general and unique class of decomposable divergence functions in $\boldsymbol{R}_{+}^{n}$, the manifold of n-dimensional positive measures. This is the (ρ, τ)-divergences, introduced by Zhang [5,6], from the point of view of "representation duality". They are Bregman divergences generating a dually flat structure. The class includes the well-known Kullback-Leibler divergence, α-divergence, β-divergence and (α, β)-divergence [1,7–9] as special cases. The merit of a decomposable Bregman divergence is that the $\boldsymbol{\theta}$-$\boldsymbol{\eta}$ Legendre transformation is computationally tractable, where $\boldsymbol{\theta}$ and $\boldsymbol{\eta}$ are two affine coordinates systems coupled by the Legendre transformation. When one uses a dually flat divergence to define the center of a cluster of elements, the center is easily given by the arithmetic mean of the dual coordinates of the elements [10,11]. However, we need to calculate its primal coordinates. This is the $\boldsymbol{\theta}$-$\boldsymbol{\eta}$ transformation. Hence, our new type of divergences has an advantage of calculating $\boldsymbol{\theta}$-coordinates for clustering and related pattern matching problems. The most general class of dually flat divergences, not necessarily decomposable, is further given in $\boldsymbol{R}_{+}^{n}$. They are the $(\boldsymbol{\rho}, \boldsymbol{\tau})$ divergence.

Positive-definite (PD) matrices appear in many engineering problems, such as convex programming, diffusion tensor analysis and multivariate statistical analysis [12–20]. The manifold, PD_n, of $n \times n$ PD matrices form a cone, and its geometry is by itself an important subject of research. If we consider the submanifold consisting of only diagonal matrices, it is equivalent to the manifold of positive measures. Hence, PD matrices can be regarded as a generalization of positive measures. There are many studies on geometry and divergences of the manifold of positive-definite matrices. We introduce a general class of dually flat divergences, the (ρ, τ)-divergence. We analyze the cases when a (ρ, τ)-divergence is invariant under the general linear transformations, $Gl(n)$, and invariant under the orthogonal transformations, $O(n)$. They not only include many well-known divergences of PD matrices, but also give new important divergences.

The present paper is organized as follows. Section 2 is preliminary, giving a short introduction to a dually flat manifold and the Bregman divergence. It also defines the cluster center due to a divergence. Section 3 defines the (ρ, τ)-structure in $\boldsymbol{R}_{+}^{n}$. It gives dually flat decomposable affine coordinates and a related canonical divergence (Bregman divergence). Section 4 is devoted to the (ρ, τ)-structure of the manifold, PD_n, of PD matrices. We first study the class of divergences that are invariant under $O(n)$. We further study a decomposable divergence that is invariant under $Gl(n)$. It coincides with the invariant divergence derived from zero-mean Gaussian distributions with PD covariance matrices. They not only include various known divergences, but new remarkable ones. Section 5 discusses a general class of non-decomposable flat divergences and miscellaneous topics. Section 6 is the conclusions.

2. Preliminaries to Information Geometry of Divergence

2.1. Dually Flat Manifold

A manifold is said to have the dually flat Riemannian structure, when it has two affine coordinate systems $\boldsymbol{\theta} = (\theta^1, \cdots, \theta^n)$ and $\boldsymbol{\eta} = (\eta_1, \cdots, \eta_n)$ (with respect to two flat affine connections) together with two convex functions, $\psi(\boldsymbol{\theta})$ and $\varphi(\boldsymbol{\eta})$, such that the two coordinates are connected by the Legendre transformations:

$$\boldsymbol{\eta} = \nabla\psi(\boldsymbol{\theta}), \quad \boldsymbol{\theta} = \nabla\varphi(\boldsymbol{\eta}), \tag{1}$$

where ∇ is the gradient operator. The Riemannian metric is given by:

$$\left(g_{ij}(\boldsymbol{\theta})\right) = \nabla\nabla\psi(\boldsymbol{\theta}), \quad \left(g^{ij}(\boldsymbol{\eta})\right) = \nabla\nabla\varphi(\boldsymbol{\eta}) \tag{2}$$

in the respective coordinate systems. A curve that is linear in the $\boldsymbol{\theta}$-coordinates is called a $\boldsymbol{\theta}$-geodesic, and a curve linear in the $\boldsymbol{\eta}$-coordinates is called an $\boldsymbol{\eta}$-geodesic.

A dually flat manifold has a unique canonical divergence, which is the Bregman divergence defined by the convex functions,

$$D[P:Q] = \psi\left(\boldsymbol{\theta}_P\right) + \varphi\left(\boldsymbol{\eta}_Q\right) - \boldsymbol{\theta}_P \cdot \boldsymbol{\eta}_Q, \tag{3}$$

where $\boldsymbol{\theta}_P$ is the $\boldsymbol{\theta}$-coordinates of P, $\boldsymbol{\eta}_Q$ is the $\boldsymbol{\eta}$-coordinates of Q and $\boldsymbol{\theta}_P \cdot \boldsymbol{\eta}_Q = \sum_i \left(\theta_P^i\right)\left(\eta_{Qi}\right)$, where θ_P^i and η_{Qi} are components of $\boldsymbol{\theta}_p$ and $\boldsymbol{\eta}_Q$, respectively. The Pythagorean and projection theorems hold in a dually flat manifold:

Pythagorean Theorem Given three points, P, Q, R, when the $\boldsymbol{\eta}$-geodesic connecting P and Q is orthogonal to the $\boldsymbol{\theta}$-geodesic connecting Q and R with respect to the Riemannian metric,

$$D[P:Q] + D[Q:R] = D[P:R]. \tag{4}$$

Projection Theorem Given a smooth submanifold, S, let P_S be the minimizer of divergence from P to S,

$$P_S = \min_{Q \in S} D[P : Q]. \tag{5}$$

Then, P_S is the $\boldsymbol{\eta}$-geodesic projection of P to S, that is the $\boldsymbol{\eta}$-geodesic connecting P and P_S is orthogonal to S.

We have the dual of the above theorems where $\boldsymbol{\theta}$- and $\boldsymbol{\eta}$-geodesics are exchanged and $D[P : Q]$ is replaced by its dual $D[Q : P]$.

2.2. Decomposable Divergence

A divergence, $D[P : Q]$, is said to be decomposable, when it is written as a sum of component-wise divergences,

$$D[P : Q] = \sum_{i=1}^{n} d\left(\theta_P^i, \theta_Q^i\right), \tag{6}$$

where θ_P^i and θ_Q^i are the components of $\boldsymbol{\theta}_P$ and $\boldsymbol{\theta}_Q$ and $d\left(\theta_P^i, \theta_Q^i\right)$ is a scalar divergence function.

An f-divergence:

$$D_f[P : Q] = \sum p_i f\left(\frac{q_i}{p_i}\right) \tag{7}$$

is a typical example of decomposable divergence in the manifold of probability distributions, where $P = (\boldsymbol{p})$ and $Q = (\boldsymbol{q})$ are two probability vectors with $\sum p_i = \sum q_i = 1$. A convex function, $\psi(\boldsymbol{\theta})$, is said to be decomposable, when it is written as:

$$\psi(\boldsymbol{\theta}) = \sum_{i=1}^{n} \tilde{\psi}\left(\theta^i\right) \tag{8}$$

by using a scalar convex function, $\tilde{\psi}(\theta)$. The Bregman divergence derived from a decomposable convex function is decomposable.

When $\psi(\boldsymbol{\theta})$ is a decomposable convex function, its Legendre dual is also decomposable. The Legendre transformation is given componentwise as:

$$\eta_i = \tilde{\psi}'\left(\theta_i\right), \tag{9}$$

where $'$ is the differentiation of a function, so that it is computationally tractable. Its inverse transformation is also componentwise,

$$\theta_i = \tilde{\varphi}'\left(\eta_i\right), \tag{10}$$

where $\tilde{\varphi}$ is the Legendre dual of $\tilde{\psi}$.

2.3. Cluster Center

Consider a cluster of points $P_1, \cdots, P_m$ of which $\boldsymbol{\theta}$-coordinates are $\boldsymbol{\theta}_1, \cdots, \boldsymbol{\theta}_m$ and $\boldsymbol{\eta}$-coordinates are $\boldsymbol{\eta}_1, \cdots, \boldsymbol{\eta}_m$. The center, R, of the cluster with respect to the divergence, $D[P : Q]$, is defined by:

$$R = \arg\min_{Q} \sum_{i=1}^{m} D\left[Q : P_i\right]. \tag{11}$$

By differentiating $\sum D\left[Q : P_i\right]$ by $\boldsymbol{\theta}$ (the $\boldsymbol{\theta}$-coordinates of Q), we have:

$$\nabla\psi\left(\boldsymbol{\theta}_R\right) = \frac{1}{m} \sum_{i=1}^{m} \boldsymbol{\eta}_i. \tag{12}$$

Hence, the cluster-center theorem due to Banerjee *et al.* [10] follows; see also [11]:

Cluster-Center Theorem The $\boldsymbol{\eta}$-coordinates $\boldsymbol{\eta}_R$ of the cluster center are given by the arithmetic average of the $\boldsymbol{\eta}$-coordinates of the points in the cluster:

$$\boldsymbol{\eta}_R = \frac{1}{m}\sum_{i=1}^{m}\boldsymbol{\eta}_i. \tag{13}$$

When we need to obtain the $\boldsymbol{\theta}$-coordinates of the cluster center, it is given by the $\boldsymbol{\theta}$-$\boldsymbol{\eta}$ transformation from $\boldsymbol{\eta}_R$,

$$\boldsymbol{\theta}_R = \nabla\varphi\left(\boldsymbol{\eta}_R\right). \tag{14}$$

However, in many cases, the transformation is computationally heavy and intractable when the dimensions of a manifold is large. The transformation is easy in the case of a decomposable divergence. This is motivation for considering a general class of decomposable Bregman divergences.

3. (ρ, τ) Dually Flat Structure in $\boldsymbol{R}^n_+$

3.1. (ρ, τ)-Coordinates of $\boldsymbol{R}^n_+$

Let $\boldsymbol{R}^n_+$ be the manifold of positive measures over n elements $x_1, \cdots, x_n$. A measure (or a weight) of x_i is given by:

$$\xi_i = m\left(x_i\right) > 0 \tag{15}$$

and $\boldsymbol{\xi} = (\xi_1, \cdots, \xi_n)$ is a distribution of measures. When $\sum \xi_i = 1$ is satisfied, it is a probability measure. We write:

$$\boldsymbol{R}^+_n = \{\boldsymbol{\xi}\,|\xi_i > 0\} \tag{16}$$

and $\boldsymbol{\xi}$ forms a coordinate system of $\boldsymbol{R}^n_+$.

Let $\rho(\xi)$ and $\tau(\xi)$ be two monotonically increasing differentiable functions. We call:

$$\theta = \rho(\xi), \quad \eta = \tau(\xi) \tag{17}$$

the ρ- and τ-representations of positive measure ξ. This is a generalization of the $\pm\alpha$ representations [1] and was introduced in [5] for a manifold of probability distributions. See also [6].

By using these functions, we construct new coordinate systems $\boldsymbol{\theta}$ and $\boldsymbol{\eta}$ of $\boldsymbol{R}^n_+$. They are given, for $\boldsymbol{\theta} = \left(\theta^i\right)$ and $\boldsymbol{\eta} = \left(\eta_i\right)$, by componentwise relations,

$$\theta^i = \rho\left(\xi_i\right), \quad \eta_i = \tau\left(\xi_i\right). \tag{18}$$

They are called the ρ- and τ-representations of $\boldsymbol{\xi} \in \boldsymbol{R}^n_+$, respectively. We search for convex functions, $\psi_{\rho,\tau}(\boldsymbol{\theta})$ and $\varphi_{\rho,\tau}(\boldsymbol{\eta})$, which are Legendre duals to each other, such that $\boldsymbol{\theta}$ and $\boldsymbol{\eta}$ are two dually coupled affine coordinate systems.

3.2. Convex Functions

We introduce two scalar functions of θ and η by:

$$\tilde{\psi}_{\rho,\tau}(\theta) = \int_0^{\rho^{-1}(\theta)} \tau(\xi)\rho'(\xi)d\xi, \tag{19}$$

$$\tilde{\varphi}_{\rho,\tau}(\eta) = \int_0^{\tau^{-1}(\eta)} \rho(\xi)\tau'(\xi)d\xi. \tag{20}$$

Then, the first and second derivatives of $\tilde{\psi}_{\rho,\tau}$ are:

$$\tilde{\psi}'_{\rho,\tau}(\theta) = \tau(\xi), \tag{21}$$

$$\tilde{\psi}''_{\rho,\tau}(\theta) = \frac{\tau'(\xi)}{\rho'(\xi)}. \tag{22}$$

Since $\rho'(\xi) > 0$, $\tau'(\xi) > 0$, we see that $\tilde{\psi}_{\rho,\tau}(\theta)$ is a convex function. So is $\tilde{\varphi}_{\rho,\tau}(\eta)$. Moreover, they are Legendre duals, because:

$$\begin{aligned} \tilde{\psi}_{\rho,\tau}(\theta) + \tilde{\varphi}_{\rho,\tau}(\eta) - \theta\eta &= \int_0^{\xi} \tau(\xi)\rho'(\xi)d\xi + \int_0^{\xi} \rho(\xi)\tau'(\xi)d\xi - \rho(\xi)\tau(\xi) && (23)\\ &= 0. && (24) \end{aligned}$$

We then define two decomposable convex functions of $\boldsymbol{\theta}$ and $\boldsymbol{\eta}$ by:

$$\begin{aligned} \psi_{\rho,\tau}(\boldsymbol{\theta}) &= \sum \tilde{\psi}_{\rho,\tau}\left(\theta^i\right), && (25)\\ \varphi_{\rho,\tau}(\boldsymbol{\eta}) &= \sum \tilde{\varphi}_{\rho,\tau}\left(\eta_i\right). && (26) \end{aligned}$$

They are Legendre duals to each other.

3.3. (ρ, τ)-*Divergence*

The (ρ, τ)-divergence between two points, $\boldsymbol{\xi}, \boldsymbol{\xi}' \in \boldsymbol{R}_n^+$, is defined by:

$$\begin{aligned} D_{\rho,\tau}\left[\boldsymbol{\xi} : \boldsymbol{\xi}'\right] &= \psi_{\rho,\tau}(\boldsymbol{\theta}) + \varphi_{\rho,\tau}(\boldsymbol{\eta}') - \boldsymbol{\theta}\cdot\boldsymbol{\eta}' && (27)\\ &= \sum_{i=1}^{n}\left[\int_0^{\xi_i} \tau(\xi)\rho'(\xi)d\xi + \int_0^{\xi_i'} \rho(\xi)\tau'(\xi)d\xi - \rho\left(\xi_i\right)\tau\left(\xi_i'\right)\right], && (28) \end{aligned}$$

where $\boldsymbol{\theta}$ and $\boldsymbol{\eta}'$ are ρ- and τ-representations of $\boldsymbol{\xi}$ and $\boldsymbol{\xi}'$, respectively.

The (ρ, τ)-divergence gives a dually flat structure having $\boldsymbol{\theta}$ and $\boldsymbol{\eta}$ as affine and dual affine coordinate systems. This is originally due to Zhang [5] and a generalization of our previous results concerning the q and deformed exponential families [4]. The transformation between $\boldsymbol{\theta}$ and $\boldsymbol{\eta}$ is simple in the (ρ, τ)-structure, because it can be done componentwise,

$$\begin{aligned} \theta^i &= \rho\left\{\tau^{-1}\left(\eta_i\right)\right\}, && (29)\\ \eta_i &= \tau\left\{\rho^{-1}\left(\theta^i\right)\right\}. && (30) \end{aligned}$$

The Riemannian metric is:

$$g_{ij}(\boldsymbol{\xi}) = \frac{\tau'\left(\xi_i\right)}{\rho'\left(\xi_i\right)}\delta_{ij}, \tag{31}$$

and hence Euclidean, because the Riemann-Christoffel curvature due to the Levi-Civita connection vanishes, too.

The following theorem is new, characterizing the (ρ, τ)-divergence.

Theorem 1. The (ρ, τ)-divergences form a unique class of divergences in $\boldsymbol{R}_+^n$ that are dually flat and decomposable.

3.4. *Biduality:* α-(ρ, τ) *Divergence*

We have dually flat connections, $\left(\nabla_{\rho,\tau}, \nabla_{\rho,\tau}^*\right)$, represented in terms of covariant derivatives, which are derived from $D_{\rho,\tau}$. This is called the representation duality by Zhang [5]. We further have the α-(ρ, τ) connections defined by:

$$\nabla_{\rho,\tau}^{(\alpha)} = \frac{1+\alpha}{2}\nabla_{\rho,\tau} + \frac{1-\alpha}{2}\nabla_{\rho,\tau}^*. \tag{32}$$

The α-$(-\alpha)$ duality is called the reference duality [5]. Therefore, $\nabla_{\rho,\tau}^{(\alpha)}$ possesses the biduality, one concerning α and $(-\alpha)$, and the other with respect to ρ and τ.

The Riemann-Christoffel curvature of $\nabla^{(\alpha)}_{\rho,\tau}$ is:

$$R^{(\alpha)}_{\rho,\tau} = \frac{1-\alpha^2}{4} R^{(0)}_{\rho,\tau} = 0 \tag{33}$$

for any α. Hence, there exists unique canonical divergence $D^{(\alpha)}_{\rho,\tau}$ and α-(ρ,τ) affine coordinate systems. It is an interesting future problem to obtain their explicit forms.

3.5. Various Examples

As a special case of the (ρ,τ)-divergence, we have the (α,β)-divergence obtained from the following power functions,

$$\rho(\xi) = \frac{1}{\alpha}\xi^{\alpha},\ \tau(\xi) = \frac{1}{\beta}\xi^{\beta}. \tag{34}$$

This was introduced by Cichocki, Cruse and Amari in [7,8].

The affine and dual affine coordinates are:

$$\theta^i = \frac{1}{\alpha}\left(\xi_i\right)^{\alpha}, \quad \eta_i = \frac{1}{\beta}\left(\xi_i\right)^{\beta} \tag{35}$$

and the convex functions are:

$$\psi(\boldsymbol{\theta}) = c_{\alpha,\beta} \sum \theta_i^{\frac{\alpha+\beta}{\alpha}}, \quad \varphi(\boldsymbol{\eta}) = c_{\beta,\alpha} \sum \eta_i^{\frac{\alpha+\beta}{\beta}}, \tag{36}$$

where:

$$c_{\alpha,\beta} = \frac{1}{\beta(\alpha+\beta)} \alpha^{\frac{\alpha+\beta}{\alpha}}. \tag{37}$$

The induced (α,β)-divergence has a simple form,

$$D_{\alpha,\beta}[\boldsymbol{\xi} : \boldsymbol{\xi}'] = \frac{1}{\alpha\beta\left(\alpha+\beta\right)} \sum \left\{ \alpha\xi_i^{\alpha+\beta} + \beta\xi_i'^{\alpha+\beta} - (\alpha+\beta)\xi_i^{\alpha}\xi_i'^{\beta} \right\}, \tag{38}$$

for $\boldsymbol{\xi}, \boldsymbol{\xi}' \in \boldsymbol{R}^n_+$. It is defined similarly in the manifold, S_n, of probability distributions, but it is not a Bregman divergence in S_n. This is because the total mass constraint $\sum \xi_i = 1$ is not linear in $\boldsymbol{\theta}$- or $\boldsymbol{\eta}$-coordinates in general.

The α-divergence is a special case of the (α,β)-divergence, so that it is a (ρ,τ)-divergence. By putting:

$$\rho(\xi) = \frac{2}{1-\alpha}\xi^{\frac{1-\alpha}{2}}, \quad \tau(\xi) = \frac{2}{1+\alpha}\xi^{\frac{1+\alpha}{2}}, \tag{39}$$

we have:

$$D_{\alpha}\left[\boldsymbol{\xi} : \boldsymbol{\xi}'\right] = \frac{4}{1-\alpha^2} \sum \left\{ \frac{1-\alpha}{2}\xi_i + \frac{1+\alpha}{2}\xi_i^{\frac{1-\alpha}{2}} - \xi_i^{\alpha}\left(\xi_i'\right)^{\frac{1+\alpha}{2}} \right\}. \tag{40}$$

The β-divergence [19] is obtained from:

$$\rho(\xi) = \xi, \quad \tau(\xi) = \frac{1}{\beta}\xi^{1+\beta}. \tag{41}$$

It is written as:

$$D_{\beta}\left[\boldsymbol{\xi} : \boldsymbol{\xi}'\right] = \frac{1}{\beta(\beta+1)} \sum_i \left[\xi_i^{\beta+1} + (\beta+1)\xi_i' - \left(\xi_i'\right)^{\beta+1} - (\beta+1)\xi_i\left(\xi_i'\right)^{\beta} \right]. \tag{42}$$

The β-divergence is special in the sense that it gives a dually flat structure, even in S_n. This is because $u(\xi)$ is linear in ξ.

The classes of α-divergences and β-divergences intersect at the KL-divergence, and their duals are different in general. They are the only intersecting points of the two classes.

When $\rho(\xi) = \xi$ and $\tau(\xi) = U'(\xi)$ where U is a convex function, (ρ, τ)-divergence is Eguchi's U-divergence [21].

Zhang already introduced the (α, β)-divergence in [5], which is not a (ρ, τ)-divergence in $\boldsymbol{R}^n_+$ and different from ours. We regret for our confusing the naming of the (α, β)-divergence.

4. Invariant, Flat Decomposable Divergences in the Manifold of Positive-Definite Matrices

4.1. Invariant and Decomposable Convex Function

Let $\mathbf{P}$ be a positive-definite matrix and $\psi(\mathbf{P})$ be a convex function. Then, a Bregman divergence is defined between two positive definite matrices, $\mathbf{P}$ and $\mathbf{Q}$, by:

$$D[\mathbf{P} : \mathbf{Q}] = \psi(\mathbf{P}) - \blacksquare(\mathbf{Q}) - \nabla\blacksquare(\mathbf{P}) \cdot (\mathbf{P} - \mathbf{Q}) \tag{43}$$

where ∇ is the gradient operator with respect to matrix $\mathbf{P} = (P_{ij})$, so that $\nabla\psi(\mathbf{P})$ is a matrix and the inner product of two matrices is defined by:

$$\nabla\psi(\mathbf{Q}) \cdot \mathbf{P} = \mathrm{tr}\left\{\nabla\psi(\mathbf{Q})\mathbf{P}\right\}, \tag{44}$$

where tr is the trace of a matrix.

It induces a dually flat structure to the manifold of positive-definite matrices, where the affine coordinate system ($\boldsymbol{\theta}$-coordinates) is $\blacksquare = \mathbf{P}$ and the dual affine coordinate system ($\boldsymbol{\eta}$-coordinates) is:

$$\mathbf{H} = \nabla\psi(\mathbf{P}). \tag{45}$$

A convex function, $\psi(\mathbf{P})$, is said to be invariant under the orthogonal group $O(n)$, when:

$$\psi(\mathbf{P}) = \psi\left(\mathbf{O}^T\mathbf{P}\mathbf{O}\right) \tag{46}$$

holds for any orthogonal transformation $\boldsymbol{O}$, where $\boldsymbol{O}^T$ is the transpose of $\boldsymbol{O}$. An invariant function is written as a symmetric function of n eigenvalues $\lambda_1, \cdots, \lambda_n$ of $\mathbf{P}$. See Dhillon and Tropp [12]. When an invariant convex function of $\mathbf{P}$ is written, by using a convex function, f, of one variable, in the additive form:

$$\psi(\mathbf{P}) = \sum f(\lambda_i), \tag{47}$$

it is said to be decomposable. We have:

$$\psi(\boldsymbol{P}) = \mathrm{tr} f(\mathbf{P}). \tag{48}$$

4.2. Invariant, Flat and Decomposable Divergence

A divergence $D[\mathbf{P} : \mathbf{Q}]$ is said to be invariant under $O(n)$, when it satisfies:

$$D[\mathbf{P} : \mathbf{Q}] = D\left[\mathbf{O}^T\mathbf{P}\mathbf{O} : \mathbf{O}^T\mathbf{Q}\mathbf{O}\right]. \tag{49}$$

When it is derived from a decomposable convex function, $\psi(\mathbf{P})$, it is invariant, flat and decomposable.

We give well-known examples of decomposable convex functions and the divergences derived from them:

(1) For $f(\lambda) = (1/2)\lambda^2$, we have:

$$\psi(\mathbf{P}) = \frac{1}{2}\sum \lambda_i^2, \tag{50}$$

$$D[\mathbf{P}:\mathbf{Q}] = \frac{1}{2}\|\mathbf{P}-\mathbf{Q}\|^2, \tag{51}$$

where $\|\mathbf{P}\|^2$ is the Frobenius norm:

$$\|\mathbf{P}\|^2 = \sum P_{ij}^2. \tag{52}$$

(2) For $f(\lambda) = -\log\lambda$

$$\psi(\mathbf{P}) = -\log(\det|\mathbf{P}|), \tag{53}$$

$$D[\mathbf{P}:\mathbf{Q}] = \mathrm{tr}\left(\mathbf{P}\mathbf{Q}^{-1}\right) - \log\left(\det\left|\mathbf{P}\mathbf{Q}^{-1}\right|\right) - n. \tag{54}$$

The affine coordinate system is $\mathbf{P}$, and the dual coordinate system is $\mathbf{P}^{-1}$. The derived geometry is the same as that of multivariate Gaussian probability distributions with mean zero and covariance matrix $\mathbf{P}$.

(3) For $f(\lambda) = \lambda\log\lambda - \lambda$,

$$\psi(\mathbf{P}) = \mathrm{tr}(\mathbf{P}\log\mathbf{P} - \mathbf{P}), \tag{55}$$

$$D[\mathbf{P}:\mathbf{Q}] = \mathrm{tr}(\mathbf{P}\log\mathbf{P} - \mathbf{P}\log\mathbf{Q} - \mathbf{P} + \mathbf{Q}). \tag{56}$$

This divergence is used in quantum information theory. The affine coordinate system is $\mathbf{P}$, and the dual affine coordinate system is $\log\mathbf{P}$; and, $\psi(\mathbf{P})$ is called the negative von Neuman entropy.

4.3. (ρ,τ)-*Structure in Positive Definite Matrices*

We extend the (ρ,τ)-structure in the previous section to the matrix case and introduce a general dually flat invariant decomposable divergence in the manifold of positive-definite matrices. Let:

$$\mathbf{\Theta} = \rho(\mathbf{P}), \quad \mathbf{H} = \tau(\mathbf{P}) \tag{57}$$

be ρ- and τ-representations of matrices. We use two functions, $\tilde{\psi}_{\rho,\tau}(\theta)$ and $\tilde{\varphi}_{\rho,\tau}(\eta)$, defined in Equations (19) and (20), for defining a pair of dually coupled invariant and decomposable convex functions,

$$\psi(\mathbf{\Theta}) = \mathrm{tr}\,\tilde{\psi}_{\rho,\tau}\{\mathbf{\Theta}\}, \tag{58}$$

$$\varphi(\mathbf{H}) = \mathrm{tr}\,\tilde{\varphi}_{\rho,\tau}\{\mathbf{H}\}. \tag{59}$$

They are not convex with respect to $\mathbf{P}$, but are convex with respect to $\mathbf{\Theta}$ and $\mathbf{H}$, respectively. The derived Bregman divergence is:

$$D[\mathbf{P}:\mathbf{Q}] = \psi\{\mathbf{\Theta}(\mathbf{P})\} + \varphi\{\mathbf{H}(\mathbf{Q})\} - \mathbf{\Theta}(\mathbf{P})\cdot\mathbf{H}(\mathbf{Q}). \tag{60}$$

Theorem 2. The (ρ,τ)-divergences form a unique class of invariant, decomposable and dually flat divergences in the manifold of positive matrices.

The Euclidean, Gaussian and von Neuman divergences given in Equations (51), (54) and (56) are special examples of (ρ, τ)-divergences. They are given, respectively, by:

$$(1) \qquad \rho(\xi) = \tau(\xi) = \xi, \tag{61}$$

$$(2) \qquad \rho(\xi) = \xi, \quad \tau(\xi) = -\frac{1}{\xi}, \tag{62}$$

$$(3) \qquad \rho(\xi) = \xi, \quad \tau(\xi) = \log \xi. \tag{63}$$

When ρ and τ are power functions, we have the (α, β)-structure in the manifold of positive-definite matrices.

(4) $(\alpha$-$\beta)$-divergence.

By using the (α, β) power functions given by Equation (34), we have:

$$\psi(\mathbf{\Theta}) = \frac{\alpha}{\alpha+\beta} \operatorname{tr} \mathbf{\Theta}^{\frac{\alpha+\beta}{\alpha}} = \frac{\alpha}{\alpha+\beta} \operatorname{tr} \mathbf{P}^{\alpha+\beta}, \tag{64}$$

$$\varphi(\mathbf{H}) = \frac{\beta}{\alpha+\beta} \operatorname{tr} \mathbf{H}^{\frac{\alpha+\beta}{\beta}} = \frac{\beta}{\alpha+\beta} \operatorname{tr} \mathbf{P}^{\alpha+\beta} \tag{65}$$

so that the (α, β)-divergence of matrices is:

$$D[\mathbf{P} : \mathbf{Q}] = \operatorname{tr} \left\{ \frac{\alpha}{\alpha+\beta} \mathbf{P}^{\alpha+\beta} + \frac{\beta}{\alpha+\beta} \mathbf{Q}^{\alpha+\beta} - \mathbf{P}^{\alpha} \mathbf{Q}^{\beta} \right\}. \tag{66}$$

This is a Bregman divergence, where the affine coordinate system is $\mathbf{\Theta} = \mathbf{P}^{\alpha}$ and its dual is $\mathbf{H} = \mathbf{P}^{\beta}$.

(5) The α-divergence is derived as:

$$\mathbf{\Theta}(P) = \frac{2}{1-\alpha} P^{\frac{1-\alpha}{2}}, \tag{67}$$

$$\psi(\mathbf{\Theta}) = \frac{2}{1+\alpha} P, \tag{68}$$

$$D_{\alpha}[\mathbf{P} : \mathbf{Q}] = \frac{4}{1-\alpha^2} \operatorname{tr} \left(-\mathbf{P}^{\frac{1-\alpha}{2}} \mathbf{Q}^{\frac{1+\alpha}{2}} + \frac{1-\alpha}{2} \mathbf{P} + \frac{1+\alpha}{2} \mathbf{Q} \right). \tag{69}$$

The affine coordinate system is $\frac{2}{1-\alpha} \mathbf{P}^{\frac{1-\alpha}{2}}$, and its dual is $\frac{2}{1+\alpha} \mathbf{P}^{\frac{1+\alpha}{2}}$.

(6) The β-divergence is derived from Equation (41) as:

$$D_{\beta}[\mathbf{P} : \mathbf{Q}] = \frac{1}{\beta(\beta+1)} \operatorname{tr} \left[\mathbf{P}^{\beta+1} + (\beta+1)\mathbf{Q} - \mathbf{Q}^{\beta+1} - (\beta+1)\mathbf{P}\mathbf{Q}^{\beta} \right]. \tag{70}$$

4.4. Invariance Under $Gl(n)$

We extend the concept of invariance under the orthogonal group to that under the general linear group, $Gl(n)$, that is the set of invertible matrices, $L, \det|L| \neq 0$. This is a stronger condition. A divergence is said to be invariant under $Gl(n)$, when:

$$D[\mathbf{P} : \mathbf{Q}] = D\left[\mathbf{L}^T \mathbf{P} \mathbf{L} : \mathbf{L}^T \mathbf{Q} \mathbf{L}\right] \tag{71}$$

holds for any $\mathbf{L} \in Gl(n)$.

We identify matrix $\mathbf{P}$ with the zero-mean Gaussian distribution:

$$p(x, \mathbf{P}) = \exp\left\{ -\frac{1}{2} x^T \mathbf{P}^{-1} x - \frac{1}{2} \log\det|\mathbf{P}| - c \right\}, \tag{72}$$

where c is a constant. We know that an invariant divergence belongs to the class of f-divergences in the case of a manifold of probability distributions, where the invariance means the geometry does not change under a one-to-one mapping of $\boldsymbol{x}$ to $\boldsymbol{y}$. Moreover, the only invariant flat divergence is the KL-divergence [22]. These facts suggest the following conjecture.

Proposition. The invariant, flat and decomposable divergence under $Gl(n)$ is the KL-divergence given by:

$$D_{KL}[\boldsymbol{P}:\boldsymbol{Q}] = \operatorname{tr}\left(\boldsymbol{P}\boldsymbol{Q}^{-1}\right) - \log\left(\det\left|\boldsymbol{P}\boldsymbol{Q}^{-1}\right|\right) - n. \tag{73}$$

5. Non-Decomposable Divergence

We have focused on flat and decomposable divergences. There are many interesting non-decomposable divergences. We first discuss a general class of flat divergences in $\boldsymbol{R}^n_+$ and then touch upon interesting flat and non-flat divergences in the manifold of positive-definite matrices.

5.1. General Class of Flat Divergences in $\boldsymbol{R}^n_+$

We can describe a general class of flat divergence in $\boldsymbol{R}^n_+$, which are not necessarily decomposable. This is introduced in [23], which studies the conformal structure of general total Bregman divergences ([11,13]). When $\boldsymbol{R}^n_+$ is endowed with a dually flat structure, it has a $\boldsymbol{\theta}$-coordinate system given by:

$$\boldsymbol{\theta} = \boldsymbol{\rho}(\boldsymbol{\xi}) \tag{74}$$

which is not necessarily a componentwise function. Any pair of invertible $\boldsymbol{\theta} = \boldsymbol{\rho}(\boldsymbol{\xi})$ and convex function $\psi(\boldsymbol{\theta})$ defines a dually flat structure and, hence, a Bregman divergence in $\boldsymbol{R}^n_+$.

The dual coordinates $\boldsymbol{\eta} = \boldsymbol{\tau}(\boldsymbol{\xi})$ are given by:

$$\boldsymbol{\eta} = \nabla\psi(\boldsymbol{\theta}) \tag{75}$$

so that we have:

$$\boldsymbol{\eta} = \boldsymbol{\tau}(\boldsymbol{\xi}) = \nabla\psi\left\{\boldsymbol{\rho}(\boldsymbol{\xi})\right\}. \tag{76}$$

This implies that a pair $(\boldsymbol{\rho}, \boldsymbol{\tau})$ of coordinate systems can define dually coupled affine coordinates and, hence, a dually flat structure, when and only when $\boldsymbol{\eta} = \boldsymbol{\tau}\left\{\boldsymbol{\rho}^{-1}(\boldsymbol{\theta})\right\}$ is a gradient of a convex function.

This is different from the case of decomposable divergence, where any monotone pair of $\rho(\xi)$ and $\tau(\xi)$ gives a dually flat structure.

5.2. Non-Decomposable Flat Divergence in PD_n

Ohara and Eguchi [15,16] introduced the following function:

$$\psi_V(\mathbf{P}) = V\left(\det|\mathbf{P}|\right), \tag{77}$$

where $V(\xi)$ is a monotonically decreasing scalar function. ψ_V is convex when and only when:

$$1 + \frac{V''(\xi)\xi^2}{V'(\xi)} < \frac{1}{n}. \tag{78}$$

In such a case, we can introduce dually flat structure to PD_n, where $\mathbf{P}$ is an affine coordinate system with convex $\psi_V(\mathbf{P})$, and the dual affine coordinate system is:

$$\mathbf{H} = V'(\det\|P\|)\mathbf{P}^{-1}. \tag{79}$$

The derived divergence is:

$$D_V[\mathbf{P}:\mathbf{Q}] = V(\det|\mathbf{P}) - V(\det|\mathbf{Q})| \tag{80}$$
$$+ V'(\det|\mathbf{Q}|)\mathrm{tr}\left\{\mathbf{Q}^{-1}(\mathbf{Q}-\mathbf{P})\right\}. \tag{81}$$

When $V(\xi) = -\log\xi$, it reduces to the case of Equation (54), which is invariant under $Gl(n)$ and decomposable. However, the divergence $D_V[\mathbf{P}:\mathbf{Q}]$ is not decomposable. It is invariant under $O(n)$ and more strongly so under $SGl(n) \subset Gl(n)$, defined by $\det|\mathbf{L}| = \pm 1$.

5.3. *Flat Structure Derived from q-Escort Distribution*

A dually flat structure is introduced in the manifold of probability distributions [4] as:

$$\tilde{D}_\alpha[\boldsymbol{p}:\boldsymbol{q}] = \frac{1}{1-q}\frac{1}{H_q(\boldsymbol{p})}\left(1-\sum p_i^{1-q}q_i^q\right), \tag{82}$$

where:

$$H_q(\boldsymbol{p}) = \sum p_i^q, \tag{83}$$
$$q = \frac{1+\alpha}{2}. \tag{84}$$

The dual affine coordinates are the q-escort distribution: [4]

$$\eta_i = \frac{1}{H_q(\boldsymbol{p})}p_i^q. \tag{85}$$

The divergence, $\tilde{D}_q$, is flat, but not decomposable.

We can generalize it to the case of PD_n,

$$\tilde{D}_q[\mathbf{P}:\mathbf{Q}] = \frac{1}{1-q}\frac{1}{\mathrm{tr}\,\mathbf{P}^q}\left\{(1-q)\,\mathrm{tr}\,(\mathbf{P}) + q\,\mathrm{tr}\,(\mathbf{Q}) - \mathrm{tr}\,\left(\mathbf{P}^{1-q}\mathbf{Q}^q\right)\right\}. \tag{86}$$

This is flat, but not decomposable.

5.4. *γ-Divergence in PD_n*

The γ-divergence is introduced by Fujisawa and Eguchi [24]. It gives a super-robust estimator. It is interesting to generalize it to PD_n,

$$D_\gamma[\mathbf{P}:\mathbf{Q}] = \frac{1}{\gamma(\gamma-1)}\left\{\log\mathrm{tr}\,\mathbf{P}^\gamma - (\gamma-1)\log\mathrm{tr}\,\mathbf{Q}^{\gamma-1} - \gamma\log\mathrm{tr}\,\mathbf{P}\mathbf{Q}^{\gamma-1}\right\}. \tag{87}$$

This is not flat nor decomposable. This is a projective divergence in the sense that, for any $c, c' > 0$,

$$D_\gamma\left[c\mathbf{P}:c'\mathbf{Q}\right] = D_\gamma[\mathbf{P}:\mathbf{Q}]. \tag{88}$$

Therefore, it can be defined in the submanifold of $\mathrm{tr}\,\mathbf{P} = 1$.

6. Concluding Remarks

We have shown that the (ρ,τ)-divergence introduced by Zhang [5] is a general dually flat decomposable structure of the manifold of positive measures. We then extended it to the manifold of positive-definite matrices, where the criterion of invariance under linear transformations (in particular, under orthogonal transformations) were added. The decomposability is useful from the computational point of view, because the θ-η transformation is tractable. This is the motivation for studying decomposable flat divergences.

When we treat the manifold of probability distributions, it is a submanifold of the manifold of positive measures, where the total sum of measures are restricted to one. This is a nonlinear constraint in the θ or η coordinates, so that the manifold is not flat, but curved in general. Hence, our arguments hold in this case only when at least one of the ρ and τ functions are linear. The U-divergence [21] and β-divergence [19] are such cases. However, for clustering, we can take the average of the η-coordinates of member probability distributions in the larger manifold of positive measures and then project it to the manifold of probability distributions. This is called the exterior average, and the projection is simply a normalization of the result. Therefore, the (ρ, τ)-structure is useful in the case of probability distributions. The same situation holds in the case of positive-definite matrices.

Quantum information theory deals with positive-definite Hermitian matrices of trace one [25,26]. We need to extend our discussions to the case of complex matrices. The trace one constraint is not linear with respect to θ- or η-coordinates, as is the same in the case of probability distributions. Many interesting divergence functions have been introduced in the manifold of positive-definite Hermitian matrices. It is an interesting future problem to apply our theory to quantum information theory.

Conflicts of Interest: The author declares no conflicts of interest.

References

1. Amari, S.; Nagaoka, H. *Methods of Information Geometry*; American Mathematical Society and Oxford University Press: Rhode Island, RI, USA, 2000.
2. Tsallis, C. *Introduction to Nonextensive Statistical Mechanics: Approaching a Complex World*; Springer: Berlin/Heidelberg, Germany, 2009.
3. Naudts, J. *Generalized Thermostatistics*; Springer: Berlin/Heidelberg, Germany, 2011.
4. Amari, S.; Ohara, A.; Matsuzoe, H. Geometry of deformed exponential families: Invariant, dually-flat and conformal geometries. *Physica A* **2012**, *391*, 4308–4319.
5. Zhang, J. Divergence function, duality, and convex analysis. *Neural Comput.* **2004**, *16*, 159–195.
6. Zhang, J. Nonparametric information geometry: From divergence function to referential-representational biduality on statistical manifolds. *Entropy* **2013**, *15*, 5384–5418.
7. Cichocki, A.; Amari, S. Families of alpha- beta- and gamma-divergences: Flexible and robust measures of similarities. *Entropy* **2010**, *12*, 1532–1568.
8. Cichocki, A.; Cruces, S.; Amari, S. Generalized alpha-beta divergences and their application to robust nonnegative matrix factorization. *Entropy* **2011**, *13*, 134–170.
9. Minami, M.; Eguchi, S. Robust blind source separation by beta-divergence. *Neural Comput.* **2002** *14*, 1859–1886.
10. Banerjee, A.; Merugu, S.; Dhillon I.; Ghosh, J. Clustering with Bregman Divergences. *J. Mach. Learn. Res.* **2005**, *6*, 1705–1749.
11. Liu, M.; Vemuri, B.C.; Amari, S.; Nielsen, F. Shape retrieval using hierarchical total Bregman soft clustering. *IEEE Trans. Pattern Anal. Mach. Learn.* **2012**, *24*, 3192–3212.
12. Dhillon, I.S.; Tropp, J.A. Matrix nearness problems with Bregman divergences. *SIAM J. Matrix Anal. Appl.* **2007**, *29*, 1120–1146.
13. Vemuri, B.C.; Liu, M.; Amari, S.; Nielsen, F. Total Bregman divergence and its applications to DTI analysis. *IEEE Trans. Med. Imaging* **2011**, *30*, 475–483.
14. Ohara, A.; Suda, N.; Amari, S. Dualistic differential geometry of positive definite matrices and its applications to related problems. *Linear Algebra Appl.* **1996** *247*, 31–53.
15. Ohara, A.; Eguchi, S. Group invariance of information geometry on q-Gaussian distributions induced by beta-divergence. *Entropy* **2013**, *15*, 4732–4747.
16. Ohara, A.; Eguchi, S. Geometry on positive definite matrices induced from V-potential functions. In *Geometric Science of Information*; Nielsen, F., Barbaresco, F., Eds.; Springer: Berlin/Heidelberg, Germany, 2013; pp. 621–629.

17. Chebbi, Z.; Moakher, M. Means of Hermitian positive-definite matrices based on the log-determinant alpha-divergence function. *Linear Algebra Appl.* **2012**, *436*, 1872–1889.
18. Tsuda, K.; Ratsch, G.; Warmuth, M.K. Matrix exponentiated gradient updates for on-line learning and Bregman projection. *J. Mach. Learn. Res.* **2005**, *6*, 995–1018.
19. Nock, R.; Magdalou, B.; Briys, E.; Nielsen, F. Mining matrix data with Bregman matrix divergences for portfolio selection. In *Matrix Information Geometry*; Nielsen, F., Bhatia, R., Eds.; Springer: Berlin/Heidelberg, Germany, 2013; Chapter 15, pp. 373–402.
20. Nielsen, F., Bhatia, R., Eds. *Matrix Information Geometry*; Springer: Berlin/Heidelberg, Germany, 2013.
21. Eguchi, S. Information geometry and statistical pattern recognition. *Sugaku Expo.* **2006**, *19*, 197–216.
22. Amari, S. α-divergence is unique, belonging to both f-divergence and Bregman divergence classes. *IEEE Trans. Inf. Theory* **2009**, *55*, 4925–4931.
23. Nock, R.; Nielsen, F.; Amari, S. On conformal divergences and their population minimizers. *IEEE Trans. Inf. Theory* **2014**, submitted for publication.
24. Fujisawa, H.; Eguchi, S. Robust parameter estimation with a small bias against heavy contamination. *J. Multivar. Anal.* **2008**, *99*, 2053–2081.
25. Petz, P. Monotone metrics on matrix spaces. *Linear Algebra Appl.* **1996**, *244*, 81–96.
26. Hasegawa, H. α-divergence of the non-commutative information geometry. *Rep. Math. Phys.* **1993**, *33*, 87–93.

entropy

MDPI

Article

F-Geometry and Amari's $\alpha-$Geometry on a Statistical Manifold

Harsha K. V. * and Subrahamanian Moosath K S *

Indian Institute of Space Science and Technology, Department of Space, Government of India, Valiamala P.O, Thiruvananthapuram-695547, Kerala, India

* E-Mails: harsha.11@iist.ac.in (K.V.H.); smoosath@iist.ac.in (K.S.S.M.); Tel.: +91-95-6736-0425 (K.V.H.); +91-94-9574-3148 (K.S.S.M.).

Received: 13 December 2013; in revised form: 21 April 2014 / Accepted: 25 April 2014 / Published: 6 May 2014

Abstract: In this paper, we introduce a geometry called F-geometry on a statistical manifold $\mathcal{S}$ using an embedding F of $\mathcal{S}$ into the space $\mathbb{R}_{\mathbb{X}}$ of random variables. Amari's $\alpha-$geometry is a special case of $F-$geometry. Then using the embedding F and a positive smooth function G, we introduce $(F,G)-$metric and $(F,G)-$connections that enable one to consider weighted Fisher information metric and weighted connections. The necessary and sufficient condition for two $(F,G)-$connections to be dual with respect to the $(F,G)-$metric is obtained. Then we show that Amari's $0-$connection is the only self dual $F-$connection with respect to the Fisher information metric. Invariance properties of the geometric structures are discussed, which proved that Amari's $\alpha-$connections are the only $F-$connections that are invariant under smooth one-to-one transformations of the random variables.

Keywords: embedding; Amari's $\alpha-$connections; $F-$metric; $F-$connections; $(F,G)-$metric; $(F,G)-$connections; invariance

1. Introduction

Geometric study of statistical estimation has opened up an interesting new area called the Information Geometry. Information geometry achieved a remarkable progress through the works of Amari [1,2], and his colleagues [3,4]. In the last few years, many authors have considerably contributed in this area [5–9]. Information geometry has a wide variety of applications in other areas of engineering and science, such as neural networks, machine learning, biology, mathematical finance, control system theory, quantum systems, statistical mechanics, *etc.*

A statistical manifold of probability distributions is equipped with a Riemannian metric and a pair of dual affine connections [2,4,9]. It was Rao [10] who introduced the idea of using Fisher information as a Riemannian metric in the manifold of probability distributions. Chentsov [11] introduced a family of affine connections on a statistical manifold defined on finite sets. Amari [2] introduced a family of affine connections called $\alpha-$connections using a one parameter family of functions, the $\alpha-$embeddings. These $\alpha-$connections are equivalent to those defined by Chentsov. The Fisher information metric and these affine connections are characterized by invariance with respect to the sufficient statistic [4,12] and play a vital role in the theory of statistical estimation. Zhang [13] generalized Amari's $\alpha-$representation and using this general representation together with a convex function he defined a family of divergence functions from the point of view of representational and referential duality. The Riemannian metric and dual connections are defined using these divergence functions.

In this paper, Amari's idea of using $\alpha-$embeddings to define geometric structures is extended to a general embedding. This paper is organized as follows. In Section 2, we define an affine connection called $F-$connection and a Riemannian metric called $F-$metric using a general embedding F of a statistical manifold $\mathcal{S}$ into the space of random variables. We show that $F-$metric is the Fisher information metric and Amari's $\alpha-$geometry is a special case of $F-$geometry. Further, we introduce $(F,G)-$metric and $(F,G)-$connections using the embedding F and a positive smooth function G.

In Section 3, a necessary and sufficient condition for two $(F,G)-$connections to be dual with respect to the $(F,G)-$metric is derived and we prove that Amari's $0-$connection is the only self dual $F-$connection with respect to the Fisher information metric. Then we prove that the set of all positive finite measures on X, for a finite X, has an $F-$affine manifold structure for any embedding F. In Section 4, invariance properties of the geometric structures are discussed. We prove that the Fisher information metric and Amari's $\alpha-$connections are invariant under both the transformation of the parameter and the transformation of the random variable. Further we show that Amari's $\alpha-$connections are the only $F-$connections that are invariant under both the transformation of the parameter and the transformation of the random variable.

Let $(X,\mathcal{B})$ be a measurable space, where X is a non-empty subset of $\mathbb{R}$ and $\mathcal{B}$ is the σ-field of subsets of X. Let $\mathbb{R}_X$ be the space of all real valued measurable functions defined on $(X,\mathcal{B})$. Consider an $n-$dimensional statistical manifold $\mathcal{S}=\{p(x;\xi)\ /\ \xi=[\xi^1,...,\xi^n]\in\mathbb{E}\subseteq\mathbb{R}^n\}$, with coordinates $\xi=[\xi^1,...,\xi^n]$, defined on X. $\mathcal{S}$ is a subset of $\mathcal{P}(X)$, the set of all probability measures on X given by

$$\mathcal{P}(X):=\{p:X\longrightarrow\mathbb{R}\ /\ p(x)>0\ (\forall\, x\in X);\int_X p(x)dx=1\}. \tag{1}$$

The tangent space to $\mathcal{S}$ at a point p_ξ is given by

$$T_\xi(\mathcal{S})=\{\sum_{i=1}^n\alpha^i\partial_i\ /\ \alpha^i\in\mathbb{R}\}\quad\text{where }\partial_i=\frac{\partial}{\partial\xi^i}. \tag{2}$$

Define $\ell(x;\xi)=\log p(x;\xi)$ and consider the partial derivatives $\{\frac{\partial\ell}{\partial\xi^i}=\partial_i\ell\ ;i=1,....,n\}$ which are called scores. For the statistical manifold $\mathcal{S}$, $\partial_i\ell$'s are linearly independent functions in x for a fixed ξ. Let $T^1_\xi(\mathcal{S})$ be the n-dimensional vector space spanned by n functions $\{\partial_i\ell\ ;\ i=1,....,n\}$ in x. So

$$T^1_\xi(\mathcal{S})=\{\sum_{i=1}^n A^i\partial_i\ell\ /\ A^i\in\mathbb{R}\}. \tag{3}$$

Then there is a natural isomorphism between these two vector spaces $T_\xi(\mathcal{S})$ and $T^1_\xi(\mathcal{S})$ given by

$$\partial_i\in T_\xi(\mathcal{S})\longleftrightarrow\partial_i\ell(x;\xi)\in T^1_\xi(\mathcal{S}). \tag{4}$$

Obviously, a tangent vector $A=\sum_{i=1}^n A^i\partial_i\in T_\xi(\mathcal{S})$ corresponds to a random variable $A(x)=\sum_{i=1}^n A^i\partial_i\ell(x;\xi)\in T^1_\xi(\mathcal{S})$ having the same components A^i. Note that $T_\xi(\mathcal{S})$ is the differentiation operator representation of the tangent space, while $T^1_\xi(\mathcal{S})$ is the random variable representation of the same tangent space. The space $T^1_\xi(\mathcal{S})$ is called the 1-representation of the tangent space.
Let A and B be two tangent vectors in $T_\xi(\mathcal{S})$ and $A(x)$ and $B(x)$ be the $1-$representations of A and B respectively. We can define an inner product on each tangent space $T_\xi(\mathcal{S})$ by

$$g_\xi(A,B)=<A,B>_\xi=E_\xi[A(x)B(x)]=\int A(x)B(x)p(x;\xi)dx. \tag{5}$$

Especially the inner product of the basis vectors ∂_i and ∂_j is

$$g_{ij}(\xi)=<\partial_i,\partial_j>_\xi=E_\xi[\partial_i\ell\ \partial_j\ell]=\int\partial_i\ell(x;\xi)\partial_j\ell(x;\xi)p(x;\xi)dx. \tag{6}$$

Note that $g =<,>$ defines a Riemannian metric on $\mathcal{S}$ called the **Fisher information metric**.
On the Riemannian manifold $(\mathcal{S}, g =<,>)$, define n^3 functions Γ_{ijk} by

$$\Gamma_{ijk}(\xi) = E_\xi[(\partial_i\partial_j\ell(x;\xi))(\partial_k\ell(x;\xi))]. \tag{7}$$

These functions Γ_{ijk} uniquely determine an affine connection ∇ on $\mathcal{S}$ by

$$\Gamma_{ijk}(\xi) =< \nabla_{\partial_i}\partial_j, \partial_k >_\xi . \tag{8}$$

∇ is called the $1-$connection or the exponential connection.
Amari [2] defined a one parameter family of functions called the $\alpha-$embeddings given by

$$L_\alpha(p) = \begin{cases} \frac{2}{1-\alpha} p^{\frac{1-\alpha}{2}} & \alpha \neq 1 \\ \log p & \alpha = 1 \end{cases} \tag{9}$$

Using these, we can define n^3 functions Γ^α_{ijk} by

$$\Gamma^\alpha_{ijk} = \int \partial_i\partial_j L_\alpha(p(x;\xi))\partial_k L_{-\alpha}(p(x;\xi))dx \tag{10}$$

These Γ^α_{ijk} uniquely determine affine connections ∇^α on the statistical manifold $\mathcal{S}$ by

$$\Gamma^\alpha_{ijk} = < \nabla^\alpha_{\partial_i}\partial_j, \partial_k > \tag{11}$$

which are called **ff−connections**.

2. $F-$Geometry of a Statistical Manifold

On a statistical manifold $\mathcal{S}$, the Fisher information metric and exponential connection are defined using the log embedding. In a similar way, $\alpha-$connections are defined using a one parameter family of functions, the $\alpha-$embeddings. In general, we can give other geometric structures on $\mathcal{S}$ using different embeddings of the manifold $\mathcal{S}$ into the space of random variables $\mathbb{R}_X$.
Let $F : (0,\infty) \longrightarrow \mathbb{R}$ be an injective function that is at least twice differentiable. Thus we have $F'(u) \neq 0,\ \forall\ u \in (0,\infty)$. F is an embedding of $\mathcal{S}$ into $\mathbb{R}_X$ that takes each $p(x;\xi) \longmapsto F(p(x;\xi))$. Denote $F(p(x;\xi))$ by $F(x;\xi)$ and $\partial_i F$ can be written as

$$\partial_i F(x;\xi) = p(x;\xi)F'(p(x;\xi))\partial_i\ell(p(x;\xi)). \tag{12}$$

It is clear that $\partial_i F(x;\xi);\ i = 1,...,n$ are linearly independent functions in x for fixed ξ since $\partial_i\ell(p(x;\xi));\ i = 1,..,n$ are linearly independent. Let $T_{F(p_\xi)}F(\mathcal{S})$ be the n-dimensional vector space spanned by n functions $\partial_i F;\ i = 1,....,n$ in x for fixed ξ. So

$$T_{F(p_\xi)}F(\mathcal{S}) = \{\sum_{i=1}^{n} A^i\partial_i F \ /\ A^i \in \mathbb{R}\} \tag{13}$$

Let the tangent space $T_{F(p_\xi)}(F(\mathcal{S}))$ to $F(\mathcal{S})$ at the point $F(p_\xi)$ be denoted by $T^F_\xi(\mathcal{S})$. There is a natural isomorphism between the two vector spaces $T_\xi(\mathcal{S})$ and $T^F_\xi(\mathcal{S})$ given by

$$\partial_i \in T_\xi(\mathcal{S}) \longleftrightarrow \partial_i F(x;\xi) \in T^F_\xi(\mathcal{S}). \tag{14}$$

$T^F_\xi(\mathcal{S})$ is called the $F-$representation of the tangent space $T_\xi(\mathcal{S})$.

For any $A = \sum_{i=1}^{n} A^i \partial_i \in T_\xi(\mathcal{S})$, the corresponding $A(x) = \sum_{i=1}^{n} A^i \partial_i F \in T_\xi^F(\mathcal{S})$ is called the $F-$representation of the tangent vector A and is denoted by $A^F(x)$. Note that $T_\xi^F(\mathcal{S}) \subseteq T_{F(p_\xi)}(\mathbb{R}_X)$. Since $\mathbb{R}_X$ is a vector space, its tangent space $T_{F(p_\xi)}(\mathbb{R}_X)$ can be identified with $\mathbb{R}_X$. So $T_\xi^F(\mathcal{S}) \subseteq \mathbb{R}_X$.

Definition 1. *$F-$expectation of a random variable f with respect to the distribution $p(x;\xi)$ is defined as*

$$E_\xi^F(f) = \int f(x) \frac{1}{p(F'(p))^2} dx. \tag{15}$$

We can use this $F-$expectation to define an inner product in $\mathbb{R}_X$ by

$$< f, g >_\xi^F = E_\xi^F[f(x)g(x)], \tag{16}$$

which induces an inner product on $T_\xi(\mathcal{S})$ by

$$< A, B >_\xi^F = E_\xi^F[A^F(x)B^F(x)] \; ; \; A, B \in T_\xi(\mathcal{S}). \tag{17}$$

Proposition 1. *The induced metric $<,>^F$ on $\mathcal{S}$ is the Fisher information metric $g =<,>$ on $\mathcal{S}$.*

Proof. For any basis vectors $\partial_i, \partial_j \in T_\xi(\mathcal{S})$

$$\begin{aligned} < \partial_i, \partial_j >_\xi^F &= E_\xi^F[\partial_i F \, \partial_j F] \\ &= \int \partial_i F \, \partial_j F \frac{1}{p(F'(p))^2} dx \\ &= \int (p \, F'(p) \, \partial_i \ell) \, (p \, F'(p) \, \partial_j \ell) \, \frac{1}{p(F'(p))^2} dx \\ &= \int \partial_i \ell \, \partial_j \ell \, p(x;\xi) \, dx \\ &= E_\xi[\partial_i \ell \, \partial_j \ell] \\ &= g_{ij}(\xi) \\ &= < \partial_i, \partial_j >_\xi . \end{aligned} \tag{18}$$

So the metric $<,>^F$ on $\mathcal{S}$ induced by the embedding F of $\mathcal{S}$ into $\mathbb{R}_X$ is the Fisher information metric $g =<,>$ on $\mathcal{S}$. □

We can induce a connection on $\mathcal{S}$ using the embedding F.
Let $\pi^F_{|p_\xi} : \mathbb{R}_X \longrightarrow T_\xi^F(\mathcal{S})$ be the projection map.

Definition 2. *The connection induced by the embedding F on $\mathcal{S}$, the $F-$connection, is defined as*

$$\begin{aligned} \nabla^F_{\partial_i} \partial_j &= \pi^F_{|p_\xi}(\partial_i \partial_j F) \\ &= \sum_n \sum_m g^{mn} < \partial_i \partial_j F, \partial_m F >_\xi^F \partial_n . \end{aligned} \tag{19}$$

where $[g^{mn}(\xi)]$ is the inverse of the Fisher information matrix $G(\xi) = [g_{mn}(\xi)]$. Note that the $F-$connections are symmetric.

Lemma 1. *The $F-$connection and its components can be written in terms of scores as*

$$\nabla^F_{\partial_i} \partial_j = \sum_n \sum_m g^{mn} E_\xi \left[(\partial_i \partial_j \ell + (1 + \frac{pF''(p)}{F'(p)}) \partial_i \ell \, \partial_j \ell)(\partial_m \ell) \right] \partial_n \tag{20}$$

and

$$\Gamma^F_{ijk}(\xi) = E_\xi\left[(\partial_i\partial_j\ell + (1+\frac{pF''(p)}{F'(p)})\partial_i\ell\,\partial_j\ell)(\partial_k\ell)\right] \tag{21}$$

Proof. From Equation (12), we have

$$\partial_i\partial_j F = pF'(p)\partial_i\partial_j\ell + [pF'(p)+p^2F''(p)]\,\partial_i\ell\,\partial_j\ell. \tag{22}$$

Therefore

$$\begin{aligned} <\partial_i\partial_jF, \partial_mF>^F_\xi &= \int \partial_i\partial_jF\,\partial_mF\frac{1}{p(F'(p))^2}dx \\ &= \int\left(pF'(p)\partial_i\partial_j\ell + [pF'(p)+p^2F''(p)]\,\partial_i\ell\,\partial_j\ell\right)\frac{\partial_m\ell}{F'(p)}dx \\ &= \int\left(\partial_i\partial_j\ell\,\partial_m\ell + (1+\frac{pF''(p)}{F'(p)})\partial_i\ell\,\partial_j\ell\,\partial_m\ell\right)pdx \\ &= E_\xi\left[(\partial_i\partial_j\ell + (1+\frac{pF''(p)}{F'(p)})\partial_i\ell\,\partial_j\ell)(\partial_m\ell)\right]. \end{aligned} \tag{23}$$

Hence we can write

$$\begin{aligned} \nabla^F_{\partial_i}\partial_j &= \pi^F_{|p_\xi}(\partial_i\partial_jF) \\ &= \sum_n\sum_m g^{mn}E_\xi\left[(\partial_i\partial_j\ell + (1+\frac{pF''(p)}{F'(p)})\partial_i\ell\,\partial_j\ell)(\partial_m\ell)\right]\partial_n. \end{aligned} \tag{24}$$

Then we have the Christoffel symbols of the $F-$connection

$$\Gamma^n_{ij} = \sum_m g^{mn}E_\xi\left[(\partial_i\partial_j\ell + (1+\frac{pF''(p)}{F'(p)})\partial_i\ell\,\partial_j\ell)(\partial_m\ell)\right] \tag{25}$$

and components of the $F-$connection are given by

$$\Gamma^F_{ijk}(\xi) = <\nabla^F_{\partial_i}\partial_j, \partial_k>_\xi = E_\xi\left[(\partial_i\partial_j\ell + (1+\frac{pF''(p)}{F'(p)})\partial_i\ell\,\partial_j\ell)(\partial_k\ell)\right]. \tag{26}$$

□

Theorem 1. *Amari's $\alpha-$geometry is a special case of the $F-$geometry.*

Proof. Let $F(p) = L_\alpha(p)$, $L_\alpha(p)$ is the $\alpha-$embedding of Amari.
The components Γ^α_{ijk} of the $\alpha-$connection are given by

$$\begin{aligned} \Gamma^\alpha_{ijk}(\xi) &= <\nabla^\alpha_{\partial_i}\partial_j, \partial_k>_\xi \\ &= E_\xi\left[(\partial_i\partial_j\ell + \frac{1-\alpha}{2}\partial_i\ell\,\partial_j\ell)(\partial_k\ell)\right]. \end{aligned} \tag{27}$$

From Equation (26), when $F(p) = L_\alpha(p)$
we have

$$F'(p) = L'_\alpha(p) = p^{-(\frac{1+\alpha}{2})} \tag{28}$$

$$F''(p) = L''_\alpha(p) = -\frac{1+\alpha}{2}p^{-(\frac{3+\alpha}{2})}. \tag{29}$$

Then we get

$$1 + \frac{pF''(p)}{F'(p)} = 1 + \frac{pL''_\alpha(p)}{L'_\alpha(p)} = \frac{1-\alpha}{2} \tag{30}$$

Hence

$$\begin{aligned} \Gamma^F_{ijk}(\xi) = <\nabla^F_{\partial_i}\partial_j, \partial_k>_\xi &= E_\xi\left[(\partial_i\partial_j\ell + (1+\frac{pF''(p)}{F'(p)})\partial_i\ell\, \partial_j\ell)(\partial_k\ell)\right] \\ &= E_\xi\left[(\partial_i\partial_j\ell + \frac{1-\alpha}{2}\partial_i\ell\, \partial_j\ell)(\partial_k\ell)\right] \\ &= \Gamma^\alpha_{ijk}(\xi) \end{aligned} \tag{31}$$

which are the components of the $\alpha-$connection. Hence $F-$connection reduces to $\alpha-$connection. Thus we obtain that $\alpha-$geometry is a special case of $F-$geometry. □

Remark 1. *Burbea [14] introduced the concept of weighted Fisher information metric using a positive continuous function. We use this idea to define weighted $F-$metric and weighted $F-$connections. Let $G : (0, \infty) \longrightarrow \mathbb{R}$ be a positive smooth function and F be an embedding, define $(F,G)-$expectation of a random variable with respect to the distribution p_ξ as*

$$E^{F,G}_\xi(f) = \int f(x)\frac{G(p)}{p(F'(p))^2}dx. \tag{32}$$

Define $(F,G)-$metric $<,>^{F,G}_\xi$ in $T_{p_\xi}(\mathcal{S})$ by

$$\begin{aligned} <\partial_i, \partial_j>^{F,G}_\xi &= E^{F,G}_\xi[\partial_i F\, \partial_j F] \\ &= \int \partial_i F\, \partial_j F \frac{G(p)}{p(F'(p))^2}dx \\ &= \int \partial_i\ell\, \partial_j\ell\, G(p)\, p\, dx \\ &= E_\xi[G(p)\, \partial_i\ell\, \partial_j\ell]. \end{aligned} \tag{33}$$

Define $(F,G)-$connection as

$$\begin{aligned} \Gamma^{F,G}_{ijk} &= <\nabla^{F,G}_{\partial_i}\partial_j, \partial_k>_\xi \\ &= E_\xi\left[\left((\partial_i\partial_j\ell + (1+\frac{pF''(p)}{F'(p)})\partial_i\ell\, \partial_j\ell)(\partial_k\ell)\right)(G(p))\right]. \end{aligned} \tag{34}$$

When $G(p) = 1$, $(F,G)-$connection reduces to the $F-$connection and the metric $<,>^{F,G}$ reduces to the Fisher information metric. This is a more general way of defining Riemannian metrics and affine connections on a statistical manifold.

3. Dual Affine Connections

Definition 3. *Let M be a Riemannian manifold with a Riemannian metric g. Two affine connections, ∇ and ∇^* on the tangent bundle are said to be **dual connections** with respect to the metric g if*

$$Zg(X,Y) = g(\nabla_Z X, Y) + g(X, \nabla^*_Z Y) \tag{35}$$

holds for any vector fields X, Y, Z on M.

Theorem 2. *Let F, H be two embeddings of statistical manifold $\mathcal{S}$ into the space $\mathbb{R}_{\mathbb{X}}$ of random variables. Let G be a positive smooth function on $(0, \infty)$. Then the $(F, G)-$connection $\nabla^{F,G}$ and the $(H, G)-$connection $\nabla^{H,G}$ are dual connections with respect to the $(F, G)-$metric iff the functions F and H satisfy*

$$H'(p) = \frac{G(p)}{pF'(p)}. \tag{36}$$

We call such an embedding H as a $G-$dual embedding of F.
The components of the dual connection $\nabla^{H,G}$ can be written as

$$\Gamma^{H,G}_{ijk} = \int \left(\partial_i\partial_j\ell + (\frac{pG'(p)}{G(p)} - \frac{pF''(p)}{F'(p)})\partial_i\ell\, \partial_j\ell \right) \partial_k\ell\, G(p)p\, dx. \tag{37}$$

Proof. $\nabla^{F,G}$ and $\nabla^{H,G}$ are dual connections with respect to the $G-$metric means,

$$\partial_k < \partial_i, \partial_j >^{F,G} = < \nabla^{F,G}_{\partial_k}\partial_i, \partial_j >^{F,G} + < \partial_i, \nabla^{H,G}_{\partial_k}\partial_j >^{F,G}. \tag{38}$$

for any basis vectors $\partial_i, \partial_j, \partial_k \in T_\xi(\mathcal{S})$.

$$\begin{aligned} \partial_k < \partial_i, \partial_j >^{F,G} &= \int \partial_k\partial_j\ell\, \partial_i\ell\, pG(p)dx + \int \partial_k\partial_i\ell\, \partial_j\ell\, pG(p)dx \\ &+ \int (1 + \frac{pG'(p)}{G(p)})\partial_i\ell\, \partial_j\ell\, \partial_k\ell\, pG(p)dx. \end{aligned} \tag{39}$$

$$\begin{aligned} < \nabla^{F,G}_{\partial_k}\partial_i, \partial_j >^{F,G} + < \partial_i, \nabla^{H,G}_{\partial_k}\partial_j >^{F,G} &= \int \partial_k\partial_i\ell\, \partial_j\ell\, pG(p)dx \\ &+ \int 1 + \frac{pF''(p)}{F'(p)}\partial_i\ell\, \partial_j\ell\, \partial_k\ell\, pG(p)dx \\ &+ \int 1 + \frac{pH''(p)}{H'(p)}\partial_i\ell\, \partial_j\ell\, \partial_k\ell\, pG(p)dx \\ &+ \int \partial_k\partial_j\ell\, \partial_i\ell\, pG(p)dx \end{aligned} \tag{40}$$

Then the condition (38) holds iff

$$\begin{aligned} &\int [2 + \frac{pF''(p)}{F'(p)} + \frac{pH''(p)}{H'(p)}]\partial_i\ell\, \partial_j\ell\, \partial_k\ell\, pG(p)dx = \\ &\qquad \int [1 + \frac{pG'(p)}{G(p)}]\partial_i\ell\, \partial_j\ell\, \partial_k\ell\, pG(p)dx \end{aligned} \tag{41}$$

$$\Longleftrightarrow [2 + \frac{pF''(p)}{F'(p)} + \frac{pH''(p)}{H'(p)}] = 1 + \frac{pG'(p)}{G(p)}. \tag{42}$$

$$\Longleftrightarrow 1 + \frac{pH''(p)}{H'(p)} = \frac{pG'(p)}{G(p)} - \frac{pF''(p)}{F'(p)} \tag{43}$$

$$\Longleftrightarrow \frac{H''(p)}{H'(p)} = \frac{G'(p)}{G(p)} - \frac{F''(p)}{F'(p)} - \frac{1}{p} \Longleftrightarrow H'(p) = \frac{G(p)}{pF'(p)}. \tag{44}$$

Hence $\nabla^{F,G}$ and $\nabla^{H,G}$ are dual connections with respect to the $(F, G)-$metric iff Equation (36) holds. From Equation (43), we can rewrite the components of dual connection $\nabla^{H,G}$ as

$$\Gamma^{H,G}_{ijk} = \int \left(\partial_i\partial_j\ell + (\frac{pG'(p)}{G(p)} - \frac{pF''(p)}{F'(p)})\partial_i\ell\, \partial_j\ell \right) \partial_k\ell\, G(p)p\, dx. \tag{45}$$

□

Corollary 1. *Amari's* $0-$*connection is the only self dual* $F-$*connection with respect to the Fisher information metric.*

Proof. From Theorem 2, for $G(p) = 1$ the $F-$connection ∇^F and the $H-$connection ∇^H are dual connections with respect to the Fisher information metric iff the functions F and H satisfy

$$H'(p) = \frac{1}{pF'(p)} \tag{46}$$

Thus the $F-$connection ∇^F is self dual iff the embedding F satisfies the condition

$$F'(p) = \frac{1}{pF'(p)} \iff F'(p) = p^{-(\frac{1}{2})} \iff F(p) = 2p^{\frac{1}{2}} = L_0(p). \tag{47}$$

That is, Amari's $0-$connection is the only self dual $F-$connection with respect to the Fisher information metric. □

So far, we have considered the statistical manifold $\mathcal{S}$ as a subset of $\mathcal{P}(X)$, the set of all probability measures on X. Now we relax the condition $\int p(x)dx = 1$, and consider $\mathcal{S}$ as a subset of $\tilde{\mathcal{P}}(X)$, which is defined by

$$\tilde{\mathcal{P}}(X) := \{p : X \longrightarrow \mathbb{R} \;/\; p(x) > 0 \;(\forall\, x \in X); \int_X p(x)dx < \infty\}. \tag{48}$$

Definition 4. *Let* M *be a Riemannian manifold with a Riemannian metric* g. *Let* ∇ *be an affine connection on* M. *If there exists a coordinate system* $[\theta^i]$ *of* M *such that* $\nabla_{\partial_i}\partial_j = 0$ *then we say that* ∇ *is flat, or alternatively* M *is flat with respect to* ∇, *and we call such a coordinate system* $[\theta^i]$ *an affine coordinate system for* ∇.

Definition 5. *Let* $\mathcal{S} = \{p(x;\xi) \;/\; \xi = [\xi^1, ..., \xi^n] \in \mathbb{E} \subseteq \mathbb{R}^n\}$ *be an* $n-$*dimensional statistical manifold. If for some coordinate system* $[\theta^i]$; $i = 1, ..., n$

$$\partial_i\partial_j F(p(x;\theta)) = 0 \tag{49}$$

then we can see from Equation (19) that $[\theta^i]$ *is an* $F-$*affine coordinate system and that* $\mathcal{S} = \{p_\theta\}$ *is* $F-$*flat. We call such* $\mathcal{S}$ *as an* $F-$*affine manifold.*
The condition (49) is equivalent to the existence of the functions $C, F_1, .., F_n$ *on* X *such that*

$$F(p(x;\theta)) = C(x) + \sum_{i=1}^{n} \theta^i F_i(x) \tag{50}$$

Theorem 3. *For any embedding* F, $\tilde{\mathcal{P}}(X)$ *is an* $F-$*affine manifold for finite* X.

Proof. Let $X = \{x_1, ..., x_n\}$ be a finite set constituted by n elements. Let $F_i : X \longrightarrow \mathbb{R}$ be the functions defined by $F_i(x_j) = \delta_{ij}$ for $i, j = 1, .., n$. Let us define n coordinates $[\theta^i]$ by

$$\theta^i = F(p(x_i)) \tag{51}$$

Then we get $F(p(x)) = \sum_{i=1}^{n} \theta^i F_i(x)$. Therefore $\tilde{\mathcal{P}}(X)$ is an $F-$affine manifold for any embedding $F(p)$. □

Remark 2. *Zhang [13] introduced* ρ*-representation, which is a generalization of* α*-representation of Amari. Zhang's geometry is defined using this* ρ*-representation together with a convex function. Zhang also defined the* ρ*-affine family of density functions and discussed its dually flat structure. The* $F-$*geometry defined using a* *general F-representation is different from the Zhang's geometry. The metric defined in the F-embedding approach is the Fisher information metric and the Riemannian metric defined using the ρ-representation is different from the Fisher information metric. The F-connections defined are not in general dually flat and are different from the dual connections defined by Zhang.*

Remark 3. *On a statistical manifold $\mathcal{S}$, we introduced a dualistic structure (g, ∇^F, ∇^H), where g is the Fisher information metric and ∇^F, ∇^H are the dual connections with respect to the Fisher information metric. Since F-connections are symmetric, the manifold $\mathcal{S}$ is flat with respect to ∇^F iff $\mathcal{S}$ is flat with respect to ∇^H. Thus if $\mathcal{S}$ is flat with respect to ∇^F, then $(\mathcal{S}, g, \nabla^F, \nabla^H)$ is a dually flat space. The dually flat spaces are important in statistical estimation [4].*

4. Invariance of the Geometric Structures

For the statistical manifold $\mathcal{S} = \{p(x;\xi) \mid \xi \in \mathbb{E} \subseteq \mathbb{R}^n\}$, the parameters are merely labels attached to each point $p \in \mathcal{S}$, hence the intrinsic geometric properties should be independent of these labels. Consequently, it is natural to consider the invariance properties of the geometric structures under suitable transformations of the variables in a statistical manifold. Here we can consider two kinds of invariance of the geometric structures; covariance under re-parametrization of the parameter of the manifold and invariance under the transformations of the random variable [15]. Now let us investigate the invariance properties of the F-geometric structures defined in Section 2.

4.1. Covariance under Re-Parametrization

Let $[\theta^i]$ and $[\eta_j]$ be two coordinate systems on S, which are related by an invertible transformation $\eta = \eta(\theta)$. Let us denote $\partial_i = \frac{\partial}{\partial \theta^i}$ and $\partial^j = \frac{\partial}{\partial \eta_j}$. Let the coordinate expressions of the metric g be given by $g_{ij} =< \partial_i, \partial_j >$ and $\tilde{g}_{ij} =< \partial^i, \partial^j >$. Let the components of the connection ∇ with respect to the coordinates $[\theta^i]$ and $[\eta_j]$ be given by $\Gamma_{ijk}, \tilde{\Gamma}_{ijk}$ respectively.

Then the covariance of the metric g and the connection ∇ under the re-parametrization means,

$$\tilde{g}_{ij} = \sum_m \sum_n \frac{\partial \theta^m}{\partial \eta_i} \frac{\partial \theta^n}{\partial \eta_j} g_{mn} \tag{52}$$

$$\tilde{\Gamma}_{ijk} = \sum_{m,n,h} \frac{\partial \theta^m}{\partial \eta_i} \frac{\partial \theta^n}{\partial \eta_j} \frac{\partial \theta^h}{\partial \eta_k} \Gamma_{mnh} + \sum_{m,h} \frac{\partial \theta^h}{\partial \eta_k} \frac{\partial^2 \theta^m}{\partial \eta_i \partial \eta_j} g_{mh} \tag{53}$$

Lemma 2. *The Fisher information metric g is covariant under re-parametrization.*

Proof. The components of the Fisher information metric with respect to the coordinate system $[\theta^i]$ are given by

$$g_{ij}(\theta) = < \partial_i, \partial_j >_\theta = \int \partial_i p(x;\theta) \partial_j p(x;\theta) \frac{1}{p(x;\theta)} dx. \tag{54}$$

Let $\tilde{p}(x;\eta) = p(x;\theta(\eta))$. Then the components of the Fisher information metric with respect to the coordinate system $[\eta_j]$ are given by

$$\tilde{g}_{ij}(\eta) = < \partial^i, \partial^j >_\eta = \int \partial^i \tilde{p}(x;\eta) \partial^j \tilde{p}(x;\eta) \frac{1}{\tilde{p}(x;\eta)} dx. \tag{55}$$

Since

$$\partial^i \tilde{p}(x;\eta) = \sum_m \frac{\partial \theta^m}{\partial \eta_i} \frac{\partial p(x;\theta(\eta))}{\partial \theta^m} \tag{56}$$

we can write

$$\begin{aligned}
\tilde{g}_{ij}(\eta) &= \int \partial^i \tilde{p}(x;\eta)\partial^j \tilde{p}(x;\eta)\frac{1}{\tilde{p}(x;\eta)}dx \\
&= \int \sum_m \frac{\partial\theta^m}{\partial\eta_i}\frac{\partial p(x;\theta)}{\partial\theta^m}\sum_n \frac{\partial\theta^n}{\partial\eta_j}\frac{\partial p(x;\theta)}{\partial\theta^n}\frac{1}{p(x;\theta)}dx \\
&= \sum_m\sum_n \frac{\partial\theta^m}{\partial\eta_i}\frac{\partial\theta^n}{\partial\eta_j}\int \partial_m p(x;\theta)\partial_n p(x;\theta)\frac{1}{p(x;\theta)}dx. \\
&= \left[\sum_m\sum_n \frac{\partial\theta^m}{\partial\eta_i}\frac{\partial\theta^n}{\partial\eta_j} g_{mn}(\theta)\right]_{\theta=\theta(\eta)}
\end{aligned} \tag{57}$$

□

Lemma 3. *The $F-$connection ∇^F is covariant under re-parametrization.*

Proof. Let the components of ∇^F with respect to the coordinates $[\theta^i]$ and $[\eta_j]$ be given by Γ_{ijk}, $\tilde{\Gamma}_{ijk}$ respectively.
Let $\tilde{p}(x;\eta) = p(x;\theta(\eta))$. Let us denote $\log p(x;\theta)$ by $\ell(x;\theta)$ and $\log \tilde{p}(x;\eta)$ by $\tilde{\ell}(x;\eta)$.
The components of the $F-$connection ∇^F with respect to the coordinate system $[\theta^i]$ are given by

$$\Gamma_{ijk} = \int \left(\partial_i\partial_j\ell(x;\theta) + (1+\frac{pF''(p)}{F'(p)})\partial_i\ell(x;\theta)\,\partial_j\ell(x;\theta)\right)\partial_k\ell(x;\theta)p(x;\theta)dx \tag{58}$$

The components of ∇^F with respect to the coordinate system $[\eta_j]$ are given by

$$\tilde{\Gamma}_{ijk} = \int \left(\partial^i\partial^j\tilde{\ell}(x;\eta) + (1+\frac{\tilde{p}F''(\tilde{p})}{F'(\tilde{p})})\partial^i\tilde{\ell}(x;\eta)\,\partial^j\tilde{\ell}(x;\eta)\right)\partial^k\tilde{\ell}(x;\eta)\tilde{p}(x;\eta)dx \tag{59}$$

We can write

$$\partial^i\tilde{\ell}(x;\eta) = \sum_m \frac{\partial\theta^m}{\partial\eta_i}\frac{\partial\ell(x;\theta(\eta))}{\partial\theta^m} \tag{60}$$

Then

$$\partial^i\partial^j\tilde{\ell}(x;\eta) = \sum_{m,n}\frac{\partial\theta^m}{\partial\eta_i}\frac{\partial\theta^n}{\partial\eta_j}\frac{\partial^2\ell(x;\theta(\eta))}{\partial\theta^m\partial\theta^n} + \sum_m \frac{\partial^2\theta^m}{\partial\eta_i\partial\eta_j}\frac{\partial\ell(x;\theta(\eta))}{\partial\theta^m} \tag{61}$$

$$\partial^i\tilde{\ell}(x;\eta)\,\partial^j\tilde{\ell}(x;\eta) = \sum_{m,n}\frac{\partial\theta^m}{\partial\eta_i}\frac{\partial\theta^n}{\partial\eta_j}\frac{\partial\ell(x;\theta(\eta))}{\partial\theta^m}\frac{\partial\ell(x;\theta(\eta))}{\partial\theta^n} \tag{62}$$

$$\partial^k\tilde{\ell}(x;\eta) = \sum_h \frac{\partial\theta^h}{\partial\eta_k}\frac{\partial\ell(x;\theta(\eta))}{\partial\theta^h} \tag{63}$$

Hence we get

$$\begin{aligned}
\tilde{\Gamma}_{ijk} &= \int \sum_{m,n,h}\frac{\partial\theta^m}{\partial\eta_i}\frac{\partial\theta^n}{\partial\eta_j}\frac{\partial\theta^h}{\partial\eta_k}\frac{\partial^2\ell(x;\theta(\eta))}{\partial\theta^m\partial\theta^n}\frac{\partial\ell(x;\theta(\eta))}{\partial\theta^h}p(x;\theta(\eta))dx + \\
&\int \sum_{m,h}\frac{\partial^2\theta^m}{\partial\eta_i\partial\eta_j}\frac{\partial\theta^h}{\partial\eta_k}\frac{\partial\ell(x;\theta(\eta))}{\partial\theta^m}\frac{\partial\ell(x;\theta(\eta))}{\partial\theta^h}p(x;\theta(\eta))dx + \\
&\int (1+\frac{pF''(p)}{F'(p)})\sum_{m,n,h}\frac{\partial\theta^m}{\partial\eta_i}\frac{\partial\theta^n}{\partial\eta_j}\frac{\partial\theta^h}{\partial\eta_k}\frac{\partial\ell(x;\theta(\eta))}{\partial\theta^m}\frac{\partial\ell(x;\theta(\eta))}{\partial\theta^n}\frac{\partial\ell(x;\theta(\eta))}{\partial\theta^h}p(x;\theta(\eta))dx
\end{aligned} \tag{64}$$

$$
\begin{aligned}
&= \sum_{m,n,h} \frac{\partial\theta^m}{\partial\eta_i}\frac{\partial\theta^n}{\partial\eta_j}\frac{\partial\theta^h}{\partial\eta_k}\int \frac{\partial^2\ell(x;\theta(\eta))}{\partial\theta^m\partial\theta^n}\frac{\partial\ell(x;\theta(\eta))}{\partial\theta^h}p(x;\theta(\eta))dx + \\
&\sum_{m,h} \frac{\partial^2\theta^m}{\partial\eta_i\partial\eta_j}\frac{\partial\theta^h}{\partial\eta_k}\int \frac{\partial\ell(x;\theta(\eta))}{\partial\theta^m}\frac{\partial\ell(x;\theta(\eta))}{\partial\theta^h}p(x;\theta(\eta))dx + \\
&\sum_{m,n,h} \frac{\partial\theta^m}{\partial\eta_i}\frac{\partial\theta^n}{\partial\eta_j}\frac{\partial\theta^h}{\partial\eta_k}\int (1+\frac{pF''(p)}{F'(p)})\frac{\partial\ell(x;\theta(\eta))}{\partial\theta^m}\frac{\partial\ell(x;\theta(\eta))}{\partial\theta^n}\frac{\partial\ell(x;\theta(\eta))}{\partial\theta^h}p(x;\theta(\eta))dx \\
&= \sum_{m,n,h} \frac{\partial\theta^m}{\partial\eta_i}\frac{\partial\theta^n}{\partial\eta_j}\frac{\partial\theta^h}{\partial\eta_k}\Gamma_{mnh} + \sum_{m,h}\frac{\partial\theta^h}{\partial\eta_k}\frac{\partial^2\theta^m}{\partial\eta_i\partial\eta_j}g_{mh}
\end{aligned}
$$

□

Hence we showed that $F-$connections are covariant under re-parametrization of the parameter. The covariance under re-parametrization actually means that the metric and connections are coordinate independent. Hence we obtained that the $F-$geometry is coordinate independent.

4.2. Invariance Under the Transformation of the Random Variable

Amari and Nagaoka [4] defined the invariance of Riemannian metric and connections on a statistical manifold under a transformation of the random variable as follows,

Definition 6. *Let $\mathcal{S} = \{p(x;\xi) \mid \xi \in \mathbb{E} \subseteq \mathbb{R}^n\}$ be a statistical manifold defined on a sample space X. Let x, y be random variables defined on sample spaces X, Y respectively and ϕ be a transformation of x to y. Assume that this transformation induces a model $\mathcal{S}' = \{q(y;\xi) \mid \xi \in \mathbb{E} \subseteq \mathbb{R}^n\}$ on Y. Let $\lambda : \mathcal{S} \longrightarrow \mathcal{S}'$ be a diffeomorphism defined as*

$$\lambda(p_\xi) = q_\xi \tag{65}$$

Let $g = <>, g' = <>'$ be two Riemannian metrics defined on $\mathcal{S}$ and $\mathcal{S}'$ respectively. Let ∇, ∇' be two affine connections on $\mathcal{S}$ and $\mathcal{S}'$ respectively. Then the invariance properties are given by

$$
\begin{aligned}
< X, Y >_p &= < \lambda_*(X), \lambda_*(Y) >'_{\lambda(p)} \quad \forall\, X, Y \in T_p(\mathcal{S}) \quad (66)\\
\lambda_*(\nabla_X Y) &= \nabla'_{\lambda_*(X)}\lambda_*(Y) \quad (67)
\end{aligned}
$$

where λ_* is the push forward map associated with the map λ, which is defined by

$$\lambda_*(X)_{\lambda(p)} = (d\lambda)_p(X) \tag{68}$$

Now we discuss the invariance properties of the $F-$geometry under suitable transformations of the random variable. Let us restrict ourselves to the case of smooth one-to-one transformations of the random variable that are in fact statistically interesting. Amari and Nagaoka [4] mentioned a transformation, the sufficient statistic of the parameter of the statistical model, which is widely used in statistical estimation. In fact the one-to-one transformations of the random variable are trivial examples of sufficient statistic.

Consider a statistical manifold $\mathcal{S} = \{p(x;\xi) \mid \xi \in \mathbb{E} \subseteq \mathbb{R}^n\}$ defined on a sample space X. Let ϕ be a smooth one-to-one transformation of the random variable x to y. Then the density function $q(y;\xi)$ of the induced model $\mathcal{S}'$ takes the form

$$q(y:\xi) = p(w(y);\xi)w'(y) \tag{69}$$

where w is a function such that $x = w(y)$ and $\phi'(x) = \frac{1}{w'(\phi(x))}$.
Let us denote $\log q(y;\xi)$ by $\ell(q_y)$ and $\log p(x;\xi)$ by $\ell(p_x)$.

Lemma 4. *The Fisher information metric and Amari's α-connections are invariant under smooth one-to-one transformations of the random variable.*

Proof. Let ϕ be a smooth one-to-one transformation of the random variable x to y.
From Equation (69)

$$\begin{aligned} p(x;\xi) &= q(\phi(x);\xi)\phi'(x) & (70)\\ \partial_i \ell(q_y) &= \partial_i \ell(p_{w(y)}) & (71)\\ \partial_i \ell(q_{\phi(x)}) &= \partial_i \ell(p_x) & (72) \end{aligned}$$

The Fisher information metric g' on the induced manifold S' is given by

$$\begin{aligned} g'_{ij}(q_\xi) &= \int_Y \partial_i \ell(q_y)\, \partial_j \ell(q_y)\, q(y;\xi) dy \\ &= \int_X \partial_i \ell(q_{\phi(x)})\, \partial_j \ell(q_{\phi(x)})\, q(\phi(x);\xi)\, \phi'(x) dx \\ &= \int_X \partial_i \ell(p_x)\, \partial_j \ell(p_x)\, p(x;\xi) dx \\ &= g_{ij}(p_\xi) \end{aligned} \tag{73}$$

which is the Fisher information metric on S.
The components of Amari's $\alpha-$connections on the induced manifold S' are given by

$$\begin{aligned} \acute{\Gamma}^{\alpha}_{ijk}(q_\xi) &= \int_Y \partial_i\partial_j \ell(q_y)\, \partial_k \ell(q_y)\, q(y;\xi) dy\ + \\ &\quad \int_Y \frac{1-\alpha}{2} \partial_i \ell(q_y)\, \partial_j \ell(q_y)\, \partial_k \ell(q_y)\, q(y;\xi) dy \\ &= \int_X \partial_i\partial_j \ell(q_{\phi(x)})\, \partial_k \ell(q_{\phi(x)})\, q(\phi(x);\xi)\phi'(x) dx\ + \\ &\quad \int_X \frac{1-\alpha}{2} \partial_i \ell(q_{\phi(x)})\, \partial_j \ell(q_{\phi(x)})\, \partial_k \ell(q_{\phi(x)})\, q(\phi(x);\xi)\phi'(x) dx \\ &= \int_X \partial_i\partial_j \ell(p_x)\, \partial_k \ell(p_x)\, p(x;\xi) dx\ + \\ &\quad \int_X \frac{1-\alpha}{2} \partial_i \ell(p_x)\, \partial_j \ell(p_x)\, \partial_k \ell(p_x)\, p(x;\xi) dx \\ &= \Gamma^{\alpha}_{ijk}(p_\xi) \end{aligned} \tag{74}$$

which are the components of Amari's $\alpha-$connections on the manifold S. Thus we obtained that the Fisher information metric and Amari's α-connections are invariant under smooth one-to-one transformations of the random variable. □

Now we prove that α-connections are the only $F-$connections that are invariant under smooth one-to-one transformations of the random variable.

Theorem 4. *Amari's α-connections are the only $F-$connections that are invariant under smooth one-to-one transformations of the random variable.*

Proof. Let ϕ be a smooth one-to-one transformation of the random variable x to y.
The components of the $F-$connection of the induced manifold $\mathcal{S}'$ are

$$\begin{aligned}
\acute{\Gamma}^F_{ijk}(q_\xi) &= \int_Y \left(\partial_i\partial_j\ell(q_y) + (1+\frac{qF''(q)}{F'(q)})\partial_i\ell(q_y)\;\partial_j\ell(q_y)\right)\partial_k\ell(q_y)\;q(y;\xi)dy \\
&= \int_X \partial_i\partial_j\ell(p_x)\;\partial_k\ell(p_x)\;p(x;\xi)dx + \\
&\quad \int_X (1+\frac{q(\phi(x);\xi)F''(q(\phi(x);\xi))}{F'(q(\phi(x);\xi))})\partial_i\ell(p_x)\;\partial_j\ell(p_x)\;\partial_k\ell(p_x)\;p(x;\xi)dx.
\end{aligned} \tag{75}$$

and the components of the $F-$connection of the manifold $\mathcal{S}$ are

$$\begin{aligned}
\Gamma^F_{ijk}(p_\xi) &= \int_X \partial_i\partial_j\ell(p_x)\;\partial_k\ell(p_x)\;p(x;\xi)dx + \\
&\quad \int_X (1+\frac{p(\phi(x);\xi)F''(p(x;\xi))}{F'(p(x;\xi))})\partial_i\ell(p_x)\;\partial_j\ell(p_x)\;\partial_k\ell(p_x)\;p(x;\xi)dx.
\end{aligned} \tag{76}$$

Then by equating the components $\acute{\Gamma}^F_{ijk}(q_\xi), \Gamma^F_{ijk}(p_\xi)$ of the $F-$connection, we get

$$\begin{aligned}
\int \frac{q(\phi(x);\xi)F''(q(\phi(x);\xi))}{F'(q(\phi(x);\xi))}\partial_i\ell(p_x)\;\partial_j\ell(p_x)\;\partial_k\ell(p_x)\;p(x;\xi)dx = \\
\int \frac{p(x;\xi)F''(p(x;\xi))}{F'(p(x;\xi))}\partial_i\ell(p_x)\;\partial_j\ell(p_x)\;\partial_k\ell(p_x)\;p(x;\xi)dx
\end{aligned} \tag{77}$$

Then it follows that the condition for $F-$connection to be invariant under the transformation ϕ is given by

$$\frac{pF''(p)}{F'(p)} = k, \tag{78}$$

where k is a real constant.
Hence it follows from the Euler's homogeneous function theorem that the function F' is a positive homogeneous function in p of degree k. So

$$F'(\lambda p) = \lambda^k F'(p) \;\text{ for } \lambda > 0. \tag{79}$$

Since F' is a positive homogeneous function in the single variable p, without loss of generality we can take,

$$F'(p) = p^k. \tag{80}$$

Therefore

$$F(p) = \begin{cases} \frac{p^{k+1}}{k+1} & k \neq -1 \\ \log p & k = -1 \end{cases} \tag{81}$$

Let

$$k = \frac{-(1+\alpha)}{2},\; \alpha \in \mathbb{R}. \tag{82}$$

we get

$$F(p) = \begin{cases} \frac{2}{1-\alpha}p^{\frac{1-\alpha}{2}} & \alpha \neq 1 \\ \log p & \alpha = 1 \end{cases} \tag{83}$$

which is nothing but Amari's $\alpha-$embeddings $L_\alpha(p)$. Hence we obtain that Amari's $\alpha-$connections are the only $F-$connections that are invariant under smooth one-to-one transformations of the random variable. □

Remark 4. *In Section 2, we defined (F, G)-connections using a general embedding function F and a positive smooth function G. We can show that (F, G)-connection is invariant under smooth one-to-one transformation of the random variable when $G(p) = c$, where c is a real constant and $F(p) = L_\alpha(p)$ (proof is similar to that of Theorem 4). The notion of $(F, G)-$metric and $(F, G)-$connection provides a more general way of introducing geometric structures on a manifold. We were able to show that the Fisher information metric (up to a constant) and Amari's $\alpha-$connections are the only metric and connections belonging to this class that are invariant under both the transformation of the parameter and the one-to-one transformation of the random variable.*

5. Conclusions

The Fisher information metric and Amari's $\alpha-$connections are widely used in the theory of information geometry and have an important role in the theory of statistical estimation. Amari's $\alpha-$connections are defined using a one parameter family of functions, the $\alpha-$embeddings. We generalized this idea to introduce geometric structures on a statistical manifold $\mathcal{S}$. We considered a general embedding function F of $\mathcal{S}$ into $\mathbb{R}_X$ and obtained a geometric structure on $\mathcal{S}$ called the $F-$geometry. Amari's $\alpha-$geometry is a special case of $F-$geometry. A more general way of defining Riemannian metrics and affine connections on a statistical manifold $\mathcal{S}$ is given using a positive continuous function G and the embedding F.

Amari's $\alpha-$geometry is the only $F-$geometry that is invariant under both the transformation of the parameter and the random variable or equivalently under the sufficient statistic. We can relax the condition of invariance under the sufficient statistic and can consider other statistically significant transformations as well, which then gives an $F-$geometry other than $\alpha-$geometry that is invariant under these statistically significant transformations. We believe that the idea of $F-$geometry can be used in the further development of the geometric theory of q-exponential families. We look forward to studying these problems in detail later.

Acknowledgments: We are extremely thankful to Shun-ichi Amari for reading this article and encouraging our learning process. We would like to thank the reviewer who mentioned the references [13,16] that are of great importance in our future work.

Author Contributions: The authors contributed equally to the presented mathematical framework and the writing of the paper.

Conflicts of Interest: The authors declare no conflicts of interest.

References

1. Amari, S. Differential geometry of curved exponential families-curvature and information loss. *Ann. Statist.* **1982**, *10*, 357–385.
2. Amari, S. *Differential-Geometrical Methods in Statistics*; Lecture Notes in Statistics, Volume 28; Springer-Verlag: New York, NY, USA, 1985.
3. Amari, S.; Kumon, M. Differential geometry of Edgeworth expansions in curved exponential family. *Ann. Inst. Statist. Math.* **1983**, *35*, 1–24.
4. Amari, S.; Nagaoka, H. *Methods of Information Geometry, Translations of Mathematical Monographs*; Oxford University Press: Oxford, UK, 2000.
5. Barndorff-Nielsen, O.E.; Cox, D.R.; Reid, N. The role of differential geometry in statistical theory. *Internat. Statist. Rev.* **1986**, *54*, 83–96.
6. Dawid, A.P. A Discussion to Efron's paper. *Ann. Statist.* **1975**, *3*, 1231–1234.
7. Efron, B. Defining the curvature of a statistical problem (with applications to second order efficiency). *Ann. Statist.* **1975**, *3*, 1189–1242.
8. Efron, B. The geometry of exponential families. *Ann. Statist.* **1978**, *6*, 362–376.
9. Murray, M.K.; Rice, R.W. *Differential Geometry and Statistics*; Chapman & Hall: London, UK, 1995.
10. Rao, C.R. Information and accuracy attainable in the estimation of statistical parameters. *Bull. Calcutta. Math. Soc.* **1945**, *37*, 81–91.

11. Chentsov, N.N. *Statistical Decision Rules and Optimal Inference;* Transted in English, Translation of the Mathematical Monographs; American Mathematical Society: Providence, RI, USA, 1982.
12. Corcuera, J.M.; Giummole, F. A characterization of monotone and regular divergences. *Ann. Inst. Statist. Math.* **1998**, *50*, 433–450.
13. Zhang, J. Divergence function, duality and convex analysis. *Neur. Comput.* **2004**, *16*, 159–195.
14. Burbea, J. Informative geometry of probability spaces. *Expo Math.* **1986**, *4*, 347–378.
15. Wagenaar, D.A. Information Geometry for Neural Networks. Available online: http://www.danielwagenaar.net/res/papers/98-Wage2.pdf (accessed on 13 December 2013).
16. Amari, S.; Ohara, A.; Matsuzoe, H. Geometry of deformed exponential families: Invariant, dually flat and conformal geometries. *Physica A* **2012**, *391*, 4308–4319.

entropy

MDPI

Article

Computational Information Geometry in Statistics: Theory and Practice

Frank Critchley [1] and Paul Marriott [2,*]

[1] Department of Mathematics and Statistics, The Open University, Walton Hall, Milton Keynes, Buckinghamshire MK7 6AA, UK; E-Mail: f.critchley@open.ac.uk

[2] Department of Statistics and Actuarial Science, University of Waterloo, 200 University Avenue West, Waterloo, ON N2L 3G1, Canada

* E-Mail: pmarriot@uwaterloo.ca; Tel.: +1-519-888-4567.

Received: 27 March 2014; in revised form: 25 April 2014 / Accepted: 29 April 2014 / Published: 2 May 2014

Abstract: A broad view of the nature and potential of computational information geometry in statistics is offered. This new area suitably extends the manifold-based approach of classical information geometry to a simplicial setting, in order to obtain an operational universal model space. Additional underlying theory and illustrative real examples are presented. In the infinite-dimensional case, challenges inherent in this ambitious overall agenda are highlighted and promising new methodologies indicated.

Keywords: information geometry; computational geometry; statistical foundations

1. Introduction

The application of geometry to statistical theory and practice has seen a number of different approaches developed. One of the most important can be defined as starting with Efron's seminal paper [?] on statistical curvature and subsequent landmark references, including the book by Kass and Vos [?]. This approach, a major part of which has been called information geometry, continues today, a primary focus being invariant higher-order asymptotic expansions obtained through the use of differential geometry. A somewhat representative example of the type of result it generates is taken from [?], where the notation is defined:

Example 1. *The bias correction of a first-order efficient estimator, $\hat{\beta}$, is defined by:*

$$b^a(\beta) = -\frac{1}{2n} g^{aa'} \left\{ g^{bc} \Gamma^{(-1)}_{a'bc} + g^{\kappa\lambda} h^{(-1)}_{\kappa\lambda a'} \right\},$$

and has the property that if $\hat{\beta}^* := \hat{\beta} - b(\beta)$ then:

$$E_\beta(\hat{\beta}^* - \beta) = O(n^{-3/2}).$$

The strengths usually claimed of such a result are that, for a worker fluent in the language of information geometry, it is explicit, insightful as to the underlying structure and of clear utility in statistical practice. We agree entirely. However, the overwhelming evidence of the literature is that, while the benefits of such inferential improvements are widely acknowledged in principle, in practice, the overhead of first becoming fluent in information geometry prevents their routine use. As a result, a great number of powerful results of practical importance lay severely underused, locked away behind notational and conceptual bars.

This paper proposes that this problem can be addressed computationally by the development of what we call computational information geometry. This gives a mathematical and numerical computational framework in which the results of information geometry can be encoded as "black-box" numerical algorithms, allowing direct access to their power. Essentially, this works by exploiting the structural properties of information geometry, which are such that all formulae can be expressed in terms of four fundamental building blocks: defined and detailed in Amari [?], these are the $+1$ and -1 geometries, the way that these are connected via the Fisher information and the foundational duality theorem. Additionally, computational information geometry enables a range of methodologies and insights impossible without it; notably, those deriving from the operational, universal model space, which it affords; see, for example, [? ? ?].

The paper is structured as follows. Section 2 looks at the case of distributions on a finite number of categories where the extended multinomial family provides an exhaustive model underlying the corresponding information geometry. Since the aim is to produce a computational theory, a finite representation is the ultimate aim, making the results of this section of central importance. The paper also emphasises how the simplicial structures introduced here are foundational to a theory of computational information geometry. Being intrinsically constructive, a simplicial approach is useful both theoretically and computationally. Section 3 looks at how simplicial structures, defined for finite dimensions, can be extended to the infinite dimensional case.

2. Finite Discrete Case

2.1. Introduction

This section shows how the results of classical information geometry can be applied in a purely computational way. We emphasise that the framework developed here can be implemented in a purely algorithmic way, allowing direct access to a powerful information geometric theory of practical importance.

The key tool, as explained in [?], is the simplex:

$$\Delta^k := \left\{ \mathbf{\beta} = (\mathbf{\beta}_0, \mathbf{\beta}_1, \ldots, \mathbf{\beta}_k)^\top \;:\; \mathbf{\beta}_i \geq \mathbf{0}\,,\, \sum_{i=0}^{k} \mathbf{\beta}_i = \mathbf{1} \right\}, \tag{1}$$

with a label associated with each vertex. Here, k is chosen to be sufficiently large, so that any statistical model—by which we mean a sample space, a set of probability distributions and selected inference problem—can be embedded. The embedding is done in such a way that all the building blocks of information geometry (*i.e.*, manifold, affine connections and metric tensor) can be numerically computed explicitly. Within such a simplex, we can embed a large class of regular exponential families; see [?] for details. This class includes exponential family random graph models, logistic regression, log-linear and other models for categorical data analysis. Furthermore, the multinomial family on $k+1$ categories is naturally identified with the relative interior of this space, $int(\Delta^k)$, while the extended family, Equation (**??**), is a union of distributions with different support sets.

This paper builds on the theory of information geometry following that introduced by [?] via the affine space construction introduced by [?] and extended by [?]. Since this paper concentrates on categorical random variables, the following definitions are appropriate. Consider a finite set of disjoint categories or bins $\mathcal{B} = \{B_i\}_{i\in A}$. Any distribution over this finite set of categories is defined by a set, $\{\pi_i\}_{i\in A}$, which defines the corresponding probabilities. With "mix" connoting mixtures of distributions, we have:

Definition 1. *The -1-affine space structure over distributions on $\mathcal{B} := \{B_i\}_{i\in A}$ is $(X_{mix}, V_{mix}, +)$ where:*

$$X_{mix} = \left\{ \{x_i\}_{i\in A} | \sum_{i\in A} x_i = 1 \right\}, V_{mix} = \left\{ \{v_i\}_{i\in A} | \sum_{i\in A} v_i = 0 \right\}$$

and the addition operator, $+$, is the usual addition of sequences.

In Definition **??**, the space of (discretised) distributions is a -1-convex subspace of the affine space, $(X_{mix}, V_{mix}, +)$. A similar affine structure for the $+1$-geometry, once the support has been fixed, can be derived from the definitions in [**?**].

2.2. Examples

Examples **??** and **??** are used for illustration. The second of these is a moderately high dimensional family, where the way that the boundaries of the simplex are attached to the model is of great importance for the behaviour of the likelihood and of the maximum likelihood estimate. In general, working in a simplex, boundary effects mean that standard first order asymptotic results can fail, while the much more flexible higher order methods can be very effective. The other example is a continuous curved exponential family, where both higher order asymptotic sampling theory results and geometrically-based dimension reduction are described.

Example 2. *The paper [**?**] models survival times for leukaemia patients. These times, recorded in days, start at the time of diagnosis, and there are 43 observations; see [**?**] for details. We further assume that the data is censored at a fixed value. It was observed that a censored exponential distribution gives a reasonable, but not exact, fit. As discussed in [**?**], this gives a one-dimensional curved exponential family inside a two-dimensional regular exponential family of the form:*

$$\exp\left[\lambda_1 x + \lambda_2 y - \log\left\{\frac{1}{\lambda_2}\left(e^{\lambda_2 t} - 1\right) + e^{\lambda_1 + \lambda_2 t}\right\}\right], \tag{2}$$

where $y = \min(z, t)$ and $x = I(z \geq t)$, and the embedding map is given by $(\lambda_1(\theta), \lambda_2(\theta)) = (-\log\theta, -\theta)$.

*As shown in [**?**], the loss due to discretisation can be made arbitrarily small for all information geometry objects. Thus, for example, using this computational approach, it is straightforward to compute the bias correction described in Example **??**. Each of the terms in the asymptotic bias, i.e., the metric, g_{ij}, its inverse, g^{ij}, the Christoffel symbols, $\Gamma^{(-1)}_{ijk}$, and curvature term, $h^{(-1)}$, can be directly numerically coded as appropriate finite difference approximations to derivatives. Thus, "black-box" code can directly calculate the numerical value of the asymptotic bias, and this numerical value can then be used by those who are not familiar with information geometry. For example this calculation establishes the fact that, with this particular data set, the sample size is such that the bias is inferentially unimportant.*

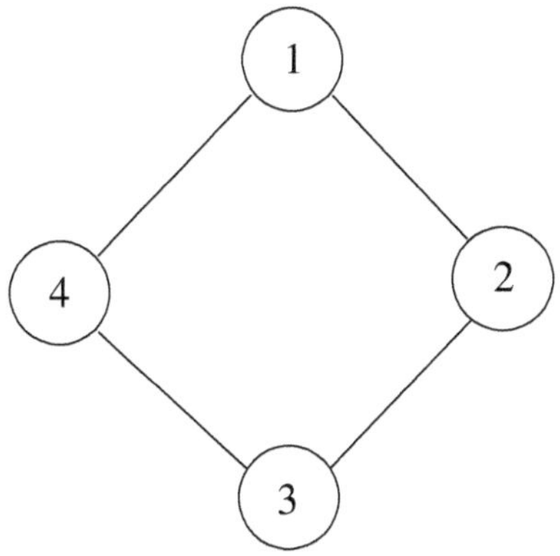

Figure 1. Undirected graphical model showing the cyclic graph of order four.

Example 3. *The paper [**?**] discusses an undirected graphical model based on the cyclic graph of order four, shown in Figure **??**, with binary random variables at each node. Without any constraints, there are 16 possible values for the graph, so model space can be thought of as a 15-dimensional simplex, including the relative* *boundary. However, the conditional independence relations encoded by the graph impose linear constraints in the natural parameters of the exponential family. Thus, the resultant model is a lower dimensional full exponential family and its closure.*

As described in [?], the four cycle model is a seven dimensional exponential family, which is a $+1$*-affine subspace of the* $+1$*-affine structure of the* 15*-dimensional simplex. The model can be written in the form:*

$$\left(\frac{\pi_i \exp\left\{ \sum_{h=1}^{8} \eta_h v_{hi} \right\}}{\sum_{j=0}^{15} \pi_j \exp\left\{ \sum_{h=1}^{8} \eta_h v_{hj} \right\}} \right)_{i=0}^{15} \tag{3}$$

for a given set of linearly independent vectors $\{v_h\}_{h=1}^{8}$*. The existence of the maximum likelihood estimate for* $\eta = (\eta_h)$ *will depend on how the limit points of Model (??) meet the observed face of* Δ^{15}*; that is, the span of the vertices (bins) having positive counts. Thus, a key computational task is to learn how a full exponential family, defined by a representation of the form of (??), is attached to boundary sub-simplices of the high-dimensional embedding simplex.*

In order to visualise the geometric aspects of this problem, consider a lower dimensional version. Define a two-dimensional full exponential family by the vectors $v_1 = (1,2,3,4), v_2 = (1,4,9,-1)$ *and the uniform distribution base point,* π_i*, embedded in the three-dimensional simplex. The two-dimensional family is defined by the* $+1$*-affine space through* $(0.25, 0.25, 0.25, 0.25)$ *spanned by the space of vectors of the form:*

$$\alpha(1,2,3,4) + \beta(1,4,9,-1) = (\alpha + \beta, 2\alpha + 4\beta, 3\alpha + 9\beta, 4\alpha - \beta).$$

Consider directions from the origin obtained by writing $\alpha = \theta\beta$*, giving, for each* θ*, a one-dimensional, full exponential family parameterized by* β *in the direction* $\beta(\theta + 1, 2\theta + 4, 3\theta + 9, 4\theta - 1)$*. The aspect of this vector, which determines the connection to the boundary, is the rank order of its elements. For example, suppose the first component was the maximum and the last the minimum. Then, as* $\beta \to \pm\infty$*, this one-dimensional family will be connected to the first and fourth vertex of the embedding four simplex, respectively. Note that changing the value of* θ *changes the rank structure, as illustrated in Figure ??. This plot shows the four element-wise linear functions of* θ *(dashed lines) and the salient overall feature of their rank order; that is, their upper and lower envelopes (solid lines). From this analysis of the envelopes of a set of linear functions, it can be seen that the function* $2\theta + 4$ *is redundant. The consequence of this is shown in Figure ??, which shows a direct computation of the two-dimensional family. It is clear that, indeed, only three of the four vertexes have been connected by the model.*

In general, the problem of finding the limit points in full exponential families inside simplex models is a problem of finding redundant linear constraints. As shown in [?], this can be converted, via convex duality, into the problem of finding extremal points in a finite dimensional affine space. In the four-cycle model, this technique can construct all sub-simplices containing limit points of the four-cycle model. For example, it can be shown that all of the 16 vertices are part of the boundary. Once the boundary points have been identified as necessary and sufficient, conditions for the existence of the maximum likelihood in the $+1$*-parameters can easily be found computationally [?].*

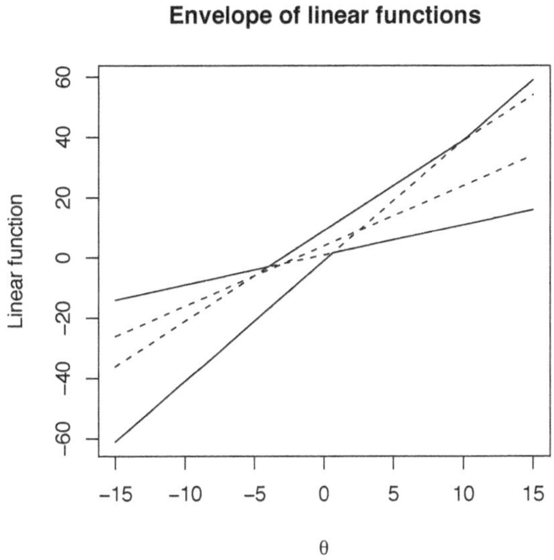

Figure 2. The envelope of a set of linear functions. Functions, dashed lines; envelope, solid lines.

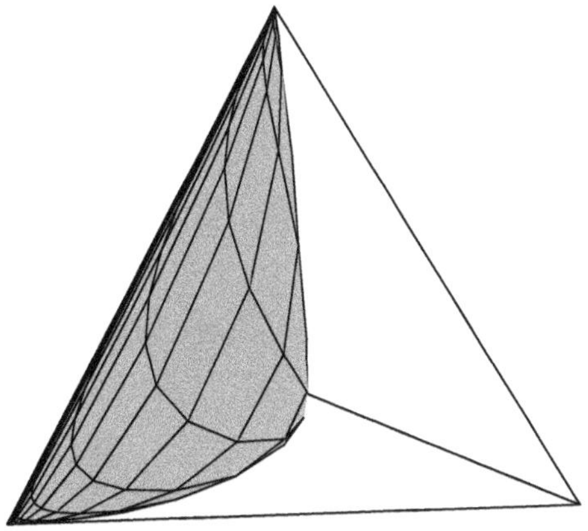

Figure 3. Attaching a two-dimensional example to the boundary of the simplex.

2.3. *Tensor Analysis and Numerical Stability*

One of the most powerful set of results from classical information geometry is the way that geometrically-based tensor analysis is perfect for use in multi-dimensional higher order asymptotic analysis; see [?] or [?]. The tensorial formulation does, however, present a couple of problems in practice. For many, its very tight and efficient notational aspects can obscure rather than enlighten, while the resulting formulae tend to have a very large number of terms, making them rather cumbersome to work with explicitly. These are not problems at all for the computational approach described in this paper. Rather, the clarity of the tensorial approach is ideal for coding, where large numbers of additive terms, of course, are easy to deal with.

Two more fundamental issues, which the global geometric approach of this paper highlights, concern numerical stability. The ability to invert the Fisher information matrix is vital in most tensorial formulae, and so understanding its spectrum, discussed in Section ??, is vital. Secondly, numerical underflow and overflow near boundaries require careful analysis, and so, understanding the way that models are attached to the boundaries of the extended multinomial models is equally important. The four-cycle model, to which we now return, illustrates computational information geometry doing this effectively.

Example 4. *The multivariate Edgeworth approximation to the sampling distribution of part of the sufficient statistic for the four-cycle model is shown in Figure ??. Using the techniques described above, a point near the boundary of the* 15*-simplex has been selected as the data generation process. For illustration, we focus on the marginal distribution of two components of the sufficient statistic, though any number could have been chosen. The boundary forces constraints on the range of the sufficient statistics, shown by the dashed line in the plot. The points, jittered for clarity, show the distribution computed by simulation. It is typical that such boundary constraints prevent standard first order methods from performing well, but the greater flexibility of higher order methods can be seen to work well here. As discussed above, methods, such as the multivariate Edgeworth expansion, can be strongly exploited in a computational framework, such as ours. Note, the discretization that can be observed in the figure is extensively discussed in [?].*

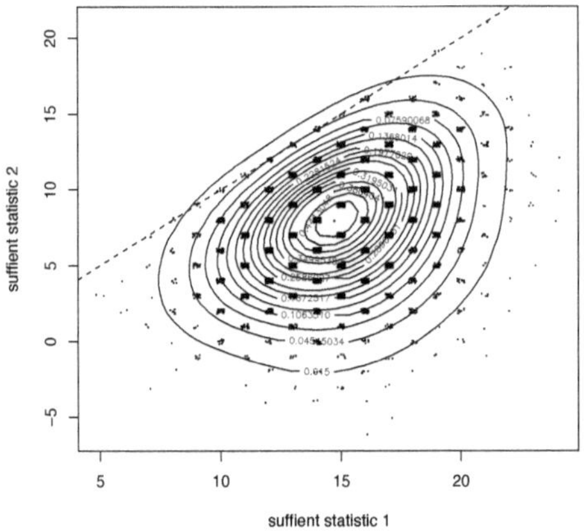

Figure 4. Using the Edgeworth expansion near the boundary of four-cycle model.

2.4. Spectrum of Fisher Information

We focus now on the second numerical issue identified above. In any multinomial, the Fisher information matrix and its inverse are explicit. Indeed, the 0-geodesics and the corresponding geodesic distance are also explicit; see [?] or [?]. However, since the simplex glues together multinomial structures with different supports and the computational theory is in high dimensions, it is a fact that the Fisher information matrix can be arbitrarily close to being singular. It is therefore of central interest that the spectral decomposition of the Fisher information itself has a very nice structure, as shown below.

Example 5. *Consider a multinomial distribution based on* 81 *equal width categories on* $[-5,5]$*, where the probability associated to a bin is proportional to that of the standard normal distribution for that bin. The Fisher information for this model is an* 80×80 *matrix, whose spectrum is shown in Figure ??. By inspection, it can be seen that there are exponentially small eigenvalues, so that while the matrix is positive definite, it is also arbitrarily close to being singular. Furthermore, it can be seen that the spectrum has the shape of a half-normal density function and that the eigenvalues seem to come in pairs. These facts are direct consequences of the general results below.*

With π_{-0} denoting the vector of all bin probabilities, except π_0, we can write the Fisher information matrix (in the +1 form) as N times:

$$I(\pi) := diag(\pi_{-0}) - \pi_{-0}\pi_{-0}^T.$$

This has an explicit spectral decomposition, which can be computed by using interlacing eigenvalue results (see for example [?], Chapter 4). In particular, if the diagonal matrices, $diag(\pi_1, \ldots, \pi_k)$ and $diag(\lambda_1 I_{m_1}|\cdots|\lambda_g I_{m_g})$, agree up to a row-and-column permutation, where $g > 1$ and $\lambda_1 > \cdots > \lambda_g > 0$, then $I(\pi)$ has ordered spectrum:

$$\lambda_1 > \tilde{\lambda}_1 > \cdots > \lambda_g > \tilde{\lambda}_g \geq 0, \tag{4}$$

with $\tilde{\lambda}_g > 0 \iff \pi_0 > 0$, each λ_i having multiplicity $m_i - 1$, while each $\tilde{\lambda}_g$ is simple.

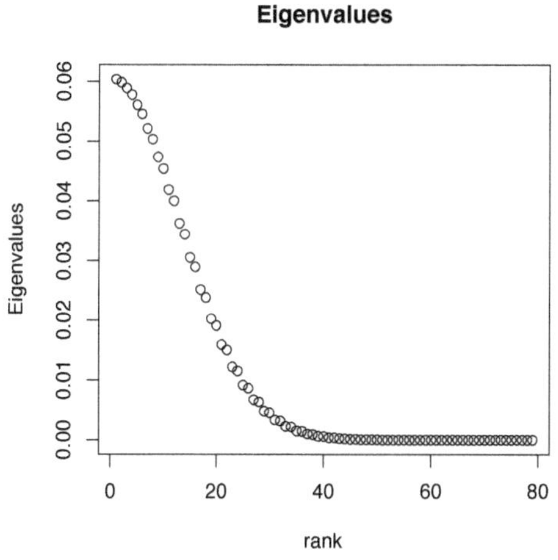

Figure 5. Spectrum of the Fisher information matrix of a discretised normal distribution.

We give a complete account of the spectral decomposition (SpD) of $I(\pi)$. There are four cases to consider, the last having the generic spectrum of (**??**). Without loss, after permutation, assume now $\pi_1 \geq \cdots \geq \pi_k$. The four cases are:

Case 1 For some $l < k$, the last $k - l$ elements of π_{-0} vanish: the sub-case $l = 0 \iff \pi_0 = 1 \iff I(\pi) = 0$ is trivial. Otherwise, writing $\pi_+ = (\pi_1, \ldots, \pi_l)^T$ and $\Pi_+ = diag(\pi_+)$, the SpD of:

$$I(\pi) = \left(\begin{array}{c|c} \Pi_+ - \pi_+\pi_+^T & 0 \\ \hline 0 & 0 \end{array} \right)$$

follows at once from that of $\Pi_+ - \pi_+\pi_+^T$, given below.

Case 2 $k = 1$: this case is trivial.

Case 3 $k > 1, \pi = \lambda 1_k, \lambda > 0$: the SpD of $I(\pi)$ is:

$$\lambda C_k + \lambda(1 - k\lambda)J_k$$

where $C_k = I_k - J_k$ and $J_k = k^{-1}1_k 1_k^T$. Here, λ has multiplicity $k - 1$ and eigenspace $[Span(1_k)]^\perp$, while $\tilde{\lambda} := \lambda(1 - k\lambda)$ has multiplicity one and eigenspace $Span(1_k)$. In particular, since $1 - \pi_0 = k\lambda$, it follows that:

$$I(\pi) \text{ is singular} \iff \pi_0 = 0.$$

Case 4 $\pi_{-0} = (\lambda_1 1_{m_1}^T | \dots | \lambda_g 1_{m_g}^T)^T$, $g > 1$ and $\lambda_1 > \cdots > \lambda_g > 0$:

This is the generic case, having the spectrum of (**??**) above. Denoting by O_m the zero matrix of order $m \times m$ and by $P(v)$ the rank one orthogonal projector onto Span(v), ($v \neq 0$), the SpD is:

$$\sum_{i=1, m_i > 1}^{g} \lambda_i diag(O_{m_{i-}}, C_{m_i}, O_{m_{i-}}) + \sum_{i=1}^{g} \tilde{\lambda}_i P\left(\left(\frac{\lambda_1}{\tilde{\lambda}_i - \lambda_1} 1_{m_1}^T, \dots, \frac{\lambda_g}{\tilde{\lambda}_i - \lambda_g} 1_{m_g}^T\right)^T\right),$$

where: $m_{i-} = \Sigma\{m_j | j < i\}$, $m_{i+} = \Sigma\{m_j | j > i\}$ and the $\tilde{\lambda}_i$ are the zeros of:

$$h(\tilde{\lambda}) := 1 + \sum_{i=1}^{g} \frac{m_i \lambda_i^2}{\tilde{\lambda} - \lambda_i} = (1 - \sum_{i=1}^{g} m_i \lambda_i) + \tilde{\lambda}\left(\sum_{i=1}^{g} \frac{m_i \lambda_i}{\tilde{\lambda} - \lambda_i}\right).$$

In particular, $\{\tilde{\lambda}_i : i = 1, \cdots, g\}$ are simple eigenvalues satisfying (**??**) while, whenever $m_i > 1$, λ_i, is also an eigenvalue having multiplicity $m_i - 1$. Further, expanding $\det(I(\pi))$, we again find:

$$I(\pi) \text{ is singular } \iff \pi_0 = 0,$$

so that $\widetilde{\lambda}_g > 0 \Leftrightarrow \pi_0 > 0$, as claimed. Finally, we note that each $\widetilde{\lambda}_i$ ($i < g$) is typically (much) closer to λ_i than to λ_{i+1}. For, considering the graph of $x \to 1/x$, $h\left((\lambda_i + \lambda_{i+1})/2 + \delta(\lambda_i - \lambda_{i+1})/2\right)$ $(-1 < \delta < +1)$ is well-approximated by:

$$1 - \frac{2m_i \lambda_i^2}{(\lambda_i - \lambda_{i+1})(1 - \delta)} + \frac{2m_{i+1} \lambda_{i+1}^2}{(\lambda_i - \lambda_{i+1})(1 + \delta)}$$

whose unique zero δ_* over $(-1, 1)$ is positive whenever, as will typically be the case, $m_i = m_{i+1}$ (both will usually be one), while $(m_i \lambda_i + m_{i+1} \lambda_{i+1}) < 1/2$. Indeed, a straightforward analysis shows that, for any m_i and m_{i+1}, $\delta_* = 1 + O(\lambda_i)$ as $\lambda_i \to 0$.

2.5. Total Positivity and Local Mixing

Mixture modelling is an exemplar of a major area of statistics in which computational information geometry enables distinctive methodological progress. The -1-convex hull of an exponential family is of great interest, mixture models being widely used in many areas of statistical science. In particular, they are explored further in [**?**]. Here, we simply state the main result, a simple consequence of the total positivity of exponential families [**?**], that, generically, convex hulls are of maximal dimension. In this result, "generic" means that the $+1$ tangent vector, which defines the exponential family as having components that are all distinct.

Theorem 1. *The -1-convex hull of an open subset of a generic one-dimensional exponential family is of full dimension.*

Proof. For any $(\pi_i) \in \Delta^k$ with each $\pi_i > 0$, $\theta_0 < \cdots < \theta_k$ and $s_0 < \cdots < s_k$, let $B = (\pi(\theta_0), ..., \pi(\theta_k))$ have general element:

$$\pi_i(\theta_j) := \pi_i \exp[s_i \theta_j - \psi(\theta_j)].$$

Further, let $\widetilde{B} = B - \pi(\theta_0) 1_{k+1}^T$, whose general column is $\pi(\theta_j) - \pi(\theta_0)$. Then, it suffices to show that $\widetilde{B}$ has rank k. However, using [**?**] (p. 33), $Rank(\widetilde{B}) = Rank(B) - 1$, so that:

$$Rank(\widetilde{B}) = k \Leftrightarrow B \text{ is nonsingular } \Leftrightarrow B^* \text{ is nonsingular},$$

where $B^* = (\exp[s_i\theta_j])$. It suffices, then, to recall [?] that $K(x,y) = \exp(xy)$ is strictly total positive (of order ∞), so that $\det B^* > 0$.

□

3. Infinite Dimensional Structure

This section will start to explore the question of whether the simplex structure, which describes the finite dimensional space of distributions, can extend to the infinite dimensional case. We examine some of the differences with the finite dimensional case, illustrating them with clear, commonly occurring examples.

3.1. Infinite Dimensional Information Geometry: A Review

In the previous sections, the underlying computational space is always finite dimensional. This section looks at issues related to an infinite dimensional extension of the theory in that paper. There is a great deal of literature concerning infinite dimensional statistical models. The discussion here concentrates on information geometric, parametrisation and boundary issues.

The information geometry theory of Amari [?] has a geometric foundation, where statistical models (typically full and curved exponential families) have a finite dimensional manifold structure. When considering the extension to infinite dimensional cases, Amari notes the problem of finding an "adequate topology" [?] (p. 93). There has to be very interesting work following up this topological challenge. By concentrating on distributions with a common support, the paper [?] uses the geometry of a Banach manifold, where local patches on the manifold are modelled by Banach spaces, via the concept of an Orlicz space. This gives a structure that is analogous to an infinite dimensional exponential family, with mean and natural parameters and including the ability to define mixed parametrisations. One drawback of this Banach structure, as pointed out in [?], is that the likelihood function with finite samples is not continuous on the manifold. Fukumizu uses a reproducing kernel Hilbert space structure rather than a Banach manifold, which is a stronger topology. There are strong connections between the approach taken in [?] and the material in Section ??, we note two issues here: (1) a focus on the finite nature of the data; and (2) using a Hilbert structure defined by a cumulant generating function. The approaches differ in that [?] uses a manifold approach rather than the simplicial complex as the fundamental geometric object. There is also other work that explicitly used infinite dimensional Hilbert spaces in statistics, a good reference being [?].

In this paper, in contrast to previous authors, a simplicial, rather than a manifold-based, approach is taken. This allows distributions with varying support, as well as closures of statistical families to be included in the geometry. Another difference in approach is the way in which geometric structures are induced by infinite dimensional affine spaces rather than by using an intrinsic geometry. This approach was introduced by [?] and extended by [?]. Spaces of distributions are convex subsets of the affine spaces, and their closure within the affine space is key to the geometry.

In exponential families, the -1-affine structure is often called the mean parametrisation, and using moments as parameters is one very important part of modelling. In the infinite dimensional case, the use of moments as a parameter system is related to the classical moment problem—when does there exist a (unique) distribution whose moments agree with a given sequence?—which has generated a vast literature in its own right; see [? ? ?]. In general terms, the existence of a solution to the moment problem is connected to positivity conditions on moment matrices. Such conditions have been used in connection to the infinite dimensional geometry of mixture models [?]. Uniqueness, however, is a much more subtle problem: sufficient conditions can be formulated in terms of the rate of growth of the moments [?]. Counter examples to general uniqueness results include the log-normal distribution [?].

The geometry of the Fisher information is also much more complex in general spaces of distributions than in exponential families. Simple mixture models, including two-component mixtures of exponential distributions [?], can have "infinite" expected Fisher information, which gives rise to non-standard inference issues. Similar results on infinitely small (and large) eigenvalues of covariance operators are also noted in [?]. Since the Fisher information is a covariance, the fact that it does not exist for certain distributions or that its spectrum can be unbounded above or arbitrarily close to zero is not a surprise. However, these observations do need to be taken into account when considering the information geometry of infinite dimensional spaces.

The rest of this section looks at the topology and geometry of the infinite dimensional simplex and gives some illustrative examples, which, in particular, show the need for specific Hilbert space structures, discussed in the final section.

3.2. Topology

For simplicity and concreteness, in this section, we will be looking at models for real valued random variables. In this paper, we restrict attention to the cases where the sample space is $\mathbb{R}^+$ or $\mathbb{R}$ and has been discretised to a countably infinite set of bins, B_i, with $i \in \mathbb{N}$ or $\mathbb{Z}$, respectively. In the finite case, the basic object is the standard simplex, Δ^k, with $k+1$ bins. We generalise this to countable unions of such objects. Of these, one is of central importance, denoted by Δ_{emp} or simply Δ, because it is the smallest object that contains all possible empirical distributions.

Definition 2. *For any finite subset of bins, indexed by $\mathcal{I} \subset \mathbb{N}$ or $\mathbb{Z}$, denote*

$$\Delta_{\mathcal{I}} = \left\{ \mathbf{x} = (x_i)_{i \in \mathcal{I}} : x_i \geq 0 \,,\ \sum_{i \in \mathcal{I}} x_i = 1 \right\}.$$

We take the union of all such sets $\bigcup_{|\mathcal{I}|<\infty} \Delta_{\mathcal{I}}$, where $|\mathcal{I}|$ denotes the number of elements of the index set. This can always be written as:

$$\Delta = \left\{ \mathbf{x} = (x_i)_{i \in \mathbb{Z}} : \sum_{i \in \mathbb{Z}} x_i = 1, x_i \geq 0 \text{ and only finitely many } x_i > 0 \right\}.$$

In what follows, it is important to note that for any given statistical inference problem, the sample size, n, is always finite, even if we frequently use asymptotic approximations, where $n \to \infty$. Thus, the data, as represented by the empirical distribution, naturally lie in the space, Δ. However, many models, used in the given inference problem, will have support over all bins, so the models most naturally lie in the "boundary" constructed using the closures of the set. These objects are subsets of sequence spaces, and the corresponding topologies can be constructed from the Banach spaces, ℓ_p, $p \in [1, \infty]$. The following results follow directly from explicit calculations, where we note that in this section, since all terms are non-negative, convergence always means absolute convergence. In particular, arbitrary rearrangements of series do not affect the existence of limits or their values.

Example 6. *Consider the sequence of "uniform distributions" $\mathbf{x}^{(n)} = (\frac{1}{n}, \ldots, \frac{1}{n}, 0, \ldots)$ as elements of Δ. This has an ℓ_p limit of the zero sequence for $p \in (1, \infty]$.*

Proposition 1. *The ℓ_p extreme points of Δ, for $p \in (1, \infty]$, are the zero sequence and the sequences, $\mathbf{ffi}_i$ $(i \in \mathbb{Z})$, with one as the $i-th$ element and zero elsewhere.*

For $p \in [1, \infty]$, let $\overline{\Delta}_p \subset \ell_p$ denote the ℓ_p closure of Δ.

Theorem 2. *(a)* $\overline{\Delta}_1 = \{\mathbf{x} = (x_i)_{i \in \mathbb{Z}} : x_i \geq 0, \sum_{i \in \mathcal{I}} x_i = 1\}$.
(b) $\overline{\Delta}_\infty = \{\mathbf{x} = (x_i)_{i \in \mathbb{Z}} : x_i \geq 0, \sum_{i \in \mathcal{I}} x_i \leq 1\}$.
(c) For $p \in (1, \infty)$, $\overline{\Delta}_p = \overline{\Delta}_\infty = \{\mathbf{x} = (x_i)_{i \in \mathbb{Z}} : x_i \geq 0, \sum_{i \in \mathcal{I}} x_i \leq 1\}$.

Proof. (a) It is immediate that $\{\mathbf{x} = (x_i)_{i\in\mathbb{Z}} : x_i \geq 0, \sum_{i\in\mathbb{Z}} x_i = 1\} \subseteq \overline{\Delta}_1$. Conversely, if $\bar{x}$ is a limit point, then all its elements must be non-negative. Finally, if $\sum_{i=1}^{\infty} \bar{x}_i$ is not bounded above by one, then there exists N, such that $\sum_{i=1}^{N} \bar{x}_i > 1+\epsilon$ for some $\epsilon > 0$. Hence, $\sum_{i=1}^{\infty} |\bar{x}_i - x_i^{(n)}| \geq \sum_{i=1}^{N} |\bar{x}_i - x_i^{(n)}| \geq \sum_{i=1}^{N} \bar{x}_i - \sum_{i=1}^{N} x_i^{(n)} > \epsilon$ for all n, which contradicts convergence. If $\sum_{i=1}^{\infty} \bar{x}_i < 1-\epsilon$, then $\sum_{i=1}^{\infty} |\bar{x}_i - x_i^{(n)}| \geq \sum_{i=1}^{\infty} x_i^{(n)} - \sum_{i=1}^{\infty} \bar{x}_i > \epsilon$, which again contradicts convergence.

(b) It is again immediate that $\{\mathbf{x} = (x_i)_{i\in\mathbb{Z}} : x_i \geq 0, \sum_{i\in\mathbb{Z}} x_i = 1\} \subseteq \overline{\Delta}_\infty$. However, by Example **??**, the zero sequence is also in $\overline{\Delta}_\infty$, so that $\{\mathbf{x} = (x_i)_{i\in\mathbb{Z}} : x_i \geq 0, \sum_{i\in\mathbb{Z}} x_i \leq 1\} \subseteq \overline{\Delta}_\infty$.

Conversely, by contradiction, it is easy to see that all elements of the closure must have non-negative elements. Finally, for any $\bar{x} \in \overline{\Delta}_\infty$, if $\sum_{i=1}^{\infty} \bar{x}_i$ is not bounded above by one, there exists N, such that $\sum_{i=1}^{N} \bar{x}_i > 1+\epsilon$ for some $\epsilon > 0$. For any sequence of points, $x^{(n)}$ in Δ, we have that $\sum_{i=1}^{N} x_i^{(n)} \leq 1$, so that, for $i = 1, \ldots, N$, the maximum value of $|x_i^{(n)} - \bar{x}_i| > \epsilon/N$. Hence, for all sequences, $x^{(n)}$, we have $\|x^{(n)} - \bar{x}\|_\infty > \epsilon/N$, which contradicts $\bar{x}$ being in the closure.

(c) This follows essentially the same argument as (b) by noting in the case where $\sum_{i=1}^{\infty} \bar{x}_i$ is not bounded above by one, we have:

$$\|x^{(n)} - \bar{x}\|_p^p \geq \sum_{i=1}^{N} |\bar{x}_i - x_i^{(n)}|^p \geq N \max_{i=1,\ldots N} |x_i^{(n)} - \bar{x}_i|^p > N^{1-p}\epsilon^p$$

for any sequences, $x^{(n)}$, which contradicts $\bar{x}$ being in the closure. □

It is immediate that the spaces, Δ and $\overline{\Delta}_1$, are convex subsets of ℓ_1 and that $\overline{\Delta}_\infty$ is a convex set in ℓ_∞.

3.3. Geometry

In the same way as for the finite case, the -1-geometry can be defined using an affine space structure using the following definition.

Definition 3. *Let $\mathcal{I}$ be a countable index set which is a subset of $\mathbb{Z}$. The -1-affine space structure over distributions is $(X_{mix}, V_{mix}, +)$, where:*

$$X_{mix} = \left\{ \mathbf{x} = (x_i) | \sum_{\mathcal{I}} x_i = 1, \sum_{\mathcal{I}} |x_i| < \infty \right\}, V_{mix} = \left\{ \mathbf{v} = v_i | \sum_{\mathcal{I}} v_i = 0, \sum_{\mathcal{I}} |v_i| < \infty, \right\},$$

and $\mathbf{x} + \mathbf{v} = (x_i + v_i)$.

In order to define the $+1$-geometric structure, we also follow the approach used in the finite case. Initially, to understand the $+1$- structure, consider the case where all distributions have a common support, *i.e.*, assume $\pi_i > 0$ for all i. We follow here the approach of [**?**].

Definition 4. *Consider the set of non-negative measures on $\mathbb{N}$ or $\mathbb{Z}$ and the equivalence relation defined by:*

$$\{a_i\} \sim \{b_i\} \iff \exists \lambda > 0\, s.t. \forall i\; a_i = \lambda b_i.$$

The equivalences classes of this are the points in the $+1$ geometry.

These points can be further partitioned into sets with the same support, i.e., $\text{supp}(< a >) = \{i : a_i > 0\}$, *where this is clearly well-defined.*

On sets of $+1$-points with the same support, we can define the $+1$-geometry in the same way as in the finite case. With "exp" connoting an exponential family distribution, we have:

Definition 5. *For a given index set, $\mathcal{I}$, define X_{exp} to be all $+1$-points whose support equals $\mathcal{I}$, and define the vector space $V_{exp} = \{v_i, i \in \mathcal{I}\}$ with the operation, $\oplus$, defined by:*

$$< x_i > \oplus v_i = \langle x_i \exp(v_i) \rangle ,$$

is an affine space. The $+1$-affine structure is then defined by $(X_{exp}, V_{exp}, \oplus)$.

Theorem 3. *If* **a** *and* **b** *lie in Δ (or $\overline{\Delta}_1$) and have the same support, then $C(\rho) = \sum(a_i^{\rho} b_i^{(1-\rho)}) < \infty$ for $\rho \in [0,1]$. Hence, $\frac{a_i^{\rho} b_i^{(1-\rho)}}{C(\rho)} \in \Delta$ (or $\overline{\Delta}_1$).*

Proof. Since a, b are absolutely convergent, the sequence, $\max(a_i, b_i)$, is also. Since we have:

$$0 \leq \min(a_i, b_i) \leq a_i^{\rho} b_i^{1-\rho} \leq \max(a_i, b_i)$$

it follows that $C(\rho) < \infty$, and we have the result. □

This result shows that sets in $\overline{\Delta}_1$ with the same support are $+1$-convex, just as the faces in the finite case are.

3.4. Examples

In order to get a sense of how the $+1$-geometry works, let us consider a few illustrative examples.

Example 7. *If we denote the discretised standard normal density by* **a** *and the discretised Cauchy density by* **b** *and consider the path:*

$$\frac{a_i^{\rho} b_i^{(1-\rho)}}{C(\rho)},$$

the normalising constant is shown in Figure ??. We see that at $\rho = 0$ (the Cauchy distribution), we have that the derivative of the normalising constant (i.e., the mean of the sufficient statistic) is tending to infinity. At the other end ($\rho = 1$), the model can be extended in the sense that the distribution exists for values greater than one.

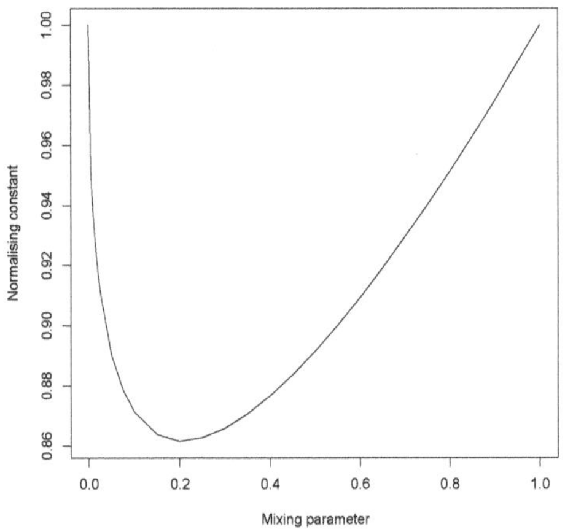

Figure 6. Normalising constant for normal-Cauchy exponential mixing example.

Thus, in this example, the path joining the two distributions is an extended, rather than natural, exponential family, since we have to include the boundary point where the mean is unbounded.

Example 8. *Let us return to Example ??, but now without the censoring. Thus, now, there is a countably infinite set of bins, and so, we can investigate its embedding in the infinite simplex. As discussed in [?], we shall discretise the continuous distribution by computing the probabilities associated to bins* $[c_i, c_{i+1}]$, $i = 1, 2, \cdots$.

For the exponential model, $Exp(\theta)$, *the bin probabilities are simply:*

$$\pi_i(\theta) = \exp(-\theta c_i) - \exp(-\theta c_{i+1}).$$

Using this, the model will lie in the infinite simplex on the positive half line with the index set $\mathcal{I} = \mathbb{N}$.

First, consider the case where we have a uniform choice of discretisation, where $c_n = n \times \epsilon$ *for some fixed,* $\epsilon > 0$. *In this case, the bin probabilities can be written as an exponential family:*

$$\pi_n(\theta) = \exp\left[-\theta\epsilon n + \log(1 - e^{-\theta\epsilon})\right]$$

for $\theta > 0$. *This gives a* $+1$*-geodesic though* $\{\pi_i(\theta_0)\}$ *in the direction* $\{\epsilon \times n\}$ *of the form:*

$$\pi_n(\theta_0)\exp\left[-\lambda\epsilon n + \log\left(\frac{1 - e^{-(\lambda+\theta_0)\epsilon}}{1 - e^{-\theta_0\epsilon}}\right)\right] \tag{5}$$

for $\lambda > -\theta_0$. *In the case where* $\lambda \to -\theta_0$, *the limiting distribution is the zero measure in* $\overline{\Delta}_\infty$, *and at the other extreme, where* $\lambda \to \infty$, *the limiting distribution is the atomic distribution in the first bin, a distribution with a different support than* $\pi_i(\theta_0)$. *However, unlike the finite case, there is no guarantee that, for a given "direction",* $\{t_i\}$, *there exists a* $+1$*-geodesic starting at* $\{\pi_i(\theta_0)\}$, *since we require the convergence of the normalising constant:*

$$\sum_{i=0}^{\infty} \pi_i(\theta_0)\exp(\lambda t_i) < \infty.$$

From this example, we see that the limit points of exponential families can lie in the space, $\overline{\Delta}_\infty$, but not in $\overline{\Delta}_1$. The next example shows that limits do not have to exist at all.

Example 9. *Consider the family whose bin probabilities,* $\pi_i \in \overline{\Delta}_\infty$, *are proportional to a discretised standard normal with bins of constant width. The exponential family, which is proportional to* $\pi_i \exp(\theta i)$, *does not have an* ℓ_∞ *limit, as it is discretised normal with mean* θ. *The natural parameter space here is* $(-\infty, \infty)$.

The last illustrative example is from [?] and shows that even for simple models, the Fisher information for the parameters of interest need not be finite.

Example 10. *Let us consider a simple example of a two-component mixture of (discretised) exponential distributions:*

$$(1-\rho)\pi_i(\theta_0 + \lambda) + \rho\pi_i(\theta_0) \tag{6}$$

the tangent vector in the ρ*-direction is:*

$$\pi_i(\theta_0) - \pi_i(\theta_0 + \lambda) = \pi_i(\theta_0)\left(1 - e^{-\lambda\epsilon n}C\right)$$

for a positive constant, C. The corresponding squared length, with respect to the Fisher information, is:

$$\sum_{n=0}^{\infty} \frac{\left(1 - e^{-\lambda\epsilon n}C\right)^2}{\pi_i(\theta_0)}.$$

As an example, consider $\theta_0 = 1$; *then, this term will be infinite for* $\lambda \leq -0.5$.

3.5. Hilbert Space Structures

Following these examples, we can consider the Hilbert space structure of exponential families inside the infinite simplex with the following results.

Definition 6. *Define the functions, $S(\cdot)$, by $S(\{v_i\}, \text{ß}) = \sup_\theta \{\theta | \sum_{\mathcal{I}} \pi_i \exp(\theta v_i) < \infty\}$, the function being set to ∞ when the set is unbounded. Furthermore, define for a given $\{\pi_i\} \in \bar{\Delta}_\infty$, the set:*

$$V(\text{ß}) = \{\{v_i\} | S(\{v_i\}, \text{ß} > 0\}, \text{ and } V^c(\text{ß}) = \{\{v_i\} | \pm \{v_i\} \in V(\text{ß})\}.$$

The spaces, $V^c(\text{ß})$, correspond to the directions in which the $+1$-geodesic and, so, the corresponding exponential families are well-defined and have particularly "nice" geometric structures.

Theorem 4. *For ß, define a Hilbert space by:*

$$H(\text{ß}) := \left\{\{v_i\} | \sum v_i^2 \pi_i < \infty\right\}$$

with inner product:

$$\langle \{v_i\}, \{w_i\} \rangle_{\text{ß}} = \sum v_i w_i \pi_i,$$

and corresponding norm $|| \cdot ||_{\text{ß}}$. Under these conditions:
(i) $V^c(\text{ß})$ is a subspace of $H(\text{ß})$, and
(ii) the set $V(\text{ß})$ is a convex cone.

Proof. (i) First, if $\{v_i\} \in V^c(\text{ß})$, then by definition, the moment generating function:

$$\sum \exp(\theta v_i) \pi_i,$$

is finite for θ in an open set containing $\theta = 0$. Hence, have both:

$$\sum v_i \pi_i < \infty, \text{ and } \sum v_i^2 \pi_i < \infty.$$

Thus, $\{v_i\} \in H(\text{ß})$. The fact that it is a subspace follows from (ii) below.

(ii) It is immediate that $V(\text{ß})$ is a cone.

Convexity follows from the Cauchy–Schwartz inequality, since for all $\{v_i\}, \{v_i^*\} \in V(\text{ß})$ and $\lambda \in [0, 1]$, it follows that:

$$\begin{aligned} \left\{\sum \pi_i e^{\frac{\theta}{2}(\lambda v_i + (1-\lambda) v_i^*)}\right\}^2 &= \left\{\sum \left(\sqrt{\pi_i} e^{\frac{\theta}{2}\lambda v_i}\right) \left(\sqrt{\pi_i} e^{\frac{\theta}{2}(1-\lambda) v_i^*}\right)\right\}^2 \\ &\leq \left\{\sum \pi_i e^{\theta \lambda v_i}\right\} \left\{\sum \pi_i e^{\theta(1-\lambda) v_i^*}\right\}, \end{aligned}$$

and, so, is finite for a strictly positive value of θ, hence $\{\lambda v_i + (1-\lambda) v_i^*\} \in V(\text{ß})$. □

Hence, this result illustrates the point above regarding the existence of "nice" geometric structure in the sense of Amari's information geometry developed for finite dimensional exponential families. Infinite dimensional families have a richer structure; for example, they include the possibility of having an infinite Fisher information; see Examples **??** and **??**.

Acknowledgments: The authors would like to thank Karim Anaya-Izquierdo and Paul Vos for many helpful discussions and the UK's Engineering and Physical Sciences Research Council (EPSRC) for the support of grant number EP/E017878/.

Author Contributions: All authors contributed to the conception and design of the study, the collection and analysis of the data and the discussion of the results. All authors read and approved the final manuscript.

Conflicts of Interest: The authors declare no conflict of interest.

References

1. Efron, B. Defining the curvature of a statistical problem (with applications to second order efficiency). *Ann. Stat.* **1975**, *3*, 1189–1242.
2. Kass, R.E.; Vos, P.W. *Geometrical Foundations of Asymptotic Inference*; John Wiley & Sons: London, UK, 1997.
3. Amari, S.-I. *Differential-Geometrical Methods in Statistics*; Lecture Notes in Statistics; Springer-Verlag Inc.: New York, NY, USA, 1985; Volume 28.
4. Anaya-Izquierdo, K.; Critchley, F.; Marriott, P.; Vos, P. Computational Information Geometry: Foundations. In *Geometric Science of Information*; Nielsen, F., Barbaresco, F., Eds.; Lecture Notes in Computer Science; Springer: Berlin/Heidelberg, Germany, 2013; Volume 8085, pp. 311–318.
5. Anaya-Izquierdo, K.; Critchley, F.; Marriott, P.; Vos, P. Computational Information Geometry: Mixture Modelling. In *Geometric Science of Information*; Nielsen, F., Barbaresco, F., Eds.; Lecture Notes in Computer Science; Springer: Berlin/Heidelberg, Germany, 2013; Volume 8085, pp. 319–326.
6. Anaya-Izquierdo, K.; Critchley, F.; Marriott, P. When are first order asymptotics adequate? A diagnostic. *Stat* **2014**, *3*, 17–22.
7. Murray, M.K.; Rice, J.W. *Differential Geometry and Statistics*; Chapman & Hall: London, UK, 1993.
8. Marriott, P. On the local geometry of mixture models. *Biometrika* **2002**, *89*, 95–97.
9. Hand, D.J.; Daly, F.; Lunn, A.D.; McConway, K.J.; Ostrowski, E. *A Handbook of Small Data Sets*; Chapman and Hall: London, UK, 1994.
10. Bryson, M.C.; Siddiqui, M.M. Survival times: Some criteria for aging. *J. Am. Stat. Assoc.* **1969**, *64*, 1472–1483.
11. Marriott, P.; West, S. On the geometry of censored models. *Calcutta Stat. Assoc. Bull.* **2002**, *52*, 567–576.
12. Geiger, D.; Heckerman, D.; King, H.; Meek, C. Stratified exponential families: Graphical models and model selection. *Ann. Stat.* **2001**, *29*, 505–529.
13. Edelsbrunner, H. *Algorithms in Combinatorial Geometry*; Springer-Verlag: NewYork, NY, USA, 1987.
14. Barndorff-Nielsen, O.E.; Cox, D.R. *Asymptotic Techniques for Use in Statistics*; Chapman & Hall: London, UK, 1989.
15. McCullagh, P. *Tensor Methods in Statistics*; Chapman & Hall: London, UK, 1987.
16. Horn, R.A.; Johnson, C.R. *Matrix Analysis*; Cambridge Universtiy Press: Cambridge, UK, 1985.
17. Karlin, S. *Total Positivity*; Stanford University Press: Stanford, CA, USA, 1968; Volume I.
18. Householder, A.S. *The Theory of Matrices in Numerical Analysis*; Dover Publications: Dover, DE, USA, 1975.
19. Pistone, G.; Rogantin, M.P. The exponential statistical manifold: Mean parameters, orthogonality and space transformations. *Bernoulli* **1999**, *5*, 571–760.
20. Fukumizu, K. Infinite dimensional exponential families by reproducing kernel Hilbert spaces. In Proceedings of the 2nd International Symposium on Information Geometry and its Applications, Tokyo, Japan, 12–16 December 2005.
21. Small, C.G.; McLeish, D.L. *Hilbert Space Methods in Probability and Statistical Inference*; John Wiley & Sons: London, UK, 1994.
22. Akhiezer, N.I. *The Classical Moment Problem*; Hafner: New York, NY, USA, 1965.
23. Stoyanov, J.M. *Counter Examples in Probability*; John Wiley & Sons: London, UK, 1987.
24. Gut, A. On the moment problem. *Bernoulli* **2002**, *8*, 407–421.
25. Lindsay, B.G. Moment matrices: Applications in mixtures. *Ann. Stat.* **1989**, *17*, 722–740.
26. Li, P.; Chen, J.; Marriott, P. Non-finite Fisher information and homogeneity: An EM approach. *Biometrika* **2009**, *96*, 411–426.

entropy

MDPI

Article

Using Geometry to Select One Dimensional Exponential Families That Are Monotone Likelihood Ratio in the Sample Space, Are Weakly Unimodal and Can Be Parametrized by a Measure of Central Tendency

Paul Vos [1] and Karim Anaya-Izquierdo [2,*]

[1] Department of Biostatistics, East Carolina University, Greenville, NC 27858, USA; E-Mail: vosp@ecu.edu
[2] Department of Mathematical Sciences, University of Bath, Bath BA27AY, UK
* E-Mail: kai21@bath.ac.uk; Tel: +44-1225-384644

Received: 30 April 2014; in revised form: 30 June 2014 / Accepted: 14 July 2014 / Published: 18 July 2014

Abstract: One dimensional exponential families on finite sample spaces are studied using the geometry of the simplex Δ_{n-1}° and that of a transformation V_{n-1} of its interior. This transformation is the natural parameter space associated with the family of multinomial distributions. The space V_{n-1} is partitioned into cones that are used to find one dimensional families with desirable properties for modeling and inference. These properties include the availability of uniformly most powerful tests and estimators that exhibit optimal properties in terms of variability and unbiasedness.

Keywords: simplex; cone; exponential family; monotone likelihood ratio; unimodal; duality

---

1. Introduction

The motivation for the constructions in this paper begins with a sample from a one dimensional space that is discrete. We allow for a continuous sample space but assume that this has been suitably discretized into n bins. The simplest underlying structure for the probability assigned to these bins is given by the multinomial distribution. The collection of all multinomial distributions can be identified with the $n-1$ simplex Δ_{n-1}. We use the geometry of the simplex along with a transformation of its interior Δ_{n-1}° to search for one dimensional subspaces that have good properties for modeling and for inference. In particular, we want families that can be parameterized by the mean, have only unimodal distributions, have desirable test characteristics (such as providing uniformly most powerful unbiased tests) and estimation properties (such as unbiasedness and small variability).

The boundary of the $(n-1)$ dimensional simplex Δ_{n-1} can be written as the union of simplexes of dimension $(n-2)$. This process can be repeated on the simplexes of lower dimension until the boundary consists of the vertices of the original simplex. This construction has statistical relevance to the possible supports for the probability distributions considered on the n bins. We obtain a dual decomposition for a transformation V_{n-1} (defined in Equation (5) in Section 5) of Δ_{n-1}°; it is dual in that the result can be obtained by replacing simplexes with cones. The statistical relevance of the conical decomposition is to the possible modes for all the distributions on the n bins. Since V_{n-1} is the natural parameter space for the distributions in Δ_{n-1}°, one dimensional exponential families are lines in V_{n-1} and these can be related to the cones that partition V_{n-1}. One result is that the limiting distribution for any one dimensional exponential family in Δ_{n-1}° is the uniform distribution whose support is determined by the cone that contains the limiting values of the line corresponding to the exponential family.

While one parameter exponential families can be defined quite generally by choosing a sufficient statistic, it can be useful to start with the sufficient statistics from well-known families such as the binomial, Poisson, negative binomial, normal, inverse Gaussian, and Gamma distribution. These exponential families have good modeling and inferential properties that we try to maintain by limiting the extent to which the sufficient statistic is modified. These restrictions lead to considering vectors in V_{n-1} that lie in a cone. Examples of how to construct these cones are given.

2. Motivating Examples

One dimensional exponential families such as the binomial or Poisson are the workhorse of parametric inference because of their excellent statistical properties. However, being one dimensional means they do not always fit data very well so an extension to a two (or higher) dimensional exponential family can be pursued in order to preserve the nice inferential structure. An issue with such extension is that, for each extra natural parameter added, we need to choose a new sufficient statistic and this choice can substantially change the shape of the corresponding density functions. For example densities can pass from being unimodal to have multiple modes for some parameter values. To see this, consider the following examples.

Example 1. *Altham [1] considered the so-called multiplicative generalization of the binomial distribution with corresponding density*

$$f(x;p,\phi) = \binom{n}{x} p^x (1-p)^{n-x} \phi^{x(x-n)} / C(p,\phi) \tag{1}$$

where C is the normalizing constant and where clearly the binomial is recovered when $\phi = 1$.

By reparametrizing using $\theta_1 = \log(p/(1-p))$ *and* $\theta_2 = \log(\phi)$ *this density can be expressed in exponential form as*

$$f(x;\theta_1,\theta_2) = h(x)\exp(\theta_1\, x + \theta_2\, T(x) - K(\theta_1,\theta_2)) \tag{2}$$

where $T(x) = x(x-n)$ *is the added sufficient statistic and* $h(x) = \binom{n}{x}$ *where dependence on n has been ignored. Note that the same family is obtained if* $T(x) = x^2$ *is added as a sufficient statistic instead of* $x(x-n)$.

If $n = 127$ *and* $(\theta_1, \theta_2) = (-0.0122, 0.018)$ *then density (2) is bimodal as shown in the left panel of Figure 1. The mean* μ *of this distribution is 50. Also plotted is the corresponding binomial density with the same mean or equivalently with* $\theta_1 = \log(50/(127-50)) = -0.4318$ *and* $\theta_2 = 0$.

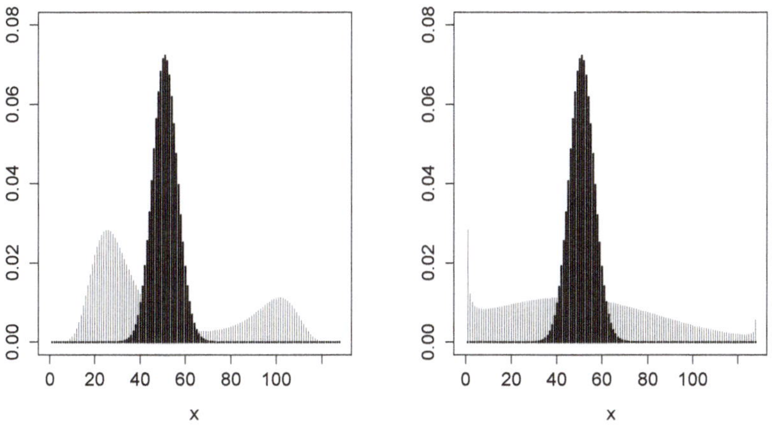

Figure 1. Binomial density (thick in both panels). Multiplicative binomial density (left panel and thin) and double binomial density (right panel and thin). All densities have the same mean $\mu = 50$ and $n = 127$. Variance of the multiplicative and double binomial densities is equal.

As explained by Lovison [2], this distribution has the feature of being under- or over-dispersed with respect to the binomial depending on θ_2 being negative or positive, respectively. Furthermore, using the mixed parametrization (μ, θ_2) (see [3] for details) it is easy to see that this distribution can be parametrized so that one parameter controls dispersion independently of the mean. In fact, for a fixed mean μ, as $\theta_2 \to -\infty$ $f(x;\theta_1,\theta_2)$ tends to a two point distribution (with support points at the extremes $x = 0$ and $x = n$) or to a degenerate distribution on $x = \mu$ when $\theta_2 \to \infty$.

Example 2. *Double exponential families [4] are two parameter exponential families that extend standard unidimensional exponential families such as the binomial and the Poisson. Similar to the multiplicative binomial in Example 1, the extra parameter involved in double exponential families controls the variance independently of the mean. The density for the so-called double binomial family can be written in the form (2) with*

$$T(x) = x \log\left(\frac{x}{n}\right) + (n-x)\log\left(1-\frac{x}{n}\right)$$

$h(x) = \binom{n}{x}$ and with the particular restriction that $\theta_2 < 1$ (see [4] for details). The range $\theta_2 < 0$ generates underdispersion and $\theta_2 \in [0,1)$ generates overdispersion with respect to the binomial. As shown on the right panel of Figure 1, the double binomial density can also be multimodal where the double binomial density shown has the same mean and variance as the multiplicative binomial shown in the left panel.

These examples show that while extending exponential families can lead to useful modeling properties such as overdispersion, the extension can also result in distributions that are not suitable for modeling. We are interested in the relationship between geometric properties of one dimensional families and the modeling properties of their distributions.

3. Sample Space and Distribution-valued Random Variables

We consider first the general case where the sample space for a single observation X_1 consists of n bins

$$S_n = \{B_1, B_2, \ldots, B_{n-1}, B_n\}.$$

We consider the space of all probability distributions $\mathcal{P}$ on this sample space S_n. Each probability distribution in $\mathcal{P}$ is defined by the n-tuple p whose i^{th} component is

$$p^i = \Pr(B_i)$$

so that $\mathcal{P}$ can be identified with the $n-1$ simplex

$$\Delta_{n-1} = \{p \in \mathbb{R}^n : p^i \geq 0 \ \forall i, 1'p = 1\}$$

where 1 in $1'p$ is the vector $1 \in \mathbb{R}^n$ each of whose components is 1. We will slightly abuse the notation by using p to name a point in Δ_{n-1}, and hence in $\mathbb{R}^n$, as well as the corresponding distribution in $\mathcal{P}$.

The sample space for a random sample of size N from a distribution $p_0 \in \Delta_{n-1}$ is

$$\mathcal{X}_n^N = \{x : x \text{ is an } n \text{ vector of nonnegative integers that sum to } N\}.$$

There is simple relationship between $\mathcal{X}_n^N$ and the simplex that we obtain by dividing each component of x by N. Although the sample space $\mathcal{X}_n^N$ can be viewed as formed by compositional data, we will follow a different approach to handle this kind of data compared with the classical approach described by Aitchison [5] because the data we consider have additional structure.

In Figure 2 the sample space for the sample of size $N = 10$ is displayed using open circles. The vertices correspond to the case where all 10 values fall in a single bin. The other points correspond to the less extreme cases. Let p_0 be any point in Δ_{n-1}. By mapping the multinomial random variable of counts X to Δ_{n-1}, we obtain the random distribution $\widehat{P} = X/N$ whose values are multinomial distributions each having number of cases N and probability vector X/N. Identifying $\mathcal{X}_n^N$-valued random variables with distribution-valued random variables provides a natural means for comparing data with probability models using the Kullback–Leibler (KL) divergence.

We can compare distributions in Δ_{n-1} using the KL divergence $D : \mathcal{P} \times \mathcal{P} \mapsto \mathbb{R}$

$$D(p_1, p_2) = \sum p_1 \log(p_1/p_2) = H(p_1, p_2) - H(p_1)$$

where $H(p_1, p_2) = -\sum p_1 \log(p_2)$ and $H(p_1) = H(p_1, p_1)$ is the entropy of p_1. Note that the arguments to D and H are distributions while the logarithm and ratios are defined on points in $\mathbb{R}^n$. Following Wu and Vos [6], the variance of the random distribution $\widehat{P}$ is defined to be

$$\mathrm{Var}_{p_0}(\widehat{P}) = \min_{p \in \Delta_{n-1}} E_{p_0} D(\widehat{P}, p)$$

and its mean is defined to be

$$\mathrm{E}_{p_0}(\widehat{P}) = \arg \min_{p \in \Delta_{n-1}} E_{p_0} D(\widehat{P}, p).$$

Note that the expectation on the right hand side of the equations above are for real-valued random variables while the expectation on the left hand side of the second equation is for a distribution-valued random variable.

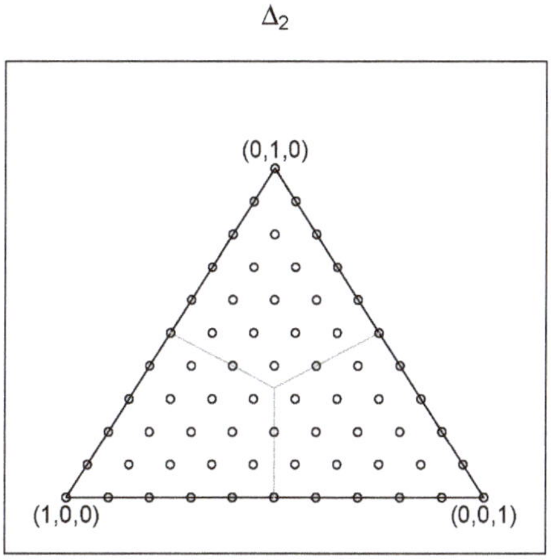

Figure 2. Simplex for $n = 3$ bins and sample space for $N = 10$ observations.

It is not difficult to show that $E_{p_0}\widehat{P} = p_0$ so that $\widehat{P}$ can be considered an unbiased estimator for p_0. Details are in [6], which also shows that the KL risk can be decomposed into bias-squared and variance terms:

$$E_{p_0} D(\widehat{P}, q) = D(p_0, q) + \mathrm{Var}_{p_0}(\widehat{P}).$$

The distributional variance is related to the entropy

$$\mathrm{Var}_{p_0}(\widehat{P}) = E_{p_0} D(\widehat{P}, p_0) = H(p_0) - E_{p_0} H(\widehat{P}).$$

Note that for $N = 1$, $H(\widehat{P}) = 0$ so that for a single observation the random distribution $\widehat{P}$ taking values on the vertices of Δ_{n-1} has variance equal to the entropy of p_0.

For inference, p_0 is unknown but we specify a subspace $M \subset \Delta_{n-1}$ that contains p_0, or at least has distributions that are not too different from p_0. Estimates can be obtained by choosing a parameterization for M, say θ, and then considering real-valued functions $\hat{\theta}$ and evaluating these in terms of bias and variance. Bias and variance are useful descriptions when θ describes a feature of the distribution that is of inherent interest. However, if θ is simply a parameterization, or if there are other features that are also of interest, then these quantities are less useful. For inference regarding the distribution p_0 we can use a distribution-valued estimator $\widehat{P}_M$ where the subscript indicates that the estimator is defined to account for the fact that $p_0 \in M$.

We will not pursue the details of distribution-valued estimators here; we mention these only because all the subspaces we consider will be exponential families and in this case the maximum likelihood estimator has important properties in terms of distribution variance and distribution bias: when M is an exponential family, the maximum likelihood estimator is distribution unbiased, and it uniquely minimizes the distribution variance among the class of all distribution unbiased estimators. Furthermore, when $p_0 \notin M$ then the maximum likelihood estimator is the unique unbiased minimum distribution variance estimator of the distribution in M that is closest (in terms of KL) to p_0. Extensions of one dimensional exponential families that do not result in exponential families will not enjoy these properties of maximum likelihood estimation. Details of these results that hold for sample spaces more general than S_n are in [7].

4. Simplices Δ_s

One dimensional exponential families on S_n are curves in Δ_{n-1} whose properties will depend on their location within various subspaces of Δ_{n-1}. An important collection of subspaces will be indexed by the subsets of S_n. For notational convenience we take B_i to the integer i. Using integers is suggestive of an ordering and a scale structure but at this point these are only being used to indicate distinct bins.

For each $s \subset S_n$,

$$\Delta_s = \left\{p \in \mathbb{R}^n : p^i \geq 0 \;\; \forall i \in s,\; p^i = 0 \;\; \forall i \in s^c,\; 1'p = 1\right\}$$

where $s^c = \{i \in S_n : i \notin s\}$. Note that $\Delta_{S_n} = \Delta_{n-1}$. The interior of Δ_s is

$$\Delta_s^\circ = \left\{p \in \Delta_s : p^i > 0 \;\; \forall i \in s\right\}.$$

As probability distributions in $\mathcal{P}$, Δ_s° corresponds to the set of all distributions having support s. There is a simple and obvious relationship between the dimension of Δ_s, $|\Delta_s|$, and the cardinality of s, $|s|$, which holds for all nonempty $s \subset S_n$

$$|\Delta_s| + 1 = |s|.$$

The boundary of Δ_s is defined as

$$\partial\Delta_s = \{p \in \Delta_s : p \notin \Delta_s^\circ\}$$

so that

$$\Delta_s = \Delta_s^\circ \uplus \partial\Delta_s$$

where $\uplus$ indicates the sets in the union are disjoint. The boundary $\partial\Delta_s$ can be written as the union of all simplices of dimension one less than that Δ_s

$$\partial\Delta_s = \bigcup_{s':s'\subset s,\; |s'|=|s|-1} \Delta_{s'} \tag{3}$$

This boundary property for Δ_s holds because the simplex $\mathcal{S}_n$ consists of all possible subsets. Each nonempty $s \in \mathcal{S}_n$ specifies one of the possible supports for distribution $P \in \mathcal{P}_n$

$$\Delta_s = \biguplus_{s':s'\subset s} \Delta_{s'}^\circ \tag{4}$$

where we set $\Delta_\emptyset = \emptyset$.

5. Cones Λ_s

The set of all nonempty subsets of the sample space provides a partition of Δ_{n-1} based on the support of the distributions in $\mathcal{P}$. The elements in the partition are simplices whose dimension is one less than the cardinality of the indexing set. In most cases we will consider models having support S_n, that is, models corresponding to Δ°_{n-1}. If we use subsets s to define the mode rather than support, we obtain a partition of $\mathcal{P}^\circ$, the distributions in $\mathcal{P}$ having support S_n. This partition can be expressed using convex cones in an $n-1$ dimensional plane V_{n-1}. The dimension of the cones are n minus the cardinality of the indexing set and the relationship between interiors of cones and their boundaries is analogous to that for simplices expressed in Equations (3) and (4).

Let

$$\mathrm{V}_{n-1} = \{v \in \mathbb{R}^n : 1'v = 0\} \tag{5}$$

be the subspace of $\mathbb{R}^n$ of dimension $n-1$ of all vectors that sum to zero. For each nonempty $s \in \mathcal{S}_n$ define

$$\Lambda_s = \left\{v \in \mathrm{V}_{n-1} : v^i \geq v^j \ \forall i \in s, \ \forall j \in S_n\right\}.$$

It is easily checked that Λ_s is a convex cone

$$v_1, v_2 \in \Lambda_s \implies a_1 v_1 + a_2 v_2 \in \Lambda_s \ \forall a_1, a_2 \in [0, \infty).$$

The dimension of Λ_s is $|\Lambda_s| = n - |s|$ since each point in $j \in s^c$ provides a basis vector b_j whose i^{th} component is 1 if $i \in s$ or $i = j$ and is zero otherwise and $|s^c| = n - |s|$. The interior of Λ_s is

$$\Lambda_s^\circ = \left\{v \in \Lambda_s : v^i > v^j \ \forall i \in s, \ \forall j \in s^c\right\},$$

the boundary is

$$\partial\Lambda_s = \{v \in \Lambda_s : v \notin \Lambda_s^\circ\},$$

so that

$$\Lambda_s = \Lambda_s^\circ \uplus \partial\Lambda_s$$

by definition. Note $\Lambda_{S_n} = \Lambda^\circ_{S_n} = 0 \in \mathrm{V}_{n-1} \subset \mathbb{R}^n$ where the first equality holds because the conditions in the definition of Λ_s° hold vacuously since $i \in S_n^c = \emptyset$ adds no restriction. Likewise, we can extend the definition of Λ_s to include $s = \emptyset$ and since $i \in \emptyset$ adds no restriction

$$\Lambda_\emptyset = \Lambda_\emptyset^\circ = \mathrm{V}_{n-1}.$$

Note that $\Lambda_\emptyset$ depends on the cardinality of the set S_n. Since we are considering n fixed, we will not show this dependence in the notation.

Corresponding to Equation (3) we have for all nonempty s that the boundary of the cone Λ_s is the union of all cones having dimension one less than the dimension of Λ_s

$$\partial\Lambda_s = \bigcup_{s' : s \subset s', \ |s'| = |s|+1} \Lambda_{s'}. \tag{6}$$

Corresponding to Equation (4) we have

$$\Lambda_s = \biguplus_{s' : s \subset s'} \Lambda_{s'}^\circ \tag{7}$$

The relationship between the simplices Δ and cones Λ is more easily seen if we suppress the sets that index these objects. Let Δ and Δ_* be any two simplices and let Λ and Λ_* be any two convex cones. We only consider cones and simplices that correspond to a nonempty subset of S_n. Then the Equations (6) and (7) for the convex cones are obtained by simply replacing Δ in Equations (3) and (4) with Λ:

$$\partial\Delta = \bigcup_{\Delta_*:|\Delta_*|=|\Delta|-1} \Delta_*, \quad \partial\Lambda = \bigcup_{\Lambda_*:|\Lambda_*|=|\Lambda|-1} \Lambda_* \tag{8}$$

$$\Delta = \biguplus_{\Delta_*\subset\Delta} \Delta_*^\circ, \quad \Lambda = \biguplus_{\Lambda_*\subset\Lambda} \Lambda_*^\circ \tag{9}$$

Equation (9) also holds for the empty set since $\Delta_\emptyset = \emptyset$ and $\Lambda_\emptyset = V_{n-1}$.

6. V_{n-1} and $\mathcal{P}^\circ$

There is a natural bijection ϕ between V_{n-1} and Δ_{n-1}° defined by

$$\phi(p) = \log(p) - m(p)1$$

where $\log(p)$ is the vector with i^{th} component $\log(p^i)$ and $m(p)$ is defined so that $1'\phi(p) = 0$. The inverse is

$$\varphi(v) = k^{-1}(v)\exp(v)$$

where $\exp(v)$ is the vector with i^{th} component $\exp(v^i)$ and $k(v)$ is defined so that $1'\exp(v) = 1$.

Each cone Λ_s° in the partition

$$V_{n-1} = \biguplus \Lambda_s^\circ$$

corresponds to one of the $2^n - 1$ possible modes for any distribution having support S_n since $v^i > v^j$ if and only if $\varphi^i(v) > \varphi^j(v)$.

7. V_{n-1} and Exponential Families in $\mathcal{P}^\circ$

We define a line by a pair of vectors $v_0, v_1 \in V_{n-1}$ with $v_1 \neq 0$

$$\ell = \ell(t) = \{v \in V_{n-1} : v = v_0 + tv_1, \ t \in \mathbb{R}\}$$

Note that v_0 and v_1 are not unique. Applying the inverse transformation φ to points in ℓ gives probability densities

$$\varphi(v_0 + tv_1) = \frac{\exp(v_0 + tv_1)}{1'\exp(v_0 + tv_1)} \tag{10}$$

which have the exponential family form with t playing the role of the natural parameter. Therefore, the space V_{n-1} is easily recognized as the natural parameter space for the distributions Δ_{n-1}° so that each line ℓ in V_{n-1} corresponds to a one dimensional exponential family.

For each line $\ell(t)$ there is a value t_{max} such that $\{\ell(t) : t \geq t_{max}\}$ is contained in one of the cones Λ_s° where s is the subset of S_n with the property that $v_1^i \geq v_1^j$ for all $i \in s$ for vectors $v_1 \in \Lambda_x^\circ$. For each line $\ell(t)$ there is a value t_{min} such that $\{\ell(t) : t \leq t_{min}\}$ is contained in one of the cones $\Lambda_{s'}^\circ$ where s' is the subset of S_n with the property that $v_1^i \leq v_1^j$ for all $i \in s'$ for vectors $v_1 \in \Lambda_x^\circ$. The cones Λ_s° and $\Lambda_{s'}^\circ$ are disjoint and will be called the *extremal* cones for ℓ. There is at least one other cone $\Lambda_{s''}^\circ$ such that $\ell \cap \Lambda_{s''}^\circ \neq \emptyset$.

Any one dimensional exponential family $\ell(t)$ can be described by an ordered sequence of disjoint cones

$$\left(\Lambda_{s_1}^\circ, \Lambda_{s_2}^\circ, \ldots, \Lambda_{s_k}^\circ\right)$$

where $k = k(\ell)$ will depend on the family. These are simply the cones that are traversed by $\ell(t)$ between its extremal cones. We take $\Lambda^{\circ}_{s_k}$ to be the cone that contains $\ell(t)$ for all sufficiently large t. Equation (6) for cones means that

$$\partial\Lambda_{s_i} \subset \Lambda_{s_j} \text{ for } j = i+1 \text{ or } j = i-1$$

The ordered sequence of cones provides an ordered sequence of unique subsets of S_n

$$(s_1, s_2, \ldots, s_k)$$

that we call the *modal profile* for ℓ as these are the modes realized by the exponential family $\ell(t)$ between its extremal cones that have modes s_1 and s_k.

Each point on a line $\ell(t)$ in V_{n-1} corresponds to a distribution having support S_n. As t goes to $-\infty$ ($+\infty$) $\varphi(\ell(t))$ goes to a distribution having support s_1 (s_k). In fact, these are the uniform distribution on these supports. For every $s \subset S_n$ other than $\emptyset$ and S_n, the uniform distribution on s is a limiting distribution for some one dimensional exponential family in $\mathcal{P}^{\circ}$.

Figure 3 shows V_{n-1} for the two dimensional simplex shown in Figure 2. The three rays are the one dimensional cones and the spaces between these cones are the two dimensional cones. The origin is the zero dimensional cone. The sample values on the boundary of Δ_2 are not in V_2. Note that the one dimensional cones are line segments in Δ_2.

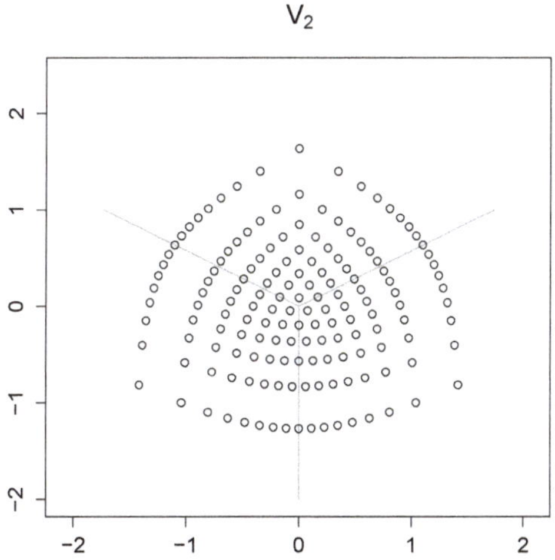

Figure 3. V_2 for $n = 3$ bins and sample space for $N = 10$ observations that are in the interior of Δ_2.

8. Ordered Bins and the Monotone Likelihood Ratio Property

Let the bins be ordered and assign the first n integers to the bins to reflect this ordering. We seek to define exponential families that have a modal profile of the form

$$(\{1\}, \{1,2\}, \{2\}, \{2,3\}, \ldots, \{n-1,n\}, \{n\}) \tag{11}$$

or a contiguous sub-collection of this profile. Extensions to three or more contiguous modes are clearly possible but not discussed here.

From the definition of modal profile, it follows that a family with modal profile (11) will have the property that the mode is a non-decreasing function of t. In addition to this property for the mode, we want the likelihood ratio for any two members of the family to provide the same ordering structure as that of the bins. A family that satisfies this condition is said to have the monotone likelihood ratio property with respect to x where x takes the values of the bin labels: $1, 2, \ldots, n$. Let p_{θ_1} and p_{θ_2} be two distributions in a one dimensional family parameterized by θ and let $p_{\theta_2}/p_{\theta_1}$ be the n-vector with components $p^j_{\theta_2}/p^j_{\theta_1}$ for $1 \le j \le n$. This family has monotone likelihood ratio if for all $\theta_1 < \theta_2$ and $j < j'$

$$\frac{p^j_{\theta_2}}{p^j_{\theta_1}} < \frac{p^{j'}_{\theta_2}}{p^{j'}_{\theta_1}}.$$

A family with this property avoids the problem situation where in general the data in the higher numbered bins are evidence for p_{θ_2} but in going from a particular bin, say j_0 to $j_0 + 1$, the likelihood ratio actually decreases. Exponential families such as the binomial and Poisson have this monotone likelihood ratio property for the bin labels. The monotone likelihood ratio property can be extended to allow for likelihood ratios that are monotone in some function of x. An important advantage of families with the monotone likelihood ratio property is the existence of uniformly most powerful tests.

To ensure that our exponential families have the monotone likelihood ratio property we consider vectors in the cone $\Lambda^\uparrow \subset \Lambda_n$

$$\Lambda^\uparrow = \left\{ v : v^i < v^j, i < j \right\}.$$

From Equation (10), the exponential family indexed by θ is $k(\theta)\exp(v_0 + \theta v_1)$

$$\frac{p^j_{\theta_2}}{p^j_{\theta_1}} = \frac{k(\theta_2)}{k(\theta_1)} \exp\left\{ (\theta_2 - \theta_1)\, v^j_1 \right\}$$

so that the likelihood ratio is monotone in j if $v_1 \in \Lambda^\uparrow$.

9. Selecting Vectors in $\Lambda^\uparrow$

In order to choose n-dimensional vectors $v \in \Lambda^\uparrow$ we will consider a set of infinite dimensional vectors f. Let $\bar{f} : \mathbb{R} \mapsto \mathbb{R}$ and consider $f = \bar{f}|_{\mathbb{Z}}$ where $\mathbb{Z}$ is the set of integers. The function f is represented by a doubly infinite sequence

$$f = \ldots, f^{j-1}, f^j, f^{j+1}, \ldots$$

and we denote the set of all such functions as

$$\mathcal{F} = \left\{ f : f^j \in \mathbb{R}\, \forall\, j \in \mathbb{Z} \right\}.$$

While it is not necessary to consider functions $\bar{f}$ to define f, these functions are useful to describe properties of f, which can be thought of as a discretized version of $\bar{f}$.

Define the gradient of f as the function ∇ whose j^{th} component is

$$(\nabla f)^j = f^j - f^{j-1}$$

The simplest functions in $\mathcal{F}$ are the constant functions

$$\mathcal{F}_0 = \left\{ f \in \mathcal{F} : f^j = f^{j'}\ \forall j, j' \in \mathbb{Z} \right\}.$$

The next simplest functions are those whose gradient is constant. We call these first order functions and denote the set of these as

$$\mathcal{F}_1 = \{ f \in \mathcal{F} : \nabla f \in \mathcal{F}_0 \}.$$

Functions in $\mathcal{F}_1$ are such that changes from one bin to the next bin is the same for all bins. That is, these functions describe constant change. We can write the functions in $\mathcal{F}_1$ explicitly as

$$\mathcal{F}_1 = \left\{ f \in \mathcal{F} : f^j = aj + b, \;\; a, b \in \mathbb{R} \right\}$$

which shows that each $f \in \mathcal{F}_1$ is the discretized version of a function $\bar{f}$ whose graph is a line in $\mathbb{R} \times \mathbb{R}$. We obtain a vector v from f by defining the jth component of v as

$$v^j = f^j - \sum_1^n f^i$$

. From this definition we see that the intercept b of f does not affect v and that the slope is a scaling factor so that the restriction to first order functions results in a single direction in $\Lambda^\uparrow$. This direction defines the one dimensional cone defined by the vector with $v^j = j - (n+1)/2$.

Additional directions can be obtained from the second order functions

$$\mathcal{F}_2 = \{ f \in \mathcal{F} : \nabla f \in \mathcal{F}_1 \}.$$

If $f \in \mathcal{F}_2$ then $(\nabla^2 f)^j = a$ for some $a \in \mathbb{R}$ and for all $j \in \mathbb{Z}$. Using the fact that

$$\begin{aligned}(\nabla^2 f)^j = (\nabla(\nabla f))^j &= (f^j - f^{j-1}) - (f^{j-1} - f^{j-2}) \\ &= f^j + f^{j-2} - 2f^{j-1}\end{aligned}$$

the second order functions can be written explicitly as

$$\mathcal{F}_2 = \left\{ f \in \mathcal{F} : f^j = \frac{a}{2} j(j+1) + bj + c, \;\; a, b, c \in \mathbb{R} \right\}$$

.

In order for the vector v obtained from $f \in \mathcal{F}_2$ to be in $\Lambda^\uparrow$ we need $(\nabla f)^j \geq 0$ for $j = 1, 2, \ldots, n$. With $f^j = (a/2)j(j+1) + bj + c$ we have $(\nabla f)^j = aj + b$ so that for $a > 0$ we require $b \geq -a$ and for $a < 0$ we require $b \geq -an$. Since we are concerned with the direction rather than the magnitude we can take $a = \pm 1$ and the value of c is chosen so the sum of the components is zero.

The second order vectors in $\Lambda^\uparrow$ consists of the cone defined by the vectors v_{20} and v_{21} having components defined by

$$\begin{aligned}(n-1)(v_{20})^j &= \frac{1}{2} j(j+1) - j - c_{20} \\ (n-1)(v_{21})^j &= -\frac{1}{2} j(j+1) + nj - c_{21}\end{aligned}$$

Notice that this cone contains v_1 since v_1 is proportional to $v_{20} + v_{21}$. Many discrete one dimensional exponential families (e.g., binomial, negative binomial, and Poisson) use the vector v_1. Furthermore, many continuous one dimensional exponential families use the continuous function f used to define v_1: normal with σ known, and the gamma and inverse Gaussian distributions with known shape parameter (the shape parameter is the non-scale parameter). The cone defined by v_{20} and v_{21} allows us to perturb the v_1 direction to obtain related exponential families that we would expect to have similar properties. Figure 4 shows v_{20} and v_{21} as well as $v_1 = 0.5v_{20} + 0.5v_{21}$.

Other vectors can be used to define cones around v_1. Looking at common exponential families we see that $\log(x)$ and x^{-1} are sufficient statistics so that these suggest taking $\bar{f}(x) = \log(x)$ or $\bar{f}(x) = 1/x$. These can be further generalized to $\bar{f}(x; \lambda)$, which can be the power family or some other family of transformations. The vectors v_{f0} and v_{f1} are defined using the discretized f with the constraints that $v_{f0}, v_{f1} \in \Lambda^\uparrow$ and $0.5v_{f0} + 0.5v_{f1} = v_1$.

An exponential family with sufficient statistic x can be modified by choosing a function $\bar{f}(x)$ and $0 \leq \alpha \leq 1$ where $\alpha = 0.5$ corresponds to the original exponential family and other values perturb this direction. We denote this vector as $v_{f\alpha}$ so that $v_0 + tv_{f\alpha}$ is the natural parameter of the modified family.

Figure 4 shows the components of the vectors v_{20} and v_{21}.

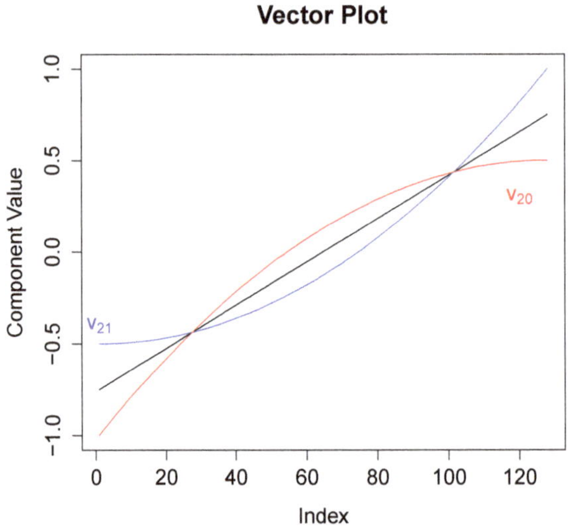

Figure 4. Components of the vectors v_{20} and v_{21} for $n = 128$ bins.

Since v_0 is common to each exponential family with natural parameter $\ell(t) = v_0 + tv_{f\alpha}$, the monotone likelihood ratio property will hold even if $v_0 \notin \Lambda^{\uparrow}$. Initial choices for v_0 are suggested by the Poisson, binomial, and negative binomial distributions:

$$
\begin{aligned}
(v_{\text{Poisson}})^j &= -\log\Gamma(j) + c \quad \notin \Lambda^{\uparrow} \\
(v_{\text{binomial}})^j &= \log\Gamma(n) - \log\Gamma(j) - \log\Gamma(n-j) + c \quad \notin \Lambda^{\uparrow} \\
(v_{\text{neg.bin.}})^j &= \log\Gamma(j+r) - \log\Gamma(j) + c \quad \in \Lambda^{\uparrow}
\end{aligned}
$$

where c is a constant chosen so that the components sum to 1, n is the number of bins, and r is a positive real constant.

Author Contributions: This paper was initiated by the first author but all sections reflect a collaborative effort. Both authors have read and approved the final manuscript.

Conflicts of Interest: The authors declare no conflict of interest.

References

1. Altham, P.M.E. Two Generalizations of the Binomial Distribution. *J. R. Stat. Soc. Ser. C (Appl. Stat.)* **1978**, *27*, 162–167.
2. Lovison, G. An alternative representation of Altham's multiplicative-binomial distribution. *Stat. Probab. Lett.* **1998**, *36*, 415–420.
3. Brown, L. *Fundamentals of Statistical Exponential Families: With Applications in Statistical Decision Theory*; IMS Lecture Notes; Institute of Mathematical Statistics: Hayward, CA, USA, 1986.
4. Efron, B. Double Exponential Families and Their Use in Generalized Linear Regression. *J. Am. Stat. Assoc.* **1986**, *81*, 709–721.
5. Aitchison, J. *The Statistical Analysis of Compositional Data*; Chapman and Hall: London, UK, 1986.

6. Wu, Q.; Vos, P. Decomposition of Kullback–Leibler risk and unbiasedness for parameter-free estimators. *J. Stat. Plan. Inference* **2012**, *142*, 1525–1536.
7. Vos, P.; Wu, Q. Maximum Likelihood Estimators Uniformly Minimize Distribution Variance among Distribution Unbiased Estimators in Exponential Families. *Bernoulli* **2014**, submitted.

entropy

MDPI

Article

On the Fisher Metric of Conditional Probability Polytopes

Guido Montúfar [1,*], Johannes Rauh [1] and Nihat Ay [1,2,3]

[1] Max Planck Institute for Mathematics in the Sciences, Inselstraße 22, Leipzig 04103, Germany; E-Mails: jrauh@mis.mpg.de (J.R.); nay@mis.mpg.de (N.A.)

[2] Department of Mathematics and Computer Science, Leipzig University, PF 10 09 20, Leipzig 04009, Germany

[3] Santa Fe Institute, 1399 Hyde Park Road, Santa Fe, NM 87501, USA

* E-Mail: montufar@mis.mpg.de; Tel.: +49-341-9959-521.

Received: 31 March 2014; in revised form: 18 May 2014 / Accepted: 29 May 2014 / Published: 6 June 2014

Abstract: We consider three different approaches to define natural Riemannian metrics on polytopes of stochastic matrices. First, we define a natural class of stochastic maps between these polytopes and give a metric characterization of Chentsov type in terms of invariance with respect to these maps. Second, we consider the Fisher metric defined on arbitrary polytopes through their embeddings as exponential families in the probability simplex. We show that these metrics can also be characterized by an invariance principle with respect to morphisms of exponential families. Third, we consider the Fisher metric resulting from embedding the polytope of stochastic matrices in a simplex of joint distributions by specifying a marginal distribution. All three approaches result in slight variations of products of Fisher metrics. This is consistent with the nature of polytopes of stochastic matrices, which are Cartesian products of probability simplices. The first approach yields a scaled product of Fisher metrics; the second, a product of Fisher metrics; and the third, a product of Fisher metrics scaled by the marginal distribution.

Keywords: Fisher information metric; information geometry; convex support polytope; conditional model; Markov morphism; isometric embedding; natural gradient

---

1. Introduction

The Riemannian structure of a function's domain has a crucial impact on the performance of gradient optimization methods, especially in the presence of plateaus and local maxima. The natural gradient [1] gives the steepest increase direction of functions on a Riemannian space. For example, artificial neural networks can often be trained by following some function's gradient on a space of probabilities. In this context, it has been observed that following the natural gradient with respect to the Fisher information metric, instead of the Euclidean metric, can significantly alleviate the plateau problem [1,2]. The Fisher information metric, which is also called Shahshahani metric [3] in biological contexts, is broadly recognized as the natural metric of probability spaces. An important argument was given by Chentsov [4], who showed that the Fisher information metric is the only metric on probability spaces for which certain natural statistical embeddings, called Markov morphisms, are isometries. More generally, Chentsov's Theorem characterizes the Fisher metric and α-connections of statistical manifolds uniquely (up to a multiplicative constant) by requiring invariance with respect to Markov morphisms. Campbell [5] gave another proof that characterizes invariant metrics on the set of non-normalized positive measures, which restrict to the Fisher metric in the case of probability measures (up to a multiplicative constant). In this paper, we explore ways of defining distinguished Riemannian metrics on spaces of stochastic matrices.

In learning theory, when modeling the policy of a system, it is often preferred to consider stochastic matrices instead of joint probability distributions. For example, in robotics applications, policies are optimized over a parametric set of stochastic matrices by following the gradient of a reward function [6,7]. The set of stochastic matrices can be parametrized in many ways, e.g., in terms of feedforward neural networks, Boltzmann machines [8] or projections of exponential families [9]. The information geometry of policy models plays an important role in these applications and has been studied by Kakade [2], Peters and co-workers [10–12], and Bagnell and Schneider [13], among others. A stochastic matrix is a tuple of probability distributions, and therefore, the space of stochastic matrices is a Cartesian product of probability simplices. Accordingly, in applications, usually a product metric is considered, with the usual Fisher metric on each factor. On the other hand, Lebanon [14] takes an axiomatic approach, following the ideas of Chentsov and Campbell, and characterizes a class of invariant metrics of positive matrices that restricts to the product of Fisher metrics in the case of stochastic matrices. We will consider three different approaches discussed in the following.

In the first part, we take another look at Lebanon's approach for characterizing a distinguished metric on polytopes of stochastic matrices. However, since the maps considered by Lebanon do not map stochastic matrices to stochastic matrices, we will use different maps. We show that the product of Fisher metrics can be characterized by an invariance principle with respect to natural maps between stochastic matrices.

In the second part, we consider an approach that allows us to define Riemannian structures on arbitrary polytopes. Any polytope can be identified with an exponential family by using the coordinates of the polytope vertices as observables. The inverse of the moment map then defines an embedding of the polytope in a probability simplex. This embedding can be used to pull back geometric structures from the probability simplex to the polytope, including Riemannian metrics, affine connections, divergences, *etc.* This approach has been considered in [9] as a way to define low-dimensional families of conditional probability distributions. More general embeddings can be defined by identifying each exponential family with a point configuration, $\mathbf{B}$, together with a weight function, ν. Given $\mathbf{B}$ and ν, the corresponding exponential family defines geometric structures on the set $(\operatorname{conv}\mathbf{B})^\circ$, which is the relative interior of the convex support of the exponential family. Moreover, we can define natural morphisms between weighted point configurations as surjective maps between the point sets, which are compatible with the weight functions. As it turns out, the Fisher metric on $(\operatorname{conv}\mathbf{B})^\circ$ can be characterized by invariance under these maps.

In the third part, we return to stochastic matrices. We study natural embeddings of conditional distributions in probability simplices as joint distributions with a fixed marginal. These embeddings define a Fisher metric equal to a weighted product of Fisher metrics. This result corresponds to the Definitions commonly used in robotics applications.

All three approaches give very similar results. In all cases, the identified metric is a product metric. This is a sensible result, since the set of $k \times m$ stochastic matrices is a Cartesian product of probability simplices $\Delta_{m-1} \times \cdots \times \Delta_{m-1} = \Delta_{m-1}^k$, which suggests using the product metric of the Fisher metrics defined on the factor simplices, Δ_{m-1}. Indeed, this is the result obtained from our second approach. The first approach yields that same result with an additional scaling factor of $1/k$. Only when stochastic matrices of different sizes are compared, the two approaches differ. The third approach yields a product of Fisher metrics scaled by the marginal distribution that defines the embedding.

Which metric to use depends on the concrete problem and whether a natural marginal distribution is defined and known. In Section 7, we do a case study using a reward function that is given as an expectation value over a joint distribution. In this simple example, the weighted product metric gives the best asymptotic rate of convergence, under the assumption that the weights are optimally chosen. In Section 8, we sum up our findings.

The contents of the paper is organized as follows. Section 2 contains basic Definitions around the Fisher metric and concepts of differential geometry. In Section 3, we discuss the Theorems of Chentsov, Campbell and Lebanon, which characterize natural geometric structures on the probability simplex, on the set of positive measures and on the cone of positive matrices, respectively. In Section 4, we study metrics on polytopes of stochastic matrices, which are invariant under natural embeddings. In Section 5, we define a Riemannian structure for polytopes, which generalizes the Fisher information metric of probability simplices and conditional models in a natural way. In Section 6, we study a class of weighted product metrics. In Section 7, we study the gradient flow with respect to an expectation value. Section 8 contains concluding Remarks. In Appendix A, we investigate restrictions on the parameters of the metrics characterized in Sections 3 and 4 that make them positive definite. Appendix B contains the proofs of the results from Section 4.

2. Preliminaries

We will consider the simplex of probability distributions on $[m] := \{1, \ldots, m\}$, $m \geq 2$, which is given by $\Delta_{m-1} := \{(p_i)_i \in \mathbb{R}^m : p_i \geq 0, \sum_i p_i = 1\}$. The relative interior of Δ_{m-1} consists of all strictly positive probability distributions on $[m]$, and will be denoted Δ_{m-1}°. This is a subset of $\mathbb{R}_+^m$, the cone of strictly positive vectors. The set of $k \times m$ row-stochastic matrices is given by $\Delta_{m-1}^k := \{(K_{ij})_{ij} \in \mathbb{R}^{k\times m} : (K_{ij})_j \in \Delta_{m-1}$ for all $i \in [k]\}$ and is equal to the Cartesian product $\times_{i\in[k]} \Delta_{m-1}$. The relative interior $(\Delta_{m-1}^k)^\circ$ is a subset of $\mathbb{R}_+^{k\times m}$, the cone of strictly positive matrices.

Given two random variables X and Y taking values in the finite sets $[k]$ and $[m]$, respectively, the conditional probability distribution of Y given X is the stochastic matrix $K = (P(y|x))_{x\in[k],y\in[m]}$ with rows $(P(y|x))_{y\in[m]} \in \Delta_{m-1}$ for all $x \in [k]$. Therefore, the polytope of stochastic matrices Δ_{m-1}^k is called a conditional polytope.

The tangent space of $\mathbb{R}_+^n$ at a point $p \in \mathbb{R}_+^n$, denoted by $T_p\mathbb{R}_+^n$, is the real vector space spanned by the vectors $\partial_1, \ldots, \partial_n$ of partial derivatives with respect to the n components. The tangent space of Δ_{n-1}° at a point $p \in \Delta_{n-1}^\circ \subset \mathbb{R}_+^n$ is the subspace $T_p\Delta_{n-1}^\circ \subset T_p\mathbb{R}_+^n$ consisting of the vectors:

$$u = \sum_i u_i \partial_i \in T_p\mathbb{R}_+^n \quad \text{with} \quad \sum_i u_i = 0. \tag{1}$$

The Fisher metric on the positive probability simplex Δ_{n-1}° is the Riemannian metric given by:

$$g_p^{(n)}(u, v) = \sum_{i=1}^n \frac{u_i v_i}{p_i}, \quad \text{for all } u, v \in T_p\Delta_{n-1}^\circ. \tag{2}$$

The same formula (2) also defines a Riemannian metric on $\mathbb{R}_+^n$, which we will denote by the same symbol. This, however, is not the only way in which the Fisher metric can be extended from Δ_{n-1}° to $\mathbb{R}_+^n$. We will discuss other extensions in the next section (see Campbell's Theorem, Theorem 2).

Consider a smoothly parametrized family of probability distributions $\mathcal{M} = \{(p(x;\theta))_{x\in[n]} : \theta \in \Omega\} \subseteq \Delta_{n-1}^\circ$, where $\Omega \subseteq \mathbb{R}^d$ is open. Then, $g^{(n)}$ induces a Riemannian metric on $\mathcal{M}$. Denote by $\partial_{\theta_i} = \frac{\partial}{\partial\theta_i}$ the tangent vector corresponding to the partial derivative with respect to θ_i, for all $i \in [d]$. Then, the Fisher matrix has coordinates:

$$g_\theta^{\mathcal{M}}(\partial_{\theta_i}, \partial_{\theta_j}) = \sum_{x\in[n]} p(x;\theta) \frac{\partial \log p(x;\theta)}{\partial\theta_i} \frac{\partial \log p(x;\theta)}{\partial\theta_j}, \quad \text{for all } i, j \in [d], \quad \text{for all } \theta \in \Omega. \tag{3}$$

Here, it is not necessary to assume that the parameters θ_i are independent. In particular, the dimension of $\mathcal{M}$ may be smaller than d, in which case the matrix is not positive definite. If the map $\Omega \to \mathcal{M}, \theta \mapsto p(\cdot;\theta)$ is an embedding (*i.e.*, a smooth injective map that is a diffeomorphism onto its image), then $g_\theta^{\mathcal{M}}$ defines a Riemannian metric on Ω, which corresponds to the pull-back of $g^{(n)}$.

Consider an embedding $f : \mathcal{E} \to \mathcal{E}'$. The pull-back of a metric g' on $\mathcal{E}'$ through f is defined as:

$$(f^* g')_p(u, v) := g'_{f(p)}(f_* u, f_* v), \quad \text{for all } u, v \in T_p\mathcal{E}, \tag{4}$$

where f_* denotes the push-forward of $T_p\mathcal{E}$ through f, which in coordinates is given by:

$$f_*\colon \quad T_p\mathcal{E} \to T_{f(p)}\mathcal{E}'; \quad \sum_i u_i \partial_{\theta_i} \mapsto \sum_j \sum_i u_i \frac{\partial f_j(p)}{\partial \theta_i} \partial_{\theta'_j}, \tag{5}$$

where $\{\partial_{\theta_i}\}_i$ spans $T_q\mathcal{E}$ and $\{\partial_{\theta'_j}\}_j$ spans $T_{f(p)}\mathcal{E}'$.

An embedding $f\colon \mathcal{E} \to \mathcal{E}'$ of two Riemannian manifolds $(\mathcal{E}, g)$ and $(\mathcal{E}', g')$ is an isometry iff:

$$g_p(u,v) = (f^*g')_p(u,v), \quad \text{for all } p \in \mathcal{E} \text{ and } u, v \in T_p\mathcal{E}. \tag{6}$$

In this case, we say that the metric g is invariant with respect to f (and g').

3. The Results of Campbell and Lebanon

One of the theoretical motivations for using the Fisher metric is provided by Chentsov's characterization [4], which states that the Fisher metric is uniquely specified, up to a multiplicative constant, by an invariance principle under a class of stochastic maps, called Markov morphisms. Later, Campbell [5] considered the characterization problem on the space $\mathbb{R}^n_+$ instead of Δ°_{n-1}. This simplifies the computations, since $\mathbb{R}^n_+$ has a more symmetric parametrization.

Definition 1. *Let $2 \le m \le n$. A (row) stochastic partition matrix (or just row-partition matrix) is a matrix $Q \in \mathbb{R}^{m\times n}$ of non-negative entries, which satisfies $\sum_{j\in A_{i'}} Q_{ij} = \delta_{ii'}$ for an m block partition $\{A_1, \ldots, A_m\}$ of $[n]$. The linear map defined by:*

$$\mathbb{R}^m_+ \to \mathbb{R}^n_+; \quad p \mapsto p \cdot Q \tag{7}$$

is called a congruent embedding by a Markov mapping of $\mathbb{R}^m_+$ to $\mathbb{R}^n_+$ or just a Markov map, for short.

An example of a 3×5 row-partition matrix is:

$$Q = \begin{pmatrix} 1/2 & 0 & 1/2 & 0 & 0 \\ 0 & 1/3 & 0 & 2/3 & 0 \\ 0 & 0 & 0 & 0 & 1 \end{pmatrix}. \tag{8}$$

Markov maps preserve the 1-norm and restrict to embeddings $\Delta^\circ_{m-1} \to \Delta^\circ_{n-1}$.

Theorem 1 (Chentsov's Theorem.).

- *Let $g^{(m)}$ be a Riemannian metric on Δ°_{m-1} for $m \in \{2,3,\ldots\}$. Let this sequence of metrics have the property that every congruent embedding by a Markov mapping is an isometry. Then, there is a constant $C > 0$ that satisfies:*

$$g^{(m)}_p(u,v) = C \sum_i \frac{u_i v_i}{p_i}. \tag{9}$$

- *Conversely, for any $C > 0$, the metrics given by Equation (9) define a sequence of Riemannian metrics under which every congruent embedding by a Markov mapping is an isometry.*

The main result in Campbell's work [5] is the following variant of Chentsov's Theorem.

Theorem 2 (Campbell's Theorem.).

- *Let $g^{(m)}$ be a Riemannian metric on $\mathbb{R}^m_+$ for $m \in \{2,3,\ldots\}$. Let this sequence of metrics have the property that every embedding by a Markov mapping is an isometry. Then:*

$$g^{(m)}_p(\partial_i, \partial_j) = A(|p|) + \delta_{ij} C(|p|) \frac{|p|}{p_i}, \tag{10}$$

where $|p| = \sum_{i=1}^{m} p_i$, δ_{ij} *is the Kronecker delta, and* A *and* C *are* C^∞ *functions on* $\mathbb{R}_+$ *satisfying* $C(\alpha) > 0$ *and* $A(\alpha) + C(\alpha) > 0$ *for all* $\alpha > 0$.

- *Conversely, if* A *and* C *are* C^∞ *functions on* $\mathbb{R}^+$ *satisfying* $C(\alpha) > 0$, $A(\alpha) + C(\alpha) > 0$ *for all* $\alpha > 0$, *then Equation (10) defines a sequence of Riemannian metrics under which every embedding by a Markov mapping is an isometry.*

The metrics from Campbell's Theorem also define metrics on the probability simplices Δ_{m-1}° for $m = 2, 3, \ldots$. Since the tangent vectors $v = \sum_i v_i \partial_i \in T_p\Delta_{m-1}^\circ$ satisfy $\sum_i v_i = 0$, for any two vectors $u, v \in T_p\Delta_{m-1}^\circ$, also $\sum_i \sum_j A u_i v_j = 0$ for any A. In this case, the choice of A is immaterial, and the metric becomes Chentsov's metric.

Remark 1. *Observe that Chentsov's Theorem is not a direct implication of Campbell's Theorem. However, it can be deduced from it by the following arguments. Suppose that we have a family of Riemannian simplices* $(\Delta_{m-1}^\circ, g^{(m)})$ *for* $m \in \{2, 3, \ldots\}$*, and suppose that they are isometric with respect to Markov maps. If we can extend every* $g^{(m)}$ *to a Riemannian metric* $\tilde{g}^{(m)}$ *on* $\mathbb{R}_+^m$ *in such a way that the resulting spaces* $(\mathbb{R}_+^m, \tilde{g}^{(m)})$ *are still isometric with respect to Markov maps, then Campbell's Theorem implies that* $g^{(m)}$ *is a multiple of the Fisher metric. Such metric extensions can be defined as follows. Consider the diffeomorphism:*

$$\Delta_{m-1}^\circ \times \mathbb{R}_+ \cong \mathbb{R}_+^m, \quad (p, r) \mapsto r \cdot p. \tag{11}$$

Any tangent vector $u \in T_{(p,r)}\mathbb{R}_+^m$ *can be written uniquely as* $u = u_p + u_r\partial_r$*, where* u_p *is tangent to* $r\Delta_{m-1}^\circ$*. Since each Markov map* f *preserves the one-norm* $|\cdot|$*, its push-forward* f_* *maps the tangent vector* $\partial_r \in T_{(p,r)}\mathbb{R}_+^m$ *to the corresponding tangent vector* $\partial_r \in T_{f(p,r)}\mathbb{R}_+^m$*; that is,* $f_*u = f_*u_p + u_r\partial_r$*. Therefore,*

$$\tilde{g}_{(p,r)}^{(m)}(u, v) := g_p^{(m)}(u_p, v_p) + u_r v_r \tag{12}$$

is a metric on $\mathbb{R}_+^m$ *that is invariant under* f.

In what follows, we will focus on positive matrices. In order to define a natural Riemannian metric, we can use the identification $\mathbb{R}_+^{k\times m} \cong \mathbb{R}_+^{km}$ and apply Campbell's Theorem. This leads to metrics of the form:

$$g_M^{(k,m)}(\partial_{ij}, \partial_{kl}) = A(|M|) + \delta_{ik}\delta_{jl}C(|M|)/M_{ij}, \tag{13}$$

where $\partial_{ij} = \frac{\partial}{\partial M_{ij}}$ and $|M| = \sum_{ij} M_{ij}$. However, a disadvantage of this approach is that the action of general Markov maps on $\mathbb{R}_+^{km}$ has no natural interpretation in terms of the matrix structure. Therefore, Lebanon [14] considered a special class of Markov maps defined as follows.

Definition 2. *Consider a* $k \times l$ *row-partition matrix* R *and a collection of* $m \times n$ *row-partition matrices* $Q = \{Q^{(1)}, \ldots, Q^{(k)}\}$*. The map:*

$$\mathbb{R}_+^{k\times m} \to \mathbb{R}_+^{l\times n}; \quad M \mapsto R^\top(M \otimes Q) \tag{14}$$

is called a congruent embedding by a Markov morphism of $\mathbb{R}_+^{k\times m}$ *to* $\mathbb{R}_+^{l\times n}$ *in [15]. We will refer to such an embedding as a Lebanon map. Here, the row product* $M \otimes Q$ *is defined by:*

$$(M \otimes Q)_{ab} = (M \cdot Q^{(a)})_{ab}, \quad \textit{for all } a \in [k], b \in [n]; \tag{15}$$

that is, the a*-th row of* M *is multiplied by the matrix* $Q^{(a)}$.

In a Lebanon map, each row of the input matrix M is mapped by an individual Markov mapping $Q^{(i)}$, and each resulting row is copied and scaled by an entry of R. This kind of map preserves the sum of all matrix entries. Therefore, with the identification $\mathbb{R}_+^{k\times m} \cong \mathbb{R}_+^{km}$, each Lebanon map restricts to a map $\Delta^\circ_{mk-1} \to \Delta^\circ_{nl-1}$. The set Δ°_{mk-1} can be identified with the set of joint distributions of two random variables. Lebanon maps can be regarded as special Markov maps that incorporate the product structure present in the set of joint probability distributions of a pair of random variables. In Section 4, we will give an interpretation of these maps.

Contrary to what is stated in [15], a Lebanon map does not map $(\Delta^k_{m-1})^\circ$ to $(\Delta^l_{n-1})^\circ$, unless $k = l$. Therefore, later, we will provide a characterization for the metrics on $(\Delta^k_{m-1})^\circ$ in terms of invariance under other maps (which are not Markov nor Lebanon maps).

The main result in Lebanon's work [15, Theorems 1 and 2] is the following.

Theorem 3 (Lebanon's Theorem.).

- *For each $k \geq 1, m \geq 2$, let $g^{(k,m)}$ be a Riemannian metric on $\mathbb{R}^{k\times m}_+$ in such a way that every Lebanon map is an isometry. Then:*

$$g^{(k,m)}_M(\partial_{ab}, \partial_{cd}) = A(|M|) + \delta_{ac}\left(\frac{B(|M|)}{|M_a|} + \delta_{bd}\frac{C(|M|)}{M_{ab}}\right) \tag{16}$$

  for some differentiable functions $A, B, C \in C^\infty(\mathbb{R}_+)$.
- *Conversely, let $\{(\mathbb{R}^{k\times m}_+, g^{(k,m)})\}$ be a sequence of Riemannian manifolds, with metrics $g^{(k,m)}$ of the form (16) for some $A, B, C \in C^\infty(\mathbb{R}_+)$. Then, every Lebanon map is an isometry.*

Lebanon does not study the question under which assumptions on $A, B, C \in C^\infty(\mathbb{R}_+)$ the formula (16) does indeed define a Riemannian metric. This question has the following simple answer, which we will prove in Appendix A:

Proposition 1. *The matrix (16) is positive definite if and only if $C(|M|) > 0$, $B(|M|) + C(|M|) > 0$ and $A(|M|) + B(|M|) + C(|M|) > 0$.*

The class of metrics (16) is larger than the class of metrics (13) derived in Campbell's Theorem. The reason is that Campbell's metrics are invariant with respect to a larger class of embeddings.

The special case with $A(|M|) = 0$, $B(|M|) = 0$ and $C(|M|) = 1$ is called product Fisher metric,

$$g^{(k,m)}_M(\partial_{ab}, \partial_{cd}) = \delta_{ac}\delta_{bd}\frac{1}{M_{ab}}. \tag{17}$$

Furthermore, if we restrict to $(\Delta^k_{m-1})^\circ$, the functions A and B do not play any role. In this case $|M| = k$, and we obtain the scaled product Fisher metric:

$$g^{(k,m)}_M(\partial_{ab}, \partial_{cd}) = \delta_{ac}\delta_{bd}\frac{C(k)}{M_{ab}}, \tag{18}$$

where $C(k) : \mathbb{N} \to \mathbb{R}_+$ is a positive function. As mentioned before, Lebanon's Theorem does not give a characterization of invariant metrics of stochastic matrices, since Lebanon maps do not preserve the stochasticity of the matrices. However, Lebanon maps are natural maps on the set Δ°_{mk-1} of positive joint distributions. In the same way as Chentsov's Theorem can be derived from Campbell's Theorem (see Remark 1), we obtain the following Corollary:

Corollary 1.

- *Let* $\{(\Delta^{\circ}_{km-1}, g^{(k,m)}) : k \geq 1, m \geq 2\}$ *be a double sequence of Riemannian manifolds with the property that every Lebanon map is an isometry. Then:*

$$g^{(k,m)}_P(u,v) = B \sum_a \sum_{b,c} \frac{u_{ab}u_{ac}}{|P_a|} + C \sum_a \sum_b \frac{u_{ab}v_{ab}}{P_{ab}}, \quad \text{for each } P \in \Delta^{\circ}_{km-1}, \tag{19}$$

  for some constants $B, C \in \mathbb{R}$ *with* $C > 0$ *and* $B + C > 0$*, where* $|P_a| = \sum_b P_{ab}$.
- *Conversely, let* $\{(\Delta^{\circ}_{km-1}, g^{(k,m)})\}$ *be a sequence of Riemannian manifolds with metrics* $g^{(k,m)}$ *of the form of Equation* (19) *for some* $B, C \in \mathbb{R}$*. Then, every Lebanon map is an isometry.*

Observe that these metrics agree with (a multiple of) the Fisher metric only if $B = 0$. The case $B = 0$ can also be characterized; note that Lebanon maps do not treat the two random variables symmetrically. Switching the two random variables corresponds to transposing the joint distribution matrix P. When exchanging the role of the two random variables, the Lebanon map becomes $P \mapsto (P^{\top} \otimes Q)^{\top} R$. We call such a map a dual Lebanon map. If we require invariance under both Lebanon maps and their duals in Theorem 3 or Corollary 1, the statements remain true with the additional restriction that $B = 0$ (as a function or constant, respectively).

4. Invariance Metric Characterizations for Conditional Polytopes

According to Chentsov's Theorem (Theorem 1), a natural metric on the probability simplex can be characterized by requiring the isometry of natural embeddings. Lebanon follows this axiomatic approach to characterize metrics on products of positive measures (Theorem 3). However, the maps considered by Lebanon dissolve the row-normalization of conditional distributions. In general, they do not map conditional polytopes to conditional polytopes. Therefore, we will consider a slight modification of Lebanon maps, in order to obtain maps between conditional polytopes.

4.1. Stochastic Embeddings of Conditional Polytopes

A matrix of conditional distributions $P(Y|X)$ in Δ^k_{m-1} can be regarded as the equivalence class of all joint probability distributions $P(X,Y) \in \Delta_{km-1}$ with conditional distribution $P(Y|X)$. Which Markov maps of probability simplices are compatible with this equivalence relation? The most obvious examples are permutations (relabelings) of the state spaces of X and Y.

In information theory, stochastic matrices are also viewed as channels. For any distribution of X, the stochastic matrix gives us a joint distribution of the pair (X,Y) and, hence, a marginal distribution of Y. If we input a distribution of X into the channel, the stochastic matrix determines what the distribution of the output Y will be.

Channels can be combined, provided the cardinalities of the state spaces fit together. If we take the output Y of the first channel $P(Y|X)$ and feed it into another channel $P(Y'|Y)$ then we obtain a combined channel $P(Y'|X)$. The composition of channels corresponds to ordinary matrix multiplication. If the first channel is described by the stochastic matrix K and the second channel by Q, then the combined channel is described by $K \cdot Q$. Observe that in this case, the joint distribution P (considered as a normalized matrix $P \in \Delta_{km-1}$) is transformed similarly; that is, the joint distribution of the pair (X,Y') is given by $P \cdot Q$.

More general maps result from compositions where the choice of the second channel depends on the input of the first channel. In other words, we have a first channel that takes as input X and gives as output Y, and we have another channel that takes as input (X,Y) and gives as output Y'; we are interested in the resulting channel from X to Y'. The second channel can be described by a collection of stochastic matrices $Q = \{Q^{(i)}\}_i$. If K describes the first channel, then the combined channel is described by the row product $K \otimes Q$ (see Definition 2). Again, the joint distribution of (X,Y') arises in a similar way as $P \otimes Q$.

We can also consider transformations of the first random variable X. Suppose that we use X as the input to a channel described by a stochastic matrix R. In this case, the joint distribution of the output X' of the channel and Y is described by $R^\top X$. However, in general, there is not much that we can say about the conditional distribution of Y given X'. The result depends in an essential way on the original distribution of X. However, this is not true in the special case that the channel is "not mixing", that is, in the case that R is a stochastic partition matrix. In this case, the conditional distribution $P(Y|X')$ is described by $\overline{R}^\top K$, where $\overline{R}$ is the corresponding partition indicator matrix, where all non-zero entries of R are replaced by one. In other words, each state of X corresponds to several states of X', and the corresponding row of K is copied a corresponding number of times.

To sum up, if we combine the transformations due to Q and R, then the joint probability distribution transforms as $P \mapsto R^\top(P \otimes Q)$ and the conditional transforms as $K \mapsto \overline{R}^\top(K \otimes Q)$. In particular, for the joint distribution, we obtain the Definition of a Lebanon map. Figure 1 illustrates the situation.

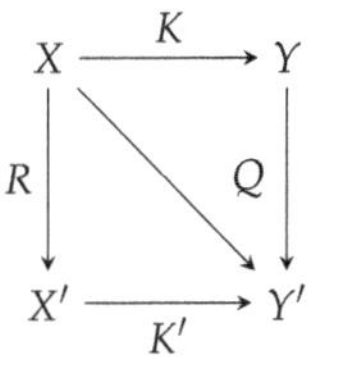

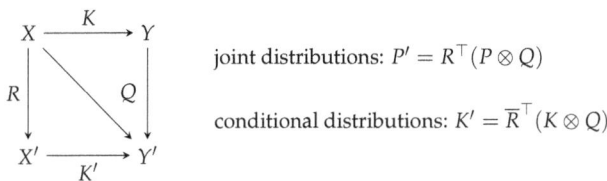

joint distributions: $P' = R^\top(P \otimes Q)$

conditional distributions: $K' = \overline{R}^\top(K \otimes Q)$

Figure 1. An interpretation for Lebanon maps and conditional embeddings. The variable X' is computed from X by R, and Y' is computed from X and Y by Q.

Finally, we will also consider the special case where the partition of R (and $\overline{R}$) is homogeneous, *i.e.*, such that all blocks have the same size. For example, this describes the case where there is a third random variable Z that is independent of Y given X. In this case, the conditional distribution satisfies $P(Y|X) = P(Y|X,Z)$, and R describes the conditional distribution of (X,Z) given X.

Definition 3. *A (row) partition indicator matrix is a matrix $\overline{R} \in \{0,1\}^{k \times l}$ that satisfies:*

$$\overline{R}_{ij} = \begin{cases} 1, & \text{if } j \in A_i, \\ 0, & \text{else}, \end{cases} \tag{20}$$

for a k block partition $\{A_1, \dots, A_k\}$ of $[l]$.

For example, the 3×5 partition indicator matrix corresponding to Equation (8) is:

$$\overline{R} = \begin{pmatrix} 1 & 0 & 1 & 0 & 0 \\ 0 & 1 & 0 & 1 & 0 \\ 0 & 0 & 0 & 0 & 1 \end{pmatrix}. \tag{21}$$

Definition 4. *Consider a $k \times l$ partition indicator matrix $\overline{R}$ and a collection of $m \times n$ stochastic partition matrices $Q = \{Q^{(i)}\}_{i=1}^k$. We call the map:*

$$f\colon \quad \mathbb{R}_+^{k \times m} \to \mathbb{R}_+^{l \times n}; \quad M \mapsto \overline{R}^\top(M \otimes Q) \tag{22}$$

a conditional embedding of $\mathbb{R}_+^{k \times m}$ in $\mathbb{R}_+^{l \times n}$. We denote the set of all such maps by $\hat{\mathcal{F}}_{k,m}^{l,n}$. If $\overline{R}$ is the partition indicator matrix of a homogeneous partition (with partition blocks of equal cardinality), then we call f a homogeneous conditional embedding. We denote the set of all such homogeneous conditional embeddings by $\mathcal{F}_{k,m}^{l,n}$ and assume that l is a multiple of k.

Conditional embeddings preserve the 1-norm of the matrix rows; that is, the elements of $\hat{\mathcal{F}}_{k,m}^{l,n}$ map $(\Delta_{m-1}^k)^\circ$ to $(\Delta_{n-1}^l)^\circ$. On the other hand, they do not preserve the 1-norm of the entire matrix. Conditional embeddings are Markov maps only when $k = l$, in which case they are also Lebanon maps.

4.2. Invariance Characterization

Considering the conditional embeddings discussed in the previous section, we obtain the following metric characterization.

Theorem 4.

- *Let $g^{(k,m)}$ denote a metric on $\mathbb{R}_+^{k\times m}$ for each $k \geq 1$ and $m \geq 2$. If every homogeneous conditional embedding $f \in \mathcal{F}_{k,m}^{l,n}$ is an isometry with respect to these metrics, then:*

$$g_M^{(k,m)}(\partial_{ab}, \partial_{cd}) = \frac{A}{k^2} + \delta_{ac}\left(k\frac{B}{k^2} + \delta_{bd}\frac{|M|}{M_{ab}}\frac{C}{k^2}\right), \quad \text{for all } M \in \mathbb{R}_+^{k\times m}, \tag{23}$$

  for some constants $A, B, C \in \mathbb{R}$, where $\partial_{ab} = \frac{\partial}{\partial M_{ab}}$ and $|M| = \sum_{ab} M_{ab}$.
- *Conversely, given the metrics defined by Equation (23) for any non-degenerate choice of constants $A, B, C \in \mathbb{R}$, each homogeneous conditional embedding $f \in \mathcal{F}_{k,m}^{l,n}$, $k \leq l$, $m \leq n$ is an isometry.*
- *Moreover, the tensors $g^{(k,m)}$ from Equation (23) are positive-definite for all $k \geq 1$ and $m \geq 2$ if and only if $C > 0$, $B + C > 0$ and $A + B + C > 0$.*

The proof of Theorem 4 is similar to the proof of the Theorems of Chentsov, Campbell and Lebanon. Due to its technical nature, we defer it to Appendix B.

Now, for the restriction of the metric $g^{(k,m)}$ to $(\Delta_{m-1}^k)^\circ$, we have the following. In this case, $|M| = k$. Since tangent vectors $v = \sum_{ab} v_{ab}\partial_{ab} \in T_M(\Delta_{m-1}^k)^\circ$ satisfy $\sum_b v_{ab} = 0$ for all a, the constants A and B become immaterial, and the metric can be written as:

$$g_M^{(k,m)}(u, v) = \sum_{ab} \frac{|M| u_{ab} v_{ab}}{M_{ab}} \frac{C}{k^2} = \sum_{ab} \frac{u_{ab} v_{ab}}{M_{ab}} \frac{C}{k}, \quad \text{for all } u, v \in T_M(\Delta_{m-1}^k)^\circ. \tag{24}$$

This metric is a specialization of the metric (18) derived by Lebanon (Theorem 3).

The statement of Theorem 4 becomes false if we consider general conditional embeddings instead of homogeneous ones:

Theorem 5. *There is no family of metrics $g^{(k,m)}$ on $\mathbb{R}_+^{k\times m}$ (or on $(\Delta_{m-1}^k)^\circ$) for each $k \geq 1$ and $m \geq 2$, for which every conditional embedding $f \in \hat{\mathcal{F}}_{k,m}^{l,n}$ is an isometry.*

This negative result will become clearer from the perspective of Section 6: as we will show in Theorem 7, although there are no metrics that are invariant under all conditional embeddings, there are families of metrics (depending on a parameter, ρ) that transform covariantly (that is, in a well-defined manner) with respect to the conditional embeddings. We defer the proof of Theorem 5 to Appendix B.

5. The Fisher Metric on Polytopes and Point Configurations

In the previous section, we obtained distinguished Riemannian metrics on $\mathbb{R}_+^{k\times m}$ and $(\Delta_{m-1}^k)^\circ$ by postulating invariance under natural maps. In this section, we take another viewpoint based on general considerations about Riemannian metrics on arbitrary polytopes. This is achieved by embedding each polytope in a probability simplex as an exponential family. We first recall the necessary background. In Section 5.2, we then present our general results, and in Section 5.3, we discuss the special case of conditional polytopes.

5.1. Exponential Families and Polytopes

Let $\mathcal{X}$ be a finite set and $A \in \mathbb{R}^{d \times \mathcal{X}}$ a matrix with columns a_x indexed by $x \in \mathcal{X}$. It will be convenient to consider the rows A_i, $i \in [d]$ of A as functions $A_i : \mathcal{X} \to \mathbb{R}$. Finally, let $\nu \colon \mathcal{X} \to \mathbb{R}_+$. The exponential family $\mathcal{E}_{A,\nu}$ is the set of probability distributions on $\mathcal{X}$ given by:

$$p(x;\theta) = \exp(\theta^\top a_x + \log(\nu(x)) - \log(Z(\theta))), \quad \text{for all } x \in \mathcal{X}, \quad \text{for all } \theta \in \mathbb{R}^d, \tag{25}$$

with the normalization function $Z(\theta) = \sum_{x' \in \mathcal{X}} \exp(\theta^\top a_{x'} + \log(\nu(x')))$. The functions A_i are called the observables and ν the reference measure of the exponential family. When the reference measure ν is constant, $\nu(x) = 1$ for all $x \in \mathcal{X}$, we omit the subscript and write $\mathcal{E}_A$.

A direct calculation shows that the Fisher information matrix of $\mathcal{E}_{A,\nu}$ at a point $\theta \in \mathbb{R}^d$ has coordinates:

$$g_\theta^{\mathcal{E}_{A,\nu}}(\partial_{\theta_i}, \partial_{\theta_j}) = \operatorname{cov}_\theta(A_i, A_j), \qquad \text{for all } i, j \in [d]. \tag{26}$$

Here, $\operatorname{cov}_\theta$ denotes the covariance computed with respect to the probability distribution $p(\cdot;\theta)$.

The convex support of $\mathcal{E}_{A,\nu}$ is defined as:

$$\operatorname{conv} A := \operatorname{conv}\{a_x \colon x \in \mathcal{X}\} = \left\{\mathbb{E}_p[A] \colon p \in \Delta_{|\mathcal{X}|-1}\right\} = \left\{\mathbb{E}_p[A] \colon p \in \overline{\mathcal{E}_{A,\nu}}\right\}, \tag{27}$$

where $\operatorname{conv} S$ is the set of all convex combinations of points in S. The moment map $\mu : p \in \Delta_{n-1} \mapsto A \cdot p \in \mathbb{R}^d$ restricts to a homeomorphism $\overline{\mathcal{E}_{A,\nu}} \to \operatorname{conv} A$; see [16]. Here, $\overline{\mathcal{E}_{A,\nu}}$ denotes the Euclidean closure of $\mathcal{E}_{A,\nu}$. The inverse of μ will be denoted by $\mu^{-1} : \operatorname{conv} A \to \overline{\mathcal{E}_{A,\nu}} \subseteq \Delta_{n-1}$. This gives a natural embedding of the polytope $\operatorname{conv} A$ in the probability simplex $\Delta_{|\mathcal{X}|-1}$. Note that the convex support is independent of the reference measure ν. See [17] for more details.

5.2. Invariance Fisher Metric Characterizations for Polytopes

Let $\mathbf{P} \in \mathbb{R}^d$ be a polytope with n vertices $a_1, \ldots, a_n$. Let $A = (a_1, \ldots, a_n)$ be the matrix with columns $a_i \in \mathbb{R}^d$ for all $i \in [n]$. Then, $\mathcal{E}_A \subseteq \Delta^\circ_{n-1}$ is an exponential family with convex support $\mathbf{P}$. We will also denote this exponential family by $\mathcal{E}_{\mathbf{P}}$. We can use the inverse of the moment map, μ^{-1}, to pull back geometric structures on Δ°_{n-1} to the relative interior $\mathbf{P}^\circ$ of $\mathbf{P}$.

Definition 5. *The Fisher metric on $\mathbf{P}^\circ$ is the pull-back of the Fisher metric on $\mathcal{E}_A \subseteq \Delta^\circ_{n-1}$ by μ^{-1}.*

Some obvious questions are: Why is this a natural construction? Which maps between polytopes are isometries between their Fisher metrics? Can we find a characterization of Chentsov type for this metric?

Affine maps are natural maps between polytopes. However, in order to obtain isometries, we need to put some additional constraints. Consider two polytopes $\mathbf{P} \in \mathbb{R}^d$, $\mathbf{P}' \in \mathbb{R}^{d'}$ and an affine map $\phi : \mathbb{R}^d \to \mathbb{R}^{d'}$ that satisfies $\phi(\mathbf{P}) \subseteq \mathbf{P}'$. A natural condition in the context of exponential families is that ϕ restricts to a bijection between the set $\operatorname{vert}(\mathbf{P})$ of vertices of $\mathbf{P}$ and the set $\operatorname{vert}(\mathbf{P}')$ of vertices of $\mathbf{P}'$. In this case, $\mathcal{E}_{\mathbf{P}'} \subseteq \mathcal{E}_{\mathbf{P}} \subseteq \Delta^\circ_{n-1}$. Moreover, the moment map μ' of $\mathbf{P}'$ factorizes through the moment map μ of $\mathbf{P}$: $\mu' = \phi \circ \mu$. Let $\phi^{-1} = \mu \circ \mu'^{-1}$. Then, the following diagram commutes:

$$\begin{array}{ccccc} \mathbf{P}^\circ & \xrightarrow{\mu^{-1}} & \mathcal{E}_{\mathbf{P}} & \searrow & \\ \phi^{-1}\big\uparrow & & \big\uparrow & & \Delta^\circ_{n-1} \\ \mathbf{P}'^\circ & \xrightarrow[\mu'^{-1}]{} & \mathcal{E}_{\mathbf{P}'} & \nearrow & \end{array} \tag{28}$$

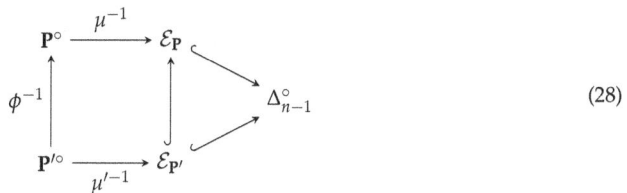

It follows that ϕ^{-1} is an isometry from $\mathbf{P}'^{\circ}$ to its image in $\mathbf{P}^{\circ}$. Observe that the inverse moment map itself arises in this way: In the diagram (28), if $\mathbf{P}$ is equal to Δ_{n-1}, then the upper moment map μ^{-1} is the identity map, and ϕ^{-1} equals the inverse moment map μ'^{-1} of $\mathbf{P}'$.

The constraint of mapping vertices to vertices bijectively is very restrictive. In order to consider a larger class of affine maps, we need to generalize our construction from polytopes to weighted point configurations.

Definition 6. *A weighted point configuration is a pair (A, ν) consisting of a matrix $A \in \mathbb{R}^{d\times n}$ with columns $a_1, \ldots, a_n$ and a positive weight function $\nu : \{1, \ldots, n\} \to \mathbb{R}_+$ assigning a weight to each column a_i. The pair (A, ν) defines the exponential family $\mathcal{E}_{A,\nu}$.*

The (A, ν)-Fisher metric on $(\operatorname{conv} A)^{\circ}$ is the pull-back of the Fisher metric on Δ°_{n-1} through the inverse of the moment map.

We recover Definition 5 as follows. For a polytope $\mathbf{P}$, let A be the point configuration consisting of the vertices of $\mathbf{P}$. Moreover, let ν be a constant function. Then, $\mathcal{E}_{\mathbf{P}} = \mathcal{E}_{A,\nu}$, and the two Definitions of the Fisher metric on $\mathbf{P}^{\circ}$ coincide.

The following are natural maps between weighted point configurations:

Definition 7. *Let (A, ν), (A', ν') be two weighted point configurations with $A = (a_i)_i \in \mathbb{R}^{d\times n}$ and $A' = (a'_j)_j \in \mathbb{R}^{d'\times n'}$. A morphism $(A, \nu) \to (A', \nu')$ is a pair (ϕ, σ) consisting of an affine map$\phi : \mathbb{R}^d \to \mathbb{R}^{d'}$ and a surjective map $\sigma : \{1, \ldots, n\} \to \{1, \ldots, n'\}$ with $\phi(a_i) = a'_{\sigma(i)}$ and$\nu'(a'_j) = \alpha \sum_{i:\sigma(i)=j} \nu(a_i)$, where $\alpha > 0$ is a constant that does not depend on j.*

Consider a morphism $(\phi, \sigma) : (A, \nu) \to (A', \nu')$. For each $j \in [n']$, let $A_j = \{i : \phi(a_i) = a'_j\}$. Then, $(A_1, \ldots, A_{n'})$ is a partition of $[n]$. Define a matrix $Q \in \mathbb{R}^{n'\times n}$ by:

$$Q_{ji} = \begin{cases} \frac{\nu(i)}{\sum_{i'\in A_j} \nu(i')}, & \text{if } i \in A_j, \\ 0, & \text{else.} \end{cases} \tag{29}$$

Then, Q is a Markov mapping, and the following diagram commutes:

$$\begin{array}{ccccc} (\operatorname{conv} A)^{\circ} & \xrightarrow{\mu^{-1}} & \mathcal{E}_{A,\nu} & \hookrightarrow & \Delta^{\circ}_{n-1} \\ {\scriptstyle \phi^{-1}}\uparrow & & \uparrow & & \uparrow{\scriptstyle Q} \\ (\operatorname{conv} A')^{\circ} & \xrightarrow[\mu'^{-1}]{} & \mathcal{E}_{A',\nu'} & \hookrightarrow & \Delta^{\circ}_{n'-1} \end{array} \tag{30}$$

By Chentsov's Theorem (Theorem 1), Q is an isometric embedding. It follows that ϕ^{-1} also induces an isometric embedding. This shows the first part of the following Theorem:

Theorem 6.

- *Let $(\phi, \sigma) : (A, \nu) \to (A', \nu')$ be a morphism of weighted point configurations. Then, $\phi^{-1} : (\operatorname{conv} A')^{\circ} \to (\operatorname{conv} A)^{\circ}$ is an isometric embedding with respect to the Fisher metrics on $(\operatorname{conv} A)^{\circ}$ and $(\operatorname{conv} A')^{\circ}$.*
- *Let $g^{A,\nu}$ be a Riemannian metric on $(\operatorname{conv} A)^{\circ}$ for each weighted point configuration (A, ν). If every morphism $(\phi, \sigma) : (A, \nu) \to (A', \nu')$ of weighted point configurations induces an isometric embedding $\phi^{-1} : (\operatorname{conv} A')^{\circ} \to (\operatorname{conv} A)^{\circ}$, then there exists a constant $\alpha \in \mathbb{R}_+$ such that $g^{A,\nu}$ is equal to α times the (A, ν)-Fisher metric.*

Proof. The first statement follows from the discussion before the Theorem. For the second statement, we show that under the given assumptions, all Markov maps are isometric embeddings. By Chentsov's Theorem (Theorem 1), this implies that the metrics $g^{\mathbf{P}}$ agree with the Fisher metric whenever $\mathbf{P}$ is a simplex. The statement then follows from the two facts that the metric on $\mathbf{P}^\circ$ or $(\operatorname{conv} A)^\circ$ is the pull-back of the Fisher metric through the inverse of the moment map and that μ^{-1} is itself a morphism.

Observe that $\Delta_{n-1} = \operatorname{conv} I_n = \operatorname{conv}\{e_1, \dots, e_n\}$ is a polytope, and Δ°_{n-1} is the corresponding exponential family. Consider a Markov embedding $Q : \Delta^\circ_{n'-1} \to \Delta^\circ_{n-1}, p \mapsto p \cdot Q$. Let $\nu(i) = \sum_j Q_{ji}$ be the value of the unique non-zero entry of Q in the i-th column. This defines a morphism and an embedding as follows:

Let A be the matrix that arises from Q by replacing each non-zero entry by one. We define ϕ as the linear map represented by the matrix A, and define $\sigma : [n] \to [n']$ by $\sigma(j) = i$ if and only if $a_j = e_i$, that is, $\sigma(j)$ indicates the row i in which the j-th column of A is non-zero. Then, (ϕ, σ) is a morphism $(I_n, \nu) \to (I_{n'}, 1)$, and by assumption, the inverse ϕ^{-1} is an isometric embedding $\Delta^\circ_{n'-1} \to \Delta^\circ_{n-1}$. However, ϕ^{-1} is equal to the Markov map Q. This shows that all Markov maps are isometric embeddings, and so, by Chentsov's Theorem, the statement holds true on the simplices. □

Theorem 6 defines a natural metric on $(\Delta^k_{m-1})^\circ$ that we want to discuss in more detail next.

5.3. Independence Models and Conditional Polytopes

Consider k random variables with finite state spaces $[n_1], \dots, [n_k]$. The independence model consists of all joint distributions $p \in \Delta_{\prod_{i\in[k]} n_i - 1}$ of these variables that factorize as:

$$p(x_1, \dots, x_k) = \prod_{i\in[k]} p_i(x_i), \quad \text{for all } x_1 \in [n_1], \dots, x_k \in [n_k], \tag{31}$$

where $p_i \in \Delta_{n_i-1}$ for all $i \in [k]$. Assuming fixed $n_1, \dots, n_k$, we denote the independence model by $\overline{\mathcal{E}_k}$. It is the Euclidean closure of an exponential family (with observables of the form δ_{iy_i}). The convex support of $\mathcal{E}_k$ is equal to the product of simplices $\mathbf{P}_k := \Delta_{n_1-1} \times \cdots \times \Delta_{n_k-1}$. The parametrization (31) corresponds to the inverse of the moment map.

We can write any tangent vector $u \in T_{(p_1,\dots,p_k)}\mathbf{P}^\circ_k$ of this open product of simplices as a linear combination $u = \sum_{i\in[k]} \sum_{x_i\in[n_i]} u_{ix_i}\partial_{i,x_i}$, where $\sum_{x_i\in[n_i]} v_{ix_i} = 0$ for all $i \in [k]$. Given two such tangent vectors, the Fisher metric is given by:

$$g^{\mathbf{P}_k}_{(p_1,\dots,p_k)}(u, v) = \sum_{i\in[k]} \sum_{x_i\in[n_i]} \frac{u_{ix_i} v_{ix_i}}{p_i(x_i)}. \tag{32}$$

Just as the convex support of the independence model is the Cartesian product of probability simplices, the Fisher metric on the independence model is the product metric of the Fisher metrics on the probability simplices of the individual variables. If $n_1 = \cdots = n_k =: n$, then $\mathbf{P}_k = \Delta^k_{n-1}$ can be identified with the set of $k \times n$ stochastic matrices.

The Fisher metric on the product of simplices is equal to the product of the Fisher metrics on the factors. More generally, if $\mathbf{P} = \mathbf{Q}_1 \times \mathbf{Q}_2$ is a Cartesian product, then the Fisher metric on $\mathbf{P}^\circ$ is equal to the product of the Fisher metrics on $\mathbf{Q}^\circ_1$ and $\mathbf{Q}^\circ_2$. In fact, in this case, the inverse of the moment map of $\mathbf{P}$ can be expressed in terms of the two moment map inverses $\mu_1 : \mathbf{Q}_1 \to \overline{\mathcal{E}_{\mathbf{Q}_1}} \subseteq \Delta_{m_1-1}$ and $\mu_2 : \mathbf{Q}_2 \to \overline{\mathcal{E}_{\mathbf{Q}_2}} \subseteq \Delta_{m_2-1}$ and the moment map $\tilde{\mu}$ of the independence model $\Delta_{m_1-1} \times \Delta_{m_2-1}$, by:

$$\mu^{-1}(q_1, q_2) = \tilde{\mu}^{-1}(\mu_1^{-1}(q_1), \mu_2^{-1}(q_2)). \tag{33}$$

Therefore, the pull-back by μ^{-1} factorizes through the pull-back by $\tilde{\mu}^{-1}$, and since the independence model carries a product metric, the product of polytopes also carries a product metric.

Let us compare the metric $g_K^{(k,m)}$ from Equation (24), with the Fisher metric $g^{\mathbf{P}_k}_{(K_1,\dots,K_k)}$ from Equation (32) on the product of simplices $\mathbf{P}^\circ = (\Delta^k_{m-1})^\circ$. In both cases, the metric is a product metric; that is, it has the form:

$$g = g_1 + \cdots + g_k, \tag{34}$$

where g_i is a metric on the i-th factor Δ°_{m-1}. For $g_K^{\Delta^k_{m-1}}$, g_i is equal to the Fisher metric on Δ°_{m-1}. However, for $g_K^{(k,m)}$, g_i is equal to $1/k$ times the Fisher metric on Δ°_{m-1}. Since this factor only depends on k, it only plays a role if stochastic matrices of different sizes are compared. The additional factor of $1/k$ can be interpreted as the uniform distribution on k elements. This is related to another more general class of Riemannian metrics that are used in applications; namely, given a function $K \in \Delta^k_{m-1} \to \rho^K \in \mathbb{R}^k_+$, it is common to use product metrics with g_i equal to $\rho^K(i)$ times the Fisher metric on Δ°_{m-1}. When K has the interpretation of a channel or when K describes the policy by which a system reacts to some sensor values, a natural possibility is to let ρ^K be the stationary distribution of the channel input or of the sensor values, respectively. We will discuss this approach in Section 6.

6. Weighted Product Metrics for Conditional Models

In this section, we consider metrics on spaces of stochastic matrices defined as weighted sums of the Fisher metrics on the spaces of the matrix rows, similar to Equation (34). This kind of metric was used initially by Amari [1] in order to define a natural gradient in the supervised learning context. Later, in the context of reinforcement learning, Kakade [2] defined a natural policy gradient based on this kind of metric, which has been further developed by Peters *et al.* [10]. Related applications within unsupervised learning have been pursued by Zahedi *et al.* [18].

Consider the following weighted product Fisher metric:

$$g_K^{\rho,m} = \sum_a \rho^K(a) g_{K_a}^{(m),a}, \quad \text{for all } K \in (\Delta^k_{m-1})^\circ, \tag{35}$$

where $g_{K_a}^{(m),a}$ denotes the Fisher metric of Δ°_{m-1} at the a-th row of K and $\rho^K \in \Delta^\circ_{k-1}$ is a probability distribution over a associated with each $K \in (\Delta^k_{m-1})^\circ$. For example, the distribution ρ^K could be the stationary distribution of sensor values observed by an agent when operating under a policy described by K.

In the following, we will try to illuminate the properties of polytope embeddings that yield the metric (35) as the pull-back of the Fisher information metric on a probability simplex. We will focus on the case that $\rho^K = \rho$ is independent of K.

There are two direct ways of embedding Δ^k_{n-1} in a probability simplex. In Section 5, we used the inverse of the moment map of an exponential family, possibly with some reference measure. This embedding is illustrated in the left panel of Figure 2. If we have given a fixed probability distribution $\rho \in \Delta^\circ_{k-1}$, there is a second natural embedding $\psi_\rho \colon \Delta^k_{m-1} \to \Delta_{k\cdot m-1}$ defined as follows:

$$\psi_\rho(K)(x,y) = \rho(x) K_{x,y} \quad \text{for all } x \in [k], y \in [m]. \tag{36}$$

If ρ is the distribution of a random variable X and $K \in \Delta^k_{m-1}$ is the stochastic matrix describing the conditional distribution of another variable Y given X, then $\psi_\rho(K)$ is the joint distribution of X and Y. Note that ψ_ρ is an affine embedding. See the right panel of Figure 2 for an illustration.

The pull-back of the Fisher metric on Δ°_{km-1} through ψ_ρ is given by:

$$\begin{aligned} g^{(km)}_{\psi_\rho(K)}(\psi_{\rho *}u, \psi_{\rho *}v) &= \sum_{a,b}\sum_{c,d}\sum_{i,j} \rho(i) K_{ij} u_{ab} \frac{\partial \log \rho(i) K_{ij}}{\partial K_{ab}} v_{cd} \frac{\partial \log \rho(i) K_{ij}}{\partial K_{cd}} \\ &= \sum_i \rho(i) \sum_j \frac{u_{ij} v_{ij}}{K_{ij}} = \sum_i \rho(i) g^i_{K_i}(u_i, v_i) = g_K^{\rho,m}(u,v). \end{aligned} \tag{37}$$

This recovers the weighted sum of Fisher metrics from Equation (35).

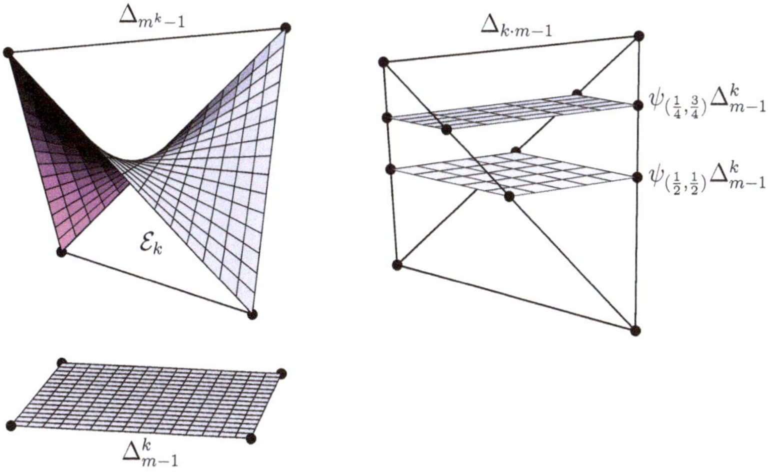

Figure 2. An illustration of different embeddings of the conditional polytope Δ^k_{m-1} in a probability simplex. The left panel shows an embedding in Δ_{m^k-1} by the inverse of the moment map μ of the independence model. The right panel shows an affine embedding in $\Delta_{k\cdot m-1}$ as a set of joint probability distributions for two different specifications of marginals.

Are there natural maps that leave the metrics $g^{\rho,m}$ invariant? Let us reconsider the stochastic embeddings from Definition 4. Let $\overline{R}$ be a $k \times l$ indicator partition matrix and R a stochastic partition matrix with the same block structure as $\overline{R}$. Observe that to each indicator partition matrix $\overline{R}$ there are many compatible stochastic partition matrices R, but the indicator partition matrix $\overline{R}$ for any stochastic partition matrix R is unique. Furthermore, let $Q = \{Q^{(a)}\}_{a\in[k]}$ be a collection of stochastic partition matrices. The corresponding conditional embedding $\overline{f}$ maps $K \in \Delta^k_{m-1}$ to $\overline{f}(K) := \overline{R}^\top(K \otimes Q) \in \Delta^l_{n-1}$.

Let $\rho \in \Delta^\circ_{k-1}$. Suppose that K describes the conditional distribution of Y given X and that $\psi_\rho(K)$ describes the joint distribution of Y and X. As explained in Section 4.1, the matrix $f(P) := R^\top(P \otimes Q)$ describes the joint distribution of a pair of random variables (X', Y'), and the conditional distribution of Y' given X' is given by $\overline{f}(K)$. In this situation, the marginal distribution of X' is given by $\rho' = \rho R$. Therefore, the following diagram commutes:

$$\begin{array}{ccc} (\Delta^k_{m-1})^\circ & \xrightarrow{\psi_\rho} & \Delta^\circ_{mk-1} \\ \Big\downarrow \overline{f} & & \Big\downarrow f \\ (\Delta^l_{n-1})^\circ & \xrightarrow[\psi_{\rho'}]{} & \Delta^\circ_{nl-1} \end{array} \tag{38}$$

The preceding discussion implies the first statement of the following result:

Theorem 7.

- *For any $k \geq 1$ and $m \geq 2$ and any $\rho \in \Delta_{k-1}^{\circ}$, the Riemannian metric $g^{\rho,m}$ on $(\Delta_{m-1}^{k})^{\circ}$ satisfies:*

$$g^{\rho,m} = \overline{f}^{*}(g^{\rho',n}), \quad \text{for } \rho' = \rho R, \tag{39}$$

for any conditional embedding $\overline{f} : K \mapsto \overline{R}(K \otimes Q)$.
- *Conversely, suppose that for any $k \geq 1$ and $m \geq 2$ and any $\rho \in \Delta_{k-1}^{\circ}$, there is a Riemannian metric $g^{(\rho,m)}$ on $(\Delta_{m-1}^{k})^{\circ}$, such that Equation (39) holds for all conditional embeddings, and suppose that $g^{(\rho,m)}$ depends continuously on ρ. Then, there is a constant $A > 0$ that satisfies $g^{(\rho,m)} = A g^{\rho,m}$.*

Proof. The first statement follows from the commutative diagram (38). For the second statement, denote by ρ^k the uniform distribution on a set of k elements. If $\overline{f}\colon K \mapsto \overline{R}(K \otimes Q)$ is a homogeneous conditional embedding of Δ_{m-1}^{k} in Δ_{n-1}^{l}, then $R = \frac{k}{l}\overline{R}$ is a stochastic partition matrix corresponding to the partition indicator matrix $\overline{R}$. Observe that $\rho^l = \rho^k R$. Therefore, the family of Riemannian metrics $g^{\rho^k,m}$ on Δ_{m-1}^{k} satisfies the assumptions of Theorem 4. Therefore, there is a constant $A > 0$ for which $g^{\rho^k,m}$ equals A/k times the product Fisher metric. This proves the statement for uniform distributions ρ.

A general distribution $\rho \in \Delta_{k-1}^{\circ}$ can be approximated by a distribution with rational probabilities. Since $g^{(\rho,m)}$ is assumed to be continuous, it suffices to prove the statement for rational ρ. In this case, there exists a stochastic partition matrix R for which $\rho' := \rho R$ is a uniform distribution, and so, $g^{(\rho',n)}$ is of the desired form. Equation (39) shows that $g^{(\rho,m)}$ is also of the desired form. □

7. Gradient Fields and Replicator Equations

In this section, we use gradient fields in order to compare Riemannian metrics on the space $(\Delta_{n-1}^{k})^{\circ}$.

7.1. Replicator Equations

We start with gradient fields on the simplex Δ_{n-1}°. A Riemannian metric g on Δ_{n-1}° allows us to consider gradient fields of differentiable functions $F\colon \Delta_{n-1}^{\circ} \to \mathbb{R}$. To be more precise, consider the differential $d_pF : T_p\Delta_{n-1}^{\circ} \to \mathbb{R}$ of F in p. It is a linear form on $T_p\Delta_{n-1}^{\circ}$, which maps each tangent vector u to $d_pF(u) = \frac{\partial F}{\partial u}(p) \in \mathbb{R}$. Using the map $u \mapsto g_p(u,\cdot)$, this linear form can be identified with a tangent vector in $T_p\Delta_{n-1}^{\circ}$, which we denote by $\mathrm{grad}_p F$. If we choose the Fisher metric $g^{(n)}$ as the Riemannian metric, we obtain the gradient in the following way. First consider a differentiable extension of F to the positive cone $\mathbb{R}_+^n$, which we will denote by the same symbol F. With the partial derivatives $\partial_i F$ of F, the Fisher gradient of F on the simplex Δ_{n-1}° is given as:

$$(\mathrm{grad}_p F)_i \;=\; p_i \left(\partial_i F(p) - \sum_{j=1}^{n} p_j \, \partial_j F(p) \right), \qquad i \in [n]. \tag{40}$$

Note that the expression on the right-hand side of Equation (40) does not depend on the particular differentiable extension of F to $\mathbb{R}_+^n$. The corresponding differential equation is well known in theoretical biology as the replicator equation; see [19,20].

$$\dot{p}_i \;=\; p_i \left(\partial_i F(p) - \sum_{j=1}^{n} p_j \, \partial_j F(p) \right), \qquad i \in [n]. \tag{41}$$

We now apply this gradient formula to functions that have the structure of an expectation value. Given real numbers F_i, $i \in [n]$, referred to as fitness values, we consider the mean fitness:

$$\bar{F}(p) := \sum_{i=1}^{n} p_i \, F_i. \tag{42}$$

Replacing the p_i by any positive real numbers leads to a differentiable extension of F, also denoted by F. Obviously, we have $\partial_i F = F_i$, which leads to the following replicator equation:

$$\dot{p}_i \; = \; p_i \left(F_i - \bar{F}(p) \right), \qquad i \in [n]. \tag{43}$$

This equation has the solution:

$$p_i(t) \; = \; \frac{p_i(0) e^{t F_i}}{\sum_{j=1}^{n} p_j(0) e^{t F_i}}, \qquad i \in [n]. \tag{44}$$

Clearly, the mean fitness will increase along this solution of the gradient field. The rate of increase can be easily calculated:

$$\frac{d}{dt} \bar{F}(p(t)) \; = \; \sum_{i=1}^{n} \dot{p}_i(t) \, F_i \; = \; \sum_{i=1}^{n} p_i \left(F_i - \bar{F}(p) \right) F_i \; = \; \sum_{i=1}^{n} p_i \left(F_i - \bar{F}(p) \right)^2 \; = \; \mathrm{var}_p(F) \; > \; 0. \tag{45}$$

As limit points of this solution, we obtain:

$$\lim_{t \to -\infty} p_i(t) \; = \; \begin{cases} \frac{p_i(0)}{\sum_{j \in \mathrm{argmin}\, F} p_j(0)}, & \text{if } i \in \mathrm{argmin}\, F \\ 0 \quad , & \text{otherwise} \end{cases}, \qquad i \in [n], \tag{46}$$

and:

$$\lim_{t \to +\infty} p_i(t) \; = \; \begin{cases} \frac{p_i(0)}{\sum_{j \in \mathrm{argmax}\, F} p_j(0)}, & \text{if } i \in \mathrm{argmax}\, F \\ 0 \quad , & \text{otherwise} \end{cases}, \qquad i \in [n]. \tag{47}$$

7.2. Extension of the Replicator Equations to Stochastic Matrices

Now, we come to the corresponding considerations of gradient fields in the context of stochastic matrices $K \in (\Delta_{n-1}^k)^\circ$. We consider a function:

$$K \; \mapsto \; F(K) = F(K_{11}, \ldots, K_{1n}; K_{21}, \ldots, K_{2n}; \ldots; K_{k1}, \ldots, K_{kn}). \tag{48}$$

One way to deal with this is to consider for each $i \in [k]$ the corresponding replicator equation:

$$\dot{K}_{ij} \; = \; K_{ij} \left(\partial_{ij} F(K) - \sum_{j'=1}^{n} K_{ij'} \, \partial_{ij'} F(K) \right), \qquad j \in [n]. \tag{49}$$

Obviously, this is the gradient field that one obtains by using the product Fisher metric on $(\Delta_{n-1}^k)^\circ$ (Equation (17)):

$$g_K^{(k,m)}(u,v) \; = \; \sum_{ij} \frac{1}{K_{ij}} \, u_{ij} v_{ij}. \tag{50}$$

If we replace the metric by the weighted product Fisher metric considered by Kakade (Equation (35)),

$$g_K^{\rho,m}(u,v) \; = \; \sum_{ij} \frac{\rho_i}{K_{ij}} \, u_{ij} v_{ij}, \tag{51}$$

then we obtain

$$\dot{K}_{ij} = \frac{K_{ij}}{\rho_i}\left(\partial_{ij}F(K) - \sum_{j'=1}^{n} K_{ij'}\,\partial_{ij'}F(K)\right), \qquad j \in [n]. \tag{52}$$

7.3. *The Example of Mean Fitness*

Next, we want to study how the gradient flows with respect to different metrics compare. We restrict to the class of metrics $g^{\rho,m}$ (Equation (35)), where $\rho \in \Delta_k^\circ$ is a probability distribution. In principle, one could drop the normalization condition $\sum_i \rho_i = 1$ and allow arbitrary coefficients ρ_i. However, it is clear that the rate of convergence can always be increased by scaling all values ρ_i with a common positive factor. Therefore, some normalization condition is needed for ρ.

With a probability distribution $p \in \Delta_{k-1}^\circ$ and fitness values F_{ij}, let us consider again the example of an expectation value function:

$$\bar{F}(K) = \sum_{i=1}^{k} p_i \sum_{j=1}^{n} K_{ij}\,F_{ij}. \tag{53}$$

With $\partial_{ij}\bar{F}(\pi) = p_i\,F_{ij}$, this leads to:

$$\dot{K}_{ij} = \frac{p_i}{\rho_i}K_{ij}\left(F_{ij} - \sum_{j'=1}^{n} K_{ij'}\,F_{ij'}\right), \qquad j \in [n]. \tag{54}$$

The corresponding solutions are given by:

$$K_{ij}(t) = \frac{K_{ij}(0)\,e^{t\frac{p_i}{\rho_i}F_{ij}}}{\sum_{j'=1}^{n} K_{ij'}(0)\,e^{t\frac{p_i}{\rho_i}F_{ij'}}}, \qquad i \in [n]. \tag{55}$$

Since $\operatorname{argmax}(\frac{p_i}{\rho_i}F_{i\cdot})$ and $\operatorname{argmin}(\frac{p_i}{\rho_i}F_{i\cdot})$ are independent of $\rho_i > 0$, the limit points are given independently of the chosen ρ as:

$$\lim_{t\to-\infty} K_{ij}(t) = \begin{cases} \frac{K_{ij}(0)}{\sum_{j'\in\operatorname{argmin}F_{i\cdot}} K_{ij'}(0)}, & \text{if } j \in \operatorname{argmin}F_{i\cdot} \\ 0 \quad , & \text{otherwise} \end{cases}, \qquad i \in [n], \tag{56}$$

and:

$$\lim_{t\to+\infty} K_{ij}(t) = \begin{cases} \frac{K_{ij}(0)}{\sum_{j'\in\operatorname{argmax}F_{i\cdot}} K_{ij'}(0)}, & \text{if } j \in \operatorname{argmax}F_{i\cdot} \\ 0 \quad , & \text{otherwise} \end{cases}, \qquad i \in [n]. \tag{57}$$

This is consistent with the fact that the critical points of gradient fields are independent of the chosen Riemannian metric. However, the speed of convergence does depend on the metric:

For each i, let $G_i = \max_j F_{ij}$ and $g_i = \max_{j\notin\operatorname{argmax}(F_{ij})} F_{ij}$ be the largest and second-largest values in the i-th row of F_{ij}, respectively. Then, as: $t \to \infty$,

$$K_{ij}(t) = \begin{cases} 1 - O(\exp(-\frac{p_i}{\rho_i}(G_i - g_i)t), & \text{if } i \in \operatorname{argmax}F_{i\cdot} \\ O(\exp(-\frac{p_i}{\rho_i}(G_i - F_{ij})t) \quad , & \text{otherwise} \end{cases} \tag{58}$$

Therefore,

$$
\begin{aligned}
\bar{F}(K(t)) &= \sum_i p_i \sum_{j \in \operatorname{argmax} F_{i\cdot}} F_{ij} + O\left(\exp(-\frac{p_i}{\rho_i}(G_i - g_i)t)\right) \\
&= \sum_i p_i \sum_{j \in \operatorname{argmax} F_{i\cdot}} F_{ij} + O\left(\exp(-\inf_i \left\{\frac{p_i}{\rho_i}(G_i - g_i)\right\} t)\right). \quad (59)
\end{aligned}
$$

Thus, in the long run, the rate of convergence is given by $\inf_i\{\frac{p_i}{\rho_i}(G_i - g_i)\}$, which depends on the parameter ρ of the metric. As a result, in this case study, the optimal choice of ρ_i, *i.e.*, with the largest convergence rate, can be computed if the numbers G_i and g_i are known.

Consider, for example, the case that the differences $G_i - g_i$ are of comparable sizes for all i. Then, we need to find the choice of ρ that maximizes $\inf_i\{\frac{p_i}{\rho_i}\}$. Clearly, $\inf_i\{\frac{p_i}{\rho_i}\} \leq 1$ (since there is always an index i with $p_i \leq \rho_i$). Equality is attained for the choice $\rho_i = p_i$. Thus, we recover the choice of Kakade.

8. Conclusions

So, which Riemannian metric should one use in practice on the set of stochastic matrices, $(\Delta_{m-1}^k)^\circ$? The results provided in this manuscript give different answers, depending on the approach. In all cases, the characterized Riemannian metrics are products of Fisher metrics with suitable factor weights. Theorem 4 suggests to use a factor weight proportional to $1/k$, and Theorem 6 suggests to use a constant weight independent of k. In many cases, it is possible to work within a single conditional polytope $(\Delta_{m-1}^k)^\circ$ and a fixed k, and then, these two results are basically equivalent. On the other hand, Theorem 7 gives an answer that allows arbitrary factor weights ρ.

Which metric performs best obviously depends on the concrete application. The first observation is that in order to use the metric $g^{\rho,m}$ of Theorem 7, it is necessary to know ρ. If the problem at hand suggests a natural marginal distribution ρ, then it is natural to make use of it and choose the metric $g^{\rho,m}$. Even if ρ is not known at the beginning, a learning system might try to learn it to improve its performance.

On the other hand, there may be situations where there is no natural choice of the weights ρ. Observe that ρ breaks the symmetry of permuting the rows of a stochastic matrix. This is also expressed by the structural difference between Theorems 4 and 6 on the one side and Theorem 7 on the other. While the first two Theorems provide an invariance metric characterization, Theorem 7 provides a "covariance" classification; that is, the metrics $g^{\rho,m}$ are not invariant under conditional embeddings, but they transform in a controlled manner. This again illustrates that the choice of a metric should depend on which mappings are natural to consider, e.g., which mappings describe the symmetries of a given problem.

For example, consider a utility function of the form $F = \sum_i \rho_i \sum_j K_{ij} F_{ij}$. Row permutations do not leave $g^{\rho,m}$ invariant (for a general ρ), but they are not symmetries of the utility function F, either, and hence, they are not very natural mappings to consider. However, row permutations transform the metric $g^{\rho,m}$ and the utility function in a controlled manner; in such a way that the two transformations match. Therefore, in this case, it is natural to use $g^{\rho,m}$. On the other hand, when studying problems that are symmetric under all row permutations, it is more natural to use the invariant metric $g^{(k,m)}$.

Appendix A

Appendix A Conditions for Positive Definiteness

Equation (16) in Lebanon's Theorem 3 defines a Riemannian metrics whenever it defines a positive-definite quadratic form. The next Proposition gives sufficient and necessary conditions for which this is the case.

Proposition A1. *For each pair $k \geq 1$ and $m \geq 2$, consider the tensor on $\mathbb{R}_+^{k \times m}$ defined by:*

$$g_M^{(k,m)}(\partial_{ab}, \partial_{cd}) = A(|M|) + \delta_{ac}\left(\frac{B(|M|)}{|M_a|} + \delta_{bd}\frac{C(|M|)}{M_{ab}}\right) \tag{A1}$$

for some differentiable functions $A, B, C \in C^\infty(\mathbb{R}_+)$. The tensor $g^{(k,m)}$ defines a Riemannian metric for all k and m if and only if $C(\alpha) > 0$, $B(\alpha) + C(\alpha) > 0$ and $A(\alpha) + B(\alpha) + C(\alpha) > 0$ for all $\alpha \in \mathbb{R}_+$.

Proof. The tensors are Riemannian metrics when:

$$g_M^{(k,m)}(V) = A(|M|)(\sum_{ab} V_{ab})^2 + B(|M|)\sum_a \frac{|M|}{|M_a|}(\sum_b V_{ab})^2 + C(|M|)\sum_{ab}\frac{|M|}{M_{ab}}V_{ab}^2 \tag{A2}$$

is strictly positive for all non-zero $V \in \mathbb{R}^{k \times m}$, for all $M \in \mathbb{R}_+^{k \times m}$.

We can derive necessary conditions on the functions A, B, C from some basic observations. Choosing $V = \partial_{ab}$ in Equation (A2) shows that $A(|M|) + \frac{|M|}{|M_a|}B(|M|) + \frac{|M|}{M_{ab}}C(|M|)$ has to be positive for all $a \in [k], b \in [m]$, for all $M \in \mathbb{R}_+^{k \times m}$. Since M_{ab} can be arbitrarily small for fixed $|M|$ and $|M_a|$, we see that C has to be non-negative. Since we can choose $|M_a| \approx M_{ab} \ll |M|$ for a fixed $|M|$, we find that $B + C$ has to be non-negative. Further, since we can choose $M_{ab} \approx |M_a| \approx |M|$ for a given $|M|$, we find that $A + B + C$ has to be non-negative. This shows that the quadratic form is positive definite only if $C \geq 0$, $B + C \geq 0$, $A + B + C \geq 0$. Since the cone of positive definite matrices is open, these inequalities have to be strictly satisfied. In the following, we study sufficient conditions.

For any given $M \in \mathbb{R}_+^{k \times m}$, we can write Equation (A2) as a product $V^\top GV$, for all $V \in \mathbb{R}^{km}$, where $G = G_A + G_B + G_C \in \mathbb{R}^{km \times km}$ is the sum of a matrix G_A with all entries equal to $A(|M|)$, a block diagonal matrix G_B whose a-th block has all entries equal to $\frac{|M|}{|M_a|}B(|M|)$, and a diagonal matrix G_C with diagonal entries equal to $\frac{|M|}{M_{ab}}C(|M|)$. The matrix G is obviously symmetric, and by Sylvester's criterion, it is positive definite iff all its leading principal minors are positive. We can evaluate the minors using Sylvester's determinant Theorem. That Theorem states that for any invertible $m \times m$ matrix X, an $m \times n$ matrix Y and an $n \times m$ matrix Z, one has the equality $\det(X + YZ) = \det(X)\det(I_n + ZX^{-1}Y)$.

Let us consider a leading square block G', consisting of all entries $G_{ab,cd}$ of G with row-index pairs (a, b) satisfying $b \in [m]$ for all $a < a'$ and $b \leq b'$ for $a = a'$ for some $a' \leq k$ and $b' \leq m$; and the same restriction for the column index pairs. The corresponding block $G'_A + G'_B$ can be written as the rank-a' matrix YZ, with Y consisting of columns $\mathbb{1}_a$ for all $a \leq a'$ and Z consisting of rows $A + \mathbb{1}_a \frac{|M|}{|M_a|}B$ for all $a \leq a'$. Hence, the determinant of G' is equal to:

$$\det(G') = \det(G'_C) \cdot \det(I_{a'} + Z{G'}_C^{-1}Y). \tag{A3}$$

Since ${G'}_C$ is diagonal, the first term is just:

$$\det(G'_C) = \left(\prod_{a<a'}\prod_b \tfrac{|M|}{M_{ab}}C\right)\left(\prod_{b \leq b'} \tfrac{|M|}{M_{a'b}}C\right). \tag{A4}$$

The matrix in the second term of Equation (A3) is given by:

$$I_{a'} + Z{G'}_C^{-1}Y =$$

$$\frac{1}{C}\begin{pmatrix} C+B & & & \\ & \ddots & & \\ & & C+B & \\ & & & C + \frac{\sum_{b \leq b'} M_{a'b}}{|M_{a'}|}B \end{pmatrix} + \frac{1}{C}\begin{pmatrix} \frac{|M_1|}{|M|}A & \cdots & \frac{|M_{a'-1}|}{|M|}A & \frac{\sum_{b \leq b'} M_{a'b}}{|M|}A \\ \vdots & & \vdots & \vdots \\ \frac{|M_1|}{|M|}A & \cdots & \frac{|M_{a'-1}|}{|M|}A & \frac{\sum_{b \leq b'} M_{a'b}}{|M|}A \end{pmatrix}. \tag{A5}$$

By Sylvester's determinant Theorem, we have:

$$\det(I_{a'} + ZG'^{-1}_{C}Y) = C^{-a'}(C+B)^{a'-1}(C + \frac{\sum_{b\leq b'} M_{a'b}}{|M_{a'}|}B)(1 + \sum_{a<a'} \frac{\frac{|M_a|}{|M|}A}{C+B} + \frac{\frac{\sum_{b\leq b'} M_{a'b}}{|M|}A}{C + \frac{\sum_{b\leq b'} M_{a'b}}{|M_{a'}|}B})$$

$$= \left(\prod_{a} \frac{C+B_a}{C}\right)\left(1 + \sum_{a} \frac{A_a}{C+B_a}\right), \tag{A6}$$

where $A_a = \frac{|M_a|}{|M|}A$ for $a < a'$ and $A_{a'} = \frac{\sum_{b\leq b'} M_{a'b}}{|M|}A$, and $B_a = B$ for $a < a'$ and $B_{a'} = \frac{\sum_{b\leq b'} M_{a'b}}{|M_{a'}|}B$.

This shows that the matrix G is positive definite for all M if and only if $C > 0$, $C + B > 0$ and $\left(1 + \sum_{a\leq a'} \frac{A_a}{C+B_a}\right) > 0$ for all a' and b'. The latter inequality is satisfied whenever $A + B + C > 0$. This completes the proof. □

Appendix B Proofs of the Invariance Characterization

The following Lemma follows directly from the Definition and contains all the technical details we need for the proofs.

Lemma A1. *The push-forward $f_*\colon T_M\mathbb{R}_+^{k\times m} \to T_{f(M)}\mathbb{R}_+^{l\times n}$ of a map $f \in \hat{\mathcal{F}}_{k,m}^{l,n}$ is given by:*

$$f_*(\partial_{ab}) = \sum_{i=1}^{l}\sum_{j=1}^{n} \overline{R}_{ai} Q_{bj}^{(a)} \partial'_{ij}, \tag{A7}$$

and the pull-back of a metric $g^{(l,n)}$ on $\mathbb{R}_+^{l\times n}$ through f is given by:

$$(f^* g^{(l,n)})_M(\partial_{ab}, \partial_{cd}) = g_{f(M)}^{(l,n)}(f_*\partial_{ab}, f_*\partial_{cd}) = \sum_{i=1}^{l}\sum_{j=1}^{n}\sum_{s=1}^{l}\sum_{t=1}^{n} \overline{R}_{ai}\overline{R}_{cs} Q_{bj}^{(a)} Q_{dt}^{(c)} g_{f(M)}^{(l,n)}(\partial'_{ij}, \partial'_{st}). \tag{A8}$$

Proof of Theorem 4. We follow the strategy of [5,14]. The idea is to consider subclasses of maps from the class $\mathcal{F}_{k,m}^{l,n}$ and to evaluate their push-forward and pull-back maps together with the isometry requirement. This yields restrictions on the possible metrics, eventually fully characterizing them.

First. Consider the maps $h_{\pi,\sigma} \in \mathcal{F}_{k,m}^{k,m}$, resulting from permutation matrices $Q^{(a)} = P_{\pi^a}$, $\pi^a\colon [m] \to [m]$ for all $a \in [k]$, and $\overline{R} = P_\sigma$, $\sigma\colon [k] \to [k]$. Requiring isometry yields:

$$(h_{\pi,\sigma})_*(\partial_{ab}) = \partial'_{\sigma(a)\,\pi^a(b)} \tag{A9}$$

$$g_M^{(k,m)}(\partial_{ab}, \partial_{cd}) = g_{h_{\pi,\sigma}(M)}^{(k,m)}(\partial_{\sigma(a)\,\pi^{(a)}(b)}, \partial_{\sigma(c)\,\pi^{(c)}(d)}). \tag{A10}$$

Second. Consider the maps $r_{zw} \in \mathcal{F}_{k,m}^{kz,mw}$ defined by $Q^{(1)} = \cdots = Q^{(k)} \in \mathbb{R}^{m\times mw}$ and $\overline{R} \in \mathbb{R}^{k\times kz}$ being uniform. In this case, for some permutations π and σ,

$$(r_{zw})_*(\partial_{ab}) = \frac{1}{w}\sum_{i=1}^{z}\sum_{j=1}^{w} \partial'_{\sigma^{(a)}(i)\,\pi^{(b)}(j)} \tag{A11}$$

$$(r_{zw}{}^* g^{(kz,mw)})_M(\partial_{ab}, \partial_{cd}) = \frac{1}{w^2}\sum_{i=1}^{z}\sum_{j=1}^{w}\sum_{s=1}^{z}\sum_{t=1}^{w} g_{r_{zw}(M)}^{(kz,mw)}(\partial'_{\sigma^{(a)}(i)\,\pi^{(b)}(j)}, \partial'_{\sigma^{(c)}(s)\,\pi^{(d)}(t)}). \tag{A12}$$

Third. For a rational matrix $M = \frac{1}{Z}\tilde{M}$ with $\tilde{M} \in \mathbb{N}^{k\times m}$ and row-sum $|\tilde{M}_a| = N \in \mathbb{N}$ for all $a \in [k]$, consider the map $v_M \in \mathcal{F}^{zk,N}_{k,m}$ that maps M to a constant matrix. In this case, $\overline{R} \in \mathbb{R}^{k\times kz}$ and $Q^{(a)}$ has the b-th row with $|\tilde{M}_{ab}|$ entries with value $\frac{1}{|\tilde{M}_{ab}|}$ at positions $\pi^{(ab)}([\tilde{M}_{ab}]) \subseteq [N]$, and:

$$(v_M)_*(\partial_{ab}) = \frac{1}{\tilde{M}_{ab}} \sum_{i=1}^{k} \sum_{j=1}^{\tilde{M}_{ab}} \partial'_{\sigma^{(a)}(i)\ \pi^{(ab)}(j)} \tag{A13}$$

$$(v_M{}^* g^{(kz,N)})_M(\partial_{ab}, \partial_{cd}) = \frac{1}{\tilde{M}_{ab}} \frac{1}{\tilde{M}_{cd}} \sum_{i=1}^{z} \sum_{j=1}^{\tilde{M}_{ab}} \sum_{s=1}^{z} \sum_{t=1}^{\tilde{M}_{cd}} g^{(kz,N)}_{v_M(M)}(\partial'_{\sigma^{(a)}(i)\ \pi^{(ab)}(j)}, \partial'_{\sigma^{(c)}(s)\ \pi^{(cd)}(t)}). \tag{A14}$$

Step 1: $a \neq c$. Consider a constant matrix $M = U$. Then:

$$g^{(k,m)}_U(\partial_{a_1b_1}, \partial_{c_1d_1}) = g^{(k,m)}_{h_{\pi,\sigma}(U)}(\partial_{a_2b_2}, \partial_{c_2d_2}) = g^{(k,m)}_U(\partial_{a_2b_2}, \partial_{c_2d_2}). \tag{A15}$$

This implies that $g^{(k,m)}_U(\partial_{ab}, \partial_{cd}) = \hat{A}(k,m)$ when $a \neq c$.

Using the second type of map, we get:

$$\hat{A}(k,m) = \frac{z^2 w^2}{w^2} \hat{A}(kz, mw), \tag{A16}$$

which implies $g^{(k,m)}_U(\partial_{ab}, \partial_{cd}) = \frac{A}{k^2}$, when $a \neq c$. Considering a rational matrix M and the map v_M yields:

$$g^{(k,m)}_M(\partial_{ab}, \partial_{c,d}) = \frac{A}{k^2}. \tag{A17}$$

Step 2: $b \neq d$. By similar arguments as in Part 1, $g^{(k,m)}_U(\partial_{ab}, \partial_{ad}) = \hat{B}(k,m)$. Evaluating the map r_{zw} yields:

$$\begin{aligned}\hat{B}(k,m) &= \frac{zw^2}{w^2}\hat{B}(kz,mw) + \frac{z(z-1)w^2}{w^2}\frac{A}{(kz)^2} \\ &= z\hat{B}(kz,mw) + \frac{z-1}{z}\frac{A}{k^2},\end{aligned} \tag{A18}$$

and therefore,

$$\frac{1}{z}\left(\hat{B}(k,m) - \frac{A}{k^2}\right) = \hat{B}(kz,mw) - \frac{A}{(kz)^2}, \tag{A19}$$

which implies that $\left(\hat{B}(k,m) - \frac{A}{k^2}\right)$ is independent of m and scales with the inverse of k, such that it can be written as $\frac{B}{k}$. Rearranging the terms yields $g^{(k,m)}_U(\partial_{ab}, \partial_{ad}) = \frac{A}{k^2} + \frac{B}{k}$, for $b \neq d$.

For a rational matrix M, the pull-back through v_M shows then:

$$g^{(k,m)}_M(\partial_{ab}, \partial_{cd}) = z\frac{\tilde{M}_{ab}\tilde{M}_{ad}}{\tilde{M}_{ab}\tilde{M}_{ad}}\left(\frac{A}{(kz)^2} + \frac{B}{kz}\right) + \frac{z(z-1)\tilde{M}_{ab}\tilde{M}_{ad}}{\tilde{M}_{ab}\tilde{M}_{ad}}\frac{A}{(kz)^2} = \frac{A}{k^2} + \frac{B}{k}. \tag{A20}$$

Step 3: $a = c$ and $b = d$. In this case, $g_U^{(k,m)}(\partial_{a_1b_1}, \partial_{a_1b_1}) = g_U^{(k,m)}(\partial_{a_2b_2}, \partial_{a_2b_2}) = \hat{C}(k,m)$, and:

$$\begin{aligned}\hat{C}(k,m) &= \frac{1}{w^2}zw\hat{C}(kz,mw) + \frac{1}{w^2}zw(w-1)\left(\frac{A}{(kz)^2}+\frac{B}{kz}\right) + \frac{1}{w^2}zw^2z(z-1)\frac{A}{(kz)^2} \\ &= \frac{z}{w}\hat{C}(kz,mw) + (1-\frac{1}{zw})\frac{A}{k^2} + (1-\frac{z}{zw})\frac{B}{k},\end{aligned} \tag{A21}$$

which implies:

$$\frac{k}{m}\left(\hat{C}(k,m) - \frac{A}{k^2} - \frac{B}{k}\right) = \frac{kz}{mw}\left(\tilde{C}(kz,mw) - \frac{A}{(kz)^2} - \frac{B}{kz}\right), \tag{A22}$$

such that the left-hand side is a constant C, and $g_U^{(k,m)}(\partial_{ab}, \partial_{ab}) = \frac{A}{k^2} + \frac{B}{k} + \frac{m}{k}C$. Now, for a rational matrix M, pulling back through v_M gives:

$$\begin{aligned}g_M^{(k,m)}(\partial_{ab}, \partial_{ab}) &= \frac{1}{\tilde{M}_{ab}^2}\tilde{M}_{ab}\left(\frac{A}{k^2} + \frac{B}{k} + \frac{|\tilde{M}_a|}{k}C\right) + \frac{1}{\tilde{M}_{ab}^2}\tilde{M}_{ab}(\tilde{M}_{ab}-1)\left(\frac{A}{k^2}+\frac{B}{k}\right) + 0 \\ &= \frac{A}{k^2} + \frac{B}{k} + \frac{|\tilde{M}_a|}{\tilde{M}_{ab}k}C \\ &= \frac{A}{k^2} + k\frac{B}{k^2} + \frac{|M|}{M_{ab}}\frac{C}{k^2}.\end{aligned} \tag{A23}$$

Summarizing, we found:

$$g_M^{(k,m)}(\partial_{ab}, \partial_{cd}) = \frac{A}{k^2} + \delta_{ac}\left(k\frac{B}{k^2} + \delta_{bd}\frac{|M|}{M_{ab}}\frac{C}{k^2}\right), \tag{A24}$$

which proves the first statement. The second statement follows by plugging Equation (23) into Equation (A8). Finally, the statement about the positive-definiteness is a direct consequence of Proposition 1. □

Proof of Theorem 5. Suppose, contrary to the claim, that a family of metrics $g_M^{(k,m)}$ exists, which is invariant with respect to any conditional embedding. By Theorem 4, these metrics are of the form of Equation (23). To prove the claim, we only need to show that A, B and C vanish. In the following, we study conditional embeddings where Q consists of identity matrices and evaluate the isometry requirement $(f^*g^{(l,n)})_M(\partial_{ab}, \partial_{cd}) = g_M^{(k,m)}(\partial_{ab}, \partial_{cd})$.
Step 1: In the case $a \neq c$, we obtain from the invariance requirement and Equation (A8), that:

$$\frac{A}{k^2} = |\overline{R}_a||\overline{R}_c|\frac{A}{l^2}. \tag{A25}$$

Observe that:

$$\frac{1}{k}\sum_{i=1}^{k}|\overline{R}_i| = \frac{1}{k}|\overline{R}| = \frac{l}{k}. \tag{A26}$$

In fact, $|\overline{R}_i|$ is the cardinality of the i-th block of the partition belonging to $\overline{R}$. Therefore, if we choose $\overline{R}$ to be the partition indicator matrix of a partition that is not homogeneous and in which $|\overline{R}_a| > l/k$ and $|\overline{R}_c| > l/k$, then Equation (A25) implies that $A = 0$.
Step 2: In the case $a = c$ and $b \neq d$, we obtain from invariance and Equation (A8), that:

$$\frac{B}{k} = \sum_{i=1}^{l}\sum_{s=1}^{l}\overline{R}_{ai}\overline{R}_{as}\delta_{is}\frac{B}{l} = |\overline{R}_a|\frac{B}{l}. \tag{A27}$$

Again, we may chose $\overline{R}_a$ in such a way that $|\overline{R}_a| \neq \frac{k}{l}$ and find that $B = 0$.

Step 3: Finally, in the case $a = c$ and $b = d$, we obtain from invariance and Equation (A8), that:

$$\frac{C|M|}{k^2 M_{ab}} = \sum_{i=1}^{l}\sum_{s=1}^{l} \overline{R}_{ai}\overline{R}_{as}\delta_{i,s}\frac{C|\overline{R}^{\top}M|}{l^2(\overline{R}^{\top}M)_{ib}} = |\overline{R}_a|\frac{C|\overline{R}^{\top}M|}{l^2 M_{ab}}. \tag{A28}$$

If we chose $\overline{R}_a$, such that $|\overline{R}_a| \neq \frac{|M|}{|\overline{R}^{\top}M|}$, then we see that $C = 0$. Therefore, $g^{(k,m)}$ is the zero-tensor, which is not a metric. □

Acknowledgments

The authors are grateful to Keyan Zahedi for discussions related to policy gradient methods in robotics applications. Guido Montúfar thanks the Santa Fe Institute for hosting him during the initial work on this article. Johannes Rauh acknowledges support by the VW Foundation. This work was supported in part by the DFG Priority Program, Autonomous Learning (DFG-SPP 1527).

Author Contributions

All authors contributed to the design of the research. The research was carried out by all authors, with main contributions by Guido Montúfar and Johannes Rauh. The manuscript was written by Guido Montúfar, Johannes Rauh and Nihat Ay. All authors read and approved the final manuscript.

Conflicts of Interest

The authors declare no conflict of interests.

References

1. Amari, S. Natural gradient works efficiently in learning. *Neur. Comput.* **1998**, *10*, 251–276.
2. Kakade, S. A Natural Policy Gradient. *Advances in Neural Information Processing Systems 14*; MIT Press: Cambridge, MA, USA, 2001; pp. 1531–1538.
3. Shahshahani, S. *A New Mathematical Framework for the Study of Linkage and Selection*; American Mathematical Society: Providence, RI, USA, 1979.
4. Chentsov, N. *Statistical Decision Rules and Optimal Inference*; American Mathematical Society: Providence, RI, USA, 1982.
5. Campbell, L. An extended Čencov characterization of the information metric. *Proc. Am. Math. Soc.* **1986**, *98*, 135–141.
6. Sutton, R.S.; McAllester, D.; Singh, S.; Mansour, Y. Policy Gradient Methods for Reinforcement Learning with Function Approximation. *In Advances in Neural Information Processing Systems 12*; MIT Press: Cambridge, MA, USA, 2000; pp. 1057–1063.
7. Marbach, P.; Tsitsiklis, J. Simulation-based optimization of Markov reward processes. *IEEE Trans. Autom. Control* **2001**, *46*, 191–209.
8. Montúfar, G.; Ay, N.; Zahedi, K. Expressive power of conditional restricted boltzmann machines for sensorimotor control. **2014**, arXiv:1402.3346.
9. Ay, N.; Montúfar, G.; Rauh, J. Selection Criteria for Neuromanifolds of Stochastic Dynamics. In *Advances in Cognitive Neurodynamics (III)*; Yamaguchi, Y., Ed.; Springer-Verlag: Dordrecht, The Netherlands 2013; pp. 147–154.
10. Peters, J.; Schaal, S. Natural Actor-Critic. *Neurocomputing* **2008**, *71*, 1180–1190.
11. Peters, J.; Schaal, S. Policy Gradient Methods for Robotics. In Proceedings of the IEEE International Conference on Intelligent Robotics Systems (IROS 2006), Beijing, China, 9–15 October 2006.

12. Peters, J.; Vijayakumar, S.; Schaal, S. Reinforcement learning for humanoid robotics. In Proceedings of the third IEEE-RAS international conference on humanoid robots, Karlsruhe, Germany, 29–30 September 2003; pp. 1–20.
13. Bagnell, J.A.; Schneider, J. Covariant policy search. In Proceedings of the 18th International Joint Conference on Artificial Intelligence, Acapulco, Mexico, August 9–15, 2003; Morgan Kaufmann Publishers Inc.: San Francisco, CA, USA, 2003; pp. 1019–1024.
14. Lebanon, G. Axiomatic geometry of conditional models. *IEEE Trans. Inform. Theor.* **2005**, *51*, 1283–1294.
15. Lebanon, G. An Extended Čencov-Campbell Characterization of Conditional Information Geometry. In Proceedings of the 20th Conference in Uncertainty in Artificial Intelligence (UAI 04), Banff, AL, Canada, 7–11 July 2004; Chickering, D.M., Halpern, J.Y., Eds.; AUAI Press: Arlington, VA, USA, 2004; pp. 341–345.
16. Barndorff-Nielsen, O. *Information and Exponential Families: In statistical Theory*; John Wiley & Sons, Inc.: Hoboken, NJ, USA, 1978.
17. Brown, L.D. *Fundamentals of Statistical Exponential Families with Applications in Statistical Decision Theory*; Institute of Mathematical Statistics: Hayward, CA, USA, 1986.
18. Zahedi, K.; Ay, N.; Der, R. Higher coordination with less control—A result of informaion maximiation in the sensorimotor loop. *Adapt. Behav.* **2010**, *18*.
19. Hofbauer, J.; Sigmund, K. *Evolutionary Games and Population Dynamics*; Cambridge University Press: Cambridge, United Kingdom, 1998.
20. Ay, N.; Erb, I. On a notion of linear replicator equations. *J. Dyn. Differ. Equ.* **2005**, *17*, 427–451.

entropy

MDPI

Article

Matrix Algebraic Properties of the Fisher Information Matrix of Stationary Processes

André Klein

Rothschild Blv. 123 Apt.7, 65271 Tel Aviv, Israel; A.A.B.Klein@uva.nl or klein@contact.uva.nl; Tel.: 972.5.25594723

Received: 12 February 2014; in revised form: 11 March 2014 / Accepted: 24 March 2014 /
Published: 8 April 2014

Abstract: In this survey paper, a summary of results which are to be found in a series of papers, is presented. The subject of interest is focused on matrix algebraic properties of the Fisher information matrix (FIM) of stationary processes. The FIM is an ingredient of the Cramér-Rao inequality, and belongs to the basics of asymptotic estimation theory in mathematical statistics. The FIM is interconnected with the Sylvester, Bezout and tensor Sylvester matrices. Through these interconnections it is shown that the FIM of scalar and multiple stationary processes fulfill the resultant matrix property. A statistical distance measure involving entries of the FIM is presented. In quantum information, a different statistical distance measure is set forth. It is related to the Fisher information but where the information about one parameter in a particular measurement procedure is considered. The FIM of scalar stationary processes is also interconnected to the solutions of appropriate Stein equations, conditions for the FIM to verify certain Stein equations are formulated. The presence of Vandermonde matrices is also emphasized.**MSC Classification:** 15A23, 15A24, 15B99, 60G10, 62B10, 62M20.

Keywords: Bezout matrix; Sylvester matrix; tensor Sylvester matrix; Stein equation; Vandermonde matrix; stationary process; matrix resultant; Fisher information matrix

1. Introduction

In this survey paper, a summary of results derived and described in a series of papers, is presented. It concerns some matrix algebraic properties of the Fisher information matrix (abbreviated as FIM) of stationary processes. An essential property emphasized in this paper concerns the matrix resultant property of the FIM of stationary processes. To be more explicit, consider the coefficients of two monic polynomials $p(z)$ and $q(z)$ of finite degree, as the entries of a matrix such that the matrix becomes singular if and only if the polynomials $p(z)$ and $q(z)$ have at least one common root. Such a matrix is called a resultant matrix and its determinant is called the resultant. The Sylvester, Bezout and tensor Sylvester matrices have such a property and are extensively studied in the literature, see e.g., [1,3]. The FIM associated with various stationary processes will be expressed by these matrices. The derived interconnections are obtained by developing the necessary factorizations of the FIM in terms of the Sylvester, Bezout and tensor Sylvester matrices. These factored forms of the FIM enable us to show that the FIM of scalar and multiple stationary processes fulfill the resultant matrix property. Consequently, the singularity conditions of the appropriate Fisher information matrices and Sylvester, Bezout and tensor Sylvester matrices coincide, these results are described in [4,6].

A statistical distance measure involving entries of the FIM is presented and is based on [7]. In quantum information, a statistical distance measure is set forth, see [8,10], and is related to the Fisher information but where the information about one parameter in a particular measurement procedure is considered. This leads to a challenging question that can be presented as, can the existing distance measure in quantum information be developed at the matrix level?

The matrix Stein equation, see e.g., [11], is associated with the Fisher information matrices of scalar stationary processes through the solutions of the appropriate Stein equations. Conditions for the Fisher information matrices or associated matrices to verify certain Stein equations are formulated and proved in this paper. The presence of Vandermonde matrices is also emphasized. The general and more detailed results are set forth in [12] and [13]. In this survey paper it is shown that the FIM of linear stationary processes form a class of structured matrices. Note that in [14], the authors emphasize that statistical problems related to stationary processes have been treated successfully with the aid of Toeplitz forms. This paper is organized as follows. The various stationary processes, considered in this paper, are presented in Section 2, the Fisher information matrices of the stationary processes are displayed in Section 3. Section 3 sets forth the interconnections between the Fisher information matrices and the Sylvester, Bezout, tensor Sylvester matrices, and solutions to Stein equations. A statistical distance measure is expressed in terms of entries of a FIM.

2. The Linear Stationary Processes

In this section we display the class of linear stationary processes whose corresponding Fisher information matrix shall be investigated in a matrix algebraic context. But first some basic definitions are set forth, see e.g., [15].

If a random variable X is indexed to time, usually denoted by t, the observations $\{X_t, t \in \mathbb{T}\}$ is called a time series, where $\mathbb{T}$ is a time index set (for example, $\mathbb{T} = \mathbb{Z}$, the integer set).

2.1. Definition 2.1

A stochastic process is a family of random variables $\{X_t, t \in \mathbb{T}\}$ *defined on a probability space* $\{\Omega, F, \wp\}$.

2.2. Definition 2.2

The Autocovariance function. If $\{X_t, t \in \mathbb{T}\}$ *is a process such that* $Var(X_t) < \infty$ *(variance) for each t, then the autocovariance function* $\gamma_X(\cdot, \cdot)$ *of* $\{X_t\}$ *is defined by* $\gamma_X(r, s) = Cov(X_r, X_s) = \mathbb{E}[(X_r - \mathbb{E}X_r)(X_s - \mathbb{E}X_s)]$, $r, s \in \mathbb{Z}$ *and* $\mathbb{E}$ *represents the expected value.*

2.3. Definition 2.3

Stationarity. The time series $\{X_t, t \in \mathbb{Z}\}$, *with the index set* $\mathbb{Z} = \{0, \pm\}1, \pm\}2, \ldots\}$, *is said to be stationary if*

(i) $\mathbb{E}\,|X_t|^2 < \infty$

(ii) $\mathbb{E}(X_t) = m$ *for all* $t \in \mathbb{Z}$, *m is the constant average or mean*

(iii) $\gamma_X(r, s) = \gamma_X(r + t, s + t)$ *for all* $r, s, t \in \mathbb{Z}$,

From Definition 2.3 can be concluded that the joint probability distributions of the random variables $\{X_1, X_2, \ldots X_{t_n}\}$ and $\{X_{1+k}, X_{2+k}, \ldots X_{t_n+k}\}$ are the same for arbitrary times $t_1, t_2, \ldots, t_n$ for all n and all lags or leads $k = 0, \pm\}1, \pm\}2, \ldots$. The probability distribution of observations of a stationary process is invariant with respect to shifts in time. In the next section the linear stationary processes that will be considered throughout this paper are presented.

2.4. The Vector ARMAX or VARMAX Process

We display one of the most general linear stationary process called the multivariate autoregressive, moving average and exogenous process, the VARMAX process. To be more specific, consider the vector difference equation representation of a linear system $\{y(t), t \in \mathbb{Z}\}$, of order (p, r, q),

$$\sum_{j=0}^{p} A_j\, y(t-j) = \sum_{j=0}^{r} C_j\, x(t-j) + \sum_{j=0}^{q} B_j\, e(t-j),\ t \in \mathbb{Z} \tag{1}$$

where $y(t)$ are the observable outputs, $x(t)$ the observable inputs and $\epsilon(t)$ the unobservable errors, all are n-dimensional. The acronym VARMAX stands for vector autoregressive-moving average with exogenous variables. The left side of (1) is the autoregressive part the second term on the right is the moving average part and $x(t)$ is exogenous. If $x(t)$ does not occur the system is said to be (V)ARMA. Next to exogenous, the input $x(t)$ is also named the control variable, depending on the field of application, in econometrics and time series analysis, e.g., [15], and in signal processing and control, e.g., [16,17]. The matrix coefficients, $A_j \in \mathbb{R}^{n\times n}$, $C_j \in \mathbb{R}^{n\times n}$, and $B_j \in \mathbb{R}^{n\times n}$ are the associate parameter matrices. We have the property $A_0 \equiv B_0 \equiv C_0 \equiv I_n$.

Equation (1) can compactly be written as

$$A(z)\,y(t) = C(z)\,x(t) + B(z)\,e(t) \tag{2}$$

where

$$A(z) = \sum_{j=0}^{p} A_j\, z^j;\; C(z) = \sum_{j=0}^{r} C_j\, z^j;\; B(z) = \sum_{j=0}^{q} B_j\, z^j$$

we use z to denote the backward shift operator, for example $z\ x_t = x_{t-1}$. The matrix polynomials $A(z)$, $B(z)$ and $C(z)$ are the associated autoregressive, moving average matrix polynomials, and the exogenous matrix polynomial respectively of order p, q and r respectively. Hence the process described by (2) is denoted as a VARMAX(p, r, q) process. Here $z \in \mathbb{C}$ with a duplicate use of z as an operator and as a complex variable, which is usual in the signal processing and time series literature, e.g., [15,16,18]. The assumptions Det($A(z)$) $\neq 0$, such that $|z| \leq 1$ and Det($B(z)$) $\neq 0$, such that $|z| < 1$ for all $z \in \mathbb{C}$, is imposed so that the VARMAX(p, r, q) process (2) has exactly one stationary solution and the condition Det($B(z)$) $\neq 0$ implies the invertibility condition, see e.g., [15] for more details. Under these assumptions, the eigenvalues of the matrix polynomials $A(z)$ and $B(z)$ lie outside the unit circle. The eigenvalues of a matrix polynomial $Y(z)$ are the roots of the equation Det($Y(z)$) = 0, Det(X) is the determinant of X. The VARMAX(p, r, q) stationary process (2) is thoroughly discussed in [15,18,19].

The error $\{\epsilon(t), t \in \mathbb{Z}\}$ is a collection of uncorrelated zero mean n-dimensional random variables each having positive definite covariance matrix Σ and we assume, for all s, t, $\mathbb{E}_\vartheta\,\{x(s)\,\epsilon^{\mathrm{T}}(t)\} = 0$, where X^{T} denotes the transposition of matrix X and $\mathbb{E}_\vartheta$ represents the expected value under the parameter ϑ. The matrix ϑ represents all the VARMAX(p, r, q) parameters, with the total number of parameters being $n^2(p+q+r)$. For different purposes which will be specified in the next sections, two choices of the parameter structure are considred. First, the parameter vector $\vartheta \in \mathbb{R}^{n^2(p+q+r)\times 1}$ is defined by

$$\vartheta = \mathrm{vec}\,\{A_1, A_2, \ldots, A_p, C_1, C_2, \ldots, C_r, B_1, B_2, \ldots, B_q\} \tag{3}$$

The vec operator transforms a matrix into a vector by stacking the columns of the matrix one underneath the other according to vec $X = \mathrm{col}(\mathrm{col}(X_{ij})_{i=1}^{n})_{j=1}^{n}$, see e.g., [2,20]. A different choice is set forth, when the parameter matrix $\vartheta \in \mathbb{R}^{n\times n(p+q+r)}$ is of the form

$$\vartheta = (\vartheta_1\,\vartheta_2\ldots\vartheta_p\,\vartheta_{p+1}\,\vartheta_{p+2}\ldots\vartheta_{p+r}\,\vartheta_{p+r+1}\,\vartheta_{p+r+2}\ldots\vartheta_{p+r+q}) \tag{4}$$

$$= (A_1\,A_2\ldots A_p\,C_1\,C_2\ldots C_r\,B_1\,B_2\ldots B_q) \tag{5}$$

Representation (5) of the parameter matrix has been used in [21]. The estimation of the matrices A_1, A_2,. . ., A_p, C_1, C_2,. . ., C_r, B_1, B_2, . . ., B_q and Σ has received considerable attention in the time series and statistical signal processing literature, see e.g., [15,17,19]. In [19], the authors study the asymptotic properties of maximum likelihood estimates of the coefficients of VARMAX(p, r, q) processes, stored in a $(\ell \times 1)$ vector ϑ, where $\ell = n^2(p+q+r)$.

Before describing the control-exogenous variable $x(t)$ used in this survey paper, we shall present the different special cases of the model described in 1 and 2.

2.5. The Vector ARMA or VARMA Process

When the process (2) does not contain the control process $x(t)$ it yields

$$A(z)y(t) = B(z)e(t) \tag{6}$$

which is a vector autoregressive and moving average process, VARMA(p, q) process, see e.g., [15]. The matrix ϑ represents now all the VARMA parameters, with the total number of parameters being $n^2(p+q)$. The VARMA(p, q) version of the parameter vector ϑ defined in (3) is then given by

$$\vartheta = \text{vec}\{A_1, A_2, \ldots, A_p, B_1, B_2, \ldots, B_q\} \tag{7}$$

A VARMA process equivalent to the parameter matrix (4) is then the $n \times n(p+q)$ parameter matrix

$$\vartheta = (\vartheta_1\, \vartheta_2 \ldots \vartheta_p\, \vartheta_{p+1}\, \vartheta_{p+2} \ldots \vartheta_{p+q}) = (A_1\, A_2 \ldots A_p\, B_1\, B_2 \ldots B_q) \tag{8}$$

A description of the input variable $x(t)$, in 2 follows. Generally, one can assume either that $x(t)$ is non stochastic or that $x(t)$ is stochastic. In the latter case, we assume $\mathbb{E}_\vartheta\{\, x(s)\, \epsilon^{\mathrm{T}}(t)\} = 0$, for all s, t, and that statistical inference is performed conditionally on the values taken by $x(t)$. In this case it can be interpreted as constant, see [22] for a detailed exposition. However, in the papers referred in this survey, like in [21] and [23], the observed input variable $x(t)$, is assumed to be a stationary VARMA process, of the form

$$\alpha(z)x(t) = \beta(z)\eta(t) \tag{9}$$

where $\alpha(z)$ and $\beta(z)$ are the autoregressive and moving average polynomials of appropriate degree and $\{\eta(t), t \in \mathbb{Z}\}$ is a collection of uncorrelated zero mean n-dimensional random variables each having positive definite covariance matrix Ω. The spectral density of the VARMA process $x(t)$ is $R_x(\cdot)/2\pi$ and for a definition, see e.g., [15,16], to obtain

$$R_x(e^{\mathrm{i}\omega}) = \alpha^{-1}(e^{\mathrm{i}\omega})\beta(e^{\mathrm{i}\omega})\Omega\beta^*(e^{\mathrm{i}\omega})\alpha^{-*}(e^{\mathrm{i}\omega}) \qquad \omega \in [-\pi, \pi] \tag{10}$$

where $\mathbf{i}$ is the imaginary unit with the property $\mathbf{i}^2 = -1$, ω is the frequency, the spectral density $R_x(e^{\mathrm{i}\omega})$ is Hermitian, and we further have, $R_x(e^{\mathrm{i}\omega}) \geq 0$ and $\int_{-\pi}^{\pi} R_x(e^{\mathrm{i}\omega})d\omega < \infty$. As mentioned above, the basic assumption, $x(t)$ and $\epsilon(t)$ are independent or at least uncorrelated processes, which corresponds geometrically with orthogonal processes, holds and X^* is the complex conjugate transpose of matrix X.

2.6. The ARMAX and ARMA Processes

The scalar equivalent to the VARMAX(p, r, q) and VARMA(p, q) processes, given by 2 and 6 respectively, shall now be displayed, to obtain for the ARMAX(p, r, q) process

$$a(z)y(t) = c(z)x(t) + b(z)e(t) \tag{11}$$

and for the ARMA(p, q) process

$$a(z)y(t) = b(z)e(t) \tag{12}$$

popularized in, among others, the Box-Jenkins type of time series analysis, see e.g., [15]. Where $a(z)$, $b(z)$ and $c(z)$ are respectively the scalar autoregressive, moving average polynomials and exogenous polynomial, with corresponding scalar coefficients a_j, b_j and c_j,

$$a(z) = \sum_{j=0}^{p} a_j\, z^j;\; c(z) = \sum_{j=0}^{r} c_j\, z^j;\; b(z) = \sum_{j=0}^{q} b_j\, z^j \tag{13}$$

Note that as in the multiple case, $a_0 = b_0 = 1$. The parameter vector, ϑ, for the processes, 11 and 12 is then

$$\vartheta = \{a_1, a_2, \ldots, a_p, c_1, c_2, \ldots, c_r, b_1, b_2, \ldots, b_q\} \tag{14}$$

and

$$\vartheta = \{a_1, a_2, \ldots, a_p, b_1, b_2, \ldots, b_q\} \tag{15}$$

respectively.

In the next section the matrix algebraic properties of the Fisher information matrix of the stationary processes (2), (6), (11) and (12) will be verified. Interconnections with various known structured matrices like the Sylvester resultant matrix, the Bezout matrix and Vandermonde matrix are set forth. The Fisher information matrix of the various stationary processes is also expressed in terms of the unique solutions to the appropriate Stein equations.

3. Structured Matrix Properties of the Asymptotic Fisher Information Matrix of Stationary Processes

The Fisher information is an ingredient of the Cramér-Rao inequality, also called by some the Cauchy-Schwarz inequality in mathematical statistics, and belongs to the basics of asymptotic estimation theory in mathematical statistics. The Cramér-Rao theorem [24] is therefore considered. When assuming that the estimators of ϑ, defined in the previuos sections, are asymptotically unbiased, the inverse of the asymptotic information matrix yields the Cramér-Rao bound, and provided that the estimators are asymptotically efficient, the asymptotic covariance matrix then verifies the inequality

$$\mathrm{Cov}\,(\hat{\vartheta}) \succcurlyeq \mathcal{I}^{-1}(\hat{\vartheta})$$

here $I\,(\widehat{\vartheta})$ is the FIM, Cov $(\widehat{\vartheta})$ is the covariance of $\widehat{\vartheta}$, the unbiased estimator of ϑ, for a detailed fundamental statistical analysis, see [25,26]. The FIM equals the Cramér-Rao lower bound, and the subject of the FIM is also of interest in the control theory and signal processing literature, see e.g., [27]. Its quantum analog was introduced immediately after the foundation of mathematical quantum estimation theory in the 1960's, see [28,29] for a rigorous exposition of the subject. More specifically, the Fisher information is also emphasized in the context of quantum information theory, see e.g., [30,31]. It is clear that the Cramér-Rao inequality takes a lot of attention because it is located on the highly exciting boundary of statistics, information and quantum theory and more recently matrix theory. In the next sections, the Fisher information matrices of linear stationary processes will be presented and its role as a new class of structured matrices will be the subject of study.

When time series models are the subject, using 2 for all $t \in \mathbb{Z}$ to determine the residual $\epsilon(t)$ or $\epsilon_t(\vartheta)$, to emphasize the dependency on the parameter vector ϑ, and assuming that $x(t)$ is stochastic and that $(y(t), x(t))$ is a Gaussian stationary process, the asymptotic FIM $F(\vartheta)$ is defined by the following $(\ell \times \ell)$ matrix which does not depend on t

$$\mathcal{F}(\vartheta) = \mathbb{E}\left\{\left(\frac{\partial e_t(\vartheta)}{\partial \vartheta^{\top}}\right)^{\top} \Sigma^{-1} \left(\frac{\partial e_t(\vartheta)}{\partial \vartheta^{\top}}\right)\right\} \tag{16}$$

where the $(v \times \ell)$ matrix $\partial(\cdot)/\partial\vartheta^{\,\mathrm{T}}$, the derivative with respect to $\vartheta^{\,\mathrm{T}}$, for any $(v \times 1)$ column vector $(\cdot)$ and ℓ is the total number of parameters. The derivative with respect to $\vartheta^{\,\mathrm{T}}$ is used for obtaining the appropriate dimensions. Equality (16) is used for computing the FIM of the various time series processes presented in the previous sections and appropriate definitions of the derivatives are used, especially for the multivariate processes (2) and (6), see [21,22].

3.1. The Fisher Information Matrix of an ARMA(p, q) Process

In this section, the focus is on the FIM of the ARMA process (12). When ϑ is given in 15, the derivatives in 16 are at the scalar level

$$\frac{\partial e_t(\vartheta)}{\partial a_j} = \frac{1}{a(z)} e_{t-j} \quad \text{for } j = 1, \ldots, p \text{ and} \frac{\partial e_t(\vartheta)}{\partial b_k} = -\frac{1}{b(z)} e_{t-k} \text{ for } k = 1, \ldots, q$$

when combined for all j and k, the FIM of the ARMA process (12) with the variance of the noise process $\epsilon_t(\vartheta)$ equal to one, yields the block decomposition, see [32]

$$\mathcal{F}(\vartheta) = \begin{pmatrix} \mathcal{F}_{aa}(\vartheta) & \mathcal{F}_{ab}(\vartheta) \\ \mathcal{F}_{ba}(\vartheta) & \mathcal{F}_{bb}(\vartheta) \end{pmatrix} \tag{17}$$

The expressions of the different blocks of the matrix $F(\vartheta)$ are

$$\mathcal{F}_{aa}(\vartheta) = \frac{1}{2\pi \mathbf{i}} \oint_{|z|=1} \frac{u_p(z) u_p^\top(z^{-1})}{a(z)a(z^{-1})} \frac{dz}{z} = \frac{1}{2\pi \mathbf{i}} \oint_{|z|=1} \frac{u_p(z) v_p^\top(z)}{a(z)\hat{a}(z)} dz \tag{18}$$

$$\mathcal{F}_{ab}(\vartheta) = -\frac{1}{2\pi \mathbf{i}} \oint_{|z|=1} \frac{u_p(z) u_q^\top(z^{-1})}{a(z)b(z^{-1})} \frac{dz}{z} = -\frac{1}{2\pi \mathbf{i}} \oint_{|z|=1} \frac{u_p(z) v_q^\top(z)}{a(z)\hat{b}(z)} dz \tag{19}$$

$$\mathcal{F}_{ba}(\vartheta) = -\frac{1}{2\pi \mathbf{i}} \oint_{|z|=1} \frac{u_q(z) u_p^\top(z^{-1})}{a(z^{-1})b(z)} \frac{dz}{z} = -\frac{1}{2\pi \mathbf{i}} \oint_{|z|=1} \frac{u_q(z) v_p^\top(z)}{\hat{a}(z)b(z)} dz \tag{20}$$

$$\mathcal{F}_{bb}(\vartheta) = \frac{1}{2\pi \mathbf{i}} \oint_{|z|=1} \frac{u_q(z) u_q^\top(z^{-1})}{b(z)b(z^{-1})} \frac{dz}{z} = \frac{1}{2\pi \mathbf{i}} \oint_{|z|=1} \frac{u_q(z) v_q^\top(z)}{b(z)\hat{b}(z)} dz \tag{21}$$

where the integration above and everywhere below is counterclockwise around the unit circle. The reciprocal monic polynomials $\hat{a}(z)$ and $\widehat{b}(z)$ are defined as $\hat{a}(z) = z^p a(z^{-1})$ and $\widehat{b}(z) = z^q b(z^{-1})$ and $\vartheta = (a_1, \ldots, a_p, b_1, \ldots, b_q)^{\mathrm{T}}$ introduced in (15). For each positive integer k we have $u_k(z) = (1, z, z^2, \ldots, z^{k-1})^{\mathrm{T}}$ and $v_k(z) = z^{k-1} u_k(z^{-1})$. Considering the stability condition of the ARMA(p, q) process implies that all the roots of the monic polynomials $a(z)$ and $b(z)$ lie outside the unit circle. Consequently, the roots of the polynomials $\hat{a}(z)$ and $\widehat{b}(z)$ lie within the unit circle and will be used as the poles for computing the integrals (18)–(21) when Cauchy's residue theorem is applied. Notice that the FIM $F(\vartheta)$ is symmetric block Toeplitz so that $\mathcal{F}_{ab}(\vartheta) = \mathcal{F}_{ba}^\top(\vartheta)$ and the integrands in (18)–(21) are Hermitian. The computation of the integral expressions, (18)–(21) is easily implementable by using the standard residue theorem. The algorithms displayed in [33] and [22] are suited for numerical computations of among others the FIM of an ARMA(p, q) process.

3.2. The Sylvester Resultant Matrix - The Fisher Information Matrix

The resultant property of a matrix is considered, in order to show that the FIM $F(\vartheta)$ has the matrix resultant property implies to show that the matrix $F(\vartheta)$ becomes singular if and only if the appropriate scalar monic polynomials $\hat{a}(z)$ and $\widehat{b}(z)$ have at least one common zero. To illustrate the subject, the following known property of two polynomials is set forth. The greatest common divisor (frequently abbreviated as GCD) of two polynomials is a polynomial, of the highest possible degree, that is a factor of both the two original polynomials, the roots of the GCD of two polynomials are the common roots of the two polynomials. Consider the coefficients of two monic polynomials $p(z)$ and $q(z)$ of finite degree, as the entries of a matrix such that the matrix becomes singular if and only if the polynomials $p(z)$ and $q(z)$ have at least one common root. Such a matrix is called a resultant matrix and its determinant is called the resultant. Therefore we present the known $(p + q) \times (p + q)$ Sylvester resultant matrix of the polynomials a and b, see e.g., [2], to obtain

$$\mathcal{S}(a,b) = \begin{pmatrix} 1 & a_1 & \cdots & a_p & 0 & \cdots & 0 \\ 0 & \ddots & \ddots & & \ddots & \ddots & \vdots \\ \vdots & \ddots & \ddots & \ddots & & \ddots & 0 \\ 0 & \cdots & 0 & 1 & a_1 & \cdots & a_p \\ 1 & b_1 & \cdots & b_q & 0 & \cdots & 0 \\ 0 & \ddots & \ddots & & \ddots & \ddots & \vdots \\ \vdots & \ddots & \ddots & \ddots & & \ddots & 0_{n\times n} \\ 0 & \cdots & 0 & 1 & b_1 & \cdots & b_q \end{pmatrix} \tag{22}$$

Consider the $q \times (p+q)$ and $p \times (p+q)$ upper and lower submatrices $\mathcal{S}_p$ (b) and $\mathcal{S}_q$ $(-a)$ of the Sylvester resultant matrix $\mathcal{S}$ $(-b, a)$ such that

$$\mathcal{S}(b,-a) = \begin{pmatrix} \mathcal{S}_p(b) \\ -\mathcal{S}_q(a) \end{pmatrix} \tag{23}$$

The matrix

$$\mathcal{S}$$

(a, b) becomes singular in the presence of one or more common zeros of the monic polynomials $\hat{a}(z)$ and $\widehat{b}(z)$, this property is assessed by the following equalities

$$\mathcal{R}(a,b) = \prod_{\substack{i=1,\ldots,p \\ j=1,\ldots,q}} (\alpha_i - \beta_j), \mathcal{R}(b,a) = (-1)^{pq} \prod_{\substack{i=1,\ldots,p \\ j=1,\ldots,q}} (\alpha_i - \beta_j) \tag{24}$$

and

$$\mathcal{R}(b,-a) = (-1)^q \prod_{\substack{i=1,\ldots,p \\ j=1,\ldots,q}} (\beta_j - \alpha_i), \text{ and } \mathcal{R}(-b,a) = (-1)^p \prod_{\substack{i=1,\ldots,p \\ j=1,\ldots,q}} (\beta_j - \alpha_i) \tag{25}$$

where $R(a, b)$ is the resultant of $\hat{a}(z)$ and $\widehat{b}(z)$, and is equal to Det includegraphics[scale=1]entropy-16-02023f6.pdf (a, b). The string of equalities in (24) and (25) hold since $R(b, a) = (-1)^{pq} R(a, b)$, $R(b, -a) = (-1)^q R(b, a)$, and $R(-b, a) = (-1)^p R(b, a)$, see [34]. The zeros of the scalar monic polynomials $\hat{a}(z)$ and $\widehat{b}(z)$ are α_i and β_j respectively and are assumed to be distinct. By this is meant, when we have $(z - \alpha_i)^{n_{\alpha_i}}$ and $(z - \beta_j)^{n_{\beta_j}}$ with the powers n_{α_i} and n_{β_j} both greater than one, that only the distinct roots will be considered free from the corresponding powers. The key property of the classical Sylvester resultant matrix $\mathcal{S}$ (a, b) is that its null space provides a complete description of the common zeros of the polynomials involved. In particular, in the scalar case the polynomials $\hat{a}(z)$ and $\widehat{b}(z)$ are coprime if and only if $\mathcal{S}$ (a, b) is non-singular. The following key property of the classical Sylvester resultant matrix $\mathcal{S}$ (a, b), is given by the well known theorem on resultants, to obtain

$$\dim \operatorname{Ker} \mathcal{S}(a,b) = \nu(a,b) \tag{26}$$

where $v(a, b)$ is the number of common roots of the polynomials $\hat{a}(z)$ and $\widehat{b}(z)$, with counting multiplicities, see e.g., [3]. The dimension of a subspace $\mathcal{V}$ is represented by dim $(\mathcal{V})$, Ker (X) is the null space or kernel of the matrix X, denoted by Null or Ker. The null space of an $n \times n$ matrix A with coefficients in a field K (typically the field of the real numbers or of the complex numbers) is the set Ker $A = \{x \in K^n: Ax = 0\}$, see e.g., [1,2,20].

In order to prove that the FIM $\mathcal{F}(\vartheta)$ fulfills the resultant matrix property, the following factorization is derived, Lemma 2.1 in [5],

$$\mathcal{F}(\vartheta) = \mathcal{S}(b, -a)\mathcal{P}(\vartheta)\mathcal{S}^{\top}(b, -a) \tag{27}$$

where the matrix $\wp(\vartheta) \in \mathbb{R}^{(p+q)\times(p+q)}$ admits the form

$$\mathcal{P}(\vartheta) = \frac{1}{2\pi\mathbf{i}} \oint\limits_{|z|=1} \frac{u_{p+q}(z)u^{\top}_{p+a}(z^{-1})}{a(z)b(z)a(z^{-1})b(z^{-1})} \frac{dz}{z} = \frac{1}{2\pi\mathbf{i}} \oint\limits_{|z|=1} \frac{u_{p+q}(z)v^{\top}_{p+q}(z)}{a(z)b(z)\hat{a}(z)\hat{b}(z)} dz \tag{28}$$

It is proved in [5] that the symmetric matrix $\wp(\vartheta)$ fulfills the property, $\wp(\vartheta) \succ O$. The factorization (27) allows us to show the matrix resultant property of the FIM, Corollary 2.2 in [5] states.

The FIM of an ARMA(p, q) process with polynomials $a(z)$ and $b(z)$ of order p, q respectively becomes singular if and only if the polynomials $\hat{a}(z)$ and $\widehat{b}(z)$ have at least one common root. From Corollary 2.2 in [5] can be concluded, the FIM of an ARMA(p, q) process and the Sylvester resultant matrix

$$\mathcal{S}$$

$(-b, a)$ have the same singularity property. By virtue of (26) and (27) we will specify the dimension of the null space of the FIM $\mathcal{F}(\vartheta)$, this is set forth in the following lemma.

3.2.1. Lemma 3.1

Assume that the polynomials $\hat{a}(z)$ and $b(z)$ have $v(a, b)$ common roots, counting multiplicities. The factorization (27) of the FIM and the property (26) enable us to prove the equality

$$dim \ \ (Ker\, \mathcal{F}(\vartheta)) = dim \ \ (Ker\, \mathcal{S}(b, -a)) = v(a, b) \tag{29}$$

Proof

The matrix $\wp(\vartheta) \in \mathbb{R}^{(p+q)\times(p+q)}$, given in (27), fulfills the property of positive definiteness, as proved in [5]. This implies that a Cholesky decomposition can be applied to $\wp(\vartheta)$, see [35] for more details, to obtain $\wp(\vartheta) = L^{\mathrm{T}}(\vartheta)L(\vartheta)$, where $L(\vartheta)$ is a $\mathbb{R}^{(p+q)\times(p+q)}$ upper triangular matrix that is unique if its diagonal elements are all positive. Consequently, all its eigenvalues are then positive so that the matrix $L(\vartheta)$ is also positive definite. Factorization of (27) now admits the representation

$$\mathcal{F}(\vartheta) = \mathcal{S}(b, -a)L^{\top}(\vartheta)L(\vartheta)\mathcal{S}^{\top}(b, -a) \tag{30}$$

and taking the property, if A is an $m\times n$ matrix, then Ker (A) = Ker $(A^{\mathrm{T}}A)$, into account, yields when applied to (30)

$$\mathrm{Ker}\, \mathcal{F}(\vartheta) = \mathrm{Ker}\, \mathcal{S}(b, -a)L^{\top}(\vartheta)L(\vartheta)\mathcal{S}^{\top}(b, -a) = \mathrm{Ker}\, L(\vartheta)\mathcal{S}^{\top}(b, -a)$$

Assume the vector $u \in$ Ker $L(\vartheta)\,\mathcal{S}^{\top}(b, -a)$, such that $L(\vartheta)\,\mathcal{S}^{\top}(b, -a)u = 0$ and set $\mathcal{S}^{\top}(b, -a)u = v = \Rightarrow$ $L(\vartheta)v = 0$, since the matrix $L(\vartheta) \succ O = \Rightarrow v = 0$, this implies $\mathcal{S}^{\top}(b, -a)u = 0 = \Rightarrow u \in$ Ker $\mathcal{S}^{\top}(b, -a)$. Consequently,

$$\operatorname{Ker}\mathcal{F}(\vartheta) = \operatorname{Ker}\mathcal{S}^{\top}(b, -a) \tag{31}$$

We will now consider the Rank-Nullity Theorem, see e.g., [1], if A is an $m \times n$ matrix, then

$$\dim\ (\operatorname{Ker} A) + \dim\ (\operatorname{Im} A) = n$$

and the property dim (Im A) = dim (Im A^{T}). When applied to the $(p + q) \times (p + q)$ matrix $\mathcal{S}$ $(b, -a)$, it yields

$$\dim\ (\operatorname{Ker}\mathcal{S}(b,-a)) = \dim\ (\operatorname{Ker}\mathcal{S}^{\top}(b,-a)) \Rightarrow \dim\ (\operatorname{Ker}\mathcal{F}(\vartheta)) = \dim\ (\operatorname{Ker}\mathcal{S}(b,-a))$$

which completes the proof.

Notice that the dimension of the null space of matrix A is called the nullity of A and the dimension of the image of matrix A, dim (Im A), is termed the rank of matrix A. An alternative proof to the one developed in Corollary 2.2 in [5], is given in a corollary to Lemma 3.1, reconfirming the resultant matrix property of the FIM $\mathcal{F}(\vartheta)$.

3.2.2. Corollary 3.2

The FIM $\mathcal{F}(\vartheta)$ *of an ARMA*(p, q) *process becomes singular if and only if the autoregressive and moving average polynomials* $\hat{a}(z)$ *and* $\widehat{b}(z)$ *have at least one common root.*

Proof

By virtue of the equality (31) combining with the property Det $\mathcal{S}^{\top}$ $(b, -a)$ = Det $\mathcal{S}$ $(b, -a)$ and the matrix resultant property of the Sylvester matrix $\mathcal{S}$ $(b, -a)$ yields, Det $\mathcal{S}^{\top}$ $(b, -a) = 0 \Leftrightarrow$ Ker $\mathcal{S}^{\top}$ $(b, -a) \neq \{0\}$ if and only if the ARMA(p, q) polynomials $\hat{a}(z)$ and $\widehat{b}(z)$ have at least one common root. Equivalently, Det $\mathcal{S}^{\top}$ $(b, -a) \neq 0 \Leftrightarrow$ Ker $\mathcal{S}^{\top}$ $(b, -a) = \{0\}$ if and only if the ARMA(p, q) polynomials $\hat{a}(z)$ and $\widehat{b}(z)$ have no common roots. Consequently, by virtue of the equality Ker $\mathcal{F}(\vartheta)$ =Ker $\mathcal{S}^{\top}$ $(b, -a)$ can be concluded, the FIM $\mathcal{F}(\vartheta)$ becomes singular if and only if the ARMA(p, q) polynomials $\hat{a}(z)$ and $\widehat{b}(z)$ have at least one common root. This completes the proof.

3.3. *The Statistical Distance Measure and the Fisher Information Matrix*

In [7] statistical distance measures are studied. Most multivariate statistical techniques are based upon the concept of distance. For that purpose a statistical distance measure is considered that is a normalized Euclidean distance measure with entries of the FIM as weighting coefficients. The measurements $x_1, x_2, \ldots, x_n$ are subject to random fluctuations of different magnitudes and have therefore different variabilities. It is then important to consider a distance that takes the variability of these variables or measurements into account when determining its distance from a fix point. A rotation of the coordinate system through a chosen angle while keeping the scatter of points given by the data fixed, is also applied, see [7] for more details. It is shown that when the FIM is positive definite, the appropriate statistical distance measure is a metric. In case of a singular FIM of an ARMA stationary process, the metric property depends on the rotation angle. The statistical distance measure, is based on m parameters unlike a statistical distance measure introduced in quantum information, see e.g., [8,9], that is also related to the Fisher information but where the information about one parameter in a particular measurement procedure is considered.

The straight-line or Euclidean distance between the stochastic vector $x = \begin{pmatrix} x_1 & x_2 & \dots & x_n \end{pmatrix}^\top$ and fixed vector $y = \begin{pmatrix} y_1 & y_2 & \dots & y_n \end{pmatrix}^\top$ where $x, y \in \mathbb{R}^n$, is given by

$$d(x,y) = \|x - y\| = \left(\sum_{j=1}^{n} (x_j - y_j)^2 \right)^{1/2} \tag{32}$$

where the metric $d(x, y) := \|x - y\|$ is induced by the standard Euclidean norm $\| \cdot \|$ on $\mathbb{R}^n$, see e.g., [2] for the metric conditions.

The observations $x_1, x_2, \dots, x_n$ are used to compute maximum likelihood estimated of the parameters $\vartheta_1, \vartheta_2, \dots, \vartheta_m$ and where $m < n$. These estimated parameters are random variables, see e.g., [15]. The distance of the estimated vector $\vartheta \in \mathbb{R}^m$, given in (15), is studied. Entries of the FIM are inserted in the distance measure as weighting coefficients. The linear transformation

$$\widetilde{\vartheta} = \mathcal{L}_i(\varphi)\vartheta \tag{33}$$

is applied, where $\mathcal{L}_i(\varphi) \in \mathbb{R}^{m \times n}$ is the Givens rotation matrix with rotation angle φ, with $0 \leq \varphi \leq 2\pi$ and $i \in \{1, \dots, m-1\}$, see e.g., [36], and is given by

$$\mathcal{L}_i(\varphi) = \begin{pmatrix} I_{i-1} & 0 & 0 & 0 \\ 0 & (\cos(\varphi))_{i,i} & (-\sin(\varphi))_{i,i+1} & 0 \\ 0 & (\sin(\varphi))_{i+1,i} & (\cos(\varphi))_{i+1,i+1} & 0 \\ 0 & 0 & 0 & I_{m-i-1} \end{pmatrix}, \quad 0 \leq \varphi \leq 2\pi \tag{34}$$

The following matrix decomposition is applied in order to obtain a transformed FIM

$$\mathcal{F}_\varphi(\vartheta) = \mathcal{L}_i(\varphi)\mathcal{F}(\vartheta)\mathcal{L}_i^\top(\varphi) \tag{35}$$

where $F_\varphi(\vartheta)$ and $\mathcal{F}(\vartheta)$ are respectively the transformed and untransformed Fisher information matrices. It is straightforward to conclude that by virtue of (35), the transformed and untransformed Fisher information matrices $\mathcal{F}_\varphi(\vartheta)$ and $\mathcal{F}(\vartheta)$, are similar since the rotation matrix $\mathcal{L}_i(\varphi)$ is orthogonal. Two matrices A and B are similar if there exists an invertible matrix X such that the equality $AX = XB$ holds. As can be seen, the Givens matrix $\mathcal{L}_i(\varphi)$ involves only two coordinates that are affected by the rotation angle φ whereas the other directions, which correspond to eigenvalues of one, are unaffected by the rotation matrix.

By virtue of (35) can be concluded that a positive definite FIM, $\mathcal{F}(\vartheta) \succ 0$, implies a positive definite transformed FIM, $\mathcal{F}_\varphi(\vartheta) \succ 0$. Consequently, the elements on the main diagonal of $\mathcal{F}(\vartheta)$, $f_{1,1}$, $f_{2,2}, \dots, f_{m,m}$, as well as the elements on the main diagonal of $\mathcal{F}_\varphi(\vartheta)$, $\widetilde{f}_{1,1}, \widetilde{f}_{2,2}, \dots, \widetilde{f}_{m,m}$ are all positive. However, the elements on the main diagonal of a singular FIM of a stationary ARMA process are also positive.

As developed in [7], combining (33) and (35) yields the distance measure of the estimated parameters $\vartheta_1, \vartheta_2, \dots, \vartheta_m$ accordingly, to obtain

$$d^2_{\mathcal{F}_\varphi}(\vartheta) = \sum_{j=1, j \neq i, i+1}^{m} \left\{ \frac{\vartheta_j^2}{f_{j,j}} \right\} + \frac{\{\vartheta_i \cos(\varphi) - \vartheta_{i+1} \sin(\varphi)\}^2}{\widetilde{f}_{i,i}(\varphi)} + \frac{\{\vartheta_{i+1} \cos(\varphi) + \vartheta_i \sin(\varphi)\}^2}{\widetilde{f}_{i+1,i+1}(\varphi)} \tag{36}$$

where

$$\widetilde{f}_{i,i}(\varphi) = f_{i,i} \cos^2(\varphi) - f_{i,i+1} \sin(2\varphi) + f_{i+1,i+1} \sin^2(\varphi) \tag{37}$$

$$\widetilde{f}_{i+1,i+1}(\varphi) = f_{i+1,i+1}\cos^2(\varphi) + f_{i,i+1}\sin(2\varphi) + f_{i,i}\sin^2(\varphi) \tag{38}$$

and $f_{j,l}$ are entries of the FIM $\mathcal{F}(\vartheta)$ whereas $\widetilde{f}_{i,i}(\phi)$ and $\widetilde{f}_{i+1,i+1}(\phi)$ are the transformed components since the rotation affects only the entries, i and $i+1$, as can be seen in matrix $\mathcal{L}_i(\varphi)$. In [7], the existence of the following inequalities is proved

$$\widetilde{f}_{i,i}(\varphi) > 0 \quad \text{and} \quad \widetilde{f}_{i+1,i+1}(\varphi) > 0$$

this guaratees the metric property of (36). When the FIM of an ARMA(p, q) process is the case, a combination of (27) and (35) for the ARMA(p, q) parameters, given in (15) yields for the transformed FIM,

$$\mathcal{F}_\varphi(\vartheta) = \mathcal{S}_\varphi(-b,a)\mathcal{P}(\vartheta)\mathcal{S}_\varphi^\top(-b,a) \tag{39}$$

where $\wp(\vartheta)$ is given by (28) and the transformed Sylvester resultant matrix is of the form

$$\mathcal{S}_\varphi(-b,a) = \mathcal{L}_i(\varphi)\mathcal{S}(-b,a) \tag{40}$$

Proposition 3.5 in [7], proves that the transformed FIM $\mathcal{F}_\varphi(\vartheta)$ and the transformed Sylvester matrix $\mathcal{S}_\phi$ $(-b, a)$ fulfill the resultant matrix property by using the equalities (40) and (39). The following property is then set forth.

3.3.1. Proposition 3.3

The properties

$$Ker\,\mathcal{F}_\varphi(\vartheta) = Ker\,\mathcal{S}_\varphi^\top(-b,a) \;\; and \;\; Ker\,\mathcal{S}_\varphi(-b,a) = Ker\,\mathcal{S}(-b,a)$$

hold true.

Proof

By virtue of the equalities (39), (40) and the orthogonality property of the rotation matrix $\mathcal{L}_i(\varphi)$ which implies that Ker $\mathcal{L}_i(\varphi) = \{0\}$ combined with the same approach as in Lemma 3.1 completes the proof.

A straightforward conclusion from Proposition 3.3 is then

$$\dim \operatorname{Ker} \mathcal{F}_\varphi(\vartheta) = \dim \operatorname{Ker} \mathcal{S}_\varphi(-b,a), \; \dim \operatorname{Ker} \mathcal{S}_\varphi(-b,a) = \dim \operatorname{Ker} \mathcal{S}(-b,a)$$

In the next section a distance measure introduced in quantum information is discussed.

Statistical Distance Measure - Fisher Information and Quantum Information

In quantum information, the Fisher information, the information about a parameter θ in a particular measurement procedure, is expressed in terms of the statistical distance s, see [8,10]. The statistical distance used is defined as a measure to distinguish two probability distributions on the basis of measurement outcomes, see [37]. The Fisher information and the statistical distance are statistical quantities, and generally refer to many measurements as it is the case in this survey. However, in the quantum information theory and quantum statistics context, the problem set up is presented as follows. There may or may not be a small phase change θ, and the question is whether it is there. In that case you can design quantum experiments that will tell you the answer unambiguously in a single measurement. The equality derived is of the form

$$\mathcal{F}(\varphi) = \left(\frac{ds}{d\theta}\right)^2 \tag{41}$$

the Fisher information is the square of the derivative of the statistical distance s with respect to θ. Contrary to (36), where the square of the statistical distance measure is expressed in terms of entries of a FIM $\mathcal{F}(\vartheta)$ which is based on information about m parameters estimated from n measurements, for $m < n$. A challenging question could therefore be formulated as follows, can a generalization of equality (41) be developed in a quantum information context but at the matrix level ? To be more specific, many observations or measurements that lead to more than one parameter such that the corresponding Fisher information matrix is interconnected to an appropriate statistical distance matrix, a matrix where entries are scalar distance measures. This question could equally be a challenge to algebraic matrix theory and to quantum information.

3.4. The Bezoutian - The Fisher Information Matrix

In this section an additional resultant matrix is presented, it concerns the Bezout matrix or Bezoutian. The notation of Lancaster and Tismenetsky [2] shall be used and the results presented are extracted from [38]. Assume the polynomials a and b given by $a(z) = \sum_{j=0}^{n} a_j z^j$ and $b(z) = \sum_{j=0}^{n} b_j z^j$, cfr. (13) but where $p = q = n$, and we further assume $a_0 = b_0 = 1$. The Bezout matrix $B(a, b)$ of the polynomials a and b is defined by the relation

$$a(z)b(w) - a(w)b(z) = (z - w)u_n^\top(z)B(a,b)u_n(z)$$

This matrix is often referred as the Bezoutian. We will display a decomposition of the Bezout matrix $B(a, b)$ developed in [38]. For that purpose the matrix U_φ and its inverse T_φ are presented, where φ is a given complex number, to obtain

$$U_\varphi = \begin{pmatrix} 1 & 0 & \cdots & \cdots & 0 \\ -\varphi & 1 & \cdots & \cdots & 0 \\ 0 & & \ddots & & \vdots \\ \vdots & & & \ddots & \vdots \\ 0 & \cdots & 0 & -\varphi & 1 \end{pmatrix}, \ T_\varphi = \begin{pmatrix} 1 & 0 & \cdots & \cdots & 0 \\ \varphi & 1 & \cdots & \cdots & 0 \\ \varphi^2 & & \ddots & & \vdots \\ \vdots & & & \ddots & \vdots \\ \varphi^{n-1} & \cdots & \varphi^2 & \varphi & 1 \end{pmatrix}$$

Let $(1 - \alpha_1 z)$ and $(1 - \beta_1 z)$ be a factor of $a(z)$ and $b(z)$ respectively and α_1 and β_1 are zeros of $\hat{a}(z)$ and $\widehat{b}(z)$. Consider the factored form of the nth order polynomials $a(z)$ and $b(z)$ of the form $a(z) = (1 - \alpha_1 z)a_{-1}(z)$ and $b(z) = (1 - \beta_1 z)b_{-1}(z)$ respectively. Proceeding this way, for $\alpha_2, \ldots, \alpha_n$ yields the recursion $a_{-(k-1)}(z) = (1 - \alpha_k z)a_{-k}(z)$, equivalently for the polynomials $b_{-k}(z)$ and $a_0(z) = a(z)$ and $b_0(z) = b(z)$. Proposition 3.1 in [38] is presented.

The following non-symmetric decomposition of the Bezoutian is derived, considering the notations above

$$B(a,b) = U_{\alpha_1} \begin{pmatrix} B(a_{-1}, b_{-1}) & 0 \\ 0 & 0 \end{pmatrix} U_{\beta_1}^\top + (\beta_1 - \alpha_1) b_{\beta_1} a_{\alpha_1}^\top \quad (42)$$

with a_{α_1} such that $a_{\alpha_1}^\top u_n(z) = a_{-1}$ similarly for b_{β_1}. Iteration gives the following expansion for the Bezout matrix

$$B(a,b) = \sum_{k=1}^{n} (\beta_k - \alpha_k) U_{\alpha_1} \ldots U_{\alpha_{k-1}} U_{\beta_{k+1}} \ldots U_{\beta_n} e_1^n (e_1^n)^\top U_{\beta_1}^\top \ldots U_{\beta_{k-1}}^\top U_{\alpha_{k+1}}^\top \ldots U_{\alpha_n}^\top$$

where e_1^n is the first unit standard basis column vector in $\mathbb{R}^n$, by e_j we denote the jth coordinate vector, $e_j = (0, \ldots, 1, \ldots, 0)^\top$, with all its components equal to 0 except the jth component which equals 1. The following corollarys to Proposition 3.1 in [38] are now presented.

Corollary 3.2 in [38] states. Let φ be a common zero of the polynomials $\hat{a}(z)$ and $\widehat{b}(z)$. Then $a(z) = (1 - \varphi z)a_{-1}(z)$ and $b(z) = (1 - \varphi z)b_{-1}(z)$ and

$$B(a,b) = U_\varphi \begin{pmatrix} B(a_{-1},b_{-1}) & 0 \\ 0 & 0 \end{pmatrix} U_\varphi^\top$$

This a direct consequence of (42) and from which can be concluded that the Bezoutian $B(a, b)$ is non-singular if and only if the polynomials $a(z)$ and $b(z)$ have no common factors. A similar conclusion is drawn for the FIM in (27) so that matrices $\mathcal{F}(\vartheta)$ and $B(a, b)$ have the same singularity property.

Related to Corollary 3.2 in[38], this is where we give a description of the kernel or nullspace of the Bezout matrix.

Corollary 3.3 in [38] is now presented. Let $\varphi_1, \ldots, \varphi_m$ be all the common zeros of the polynomials $\hat{a}(z)$ and $\widehat{b}(z)$, with multiplicities $n_1, \ldots, n_m$. Let ℓ be the last unit standard basis column vector in $\mathbb{R}^n$ and put

$$w_k^j = \left(T_{\varphi_k}^j J^{j-1}\right)^\top \ell$$

for $k = 1, \ldots, m$ and $j = 1, \ldots, n_k$ and by J we denote the forward $n \times n$ shift matrix, $J_{ij} = 1$ if $i = j + 1$. Consequently, the subspace Ker $B(a, b)$ is the linear span of the vectors w_k^j.

An alternative representation to (27) but involving the Bezoutian $B(b, a)$ and derived in Proposition 5.1 in [38] is of the form

$$\mathcal{F}(\vartheta) = \mathcal{M}^{-1}(b,a)\mathcal{H}(\vartheta)\mathcal{M}^{-\top}(b,a) \tag{43}$$

where

$$\mathcal{H}(\vartheta) = \begin{pmatrix} I & 0 \\ 0 & B(b,a) \end{pmatrix} Q(\vartheta) \begin{pmatrix} I & 0 \\ 0 & B(b,a) \end{pmatrix} \quad \text{and } \mathcal{M}(b,a) = \begin{pmatrix} P & 0 \\ PS(\hat{a})P & PS(\hat{b})P \end{pmatrix} \tag{44}$$

and

$$P = \begin{pmatrix} 0 & \cdots & 0 & 1 \\ \vdots & & 1 & 0 \\ 0 & & & \vdots \\ 1 & 0 & \cdots & 0 \end{pmatrix}, \; S(\hat{a}) = \begin{pmatrix} a_{n-1} & a_{n-2} & \cdots & a_0 \\ a_{n-2} & & a_0 & 0 \\ \vdots & & & \vdots \\ a_0 & 0 & \cdots & 0 \end{pmatrix} \quad \text{and } Q(\vartheta) \succ 0$$

The matrix $S(\hat{a})$ is the symmetrizer of the polynomial $\hat{a}(z)$, in this paper $a_0 = 1$, see [2] and P is a permutation matrix. In [38] it is shown that the matrix $Q(\vartheta)$ is the unique solution to an appropriate Stein equation and is strictly positive definite. However, in the next section an explicit form of the Stein solution $Q(\vartheta)$ is developed. Some comments concerning the property summarized in Corollary 5.2 in [38] follow.

The matrix $\mathcal{H}(\vartheta)$ is non-singular if and only if the polynomials $a(z)$ and $b(z)$ have no common factors. The proof is straightforward since the matrix $Q(\vartheta)$ is non-singular which implies that the matrix $\mathcal{H}(\vartheta)$ is only non-singular when the Bezoutian $B(b, a)$ is non-singular and this is fulfilled if and only if the polynomials $a(z)$ and $b(z)$ have no common factors.

The matrix $\mathcal{M}(b, a)$ is non-singular if $a_0 \neq 0$ and $b_0 \neq 0$, which is the case since we have $a_0 = b_0 = 1$. From (43) can be concluded that the FIM $\mathcal{F}(\vartheta)$ is non-singular only when the matrix $\mathcal{H}(\vartheta)$ is non-singular or by virtue of (44) when the Bezoutian $B(b, a)$ is non-singular. Consequently, the singularity conditions of the Bezoutian $B(b, a)$, the FIM $\mathcal{F}(\vartheta)$ and the Sylvester resultant matrix

$$\mathcal{S}$$

$(b, -a)$ are therefore equivalent. Can be concluded, by virtue of (29) proved in Lemma 3.1 and the equality dim (Ker $\mathcal{S}$ (a, b)) = dim (Ker $B(a, b)$) proved in Theorem 21.11 in [1], yields

$$\dim \left(\mathrm{Ker}\, \mathcal{S}(b, -a)\right) = \dim \left(\mathrm{Ker}\, \mathcal{F}(\vartheta)\right) = \dim \left(\mathrm{Ker}\, B(b,a)\right) = \nu(a,b)$$

3.5. The Stein Equation - The Fisher Information Matrix of an ARMA(p, q) Process

In [12], a link between the FIM of an ARMA process and an appropriate solution of a Stein equation is set forth. In this survey paper we shall present some of the results and confront some results displayed in the previous sections. However, alternative proofs will be given to some results obtained in [12,38].

The Stein matrix equation is now set forth. Let $A \in \mathbb{C}^{m\times m}$, $B \in \mathbb{C}^{n\times n}$ and $\Gamma \in \mathbb{C}^{n\times m}$ and consider the Stein equation

$$S - BSA^{\top} = \Gamma \tag{45}$$

It has a unique solution if and only if $\lambda\mu \neq 1$ for any $\lambda \in \sigma(A)$ and $\mu \in \sigma(B)$, the spectrum of D is $\sigma(D)$ = $\{\lambda \in \mathbb{C}\colon \det(\lambda I_m - D) = 0\}$, the set of eigenvalues of D. The unique solution will be given in the next theorem [11].

3.5.1. Theorem 3.4

Let A and B be, such that there is a single closed contour C with $\sigma(B)$ inside C and for each non-zero $w \in \sigma(A)$, w^{-1} is outside C. Then for an arbitrary Γ the Stein 45 has a unique solution S

$$S = \frac{1}{2\pi \mathrm{i}} \oint_{C} (\lambda I_n - B)^{-1} \Gamma (I_m - \lambda A)^{-\top} d\lambda \tag{46}$$

In this section an interconnection between the representation (27) of the FIM $\mathcal{F}(\vartheta)$ and an appropriate solution to a Stein equation of the form (45) as developed in [12] is set forth. The distinct roots of the polynomials $\hat{a}(z)$ and $\widehat{b}(z)$ are denoted by $\alpha_1, \alpha_2, \ldots, \alpha_p$ and $\beta_1, \beta_2, \ldots, \beta_q$ respectively such that the non-singularity of the FIM $\mathcal{F}(\vartheta)$ is guaranteed. The following representation of the integral expression (28) is given when Cauchy's residue theorem is applied, equation (4.8) in [12]

$$\mathcal{P}(\vartheta) = \mathcal{U}(\vartheta)\mathcal{D}(\vartheta)\hat{\mathcal{U}}(\vartheta) \tag{47}$$

where

$$\mathcal{U}(\vartheta) = \{u_{p+q}(\alpha_1), u_{p+q}(\alpha_2), \ldots, u_{p+q}(\alpha_p), u_{p+q}(\beta_1), u_{p+q}(\beta_2), \ldots, u_{p+q}(\beta_q)\}$$

$$\mathcal{D}(\vartheta) = \mathrm{diag}\left\{ \left(\frac{1}{\hat{a}(z;\alpha_i)\hat{b}(\alpha_i)a(\alpha_i)b(\alpha_i)}\right), \left(\frac{1}{\hat{a}(\beta_j)\hat{b}(z;\beta_j)a(\beta_j)b(\beta_j)}\right)\right\}, i = 1, ..., p \text{ and } j = 1, ..., q$$

and

$$\hat{\mathcal{U}}(\vartheta) = \{v_{p+q}(\alpha_1), v_{p+q}(\alpha_2), \ldots, v_{p+q}(\alpha_p), v_{p+q}(\beta_1), v_{p+q}(\beta_2), \ldots, v_{p+q}(\beta_q)\}^{\top}$$

the polynomial $p(\cdot;\beta)$ is defined accordingly, $p(z;\beta) = \frac{p(z)}{(z-\beta)}$ and $\mathcal{D}$ (ϑ) is the $(p + q) \times (p + q)$ diagonal matrix. The matrices $\mathcal{U}$ (ϑ) and $\widehat{\mathcal{U}}$ (ϑ) in (47) are the $(p + q)\times (p + q)$ Vandermonde matrices $V_{\alpha\beta}$ and $\widehat{\mathcal{U}}_{\alpha\beta}$ respectively, given by

$$V_{\alpha\beta} = \begin{pmatrix} 1 & \alpha_1 & \alpha_1^2 & \cdots & \alpha_1^{p+q-1} \\ 1 & \alpha_2 & \alpha_2^2 & \cdots & \alpha_2^{p+q-1} \\ \vdots & \vdots & \vdots & \vdots & \vdots \\ 1 & \alpha_p & \alpha_p^2 & \cdots & \alpha_p^{p+q-1} \\ 1 & \beta_1 & \beta_1^2 & \cdots & \beta_1^{p+q-1} \\ 1 & \beta_2 & \beta_2^2 & \cdots & \beta_2^{p+q-1} \\ \vdots & \vdots & \vdots & \vdots & \vdots \\ 1 & \beta_q & \beta_q^2 & \cdots & \beta_q^{p+q-1} \end{pmatrix} \text{ and } \hat{V}_{\alpha\beta} = \begin{pmatrix} \alpha_1^{p+q-1} & \alpha_1^{p+q-2} & \cdots & \alpha_1 & 1 \\ \alpha_2^{p+q-1} & \alpha_2^{p+q-2} & \cdots & \alpha_2 & 1 \\ \vdots & \vdots & \vdots & \vdots & \vdots \\ \alpha_p^{p+q-1} & \alpha_p^{p+q-2} & \cdots & \alpha_p & 1 \\ \beta_1^{p+q-1} & \beta_1^{p+q-2} & \cdots & \beta_1 & 1 \\ \beta_2^{p+q-1} & \beta_2^{p+q-2} & \cdots & \beta_2 & 1 \\ \vdots & \vdots & \vdots & \vdots & \vdots \\ \beta_q^{p+q-1} & \beta_q^{p+q-2} & \cdots & \beta_q & 1 \end{pmatrix}$$

It is clear that the $(p + q) \times (p + q)$ Vandermonde matrices $V_{\alpha\beta}$ and $\widehat{\mathcal{U}}_{\alpha\beta}$ are nonsingular when $\alpha_i \neq \alpha_j$, $\beta_k \neq \beta_h$ and $\alpha_i \neq \beta_k$ for all $i, j = 1, \ldots, p$ and $k, h = 1, \ldots, q$. A rigorous systematic evaluation of the Vandermonde determinants $\mathrm{Det}V_{\alpha\beta}$ and $\mathrm{Det}\widehat{\mathcal{U}}_{\alpha\beta}$, yields

$$\mathrm{Det}V_{\alpha\beta} = (-1)^{(p+q)\,(p+q-1)/2}\Phi\left(\alpha_i, \beta_k\right)$$

where

$$\Phi\left(\alpha_i, \beta_k\right) = \prod_{1 \leq i < j \leq p} (\alpha_i - \alpha_j) \prod_{1 \leq k < h \leq q} (\beta_k - \beta_h) \prod_{\substack{m = 1, \ldots p \\ n = 1, \ldots q}} (\alpha_m - \beta_n)$$

Since $V_{\alpha\beta} = P\hat{V}_{\alpha\beta}^{\top}$ and given the configuration of the permutation matrix, P, this leads to the equalities $\mathrm{Det}\hat{V}_{\alpha\beta}^{\top} = \mathrm{Det}P\,\mathrm{Det}V_{\alpha\beta}$ and $\mathrm{Det}P = (-1)^{(p+q)(p+q-1)/2}$ so that

$$\mathrm{Det}\hat{V}_{\alpha\beta} = (-1)^{(p+q)\,(p+q-1)}\,\Phi\left(\alpha_i, \beta_k\right) \Rightarrow \mid \mathrm{Det}V_{\alpha\beta} \mid = \mid \mathrm{Det}\hat{V}_{\alpha\beta} \mid$$

We shall now introduce an appropriate Stein equation of the form (45) such that an interconnection with $\wp(\vartheta)$ in (47) can be verified. Therefore the following $(p + q)\times (p + q)$ companion matrix is introduced,

$$\mathcal{C}_g = \begin{pmatrix} 0 & 1 & \cdots & 0 \\ \vdots & & \ddots & \vdots \\ 0 & \cdots & 0 & 1 \\ -g_{p+q} & -g_{p+q-1} & \cdots & -g_1 \end{pmatrix} \tag{48}$$

where the entries g_i are given by $z^{p+q} + \sum_{i=1}^{p+q} g_i(\vartheta) z^{p+q-i} = \hat{a}(z)\hat{b}(z) = \hat{g}(z, \vartheta)$ and $\hat{g}(\vartheta)$ is the vector $\hat{g}(\vartheta) = (g_{p+q}(\vartheta), g_{p+q-1}(\vartheta), \ldots, g_1(\vartheta))^{\top}$. Likewise is the vector $g(z, \vartheta) = a(z)b(z)$ and $g(\vartheta) = (g_1(\vartheta), g_1(\vartheta), \ldots, g_{p+q}(\vartheta))^{\top}$, for investigating the properties of a companion matrix see e.g., [36], [2]. Since all the roots of the polynomials $\hat{a}(z)$ and $\widehat{b}(z)$ are distinct and lie within the unit circle implies that the products $\alpha_i\beta_j \neq 1$, $\alpha_i\alpha_j \neq 1$ and $\beta_i\beta_j \neq 1$ hold for all $i = 1, 2, \ldots, p$ and $j = 1, 2, \ldots, q$. Consequently, the uniqueness condition of the solution of an appropriate Stein equation is verified. The following Stein equation and its solution, according to (45) and (46), are now presented

$$S - \mathcal{C}_g S \mathcal{C}_g^{\top} = \Gamma \text{ and } S = \frac{1}{2\pi \mathbf{i}} \oint_{|z|=1} (zI_{p+q} - \mathcal{C}_g)^{-1}\Gamma(I_{p+q} - z\mathcal{C}_g)^{-\top} dz$$

where the closed contour is now the unit circle $\mid z \mid = 1$ and the matrix Γ is of size $(p + q)\times (p + q)$. A more explicit expression of the solution S is of the form

$$S = \frac{1}{2\pi \mathbf{i}} \oint_{|z|=1} \frac{\text{adj}(zI_{p+q} - \mathcal{C}_g)\Gamma\,\text{adj}(I_{p+q} - z\mathcal{C}_g)^{\top}}{a(z)b(z)\hat{a}(z)\hat{b}(z)} dz \tag{49}$$

where adj(X) = X^{-1} Det(X), the adjoint of matrix X. When Cauchy's residue theorem is applied to the solution S in (49), the following factored form of S is derived, equation (4.9) in [12]

$$S = (\mathcal{C}_1, \mathcal{C}_2)\,(I_{p+q} \otimes \Gamma)\,(\mathcal{D}(\vartheta) \otimes I_{p+q})\,(\mathcal{C}_3, \mathcal{C}_4)^{\top} \tag{50}$$

where

$$\begin{aligned}
\mathcal{C}_1 &= \text{adj}(\alpha_1 I_{p+q} - \mathcal{C}_g), \text{adj}(\alpha_2 I_{p+q} - \mathcal{C}_g), \ldots, \text{adj}(\alpha_p I_{p+q} - \mathcal{C}_g) \\
\mathcal{C}_2 &= \text{adj}(\beta_1 I_{p+q} - \mathcal{C}_g), \text{adj}(\beta_2 I_{p+q} - \mathcal{C}_g), \ldots, \text{adj}(\beta_p I_{p+q} - \mathcal{C}_g) \\
\mathcal{C}_3 &= \text{adj}(I_{p+q} - \alpha_1 \mathcal{C}_g), \text{adj}(I_{p+q} - \alpha_2 \mathcal{C}_g), \ldots, \text{adj}(I_{p+q} - \alpha_p \mathcal{C}_g) \\
\mathcal{C}_4 &= \text{adj}(I_{p+q} - \beta_1 \mathcal{C}_g), \text{adj}(I_{p+q} - \beta_2 \mathcal{C}_g), \ldots, \text{adj}(I_{p+q} - \beta_p \mathcal{C}_g)
\end{aligned}$$

and $\mathcal{D}$ ϑ) is given in (47), the following matrix rule is applied

$$(A \otimes B)\,(C \otimes D) = AC \otimes BD$$

and the operator $\otimes$ is the tensor (Kronecker) product of two matrices, see e.g., [2], [20].

Combining (47) and (50) and taking the assumption, $\alpha_i \neq \alpha_j$, $\beta_k \neq \beta_h$ and $\alpha_i \neq \beta_k$, into account implies that the inverse of the (p + q)× (p + q) Vandermonde matrices $V_{\alpha\beta}$ and $\widehat{\mathcal{U}}_{\alpha\beta}$ exist, as Lemma 4.2 [12] states.

The following equality holds true

$$S = (\mathcal{C}_1, \mathcal{C}_2)\left(V_{\alpha\beta}^{-1} \mathcal{P}(\vartheta) \hat{V}_{\alpha\beta}^{-1} \otimes \Gamma\right)(\mathcal{C}_3, \mathcal{C}_4)^{\top}$$

or

$$S = (\mathcal{C}_1, \mathcal{C}_2)\left(V_{\alpha\beta}^{-1} \mathcal{S}^{-1}(b, -a) \mathcal{F}(\vartheta) \mathcal{S}^{-\top}(b, -a) \hat{V}_{\alpha\beta}^{-1} \otimes \Gamma\right)(\mathcal{C}_3, \mathcal{C}_4)^{\top} \tag{51}$$

Consequently, under the condition $\alpha_i \neq \alpha_j$, $\beta_k \neq \beta_h$ and $\alpha_i \neq \beta_k$, and by virtue of (27) and (51), an interconnection involving the FIM $\mathcal{F}(\vartheta)$, a solution to an appropriate Stein equation S, the Sylvester matrix

$$\mathcal{S}$$

(b, $-a$) and the Vandermonde matrices $V_{\alpha\beta}$ and $\widehat{\mathcal{U}}_{\alpha\beta}$ is established. It is clear that by using the expression (43), the Bezoutian B (a, b) can be inserted in equality (51).

We will formulate a Stein equation when the matrix $\Gamma = e_{p+q} e_{p+q}^{\top}$,

$$S - \mathcal{C}_g S \mathcal{C}_g^{\top} = e_{p+q} e_{p+q}^{\top} \tag{52}$$

where e_{p+q} is the last standard basis column vector in $\mathbb{R}^{p+q}$, e_i^m is the i-th unit standard basis column vector in $\mathbb{R}^m$, with all its components equal to 0 except the i-th component which equals 1. The next lemma is formulated.

3.5.2. Lemma 3.5

The symmetric matrix $\wp(\vartheta)$ defined in (28) *fulfills the Stein Equation* (52).

Proof

The unique solution of (52) is according to (46)

$$S = \frac{1}{2\pi \mathbf{i}} \oint\limits_{|z|=1} (zI_{p+q} - \mathcal{C}_g)^{-1} e_{p+q} e_{p+q}^\top (I_{p+q} - z\mathcal{C}_g)^{-\top} dz$$

more explictely written,

$$S = \frac{1}{2\pi \mathbf{i}} \oint\limits_{|z|=1} \frac{\text{adj}(zI_{p+q} - \mathcal{C}_g) e_{p+q} e_{p+q}^\top \text{adj}(I_{p+q} - z\mathcal{C}_g)^\top}{a(z)b(z)\hat{a}(z)\hat{b}(z)} dz$$

Using the property of the companion matrix $\mathcal{C}_g$, standard computation shows that the last column of $\text{adj}(zI_{p+q} - \mathcal{C}_g)$ is the basic vector $u_{p+q}(z)$ and consequently the last column of $\text{adj}(I_{p+q} - z\,\mathcal{C}_g)$ is the basic vector $v_{p+q}(z) = z^{p+q-1}u_{p+q}(z^{-1})$. This implies that $\text{adj}(zI_{p+q} - \mathcal{C}_g)e_{p+q} = u_{p+q}(z)$ and $e_{p+q}^\top \text{adj}(I_{p+q} - z\mathcal{C}_g)^\top = v_{p+q}^\top(z)$ or

$$S = \frac{1}{2\pi \mathbf{i}} \oint\limits_{|z|=1} \frac{u_{p+q}(z) v_{p+q}^\top(z)}{a(z)b(z)\hat{a}(z)\hat{b}(z)} dz = \mathcal{P}(\vartheta)$$

Consequently, the solution S to the Stein 52 coincides with the matrix $\wp(\vartheta)$ defined in (28).

The Stein equation that is verified by the FIM $\mathcal{F}(\vartheta)$ will be considered. For that purpose we display the following $p \times p$ and $q \times q$ companion matrices $\mathcal{C}_a$ and $\mathcal{C}_b$ of the form,

$$\mathcal{C}_a = \begin{pmatrix} -a_1 & -a_2 & \cdots & \cdots & -a_p \\ 1 & 0 & \cdots & \cdots & 0 \\ 0 & \ddots & & & \vdots \\ \vdots & \ddots & \ddots & & \vdots \\ 0 & \cdots & 0 & 1 & 0 \end{pmatrix}, \mathcal{C}_b = \begin{pmatrix} -b_1 & -b_2 & \cdots & \cdots & -b_q \\ 1 & 0 & \cdots & \cdots & 0 \\ 0 & \ddots & & & \vdots \\ \vdots & \ddots & \ddots & & \vdots \\ 0 & \cdots & 0 & 1 & 0 \end{pmatrix}$$

respectively. Introduce the $(p + q) \times (p + q)$ matrix $\mathcal{K}(\vartheta) = \begin{pmatrix} \mathcal{C}_a & O \\ O & \mathcal{C}_b \end{pmatrix}$ and the $(p + q) \times 1$ vector $\mathcal{B} = \begin{pmatrix} e_p^1 \\ -e_q^1 \end{pmatrix}$, where e_p^1 and e_q^1 are the first standard basis column vectors in $\mathbb{R}^p$ and $\mathbb{R}^q$ respectively. Consider the Stein equation

$$S - \mathcal{K}(\vartheta) S \mathcal{K}^\top(\vartheta) = \mathcal{B}\mathcal{B}^\top \tag{53}$$

followed by the theorem.

3.5.3. Theorem 3.6

The Fisher information matrix $\mathcal{F}(\vartheta)$ (17) coincides with the solution to the Stein 53.

Proof

The eigenvalues of the companion matrices $\mathcal{C}_a$ and $\mathcal{C}_b$ are respectively the zeros of the polynomials $\hat{a}(z)$ and $\widehat{b}(z)$ which are in absolute value smaller than one. This implies that the unique solution of the Stein 53 exists and is given by

$$S = \frac{1}{2\pi \mathbf{i}} \oint\limits_{|z|=1} (zI_{p+q} - \mathcal{K}(\vartheta))^{-1} \mathcal{B}\mathcal{B}^\top (I_{p+q} - z\mathcal{K}(\vartheta))^{-\top} dz$$

developing this integral expression in a more explicit form yields

$$S = \frac{1}{2\pi\mathbf{i}} \oint\limits_{|z|=1} \begin{pmatrix} \frac{\text{adj}(zI_p - \mathcal{C}_a)}{\hat{a}(z)} & O \\ O & \frac{\text{adj}(zI_q - \mathcal{C}_b)}{\hat{b}(z)} \end{pmatrix} \begin{pmatrix} e_p^1 \\ -e_q^1 \end{pmatrix} \left\{ \begin{pmatrix} \frac{\text{adj}(I_p - z\mathcal{C}_a)}{a(z)} & O \\ O & \frac{\text{adj}(I_q - z\mathcal{C}_b)}{b(z)} \end{pmatrix} \begin{pmatrix} e_p^1 \\ -e_q^1 \end{pmatrix} \right\}^\top dz$$

Considering the form of the companion matrices $\mathcal{C}_a$ and $\mathcal{C}_b$ leads through straightforward computation to the conclusion, the first column of adj$(zI_p - \mathcal{C}_a)$ is the basic vector $v_p(z)$ and consequently the first column of adj$(I_p - z\,\mathcal{C}_a)$ is the basic vector $u_p(z)$. Equivalently for the companion matrix $\mathcal{C}_b$, this yields

$$S = \frac{1}{2\pi\mathbf{i}} \oint\limits_{|z|=1} \begin{pmatrix} \frac{v_p(z)}{\hat{a}(z)} \\ -\frac{v_q(z)}{\hat{b}(z)} \end{pmatrix} \begin{pmatrix} \frac{u_p^\top(z)}{a(z)} & -\frac{u_q^\top(z)}{b(z)} \end{pmatrix} dz \tag{54}$$

Representation (54) is such that in order to obtain an equivalent representation to the FIM $\mathcal{F}(\vartheta)$ in (17), the transpose of the solution to the Stein 53 is therefore required, to obtain

$$S^\top = \frac{1}{2\pi\mathbf{i}} \oint\limits_{|z|=1} \begin{pmatrix} \frac{u_p(z)v_p^\top(z)}{a(z)\hat{a}(z)} & -\frac{u_p(z)v_q^\top(z)}{a(z)\hat{b}(z)} \\ -\frac{u_q(z)v_p^\top(z)}{\hat{a}(z)b(z)} & \frac{u_q(z)v_q^\top(z)}{b(z)\hat{b}(z)} \end{pmatrix} dz = \mathcal{F}(\vartheta) \tag{55}$$

or

$$S^\top = \frac{1}{2\pi\mathbf{i}} \oint\limits_{|z|=1} (I_{p+q} - z\mathcal{K}(\vartheta))^{-1}\mathcal{B}\mathcal{B}^\top(zI_{p+q} - \mathcal{K}(\vartheta))^{-\top} dz = \mathcal{F}(\vartheta)$$

The symmetry property of the FIM $\mathcal{F}(\vartheta)$, leads to $S = \mathcal{F}(\vartheta)$. From the representation (55) can be concluded that the solution S of the Stein 53 coincides with the symmetric block Toeplitz FIM $\mathcal{F}(\vartheta)$ given in (17). This completes the proof.

It is straightforward to verify that the submatrix (1,2) in (55) is the complex conjugate transpose of the submatrix (2,1), whereas each submatrix on the main diagonal is Hermitian, consequently, the integrand is Hermitian. This implies that when the standard residue theorem is applied, it yields $\mathcal{F}(\vartheta) = \mathcal{F}^\top(\vartheta)$.

An Illustrative Example of Theorem 3.6

To illustrate Theorem 3.6, the case of an ARMA(2, 2) process is considered. We will use the representation (17) for computing the FIM $\mathcal{F}(\vartheta)$ of an ARMA(2, 2) process. The autoregressive and moving average polynomials are of degree two or $p = q = 2$ and the ARMA(2, 2) process is described by,

$$y(t)a(z) = b(z)e(t) \tag{56}$$

where $y(t)$ is the stationary process driven by white noise $\epsilon(t)$, $a(z) = (1 + a_1z + a_2z^2)$ and $b(z) = (1+b_1z + b_2z^2)$ and the parameter vector is $\vartheta = (a_1, a_2, b_1, b_2)^\top$. The condition, the zeros of the polynomials

$$\hat{a}(z) = z^2a(z^{-1}) = z^2 + a_1z + a_2 \text{ and } \hat{b}(z) = z^2b(z^{-1}) = z^2 + b_1z + b_2$$

are in absolute value smaller than one, is imposed. The FIM $\mathcal{F}(\vartheta)$ of the ARMA(2, 2) process (56) is of the form

$$\mathcal{F}(\vartheta) = \begin{pmatrix} \mathcal{F}_{aa}(\vartheta) & \mathcal{F}_{ab}(\vartheta) \\ \mathcal{F}_{ab}^{\top}(\vartheta) & \mathcal{F}_{bb}(\vartheta) \end{pmatrix} \tag{57}$$

where

$$\mathcal{F}_{aa}(\vartheta) = \frac{1}{(1-a_2)\left[(1+a_2)^2-a_1^2\right]} \begin{pmatrix} 1+a_2 & -a_1 \\ -a_1 & 1+a_2 \end{pmatrix}$$
$$\mathcal{F}_{bb}(\vartheta) = \frac{1}{(1-b_2)\left[(1+b_2)^2-b_1^2\right]} \begin{pmatrix} 1+b_2 & -b_1 \\ -b_1 & 1+b_2 \end{pmatrix}$$
$$\mathcal{F}_{ab}(\vartheta) = \frac{1}{(a_2b_2-1)^2+(a_2b_1-a_1)(b_1-a_1b_2)} \begin{pmatrix} a_2b_2-1 & a_1-a_2b_1 \\ b_1-a_1b_2 & a_2b_2-1 \end{pmatrix}$$

The submatrices $\mathcal{F}_{aa}(\vartheta)$ and $\mathcal{F}_{bb}(\vartheta)$ are symmetric and Toeplitz whereas $\mathcal{F}_{ab}(\vartheta)$ is Toeplitz. One can assert that without any loss of generality, the property, symmetric block Toeplitz, holds for the class of Fisher information matrices of stationary ARMA(p, q) processes, where p and q are arbitrary, finite integers that represent the degrees of the autoregressive and moving average polynomials, respectively. The appropriate companion matrices $\mathcal{C}_a$, $\mathcal{C}_b$, the 4×4 matrices $\mathcal{K}(\vartheta)$ and $\mathcal{B}\mathcal{B}^{\top}$ are

$$\mathcal{C}_a = \begin{pmatrix} -a_1 & -a_2 \\ 1 & 0 \end{pmatrix}, \mathcal{C}_b = \begin{pmatrix} -b_1 & -b_2 \\ 1 & 0 \end{pmatrix}, \mathcal{K}(\vartheta) = \begin{pmatrix} -a_1 & -a_2 & 0 & 0 \\ 1 & 0 & 0 & 0 \\ 0 & 0 & -b_1 & -b_2 \\ 0 & 0 & 1 & 0 \end{pmatrix} \quad \text{and } \mathcal{B}\mathcal{B}^{\top} = \begin{pmatrix} 1 & 0 & -1 & 0 \\ 0 & 0 & 0 & 0 \\ -1 & 0 & 1 & 0 \\ 0 & 0 & 0 & 0 \end{pmatrix} \tag{58}$$

where $\mathcal{B} = \begin{pmatrix} 1 & 0 & -1 & 0 \end{pmatrix}^{\top}$. It can be verified that the Stein equation

$$\mathcal{F}(\vartheta) - \mathcal{K}(\vartheta)\mathcal{F}(\vartheta)\mathcal{K}^{\top}(\vartheta) = \mathcal{B}\mathcal{B}^{\top}$$

holds true, when $\mathcal{F}(\vartheta)$ is of the form (57) and the matrices $\mathcal{K}(\vartheta)$ and includegraphics[scale=1]entropy-16-02023f666.pdf^{T} are given in (58).

3.5.4. Some Additional Results

In Proposition 5.1 in [38], the matrix $Q(\vartheta)$ in (44) fulfills the Stein 59 and the property $Q(\vartheta) \succ 0$ is proved. It states that when $e_P^{\top} = \left(e_1^{\top}P, 0\right)^{\top} = (e_n, 0_n)^{\top} \in \mathbb{R}^{2n}$, where e_1 is the first unit standard basis column vector in $\mathbb{R}^n$ and e_n is the last or n-th unit standard basis column vector in $\mathbb{R}^n$, the following Stein equation admits the form

$$Q(\vartheta) = F_N(\vartheta)Q(\vartheta)F_N^{\top}(\vartheta) + e_P e_P^{\top} \tag{59}$$

where

$$F_N(\vartheta) = \begin{pmatrix} \hat{\mathcal{C}}_a & 0 \\ e_1 e_1^{\top} & \mathcal{C}_b \end{pmatrix}, \hat{\mathcal{C}}_a = \begin{pmatrix} 0 & 1 & 0 & \cdots & 0 \\ 0 & 0 & 1 & \cdots & 0 \\ \vdots & \ddots & & \ddots & \vdots \\ 0 & \ddots & \ddots & & 1 \\ -a_p & -a_{p-1} & \cdots & \cdots & -a_1 \end{pmatrix}$$

A corollary to Proposition 5.1, [38] will be set forth, the involvement of various Vandermonde matrices in the explicit solution to 59 is confirmed. For that purpose the following Vandermonde matrices are displayed,

$$V_\alpha = \begin{pmatrix} 1 & 1 & 1 \\ \alpha_1 & \alpha_2 & \alpha_n \\ \alpha_1^2 & \alpha_2^2 & \alpha_n^2 \\ \vdots & \vdots & \vdots \\ \alpha_1^{n-1} & \alpha_2^{n-1} & \alpha_n^{n-1} \end{pmatrix}, \hat{V}_\alpha = \begin{pmatrix} \alpha_1^{n-1} & \alpha_1^{n-2} & 1 \\ \alpha_2^{n-1} & \alpha_2^{n-2} & 1 \\ \alpha_3^{n-1} & \alpha_3^{n-2} & 1 \\ \vdots & \vdots & \vdots \\ \alpha_n^{n-1} & \alpha_n^{n-2} & 1 \end{pmatrix}, \hat{V}_{\alpha\beta} = \begin{pmatrix} \hat{V}_\alpha \\ \hat{V}_\beta \end{pmatrix}, \text{and } V_{\alpha\beta} = \begin{pmatrix} V_\alpha & V_\beta \end{pmatrix} \tag{60}$$

where $\widehat{\mathcal{U}}_\beta$ and V_β have the same configuration as $\widehat{\mathcal{U}}_\alpha$ and V_α respectively. A corollary to Proposition 5.1 in [38] is now formulated.

3.5.5. Corollary 3.7

An explicit expression of the solution to the Stein 59 is of the form

$$Q(\vartheta) = \begin{pmatrix} V_\alpha \mathcal{D}_{11}(\vartheta)\hat{V}_\alpha & V_\alpha \mathcal{D}_{12}(\vartheta)V_\alpha^\top \\ \hat{V}_{\alpha\beta}^\top \mathcal{D}_{21}(\vartheta)\hat{V}_{\alpha\beta} & \hat{V}_{\alpha\beta}^\top \mathcal{D}_{22}(\vartheta)V_{\alpha\beta}^\top \end{pmatrix} \tag{61}$$

where the $n \times n$ and $2n \times 2n$ diagonal matrices $\mathcal{D}_{kl}$ ϑ) shall be specified in the proof.

Proof

The condition of a unique solution of the Stein 59 is guaranteed since the eigenvalues of the companions matrices $\widehat{\mathcal{C}}_a$ and $\mathcal{C}_b$ given respectively by the zeros of the polynomials $\hat{a}$ (z) and $\widehat{b}$ (z) are in absolute value smaller than one. Consequently, the unique solution to the Stein 59 exists and is given by

$$Q(\vartheta) = \frac{1}{2\pi \mathrm{i}} \oint_{|z|=1} (zI_{2n} - F_N(\vartheta))^{-1} e_p e_p^\top (I_{2n} - zF_N(\vartheta))^{-\top} dz \tag{62}$$

in order to proceed successfully, the following matrix property is displayed, to obtain

$$\begin{pmatrix} A & O \\ B & C \end{pmatrix}^{-1} = \begin{pmatrix} A^{-1} & O \\ -C^{-1}BA^{-1} & C^{-1} \end{pmatrix}$$

When applied to the 62, it yields

$$Q(\vartheta) = \frac{1}{2\pi \mathrm{i}} \oint_{|z|=1} \begin{pmatrix} \frac{\mathrm{adj}(zI_p - \hat{\mathcal{C}}_a)}{\hat{a}(z)} & O \\ \frac{\mathrm{adj}(zI_q - \mathcal{C}_b)e_1 e_1^\top \mathrm{adj}(zI_p - \hat{\mathcal{C}}_a)}{\hat{a}(z)\hat{b}(z)} & \frac{\mathrm{adj}(zI_q - \mathcal{C}_b)}{\hat{b}(z)} \end{pmatrix} \begin{pmatrix} e_n \\ 0 \end{pmatrix} \times$$
$$\left\{ \begin{pmatrix} \frac{\mathrm{adj}(I_n - z\hat{\mathcal{C}}_a)}{\hat{a}(z)} & O \\ \frac{\mathrm{adj}(I_n - z\mathcal{C}_b)e_1 e_1^\top \mathrm{adj}(I_p - z\hat{\mathcal{C}}_a)}{\hat{a}(z)\hat{b}(z)} & \frac{\mathrm{adj}(I_n - z\mathcal{C}_b)}{\hat{b}(z)} \end{pmatrix} \begin{pmatrix} e_n \\ 0 \end{pmatrix} \right\}^\top dz$$

Considering that the last column vector of the matrices adj(zI_p −. $\widehat{\mathcal{C}}_a$) and adj($I_n - z$ $\widehat{\mathcal{C}}_a$) are the vectors $u_n(z)$ and $v_n(z)$ respectively, it then yields

$$Q(\vartheta) = \frac{1}{2\pi \mathrm{i}} \oint_{|z|=1} \begin{pmatrix} \frac{u_n(z)}{\hat{a}(z)} \\ \frac{v_n(z)}{\hat{a}(z)\hat{b}(z)} \end{pmatrix} \begin{pmatrix} \frac{v_n^\top(z)}{a(z)} & \frac{z^{n-1}u_n^\top(z)}{a(z)b(z)} \end{pmatrix} dz$$
$$= \frac{1}{2\pi \mathrm{i}} \oint_{|z|=1} \begin{pmatrix} \frac{u_n(z)v_n^\top(z)}{a(z)\hat{a}(z)} & \frac{z^{n-1}u_n(z)u_n^\top(z)}{\hat{a}(z)a(z)b(z)} \\ \frac{v_n(z)v_n^\top(z)}{\hat{a}(z)\hat{b}(z)a(z)} & \frac{z^{n-1}v_n(z)u_n^\top(z)}{\hat{a}(z)\hat{b}(z)a(z)b(z)} \end{pmatrix} dz = \begin{pmatrix} Q_{11}(\vartheta) & Q_{12}(\vartheta) \\ Q_{21}(\vartheta) & Q_{22}(\vartheta) \end{pmatrix}$$

Applying the standard residue theorem leads for the respective submatrices

$$Q_{11}(\vartheta) = \{u_n(\alpha_1), \ldots, u_n(\alpha_n)\}\mathcal{D}_{11}(\vartheta)\{v_n(\alpha_1), \ldots, v_n(\alpha_n)\}^\top$$
$$Q_{12}(\vartheta) = \{u_n(\alpha_1), \ldots, u_n(\alpha_n)\}\mathcal{D}_{12}(\vartheta)\{u_n(\alpha_1), \ldots, u_n(\alpha_n)\}^\top$$
$$Q_{21}(\vartheta) = \{v_n(\alpha_1), \ldots, v_n(\alpha_n), v_n(\beta_1), \ldots, v_n(\beta_n)\}\mathcal{D}_{21}(\vartheta)\{v_n(\alpha_1), \ldots, v_n(\alpha_n), v_n(\beta_1), \ldots, v_n(\beta_n)\}^\top$$
$$Q_{22}(\vartheta) = \{v_n(\alpha_1), \ldots, v_n(\alpha_n), v_n(\beta_1), \ldots, v_n(\beta_n)\}\mathcal{D}_{22}(\vartheta)\{u_n(\alpha_1), \ldots, u_n(\alpha_n), u_n(\beta_1), \ldots, u_n(\beta_n)\}^\top$$

where the $n \times n$ diagonal matrices are

$$\mathcal{D}_{11}(\vartheta) = \text{diag}\{1/(a(\alpha_i)\hat{a}(z;\alpha_i))\}, \mathcal{D}_{12}(\vartheta) = \text{diag}\{\alpha_i^{n-1}/(a(\alpha_i)b(\alpha_i)\hat{a}(z;\alpha_i))\} \text{ for } i = 1, \ldots, n$$

and the $2n \times 2n$ diagonal matrices are

$$\mathcal{D}_{21}(\vartheta) = \text{diag}\left\{1/\left(a(\alpha_i)\hat{b}(\alpha_i)\hat{a}(z;\alpha_i)\right), 1/\left(\hat{a}(\beta_j)a(\beta_j)\hat{b}(z;\beta_j)\right)\right\}, \text{ for } i, j = 1, \ldots, n$$
$$\mathcal{D}_{22}(\vartheta) = \text{diag}\left\{\alpha_i^{n-1}/\left(a(\alpha_i)b(\alpha_i)\hat{b}(\alpha_i)\hat{a}(z;\alpha_i)\right), \beta_j^{n-1}/\left(\hat{a}(\beta_j)a(\beta_j)b(\beta_j)\hat{b}(z;\beta_j)\right)\right\}, \text{ for } i, j = 1, \ldots, n$$

It is clear that the first and third matrices in $Q_{11}(\vartheta)$, $Q_{12}(\vartheta)$, $Q_{21}(\vartheta)$ and $Q_{22}(\vartheta)$ are the appropriate Vandermonde matrices displayed in (60), it can be concluded that the representation (61) is verified. This completes the proof.

In this section an explicit form of the solution $Q(\vartheta)$, expressed in terms of various Vandermonde matrices, is displayed. Also, an interconnection between the Fisher information $\mathcal{F}(\vartheta)$ and appropriate solutions to Stein equations and related matrices is presented. Proofs are given when the Stein equations are verified by the FIM $\mathcal{F}(\vartheta)$ and the associated matrix $\wp(\vartheta)$. These are alternative to the proofs developed in [38]. The presence of various forms of Vandermonde matrices is also emphasized. In the next section some matrix properties of the FIM $\mathcal{F}(\vartheta)$ of an ARMAX process is presented.

3.6. The Fisher Information Matrix of an ARMAX(p, r, q) Process

The FIM of the ARMAX process (11) is set forth according to [4]. The derivatives in the corresponding representation (16) are

$$\frac{\partial e_t(\vartheta)}{\partial a_j} = \frac{c(z)}{a(z)b(z)}x(t-j) + \frac{1}{a(z)}e(t-j), \frac{\partial e_t(\vartheta)}{\partial c_l} = -\frac{1}{b(z)}e(t-l) \text{ and} \frac{\partial e_t(\vartheta)}{\partial b_k} = -\frac{1}{b(z)}e_{t-k}$$

where $j = 1, \ldots, p$, $l = 1, \ldots, r$ and $k = 1, \ldots, q$. Combining all j, l and k yields the $(p + r + q) \times (p + r + q)$ FIM

$$\mathcal{G}(\vartheta) = \begin{pmatrix} \mathcal{G}_{aa}(\vartheta) & \mathcal{G}_{ac}(\vartheta) & \mathcal{G}_{ab}(\vartheta) \\ \mathcal{G}_{ac}^\top(\vartheta) & \mathcal{G}_{cc}(\vartheta) & \mathcal{G}_{cb}(\vartheta) \\ \mathcal{G}_{ab}^\top(\vartheta) & \mathcal{G}_{cb}^\top(\vartheta) & \mathcal{G}_{bb}(\vartheta) \end{pmatrix} \quad (63)$$

where the submatrices of $\mathcal{G}(\vartheta)$ are given by

$$\mathcal{G}_{aa}(\vartheta) = \frac{1}{2\pi \mathrm{i}} \oint_{|z|=1} R_x(z) \frac{u_p(z)u_p^{\top}(z^{-1})c(z)c(z^{-1})}{a(z)a(z^{-1})b(z)b(z^{-1})} \frac{dz}{z} + \frac{1}{2\pi \mathrm{i}} \oint_{|z|=1} \frac{u_p(z)u_p^{\top}(z^{-1})}{a(z)a(z^{-1})} \frac{dz}{z}$$
$$= \frac{1}{2\pi \mathrm{i}} \oint_{|z|=1} R_x(z) \frac{u_p(z)v_p^{\top}(z)c(z)\hat{c}(z)}{a(z)\hat{a}(z)b(z)\hat{b}(z)z^{r-q}} dz + \frac{1}{2\pi \mathrm{i}} \oint_{|z|=1} \frac{u_p(z)v_p^{\top}(z)}{a(z)\hat{a}(z)} dz$$
$$\mathcal{G}_{ab}(\vartheta) = -\frac{1}{2\pi \mathrm{i}} \oint_{|z|=1} \frac{u_p(z)u_q^{\top}(z^{-1})}{a(z)b(z^{-1})} \frac{dz}{z} = -\frac{1}{2\pi \mathrm{i}} \oint_{|z|=1} \frac{u_p(z)v_q^{\top}(z)}{a(z)\hat{b}(z)} dz$$
$$\mathcal{G}_{ac}(\vartheta) = -\frac{1}{2\pi \mathrm{i}} \oint_{|z|=1} R_x(z) \frac{u_p(z)u_r^{\top}(z^{-1})c(z)}{a(z)b(z)b(z^{-1})} \frac{dz}{z} = -\frac{1}{2\pi \mathrm{i}} \oint_{|z|=1} R_x(z) \frac{u_p(z)v_r^{\top}(z)c(z)}{a(z)b(z)\hat{b}(z)z^{r-q}} dz$$
$$\mathcal{G}_{cc}(\vartheta) = \frac{1}{2\pi \mathrm{i}} \oint_{|z|=1} R_x(z) \frac{u_r(z)u_r^{\top}(z^{-1})}{b(z)b(z^{-1})} \frac{dz}{z} = \frac{1}{2\pi \mathrm{i}} \oint_{|z|=1} R_x(z) \frac{u_r(z)v_r^{\top}(z)}{b(z)\hat{b}(z)z^{r-q}} dz$$
$$\mathcal{G}_{bb}(\vartheta) = \frac{1}{2\pi \mathrm{i}} \oint_{|z|=1} \frac{u_q(z)u_q^{\top}(z^{-1})}{b(z)b(z^{-1})} \frac{dz}{z} = -\frac{1}{2\pi \mathrm{i}} \oint_{|z|=1} \frac{u_q(z)v_q^{\top}(z)}{b(z)\hat{b}(z)} dz, \text{ and } \mathcal{G}_{cb}(\vartheta) = O$$

where $R_x(z)$ is the spectral density of the process $x(t)$ and is defined in (10). Let $K(z) = a(z)a(z^{-1})b(z)b(z^{-1})$, combining all the expressions in (63) leads to the following representation of $\mathcal{G}(\vartheta)$ as the sum of two matrices

$$\frac{1}{2\pi \mathrm{i}} \oint_{|z|=1} \frac{R_x(z)}{K(z)} \begin{pmatrix} c(z)u_p(z) \\ -a(z)u_r(z) \\ O \end{pmatrix} \begin{pmatrix} c(z)u_p(z) \\ -a(z)u_r(z) \\ O \end{pmatrix}^{*} \frac{dz}{z} + \frac{1}{2\pi \mathrm{i}} \oint_{|z|=1} \frac{1}{K(z)} \begin{pmatrix} b(z)u_p(z) \\ O \\ -a(z)u_q(z) \end{pmatrix} \begin{pmatrix} b(z)u_p(z) \\ O \\ -a(z)u_q(z) \end{pmatrix}^{*} \frac{dz}{z} \quad (64)$$

where $(X)^*$ is the complex conjugate transpose of the matrix $X \in \mathbb{C}^{m \times n}$. Like in (23) we set forth

$$\mathcal{S}(-c, a) = \begin{pmatrix} -\mathcal{S}_p(c) \\ \mathcal{S}_r(a) \end{pmatrix}$$

here $\mathcal{S}_p(c)$ is formed by the top p rows of $\mathcal{S}(-c, a)$. In a similar way we decompose

$$\mathcal{S}(-b, a) = \begin{pmatrix} -\mathcal{S}_p(b) \\ \mathcal{S}_q(a) \end{pmatrix}$$

The representation (64) can be expressed by the appropriate block representations of the Sylvester resultant matrices, to obtain

$$\mathcal{G}(\vartheta) = \begin{pmatrix} -\mathcal{S}_p(c) \\ \mathcal{S}_r(a) \\ O \end{pmatrix} \mathcal{W}(\vartheta) \begin{pmatrix} -\mathcal{S}_p(c) \\ \mathcal{S}_r(a) \\ O \end{pmatrix}^{\top} + \begin{pmatrix} -\mathcal{S}_p(b) \\ O \\ \mathcal{S}_q(a) \end{pmatrix} \mathcal{P}(\vartheta) \begin{pmatrix} -\mathcal{S}_p(b) \\ O \\ \mathcal{S}_q(a) \end{pmatrix}^{\top} \quad (65)$$

where the matrix $\wp(\vartheta)$ is given in (28) and the matrix $\mathcal{P}(\vartheta) \in \mathbb{R}^{(p+r)\times(p+r)}$ is of the form

$$\mathcal{W}(\vartheta) = \frac{1}{2\pi \mathrm{i}} \oint_{|z|=1} R_x(z) \frac{u_{p+r}(z)u_{p+r}^{\top}(z^{-1})}{a(z)a(z^{-1})b(z)b(z^{-1})} \frac{dz}{z} = \frac{1}{2\pi \mathrm{i}} \oint_{|z|=1} R_x(z) \frac{u_{p+r}(z)v_{p+r}^{\top}(z)}{a(z)b(z)\hat{a}(z)\hat{b}(z)} dz \quad (66)$$

It is shown in [4] that $\mathcal{P}(\vartheta) \succ O$. As can be seen in (65), the ARMAX part is explained by the first term, whereas the ARMA part is described by the second term, the combination of both terms is a summary of the Fisher information of a ARMAX(p, r, q) process. The FIM $\mathcal{G}(\vartheta)$ under form (65) allows us to prove the following property, Theorem 3.1 in [4]. The FIM $\mathcal{G}(\vartheta)$ of the ARMAX(p, r, q) process with polynomials $a(z)$, $c(z)$ and $b(z)$ of order p, r, q respectively becomes singular if and only if these polynomials have at least one common root. Consequently, the class of resultant matrices is extended by the FIM $\mathcal{G}$ (ϑ).

3.7. The Stein Equation - The Fisher Information Matrix of an ARMAX(p, r, q) Process

In Lemma 3.5 it is proved that the matrix $\wp(\vartheta)$ (28) fulfills the Stein 52. We will now consider the conditions under which the matrix $\mathcal{P}$ (ϑ) (66) verifies an appropriate Stein equation. For that purpose we consider the spectral density to be of the form $R_x(z) = (1/h(z)h(z^{-1}))$. The degree of the polynomial $h(z)$ is ℓ and we assume the distinct roots of the polynomial $h(z)$ to lie outside the unit circle, consequently, the roots of the polynomial $\hat{h}(z)$ lie within the unit circle. We therefore rewrite $\mathcal{P}$ (ϑ) accordingly

$$\mathcal{W}(\vartheta) = \frac{1}{2\pi\mathbf{i}} \oint\limits_{|z|=1} \frac{u_{p+r}(z)u_{p+r}^{\top}(z^{-1})}{h(z)h(z^{-1})a(z)a(z^{-1})b(z)b(z^{-1})} \frac{dz}{z}$$

We consider a companion matrix of the form (48) and with size $p + q + \ell$, it is denoted by $\mathcal{C}_f$ and the entries f_i are given by $z^{p+q+\ell} + \sum_{i=1}^{p+q+\ell} f_i(\vartheta) z^{p+q+q\ell-i} = \hat{a}(z)\hat{b}(z)\hat{h}(z) = \hat{f}(z,\vartheta)$ and $\widehat{f}$ (ϑ) is the vector $\widehat{f}$ $(\vartheta) = (f_{p+q+\ell}(\vartheta), f_{p+q+\ell-1}(\vartheta), \ldots, f_1(\vartheta))^{\mathrm{T}}$. Likewise for the vector $f(z, \vartheta) = a(z)b(z)h(z)$ and $f(\vartheta) = (f_1(\vartheta), f_1(\vartheta), \ldots, f_{p+q+\ell}(\vartheta))^{\mathrm{T}}$. The property $\mathrm{Det}(zI_{p+q+\ell} - \mathcal{C}_f) = \hat{a}(z)\widehat{b}(z)\hat{h}(z)$ and $\mathrm{Det}(I_{p+q+\ell} - z\,\mathcal{C}_f) = a(z)b(z)h(z)$ holds and assume

$$r = q + \ell \,\text{or}\, p + q + \ell = p + r \,\text{and}\, r > q \tag{67}$$

$\mathcal{P}$ (ϑ) is then of the form

$$\mathcal{W}(\vartheta) = \frac{1}{2\pi\mathbf{i}} \oint\limits_{|z|=1} \frac{u_{p+r}(z)v_{p+r}^{\top}(z)}{h(z)\hat{h}(z)a(z)\hat{a}(z)b(z)\hat{b}(z)} dz \tag{68}$$

We will formulate a Stein equation when the matrix $\Gamma = e_{p+r}e_{p+r}^{\top}$ and which is of the form

$$S - \mathcal{C}_f \mathcal{S} \mathcal{C}_f^{\top} = e_{p+r}e_{p+r}^{\top} \tag{69}$$

where e_{p+r} is the last standard basis column vector in $\mathbb{R}^{p+r}$. The next lemma is formulated.

3.7.1. Lemma 3.8

The matrix $\mathcal{P}$ (ϑ) *given in (68) fulfills the Stein 69.*

Proof

The unique solution of (69) is assured since the product of all the eigenvalues of $\mathcal{C}_f$ are different from one, the solution is of the form

$$S = \frac{1}{2\pi\mathbf{i}} \oint\limits_{|z|=1} (zI_{p+r} - \mathcal{C}_f)^{-1} e_{p+r}e_{p+r}^{\top}(I_{p+r} - z\mathcal{C}_f)^{-\top} dz$$

or

$$S = \frac{1}{2\pi\mathbf{i}} \oint\limits_{|z|=1} \frac{\mathrm{adj}(zI_{p+r} - \mathcal{C}_f)e_{p+r}e_{p+r}^{\top}\mathrm{adj}(I_{p+r} - z\mathcal{C}_f)^{\top}}{\hat{a}(z)\hat{b}(z)\hat{h}(z)a(z)b(z)h(z)} dz$$

taking the property of the companion matrix $\mathcal{C}_f$ into account implies that the last column vector of adj$(zI_{p+r} - \mathcal{C}_f)$ is the basic vector $u_{p+r}(z)$, consequently the last column of adj$(I_{p+r} - z\,\mathcal{C}_f)$ is the basic vector $v_{p+r}(z)$, this yields

$$S = \frac{1}{2\pi\mathbf{i}} \oint\limits_{|z|=1} \frac{u_{p+r}(z) v_{p+r}^{\top}(z)}{\hat{a}(z)\hat{b}(z)\hat{h}(z)a(z)b(z)h(z)} dz = \mathcal{W}(\vartheta)$$

Consequently, the matrix $\mathcal{P}(\vartheta)$ defined in (68) verifies the Stein 69. This completes the proof.

The matrices, $\wp(\vartheta)$ and $\mathcal{P}(\vartheta)$, in (65), verify under specific conditions appropriate Stein equations, as has been shown in Lemma 3.5 and Lemma 3.8, respectively. We will now confirm the presence of Vandermonde matrices by applying the standard residue theorem to $\mathcal{P}(\vartheta)$ in (68), to obtain

$$\mathcal{W}(\vartheta) = V_{\alpha\beta\xi} \mathcal{R}(\vartheta)\, \hat{V}_{\alpha\beta\xi} \tag{70}$$

The $(p + r) \times (p + r)$ diagonal matrix $\mathcal{R}(\vartheta)$ is of the form

$$\mathcal{R}(\vartheta) = \text{diag}\left\{ \left(1/\hat{a}(z;\alpha_i)\hat{b}(\alpha_i)\hat{h}(\alpha_i)\varphi(\alpha_i)\right), \left(1/\hat{a}(\beta_j)\hat{b}(z;\beta_j)\hat{h}(\beta_j)\varphi(\beta_j)\right), \left(1/\hat{a}(\xi_k)\hat{b}(\xi_k)\hat{h}(z;\xi_k)\varphi(\xi_k)\right) \right\}$$

where $\varphi(z) = a(z)b(z)h(z)$ and $i = 1, \ldots, p$, $j = 1, \ldots, q$ and $k = 1, \ldots, \ell$. Whereas the $(p + r) \times (p + r)$ matrices $V_{\alpha\beta\xi}$ and $\widehat{\mathcal{U}}_{\alpha\beta\xi}$ are of the form

$$V_{\alpha\beta\xi} = \begin{pmatrix} 1 & \alpha_1 & \alpha_1^2 & \cdots & \alpha_1^{p+r-1} \\ \vdots & \vdots & \vdots & \vdots & \vdots \\ 1 & \alpha_p & \alpha_p^2 & \cdots & \alpha_p^{p+r-1} \\ 1 & \beta_1 & \beta_1^2 & \cdots & \beta_1^{p+r-1} \\ \vdots & \vdots & \vdots & \vdots & \vdots \\ 1 & \beta_q & \beta_q^2 & \cdots & \beta_q^{p+r-1} \\ 1 & \xi_1 & \xi_1^2 & \cdots & \xi_1^{p+r-1} \\ \vdots & \vdots & \vdots & \vdots & \vdots \\ 1 & \xi_\ell & \xi_\ell^2 & \cdots & \xi_\ell^{p+r-1} \end{pmatrix}^{\top}, \quad \hat{V}_{\alpha\beta\xi} = \begin{pmatrix} \alpha_1^{p+r-1} & \alpha_1^{p+r-2} & \cdots & \alpha_1 & 1 \\ \vdots & \vdots & & \vdots & \vdots \\ \alpha_p^{p+r-1} & \alpha_p^{p+r-2} & \cdots & \alpha_p & 1 \\ \beta_1^{p+r-1} & \beta_1^{p+r-2} & \cdots & \beta_1 & 1 \\ \vdots & \vdots & & \vdots & \vdots \\ \beta_q^{p+r-1} & \beta_q^{p+r-2} & \cdots & \beta_q & 1 \\ \xi_1^{p+r-1} & \xi_1^{p+r-2} & \cdots & \xi_1 & 1 \\ \vdots & \vdots & & \vdots & \vdots \\ \xi_\ell^{p+r-1} & \xi_\ell^{p+r-2} & \cdots & \xi_\ell & 1 \end{pmatrix}$$

The $(p + r) \times (p + r)$ Vandermonde matrices $V_{\alpha\beta\xi}$ and $\widehat{\mathcal{U}}_{\alpha\beta\xi}$ are nonsingular when $\alpha_i \neq \alpha_j$, $\beta_k \neq \beta_h$, $\xi_m \neq \xi_n$, $\alpha_i \neq \beta_k$, $\alpha_i \neq \xi_m$, $\beta_k \neq \xi_m$ for all $i, j = 1, \ldots, p$, $k, h = 1, \ldots, q$ and $m,n = 1, \ldots, \ell$. The Vandermonde determinants Det$V_{\alpha\beta\xi}$ and Det $\widehat{\mathcal{U}}_{\alpha\beta\xi}$, are

$$\text{Det}V_{\alpha\beta\xi} = (-1)^{(p+r)\,(p+r-1)/2}\, \Psi(\alpha_i, \beta_k, \xi_m)$$

where

$$\Psi(\alpha_i, \beta_k, \xi_m) = \prod_{1\le i<j\le p} (\alpha_i - \alpha_j) \prod_{1\le k<h\le q} (\beta_k - \beta_h) \prod_{1\le m<n\le \ell} (\xi_m - \xi_n) \prod_{\substack{r=1,\ldots,p \\ s=1,\ldots,q}} (\alpha_r - \beta_s) \prod_{\substack{r=1,\ldots,p \\ w=1,\ldots,\ell}} (\alpha_r - \xi_w) \prod_{\substack{s=1,\ldots,q \\ w=1,\ldots,\ell}} (\beta_s - \xi_w)$$

Like for the Vandermonde matrices $V_{\alpha\beta}$ and $\hat{V}_{\alpha\beta}^{\top}$,

$$\text{Det}\hat{V}_{\alpha\beta\xi} = (-1)^{(p+r)\,(p+r-1)}\,\Psi\,(\alpha_i, \beta_k, \xi_m) \Rightarrow \mid \text{Det}V_{\alpha\beta\xi} \mid = \mid \text{Det}\hat{V}_{\alpha\beta\xi} \mid$$

(70) is the ARMAX equivalent to (47). A combination of both equations generates a new representation of the FIM $\mathcal{G}\,(\vartheta)$, this is set forth in the following lemma.

3.7.2. Lemma 3.9

Assume the conditions (67) to hold and consider the representations of $\wp(\vartheta)$ and $\mathcal{P}\,(\vartheta)$ in (47) and (70) respectively, leads to an alternative form to (65), it is given by

$$\mathcal{G}(\vartheta) = \begin{pmatrix} -\mathcal{S}_p(c) \\ \mathcal{S}_r(a) \\ O \end{pmatrix} V_{\alpha\beta\xi}\mathcal{R}\,(\vartheta)\,\hat{V}_{\alpha\beta\xi} \begin{pmatrix} -\mathcal{S}_p(c) \\ \mathcal{S}_r(a) \\ O \end{pmatrix}^{\top} + \begin{pmatrix} -\mathcal{S}_p(b) \\ O \\ \mathcal{S}_q(a) \end{pmatrix} V_{\alpha\beta}\mathcal{D}(\vartheta)\hat{V}_{\alpha\beta} \begin{pmatrix} -\mathcal{S}_p(b) \\ O \\ \mathcal{S}_q(a) \end{pmatrix}^{\top}$$

In Lemma 3.9, the FIM $\mathcal{G}\,(\vartheta)$ is expressed by submatrices of two Sylvester matrices and various Vandermonde matrices, both type of matrices become singular if and only if the appropriate polynomials have at least one common root.

3.8. The Fisher Information Matrix of a Vector ARMA(p, q) Process

The process (5) is summarized as,

$$A(z)y(t) = B(z)e(t)$$

and we assume that $\{y(t), t \in \mathbb{N}\}$, is a zero mean Gaussian time series and $\{\epsilon(t), t \in \mathbb{N}\}$ is a n-dimensional vector random variable, such that

$$\mathbb{E}_{\vartheta}$$

$\{\epsilon(t)\} = 0$ and $\mathbb{E}_{\vartheta}\ \{\epsilon(t)\epsilon^{\top}\ (t)\} = \sum$ and the parameter vector ϑ is of the form (7). In [6] it is shown that representation (16) for the $n^2(p+q)\times n^2(p+q)$ asymptotic FIM of the VARMA process (6) is

$$\mathbf{F}(\vartheta) = \mathbb{E}_{\vartheta}\left\{\left(\frac{\partial e}{\partial\vartheta^{\top}}\right)^{\top}\Sigma^{-1}\left(\frac{\partial e}{\partial\vartheta^{\top}}\right)\right\} \tag{71}$$

where $\partial\epsilon/\partial\vartheta^{\mathrm{T}}$ is of size $n\times n^2(p+q)$ and for convenience t is omitted from $\epsilon(t)$. Using the differential rules outlined in [6], yields

$$\frac{\partial e}{\partial\vartheta^{\top}} = \left\{\left(A^{-1}(z)B(z)e\right)^{\top} \otimes B^{-1}(z)\right\}\frac{\partial\text{vec}\,A(z)}{\partial\vartheta^{\top}} - (e^{\top} \otimes B^{-1}(z))\frac{\partial\text{vec}\,B(z)}{\partial\vartheta^{\top}} \tag{72}$$

The substitution of representation (72) of $\partial\epsilon/\partial\vartheta^{\,\mathrm{T}}$ in (71) yields the FIM of a VARMA process. The purpose is to construct a factorization of the FIM $\boldsymbol{F}(\vartheta)$ that should be a multiple variant of the factorization (27), so that a multiple resultant matrix property can be proved for $\boldsymbol{F}(\vartheta)$. As illustrated in [6], the multiple version of the Sylvester resultant matrix (22) does not fulfill the multiple resultant matrix property. In that case even when the matrix polynomials $A(z)$ and $B(z)$ have a common zero or a common eigenvalue, the multiple Sylvester matrix is not neccessarily singular. This has also been illustrated in [3]. In order to consider a multiple equivalent to the resultant matrix $\mathcal{S}\ -b, a)$, Gohberg and Lerer set forth the $n^2(p+q) \times n^2(p+q)$ tensor Sylvester matrix

$$\mathcal{S}^{\otimes}(-B,A) := \begin{pmatrix} (-I_n)\otimes I_n & (-B_1)\otimes I_n & \cdots & (-B_q)\otimes I_n & O_{n^2\times n^2} & \cdots & O_{n^2\times n^2} \\ O_{n^2\times n^2} & \ddots & \ddots & & \ddots & \ddots & \vdots \\ \vdots & \ddots & \ddots & \ddots & & \ddots & O_{n^2\times n^2} \\ O_{n^2\times n^2} & \cdots & O_{n^2\times n^2} & (-I_n)\otimes I_n & (-B_1)\otimes I_n & \cdots & (-B_q)\otimes I_n \\ I_n\otimes I_n & I_n\otimes A_1 & \cdots & I_n\otimes A_p & O_{n^2\times n^2} & \cdots & O_{n^2\times n^2} \\ O_{n^2\times n^2} & \ddots & \ddots & & \ddots & \ddots & \vdots \\ \vdots & \ddots & \ddots & \ddots & & \ddots & O_{n^2\times n^2} \\ O_{n^2\times n^2} & \cdots & O_{n^2\times n^2} & I_n\otimes I_n & I_n\otimes A_1 & \cdots & I_n\otimes A_p \end{pmatrix} \tag{73}$$

In [3], the authors prove that the tensor Sylvester matrix $\mathcal{S}^{\otimes}(-B,A)$ fulfills the multiple resultant property, it becomes singular if and only if the appropriate matrix polynomials $A(z)$ and $B(z)$ have at least one common zero. In Proposition 2.2 in [6], the following factorized form of the Fisher information $\mathbf{F}(\vartheta)$ is developed

$$\mathbf{F}(\vartheta) = \frac{1}{2\pi\mathbf{i}} \oint_{|z|=1} \Phi(z)\Theta(z)\Phi^*(z)\frac{dz}{z} \tag{74}$$

where

$$\Phi(z) = \begin{pmatrix} I_p\otimes A^{-1}(z)\otimes I_n & O_{pn^2\times qn^2} \\ O_{qn^2\times pn^2} & I_q\otimes I_n\otimes A^{-1}(z) \end{pmatrix} \mathcal{S}^{\otimes}(-B,A)\left(u_{p+q}(z)\otimes I_{n^2}\right)$$

and

$$\Theta(z) = \Sigma\otimes\sigma(z), \sigma(z) = B^{-\top}(z)\Sigma^{-1}B^{-1}(z^{-1}) \tag{75}$$

In order to obtain a multiple variant of (27), the following matrix is introduced,

$$\mathbf{M}(\vartheta) = \frac{1}{2\pi\mathbf{i}} \oint_{|z|=1} \Lambda(z)\mathcal{J}(z)\Lambda^*(z)\frac{dz}{z} = \mathcal{S}^{\otimes}(-B,A)\mathbf{P}(\vartheta)\left(\mathcal{S}^{\otimes}(-B,A)\right)^{\top} \tag{76}$$

where

$$\mathcal{J}(z) = \Phi(z)\Theta(z)\Phi^*(z) \text{ and } \Lambda(z) = \begin{pmatrix} I_p\otimes A(z)\otimes I_n & O_{pn^2\times qn^2} \\ O_{qn^2\times pn^2} & I_q\otimes I_n\otimes A(z) \end{pmatrix}$$

and the matrix $\boldsymbol{P}(\vartheta)$ is a multiple variant of the matrix $\wp(\vartheta)$ in (28), it is of the form

$$\mathbf{P}(\vartheta) = \frac{1}{2\pi\mathbf{i}} \oint_{|z|=1} \left(u_{p+q}(z)\otimes I_{n^2}\right)\Theta(z)\left(u_{p+q}(z)\otimes I_{n^2}\right)^*\frac{dz}{z} \tag{77}$$

In Lemma 2.3 in [6], it is proved that the matrix $\boldsymbol{M}(\vartheta)$ in (76) becomes singular if and only if the matrix polynomials $A(z)$ and $B(z)$ have at least one common eigenvalue-zero. The proof is a multiple equivalent of the proof of Corollary 2.2 in [5], since the equality (76) is a multiple version of (27). Consequently, the matrix $\boldsymbol{M}(\vartheta)$ like the tensor Sylvester matrix $\mathcal{S}^{\otimes}(-B,A)$, fulfills the multiple resultant matrix property. Since the matrix $\boldsymbol{M}(\vartheta)$ is derived from the FIM $\boldsymbol{F}(\vartheta)$, this enables us to prove that the matrix $\boldsymbol{F}(\vartheta)$ fulfills the multiple resultant matrix property by showing that it becomes singular if and only if the matrix $\boldsymbol{M}(\vartheta)$ is singular, this is done in Proposition 2.4 in [6]. Consequently, it can be concluded from [6] that the FIM of a VARMA process $\boldsymbol{F}(\vartheta)$ and the tensor Sylvester matrix $\mathcal{S}^{\otimes}(-B,A)$ have the same singularity conditions. The FIM of a VARMA process $\boldsymbol{F}(\vartheta)$ can therefore be added to the class of multiple resultant matrices.

A brief summary of the contribution of [6] follows, in order to show that the FIM of a VARMA process $F(\vartheta)$ is a multiple resultant matrix two new representations of the FIM are derived. To construct such representations appropriate matrix differential rules are applied. The newly obtained representations are expressed in terms of the multiple Sylvester matrix and the tensor Sylvester matrix. The representation of the FIM expressed by the tensor Sylvester matrix is used to prove that the FIM becomes singular if and only if the autoregressive and moving average matrix polynomials have at least one common eigenvalue. It then follows that the FIM and the tensor Sylvester matrix have equivalent singularity conditions. In a numerical example it is shown, however, that the FIM fails to detect common eigenvalues due to some kind of numerical instability. The tensor Sylvester matrix reveals it clearly, proving the usefulness of the results derived in this paper.

3.9. The Fisher Information Matrix of a Vector ARMAX(p, r, q) Process

The $n^2(p+q+r) \times n^2(p+q+r)$ asymptotic FIM of the VARMAX(p, r, q) process (2)

$$A(z)y(t) = C(z)x(t) + B(z)e(t)$$

is displayed according to [23] and is an extension of the FIM of the VARMA(p, q) process (6). Representation (16) of the FIM of the VARMAX(p, r, q) process is then

$$\mathbf{G}(\vartheta) = \mathbb{E}_{\vartheta}\left\{\left(\frac{\partial e}{\partial \vartheta^{\top}}\right)^{\top} \Sigma^{-1} \left(\frac{\partial e}{\partial \vartheta^{\top}}\right)\right\}$$

where

$$\begin{aligned} \frac{\partial e}{\partial \vartheta^{\top}} = & \left\{\left(A^{-1}(z)C(z)x\right)^{\top} \otimes B^{-1}(z)\right\}\frac{\partial \mathrm{vec}\, A(z)}{\partial \vartheta^{\top}} \\ & +\left\{\left(A^{-1}(z)B(z)e\right)^{\top} \otimes B^{-1}(z)\right\}\frac{\partial \mathrm{vec}\, A(z)}{\partial \vartheta^{\top}} \\ & -\{x^{\top} \otimes B^{-1}(z)\}\frac{\partial \mathrm{vec}\, C(z)}{\partial \vartheta^{\top}} \\ & -(e^{\top} \otimes B^{-1}(z))\frac{\partial \mathrm{vec}\, B(z)}{\partial \vartheta^{\top}} \end{aligned} \tag{78}$$

To obtain the term $\partial \epsilon / \partial \vartheta^{\top}$, of size $n \times n^2(p+q+r)$, the same differential rules are applied as for the VARMA(p, q) process. In Proposition 2.3 in [23], the representation of the FIM of a VARMAX process is expressed in terms of tensor Sylvester matrices, this obtained when $\partial \epsilon / \partial \vartheta^{\top}$ in (78) is substituted in (16), to obtain

$$\mathbf{G}(\vartheta) = \frac{1}{2\pi i}\int_{|z|=1} \Phi_x(z)\Theta(z)\Phi_x^{*}(z)\frac{dz}{z} + \frac{1}{2\pi i}\int_{|z|=1} \Lambda_x(z)\Psi(z)\Lambda_x^{*}(z)\frac{dz}{z} \tag{79}$$

The matrices in (79) are of the form

$$\begin{aligned} \Phi_x(z) &= \begin{pmatrix} I_p \otimes A^{-1}(z) \otimes I_n & O_{pn^2 \times rn^2} & O_{pn^2 \times qn^2} \\ O_{rn^2 \times pn^2} & O_{rn^2 \times rn^2} & O_{rn^2 \times qn^2} \\ O_{qn^2 \times pn^2} & O_{qn^2 \times rn^2} & I_q \otimes I_n \otimes A^{-1}(z) \end{pmatrix} \begin{pmatrix} -\mathcal{S}_p^{\otimes}(B) \\ O_{rn^2 \times n^2(p+q)} \\ \mathcal{S}_q^{\otimes}(A) \end{pmatrix} (u_{p+q}(z) \otimes I_{n^2}) \\ \Lambda_x(z) &= \begin{pmatrix} I_p \otimes A^{-1}(z) \otimes I_n & O_{pn^2 \times rn^2} & O_{pn^2 \times qn^2} \\ O_{rn^2 \times pn^2} & I_r \otimes I_n \otimes A^{-1}(z) & O_{rn^2 \times qn^2} \\ O_{qn^2 \times pn^2} & O_{qn^2 \times rn^2} & O_{qn^2 \times qn^2} \end{pmatrix} \begin{pmatrix} -\mathcal{S}_p^{\otimes}(C) \\ \mathcal{S}_r^{\otimes}(A) \\ O_{qn^2 \times n^2(p+r)} \end{pmatrix} (u_{p+r}(z) \otimes I_{n^2}) \\ & \mathcal{S}_{p,q}^{\otimes}(-B, A) = \begin{pmatrix} -\mathcal{S}_p^{\otimes}(B) \\ \mathcal{S}_q^{\otimes}(A) \end{pmatrix}, \; \mathcal{S}_{p,r}^{\otimes}(-C, A) = \begin{pmatrix} -\mathcal{S}_p^{\otimes}(C) \\ \mathcal{S}_r^{\otimes}(A) \end{pmatrix} \end{aligned} \tag{80}$$

additionally we have $\Psi(z) = R_x(z) \otimes \sigma(z)$ and the Hermitian spectral density matrix $R_x(z)$ is defined in (10), whereas the matrix polynomials $\Theta(z)$ and $\sigma(z)$ are presented in (75). In (80), we have the $pn^2 \times (p+q)n^2$ and $qn^2 \times (p+q)n^2$ submatrices $\mathcal{S}_p^{\otimes}(-B)$ and $\mathcal{S}_q^{\otimes}(A)$ of the tensor Sylvester resultant matrix $\mathcal{S}_{p,q}^{\otimes}(-B, A)$. Whereas the matrices $\mathcal{S}_p^{\otimes}(-C)$ and $\mathcal{S}_r^{\otimes}(A)$ are the upper and lower blocks of the $(p+r)n^2 \times (p+r)n^2$ tensor Sylvester resultant matrix $\mathcal{S}_{p,r}^{\otimes}(-C, A)$. As for the FIM of the VARMA(p, q) process, the objective is to construct a multiple version of (65), this done in [23], to obtain

$$\begin{aligned} \boldsymbol{M}_x(\vartheta) &= \tfrac{1}{2\pi i}\oint_{|z|=1} \mathcal{L}(z)\mathcal{A}(z)\mathcal{L}^*(z)\tfrac{dz}{z} + \tfrac{1}{2\pi i}\oint_{|z|=1} \mathcal{W}(z)\mathcal{B}(z)\mathcal{W}^*(z)\tfrac{dz}{z} \\ &= \begin{pmatrix} -\mathcal{S}_p^{\otimes}(B) \\ O_{rn^2\times n^2(p+q)} \\ \mathcal{S}_q^{\otimes}(A) \end{pmatrix} \boldsymbol{P}(\vartheta) \begin{pmatrix} -\mathcal{S}_p^{\otimes}(B) \\ O_{rn^2\times n^2(p+q)} \\ \mathcal{S}_q^{\otimes}(A) \end{pmatrix}^{\top} + \begin{pmatrix} -\mathcal{S}_p^{\otimes}(C) \\ \mathcal{S}_r^{\otimes}(A) \\ O_{qn^2\times n^2(p+r)} \end{pmatrix} \boldsymbol{T}(\vartheta) \begin{pmatrix} -\mathcal{S}_p^{\otimes}(C) \\ \mathcal{S}_r^{\otimes}(A) \\ O_{qn^2\times n^2(p+r)} \end{pmatrix}^{\top} \end{aligned} \tag{81}$$

The matrices involved are of the form

$$\mathcal{L}(z) = \begin{pmatrix} I_p \otimes A(z) \otimes I_n & O_{pn^2\times rn^2} & O_{pn^2\times qn^2} \\ O_{rn^2\times pn^2} & O_{rn^2\times rn^2} & O_{rn^2\times qn^2} \\ O_{qn^2\times pn^2} & O_{qn^2\times rn^2} & I_q \otimes I_n \otimes A(z) \end{pmatrix} \text{and } \mathcal{A}(z) := \Phi_x(z)\Theta(z)\Phi_x^*(z)$$

$$\mathcal{W}(z) = \begin{pmatrix} I_p \otimes A(z) \otimes I_n & O_{pn^2\times rn^2} & O_{pn^2\times qn^2} \\ O_{rn^2\times pn^2} & I_r \otimes I_n \otimes A(z) & O_{rn^2\times qn^2} \\ O_{qn^2\times pn^2} & O_{qn^2\times rn^2} & O_{qn^2\times qn^2} \end{pmatrix} \text{and } \mathcal{B}(z) := \Lambda_x(z)\Psi(z)\Lambda_x^*(z)$$

$$\boldsymbol{T}(\vartheta) = \tfrac{1}{2\pi i}\oint_{|z|=1} \left(u_{p+r}(z) \otimes I_{n^2}\right) \Psi(z) \left(u_{p+r}(z) \otimes I_{n^2}\right)^* \tfrac{dz}{z}$$

and $\boldsymbol{P}(\vartheta)$ is given in (77). Note, the matrices $\Phi_x(z)$, $\Lambda_x(z)$, $\mathcal{L}(z)$ and $\mathcal{P}$ (z) are the corrected versions of the corresponding matrices in [23].

A parallel between the scalar and multiple structures is straightforward. This is best illustrated by comparing the representations (27) and (28) with (76) and (77) respectively, confronting the FIM for scalar and vector ARMA(p, q) processes. The FIM of the scalar ARMAX(p, r, q) process contains an ARMA(p, q) part, this is confirmed by (65), through the presence of the matrix $\wp(\vartheta)$ which is originally displayed in (28). The multiple resultant matrices $\boldsymbol{M}(\vartheta)$ and $\boldsymbol{M}_x(\vartheta)$ derived from the FIM of the VARMA(p, q) and VARMAX(p, r, q) processes respectively both contain $\boldsymbol{P}(\vartheta)$, whereas the first matrix term of the matrices $\Phi(z)$ and $\Phi_x(z)$, which are of different size, consist of the same nonzero submatrices. To summarize, in [23] compact forms of the FIM of a VARMAX process expressed in terms of multiple and tensor Sylvester matrices are developed. The tensor Sylvester matrices allow us to investigate the multiple resultant matrix property of the FIM of VARMAX(p, r, q) processes. However, since no proof of the multiple resultant matrix property of the FIM $G(\vartheta)$ has been done yet, justifies the consideration of a conjecture. A conjecture that states, the FIM$G(\vartheta)$ of a VARMAX(p, r, q) process becomes singular if and only if the matrix polynomials $A(z)$, $B(z)$ and $C(z)$ have at least one common eigenvalue. A multiple equivalent to Theorem 3.1 in [4] and combined with Proposition 2.4 in [6], but based on the representations (79) and (81), can be envisaged to formulate a proof which will be a subject for future study.

4. Conclusions

In this survey paper, matrix algebraic properties of the FIM of stationary processes are discussed. The presented material is a summary of papers where several matrix structural aspects of the FIM are investigated. The FIM of scalar and multiple processes like the (V)ARMA(X) are set forth with appropriate factorized forms involving (tensor) Sylvester matrices. These representations enable us to prove the resultant matrix property of the corresponding FIM. This has been done for (V)ARMA(p, q) and ARMAX(p, r, q) processes in the papers [4,6]. The development of the stages that lead to the appropriate factorized form of the FIM $G(\vartheta)$ (79) is set forth in [23]. However, there is no proof done yet that confirms the multiple resultant matrix property of the FIM $G(\vartheta)$ of a VARMAX(p, r, q) process. This justifies the consideration of a conjecture which is formulated in the former section, this can be a subject for future study.

The statistical distance measure derived in [7], involves entries of the FIM. This distance measure can be a challenge to its quantum information counterpart (41). Because (36) involves information about m parameters estimated from n measurements. Whereas in quantum information, like in e.g., [8,10], the information about one parameter in a particular measurement procedure is considered for establishing an interconnection with the appropriate statistical distance measure. A possible approach, by combining matrix algebra and quantum information, for developing a statistical distance measure in quantum information or quantum statistics but at the matrix level, can be a subject of future research. Some results concerning interconnections between the FIM of ARMA(X) models and appropriate solutions to Stein matrix equations are discussed, the material is extracted from the papers, [12] and [13]. However, in this paper, some alternative and new proofs that emphasize the conditions under which the FIM fulfills appropriate Stein equations, are set forth. The presence of various types of Vandermonde matrices is also emphasized when an explicit expansion of the FIM is computed. These Vandermonde matrices are inserted in interconnections with appropriate solutions to Stein equations. This explains, when the matrix algebraic structures of the FIM of stationary processes are investigated, the involvement of structured matrices like the (tensor) Sylvester, Bezoutian and Vandermonde matrices is essential.

Acknowledgments: The author thanks a perceptive reviewer for his comments which significantly improved the quality and presentation of the paper.

Conflicts of Interest: The authors have declared no conflict of interest.

References

1. Dym, H. *Linear Algebra in Action*; American Mathematical Society: Providence, RI, USA, 2006; Volume 78.
2. Lancaster, P.; Tismenetsky, M. *The Theory of Matrices with Applications*, 2nd ed; Academic Press: Orlando, FL, USA, 1985.
3. Gohberg, I.; Lerer, L. Resultants of matrix polynomials. *Bull. Am. Math. Soc* **1976**, *82*, 565–567.
4. Klein, A.; Spreij, P. On Fisher's information matrix of an ARMAX process and Sylvester's resultant matrices. *Linear Algebra Appl* **1996**, *237/238*, 579–590.
5. Klein, A.; Spreij, P. On Fisher's information matrix of an ARMA process. In *Stochastic Differential and Difference Equations*; Csiszar, I., Michaletzky, Gy., Eds.; Birkhäuser: Boston: Boston, USA, 1997; Progress in Systems and Control Theory; Volume 23, pp. 273–284.
6. Klein, A.; Mélard, G.; Spreij, P. On the Resultant Property of the Fisher Information Matrix of a Vector ARMA process. *Linear Algebra Appl* **2005**, *403*, 291–313.
7. Klein, A.; Spreij, P. Transformed Statistical Distance Measures and the Fisher Information Matrix. *Linear Algebra Appl* **2012**, *437*, 692–712.
8. Braunstein, S.L.; Caves, C.M. Statistical Distance and the Geometry of Quantum States. *Phys. Rev. Lett* **1994**, *72*, 3439–3443.
9. Jones, P.J.; Kok, P. Geometric derivation of the quantum speed limit. *Phys. Rev. A* **2010**, *82*, 022107.
10. Kok, P. *Tutorial: Statistical distance and Fisher information*; Oxford: UK, 2006.
11. Lancaster, P.; Rodman, L. *Algebraic Riccati Equations*; Clarendon Press: Oxford, UK, 1995.
12. Klein, A.; Spreij, P. On Stein's equation, Vandermonde matrices and Fisher's information matrix of time series processes. Part I: The autoregressive moving average process. *Linear Algebra Appl* **2001**, *329*, 9–47.
13. Klein, A.; Spreij, P. On the solution of Stein's equation and Fisher's information matrix of an ARMAX process. *Linear Algebra Appl* **2005**, *396*, 1–34.
14. Grenander, U.; Szegő, G.P. *Toeplitz Forms and Their Applications*; University of California Press: New York, NY, USA, 1958.
15. Brockwell, P.J.; Davis, R.A. *Time Series: Theory and Methods*, 2nd ed; Springer Verlag: Berlin, Germany; New York, NY, USA, 1991.
16. Caines, P. *Linear Stochastic Systems*; John Wiley and Sons: New York, NY, USA, 1988.
17. Ljung, L.; Söderström, T. *Theory and Practice of Recursive Identification*; M.I.T. Press: Cambridge, MA, USA, 1983.
18. Hannan, E.J.; Deistler, M. *The Statistical Theory of Linear Systems*; John Wiley and Sons: New York, NY, USA, 1988.
19. Hannan, E.J.; Dunsmuir, W.T.M.; Deistler, M. Estimation of vector Armax models. *J. Multivar. Anal* **1980**, *10*, 275–295.
20. Horn, R.A.; Johnson, C.R. *Topics in Matrix Analysis*; Cambridge University Press: New York, NY, USA, 1995.
21. Klein, A.; Spreij, P. Matrix differential calculus applied to multiple stationary time series and an extended Whittle formula for information matrices. *Linear Algebra Appl* **2009**, *430*, 674–691.

22. Klein, A.; Mélard, G. An algorithm for the exact Fisher information matrix of vector ARMAX time series. *Linear Algebra Its Appl* **2014**, *446*, 1–24.
23. Klein, A.; Spreij, P. Tensor Sylvester matrices and the Fisher information matrix of VARMAX processes. *Linear Algebra Appl* **2010**, *432*, 1975–1989.
24. Rao, C.R. Information and the accuracy attainable in the estimation of statistical parameters. *Bull. Calcutta Math. Soc* **1945**, *37*, 81–91.
25. Ibragimov, I.A.; Has'minskiĭ, R.Z. Statistical Estimation. In *Asymptotic Theory*; Springer-Verlag: New York, NY, USA, 1981.
26. Lehmann, E.L. *Theory of Point Estimation*; Wiley: New York, NY, USA, 1983.
27. Friedlander, B. On the computation of the Cramér-Rao bound for ARMA parameter estimation. *IEEE Trans. Acoust. Speech Signal Process* **1984**, *32*, 721–727.
28. Holevo, A.S. *Probabilistic and Statistical Aspects of Quantum Theory*, 2nd ed; Edizioni Della Normale, SNS Pisa: Pisa, Italy, 2011.
29. Petz, T. *Quantum Information Theory and Quantum Statistics*; Springer-Verlag: Berlin Heidelberg, Germany, 2008.
30. Barndorff-Nielsen, O.E.; Gill, R.D. Fisher Information in quantum statistics. *J. Phys. A* **2000**, *30*, 4481–4490.
31. Luo, S. Wigner-Yanase skew information *vs.* quantum Fisher information. *Proc. Amer. Math. Soc* **2004**, *132*, 885–890.
32. Klein, A.; Mélard, G. On algorithms for computing the covariance matrix of estimates in autoregressive moving average processes. *Comput. Stat. Q* **1989**, *5*, 1–9.
33. Klein, A.; Mélard, G. An algorithm for computing the asymptotic Fisher information matrix for seasonal SISO models. *J. Time Ser. Anal* **2004**, *25*, 627–648.
34. Bistritz, Y.; Lifshitz, A. Bounds for resultants of univariate and bivariate polynomials. *Linear Algebra Appl* **2010**, *432*, 1995–2005.
35. Horn, R.A.; Johnson, C.R. *Matrix Analysis*; Cambridge University Press: New York, NY, USA, 1996.
36. Golub, G.H.; van Loan, C.F. *Matrix Computations*, 3rd ed; John Hopkins University Press: Baltimore, USA, 1996.
37. Kullback, S. *Information Theory and Statistics*; John Wiley and Sons: New York, NY, USA, 1959.
38. Klein, A.; Spreij, P. The Bezoutian, state space realizations and Fisher's information matrix of an ARMA process. *Linear Algebra Appl* **2006**, *416*, 160–174.

entropy

MDPI

Article

Asymptotically Constant-Risk Predictive Densities When the Distributions of Data and Target Variables Are Different

Keisuke Yano [1,*] and Fumiyasu Komaki [1,2]

[1] Department of Mathematical Informatics, Graduate School of Information Science and Technology, The University of Tokyo, 7-3-1 Hongo, Bunkyo-ku, Tokyo 113-8656, Japan; E-Mail: komaki@mist.i.u-tokyo.ac.jp

[2] RIKEN Brain Science Institute, 2-1 Hirosawa, Wako City, Saitama 351-0198, Japan

* E-Mail: Keisuke_Yano@mist.i.u-tokyo.ac.jp; Tel.: +81-3-5841-6909.

Received: 28 March 2014; in revised form: 9 May 2014 / Accepted: 22 May 2014 / Published: 28 May 2014

Abstract: We investigate the asymptotic construction of constant-risk Bayesian predictive densities under the Kullback–Leibler risk when the distributions of data and target variables are different and have a common unknown parameter. It is known that the Kullback–Leibler risk is asymptotically equal to a trace of the product of two matrices: the inverse of the Fisher information matrix for the data and the Fisher information matrix for the target variables. We assume that the trace has a unique maximum point with respect to the parameter. We construct asymptotically constant-risk Bayesian predictive densities using a prior depending on the sample size. Further, we apply the theory to the subminimax estimator problem and the prediction based on the binary regression model.

Keywords: Bayesian prediction; Fisher information; Kullback–Leibler divergence; minimax; predictive metric; subminimax estimator

1. Introduction

Let $x^{(N)} = (x_1, \cdots, x_N)$ be independent N data distributed according to a probability density, $p(x|\theta)$, that belongs to a d-dimensional parametric model, $\{p(x|\theta) : \theta \in \Theta\}$, where $\theta = (\theta^1, \cdots, \theta^d)$ is an unknown d-dimensional parameter and Θ is the parameter space. Let y be a target variable distributed according to a probability density, $q(y|\theta)$, that belongs to a d-dimensional parametric model, $\{q(y|\theta) : \theta \in \Theta\}$ with the same parameter, θ. Here, we assume that the distributions of the data and the target variables, $p(x|\theta)$ and $q(y|\theta)$, are different. For simplicity, we assume that the data and the target variables are independent, given by θ.

We construct predictive densities for target variables based on the data. We measure the performance of the predictive density, $\hat{q}(y; x^{(N)})$, by the Kullback–Leibler divergence, $D(q(\cdot|\theta), \hat{q}(\cdot; x^{(N)}))$, from the true density, $q(y|\theta)$, to the predictive density, $\hat{q}(y; x^{(N)})$:

$$D(q(\cdot|\theta), \hat{q}(\cdot; x^{(N)})) = \int q(y|\theta) \log \frac{q(y|\theta)}{\hat{q}(y; x^{(N)})} \mathrm{d}y.$$

Then, the risk function, $R(\theta, \hat{q}(y; x^{(N)}))$, of the predictive density, $\hat{q}(y; x^{(N)})$, is given by:

$$\begin{aligned} R(\theta, \hat{q}(y; x^{(N)})) &= \int p(x^{(N)}|\theta) D(q(\cdot|\theta), \hat{q}(\cdot; x^{(N)})) \mathrm{d}x^{(N)} \\ &= \int p(x^{(N)}|\theta) \int q(y|\theta) \log \frac{q(y|\theta)}{\hat{q}(y; x^{(N)})} \mathrm{d}y \mathrm{d}x^{(N)}. \end{aligned}$$

For the construction of predictive densities, we consider the Bayesian predictive density defined by:

$$\hat{q}_\pi(y|x^{(N)}) \quad = \quad \frac{\int q(y|\theta)p(x^{(N)}|\theta)\pi(\theta;N)\mathrm{d}\theta}{\int p(x^{(N)}|\theta)\pi(\theta;N)\mathrm{d}\theta},$$

where $\pi(\theta;N)$ is a prior density for θ, possibly depending on the sample size, N. Aitchison [1] showed that, for a given prior density, $\pi(\theta;N)$, the Bayesian predictive density, $\hat{q}_\pi(y|x^{(N)})$, is a Bayes solution under the Kullback–Leibler risk. Based on the asymptotics as the sample size goes to infinity, Komaki [2] and Hartigan [3] showed its superiority over any plug-in predictive density, $q(y|\hat{\theta})$, with any estimator, $\hat{\theta}$. However, there remains a problem of prior selection for constructing better Bayesian predictive densities. Thus, a prior, $\pi(\theta;N)$, must be chosen based on an optimality criterion for actual applications.

Among various criteria, we focus on a criterion of constructing minimax predictive densities under the Kullback–Leibler risk. For simplicity, we refer to the priors generating minimax predictive densities as minimax priors. Minimax priors have been previously studied in various predictive settings; see [4–8]. When the simultaneous distributions of the target variables and the data belong to the submodel of the multinomial distributions, Komaki [7] shows that minimax priors are given as latent information priors maximizing the conditional mutual information between target variables and the parameter given the data. However, the explicit forms of latent information priors are difficult to obtain, and we need asymptotic methods, because they require the maximization on the space of the probability measures on Θ.

Except for [7], these studies on minimax priors are based on the assumption that the distributions, $p(x|\theta)$ and $q(y|\theta)$, are identical. Let us consider the prediction based on the logistic regression model where the covariates of the data and the target variables are not identical. In this predictive setting, the assumption that the distributions, $p(x|\theta)$ and $q(y|\theta)$, are identical is no longer valid.

We focus on the minimax priors in predictions where the distributions, $p(x|\theta)$ and $q(y|\theta)$, are different and have a common unknown parameter. Such a predictive setting has traditionally been considered in statistical prediction and experiment design. It has recently been studied in statistical learning theory; for example, see [9]. Predictive densities where the distributions, $p(x|\theta)$ and $q(y|\theta)$, are different and have a common unknown parameter are studied by [10–13].

Let $g^X_{ij}(\theta)$ be the (i,j)-component of the Fisher information matrix of the distribution, $p(x|\theta)$, and let $g^Y_{ij}(\theta)$ be the (i,j)-component of the Fisher information matrix of the distribution, $q(y|\theta)$. Let $g^{X,ij}(\theta)$ and $g^{Y,ij}(\theta)$ denote the (i,j)-components of their inverse matrices. We adopt Einstein's summation convention: if the same indices appear twice in any one term, it implies summation over that index from one to d. For the asymptotics below, we assume that the prior densities, $\pi(\theta;N)$, are smooth.

On the asymptotics as the sample size N goes to infinity, we construct the asymptotically constant-risk prior, $\pi(\theta;N)$, in the sense that the asymptotic risk:

$$R(\theta,\hat{q}_\pi(y|x^{(N)})) = \frac{1}{N}R_1(\theta,\hat{q}_\pi(y|x^{(N)})) + \frac{1}{N\sqrt{N}}R_2(\theta,\hat{q}_\pi(y|x^{(N)})) + O(N^{-2})$$

is constant up to $\mathrm{O}(N^{-2})$. Since the proper prior with the constant risk is a minimax prior for any finite sample size, the asymptotically constant-risk prior relates to the minimax prior; in Section 4, we verify that the asymptotically constant-risk prior agrees with the exact minimax prior in binomial examples.

When we use the prior, $\pi(\theta)$, independent of the sample size, N, it is known that the N^{-1}-order term, $R_1(\theta,\hat{q}_\pi(y|x^{(N)}))$, of the Kullback–Leibler risk is equal to the trace, $g^{X,ij}(\theta)g^Y_{ij}(\theta)$. If the trace does not depend on the parameter, θ, the construction of the asymptotically constant-risk prior is parallel to [6]; see also [13].

However, we consider the settings where there exists a unique maximum point of the trace, $g^{X,ij}(\theta)g^Y_{ij}(\theta)$; for example, these settings appear in predictions based on the binary regression model, where the covariates of the data and the target variables are not identical. In the settings, there do not exist asymptotically constant-risk priors among the priors independent of the sample size, N. The reason is as follows: we consider the prior, $\pi(\theta)$, independent of the sample size, N. Then, the Kullback–Leibler risk of the Bayesian predictive density is expanded as:

$$R(\theta, \hat{q}_\pi(y|x^{(N)})) = \frac{1}{2N} g^Y_{ij}(\theta) g^{X,ij}(\theta) + \mathrm{O}(N^{-2}).$$

Since, in our settings, the first-order term, $g^Y_{ij}(\theta)g^{X,ij}(\theta)$, is not constant, the prior independent of the sample size, N, is not an asymptotically constant-risk prior.

When there exists a unique maximum point of the trace, $g^{X,ij}(\theta)g^Y_{ij}(\theta)$, we construct the asymptotically constant-risk prior, $\pi(\theta;N)$, up to $\mathrm{O}(N^{-2})$, by making the prior dependent on the sample size, N, as:

$$\frac{\pi(\theta; N)}{|g^X(\theta)|^{1/2}} \quad \propto \quad \{f(\theta)\}^{\sqrt{N}} h(\theta),$$

where $f(\theta)$ and $h(\theta)$ are the scalar functions of θ independent of N and $|g^X(\theta)|$ denotes the determinant of the Fisher information matrix, $g^X(\theta)$.

The key idea is that, if the specified parameter point has more undue risk than the other parameter points, then the more prior weights should be concentrated on that point.

Further, we clarify the subminimax estimator problem based on the mean squared error from the viewpoint of the prediction where the distributions of data and target variables are different and have a common unknown parameter. We obtain the improvement achieved by the minimax estimator over the subminimax estimators up to $\mathrm{O}(N^{-2})$. The subminimax estimator problem [14,15] is the problem that, at first glance, there seems to exist asymptotically dominating estimators of the minimax estimator. However, any relationship between such subminimax estimator problems and predictions have not been investigated, and further, in general, the improvement by the minimax estimator over the subminimax estimators has not been investigated.

2. Information Geometrical Notations

In this section, we prepare the information geometrical notations; see [16] for details. We abbreviate $\partial/\partial\theta^i$ to ∂_i, where the indices, $i, j, k, \ldots$, run from one to d. Similarly, we abbreviate $\partial^2/\partial\theta^i\partial\theta^j$, $\partial^3/\partial\theta^i\partial\theta^j\partial\theta^k$ and $\partial^4/\partial\theta^i\partial\theta^j\partial\theta^k\partial\theta^l$ to ∂_{ij}, ∂_{ijk} and ∂_{ijkl}, respectively. We denote the expectations of the random variables, X, Y and $X^{(N)}$, by $\mathrm{E}_X[\cdot]$, $\mathrm{E}_Y[\cdot]$ and $\mathrm{E}_{X^{(N)}}[\cdot]$, respectively. We denote their probability densities by $p(x|\theta)$, $q(y|\theta)$ and $p(x^{(N)}|\theta)$, respectively.

We define the predictive metric proposed by Komaki [13] as:

$$\mathring{g}_{ij}(\theta) \quad = \quad g^X_{ik}(\theta) g^{Y,kl}(\theta) g^X_{lj}(\theta).$$

When the parameter is one-dimensional, $g_{\theta\theta}(\theta)$ denotes Fisher information and $g^{\theta\theta}(\theta)$ denotes its inverse. Let $\overset{\mathrm{e}}{\Gamma}{}^X_{ij,k}(\theta)$ and $\overset{\mathrm{m}}{\Gamma}{}^X_{ij,k}(\theta)$ be the quantities given by:

$$\overset{\mathrm{e}}{\Gamma}{}^X_{ij,k}(\theta) \quad := \quad \mathrm{E}_X[\partial_{ij} \log p(x|\theta) \partial_k \log p(x|\theta)]$$

and:

$$\overset{\mathrm{m}}{\Gamma}{}^X_{ij,k}(\theta) \quad := \quad \int \frac{1}{p(x|\theta)} [\partial_{ij} p(x|\theta) \partial_k p(x|\theta)] \mathrm{d}x.$$

Using these quantities, the e-connection and m-connection coefficients with respect to the parameter, θ, for the model, $\{p(x|\theta) : \theta \in \Theta\}$, are given by:

$$\overset{\mathrm{e}}{\Gamma}{}^{X}_{ij,k}(\theta) \;:=\; g^{X,lk}(\theta)\overset{\mathrm{e}}{\Gamma}{}^{X}_{ij,l}(\theta)$$

and:

$$\overset{\mathrm{m}}{\Gamma}{}^{X}_{ij,k}(\theta) \;:=\; g^{X,kl}(\theta)\overset{\mathrm{m}}{\Gamma}{}^{X}_{ij,l}(\theta),$$

respectively.

The (0, 3)-tensor, $T^X_{ijk}(\theta)$, is defined by:

$$T^X_{ijk}(\theta) \;:=\; \mathrm{E}_X[\partial_i \log p(x|\theta)\partial_j \log p(x|\theta)\partial_k \log p(x|\theta)].$$

The tensor, $T^X_{ijk}(\theta)$, also produces a (0, 1)-tensor:

$$T^X_i(\theta) \;:=\; T^X_{ijk}(\theta)g^{X,jk}(\theta).$$

In the same manner, the information geometrical quantities, $\overset{\mathrm{e}}{\Gamma}{}^{X}_{ij,l}(\theta)$, $\overset{\mathrm{m}}{\Gamma}{}^{X}_{ij,l}(\theta)$ and $T^Y_{ijk}(\theta)$, are defined for the model, $\{q(y|\theta) : \theta \in \Theta\}$.

Let $M^k_{ij}(\theta)$ be a (1, 2)-tensor defined by:

$$M^k_{ij}(\theta) := \overset{\mathrm{m}}{\Gamma}{}^{Y,k}_{ij}(\theta) - \overset{\mathrm{m}}{\Gamma}{}^{X,k}_{ij}(\theta).$$

For a derivative, $(\partial_1 v(\theta), \cdots, \partial_d v(\theta))$, of the scalar function, $v(\theta)$, the e-covariant derivative is given by:

$$\overset{\mathrm{e}}{\nabla}_i \partial_j v(\theta) \;:=\; \partial_{ij} v(\theta) - \overset{\mathrm{e}}{\Gamma}{}^{X,k}_{ij}(\theta)\partial_k v(\theta).$$

3. Asymptotically Constant-Risk Priors When the Distributions of Data and Target Variables Are Different

In this section, we consider the settings where the trace, $g^{X,ij}(\theta)g^Y_{ij}(\theta)$, has a unique maximum point. We construct the asymptotically constant-risk prior under the Kullback–Leibler risk in the sense that the asymptotic risk up to $\mathrm{O}(N^{-2})$ is constant. We find asymptotically constant-risk priors up to $\mathrm{O}(N^{-2})$ in two steps: first, expand the Kullback–Leibler risks of Bayesian predictive densities; second, find the prior having an asymptotically constant risk using this expansion.

From now on, we assume the following two conditions for the prior, $\pi(\theta; N)$:

(C1) The prior, $\pi(\theta; N)$, has the form:

$$\frac{\pi(\theta; N)}{|g^X(\theta)|^{1/2}} \;\propto\; \exp\{\sqrt{N}\log f(\theta) + \log h(\theta)\},$$

where $f(\theta)$ and $h(\theta)$ are smooth scalar functions of θ independent of N.

(C2) The unique maximum point of the scalar function, $f(\theta)$, is equal to the unique maximum point of the trace, $g^{X,ij}(\theta)g^Y_{ij}(\theta)$.

Based on Conditions (C1) and (C2), we expand the Kullback–Leibler risk of a Bayesian predictive density up to $\mathrm{O}(N^{-2})$.

Theorem 1. *The Kullback–Leibler risk of a Bayesian predictive density based on the prior, $\pi(\theta; N)$, satisfying Condition (C1), is expanded as:*

$$
\begin{aligned}
&R(\theta, \hat{q}_\pi(y|x^{(N)})) \\
&= \frac{1}{2N} g^Y_{ij}(\theta) g^{X,ij}(\theta) + \frac{1}{2N} \tilde{g}^{ij}(\theta)\partial_i \log f(\theta) \partial_j \log f(\theta) - \frac{1}{N\sqrt{N}} T^Y_{ijk}(\theta) g^{X,ij}(\theta) g^{X,kl}(\theta) \partial_l \log f(\theta) \\
&+ \frac{1}{N\sqrt{N}} \tilde{g}^{ij}(\theta) \overset{e}{\nabla}_i \partial_j \log f(\theta) + \frac{1}{N\sqrt{N}} \tilde{g}^{ij}(\theta) g^{X,kl}(\theta) \left\{ \overset{e}{\nabla}_i \partial_k \log f(\theta) \right\} \partial_j \log f(\theta) \partial_l \log f(\theta) \\
&- \frac{1}{3N\sqrt{N}} T^Y_{ijk}(\theta) g^{X,is}(\theta) g^{X,jt}(\theta) g^{X,ku}(\theta) \partial_s \log f(\theta) \partial_t \log f(\theta) \partial_u \log f(\theta) \\
&+ \frac{1}{2N\sqrt{N}} g^Y_{kl}(\theta) M^l_{ij}(\theta) g^{X,is}(\theta) g^{X,jt}(\theta) g^{X,ku}(\theta) \partial_s \log f(\theta) \partial_t \log f(\theta) \partial_u \log f(\theta) \\
&+ \frac{1}{2N\sqrt{N}} g^{X,ij}(\theta) g^Y_{kl}(\theta) g^{X,kl}(\theta) M^m_{ij}(\theta) \partial_m \log f(\theta) + \frac{1}{N\sqrt{N}} \tilde{g}^{ij}(\theta) M^k_{ij}(\theta) \partial_k \log f(\theta) \\
&+ \frac{1}{2N\sqrt{N}} \tilde{g}^{ij}(\theta) T^X_i(\theta) \partial_j \log f(\theta) + \frac{1}{2N\sqrt{N}} g^{X,im}(\theta) g^Y_{ij}(\theta) g^{X,kl}(\theta) M^j_{kl}(\theta) \partial_m \log f(\theta) \\
&+ \frac{1}{N\sqrt{N}} \tilde{g}^{ij}(\theta) \partial_i \log f(\theta) \partial_j \log h(\theta) + \mathrm{O}(N^{-2}).
\end{aligned} \tag{1}
$$

The proof is given in the Appendix. The first term in (1) represents that the precision of the estimation is determined by the geometric quantity of the data, $g^{X,ij}(\theta)$, and the metric of the parameter is determined by the geometric quantity of the target variables, $g^Y_{ij}(\theta)$. Note that each term in (1) is invariant under the reparametrization.

Remark 1. *For the subsequent theorem, it is important that at the point, θ_f, maximizing the scalar function, $\log f(\theta)$, $R(\theta_f, \hat{q}_\pi(y|x^N))$ is given by:*

$$
\begin{aligned}
&R(\theta_f, \hat{q}_\pi(y|x^N)) \\
&= \frac{1}{2N} \sup_{\theta \in \Theta} \{ g^{X,ij}(\theta) g^Y_{ij}(\theta) \} + \frac{1}{N\sqrt{N}} \tilde{g}^{ij}(\theta_f) \partial_{ij} \log f(\theta_f) + \mathrm{O}(N^{-2}).
\end{aligned} \tag{2}
$$

The $N^{-3/2}$-order term of this risk is common whenever we use the same scalar function, $\log f(\theta)$. This term is negative because of the definition of the point, θ_f. Under Condition (C2), θ_f is equal to the unique maximum point, $\theta_{\max}$, of the trace, $g^{X,ij}(\theta) g^Y_{ij}(\theta)$.

Based on (1) and (2), we construct asymptotically constant-risk priors using the solutions of the partial differential equations.

Theorem 2. *Suppose that the scalar functions, $\log \tilde{f}(\theta)$ and $\log \tilde{h}(\theta)$, satisfy the following conditions:*

(A1) *$\log \tilde{f}(\theta)$ is the solution of the Eikonal equation given by:*

$$
\tilde{g}^{ij}(\theta) \partial_i \log \tilde{f}(\theta) \partial_j \log \tilde{f}(\theta) = g^{X,ij}(\theta_{\max}) g^Y_{ij}(\theta_{\max}) - g^{X,ij}(\theta) g^Y_{ij}(\theta), \tag{3}
$$

where $\theta_{\max}$ is the unique maximum point of the scalar function, $g^{X,ij}(\theta) g^Y_{ij}(\theta)$.

(A2) $\log \tilde{h}(\theta)$ *is the solution of the first-order linear partial equation given by:*

$$\begin{aligned}
\mathring{g}^{ij}\partial_i \log \tilde{f}(\theta)\partial_j \log \tilde{h}(\theta) = &-\mathring{g}^{ij}(\theta)\overset{\mathrm{e}}{\nabla}_i\partial_j \log \tilde{f}(\theta) \\
&- \mathring{g}^{ij}(\theta)g^{X,kl}(\theta)\left\{\overset{\mathrm{e}}{\nabla}_i\partial_k \log \tilde{f}(\theta)\right\}\partial_j \log \tilde{f}(\theta)\partial_l \log \tilde{f}(\theta) \\
&+ T^{Y}_{ijk}(\theta)g^{X,ij}(\theta)g^{X,kl}(\theta)\partial_l \log \tilde{f}(\theta) \\
&+ \frac{1}{3}T^{Y}_{ijk}(\theta)g^{X,is}(\theta)g^{X,jt}(\theta)g^{X,ku}(\theta)\partial_s \log \tilde{f}(\theta)\partial_t \log \tilde{f}(\theta)\partial_u \log \tilde{f}(\theta) \\
&- \frac{1}{2}g^{Y}_{kl}(\theta)M^{l}_{ij}(\theta)g^{X,is}(\theta)g^{X,jt}(\theta)g^{X,ku}(\theta)\partial_s \log \tilde{f}(\theta)\partial_t \log \tilde{f}(\theta)\partial_u \log \tilde{f}(\theta) \\
&- \frac{1}{2}g^{X,ij}(\theta)g^{Y}_{kl}(\theta)g^{X,kl}(\theta)M^{m}_{ij}(\theta)\partial_m \log \tilde{f}(\theta) - \mathring{g}^{ij}(\theta)M^{k}_{ij}(\theta)\partial_k \log \tilde{f}(\theta) \\
&- \frac{1}{2}\mathring{g}^{ij}(\theta)T^{X}_{i}(\theta)\partial_j \log \tilde{f}(\theta) - \frac{1}{2}g^{X,im}(\theta)g^{Y}_{ij}(\theta)g^{X,kl}(\theta)M^{j}_{kl}(\theta)\partial_m \log \tilde{f}(\theta) \\
&+ \mathring{g}^{ij}(\theta_{\max})\partial_{ij} \log \tilde{f}(\theta_{\max}). \qquad (4)
\end{aligned}$$

Let $\pi(\theta; N)$ be the prior that is constructed as:

$$\frac{\pi(\theta;N)}{|g^X(\theta)|^{1/2}} \propto \exp\{\sqrt{N}\log \tilde{f}(\theta) + \log \tilde{h}(\theta)\}.$$

Further, suppose that $\log \tilde{f}(\theta)$ satisfies Condition (C2).

Then, the Bayesian predictive density based on the prior, $\pi(\theta; N)$, has the asymptotically smallest constant risk up to $\mathrm{O}(N^{-2})$ among all priors with the form (C1).

Proof. First, we consider the prior, $\phi(\theta; N)$, constructed as:

$$\frac{\phi(\theta;N)}{|g^X(\theta)|^{1/2}} \propto \exp\{\sqrt{N}\log \tilde{f}(\theta)\}.$$

From Theorem 1, the Kullback–Leibler risk, $R(\theta, \hat{q}_\phi(y|x^{(N)}))$, based on the prior, $\phi(\theta; N)$, is given by:

$$R(\theta, \hat{q}_\phi(y|x^{(N)})) = \frac{1}{2N}g^{X,ij}(\theta_{\max})g^{Y}_{ij}(\theta_{\max}) + \mathrm{o}(N^{-1}). \qquad (5)$$

This is constant up to $\mathrm{o}(N^{-1})$.

Suppose that there exists another prior, $\varphi(\theta; N)$, constructed as:

$$\frac{\varphi(\theta;N)}{|g^X(\theta)|^{1/2}} \propto \exp\{\sqrt{N}\log f(\theta)\},$$

and the Bayesian predictive density based on the prior, $\varphi(\theta; N)$, has the asymptotically constant risk:

$$R(\theta, \hat{q}_\varphi(y|x^{(N)})) = \frac{k}{2N} + \mathrm{o}(N^{-1}).$$

From Theorem 1, the prior $\varphi(\theta; N)$ must satisfy the equation:

$$\mathring{g}^{ij}(\theta)\partial_i \log f(\theta)\partial_j \log f(\theta) = k - g^{X,ij}(\theta)g^{Y}_{ij}(\theta).$$

The left-hand side of the above equation is non-negative, because the matrix, $\mathring{g}^{ij}(\theta)$, is positive-definite. Hence, the infimum of the constant, k, is equal to $g^{X,ij}(\theta_{\max})g^{Y}_{ij}(\theta_{\max})$. From (5), the N^{-1}-order term of the risk based on the prior, $\phi(\theta; N)$, achieves the infimum, $g^{X,ij}(\theta_{\max})g^{Y}_{ij}(\theta_{\max})$. Thus, the Bayesian predictive density based on the prior, $\phi(\theta;N)$, has the asymptotically smallest constant risk up to $\mathrm{o}(N^{-1})$.

Second, we consider the prior, $\pi(\theta;N)$, constructed as:

$$\frac{\pi(\theta;N)}{|g^X(\theta)|^{1/2}} \propto \exp\{\sqrt{N}\log\tilde{f}(\theta)+\log\tilde{h}(\theta)\}.$$

The above argument ensures that the prior, $\pi(\theta;N)$, has the asymptotically smallest constant risk up to $\mathrm{o}(N^{-1})$. Thus, we only have to check if the $N^{-3/2}$-order term of the risk is the smallest constant. From (2), the $N^{-3/2}$-order term of the risk at the point, $\theta_{\max}$, is unchanged by the choice of the scalar function, $\log h(\theta)$. In other words, the constant $N^{-3/2}$-order term must agree with the quantity, $\mathring{g}^{ij}(\theta_{\max})\partial_{ij}\log\tilde{f}(\theta_{\max})$. From Theorem 1, if we choose the prior, $\pi(\theta;N)$, the $N^{-3/2}$-order term of the risk is the smallest constant, and it agrees with the quantity, $\mathring{g}^{ij}(\theta_{\max})\partial_{ij}\log\tilde{f}(\theta_{\max})$. Thus, the prior, $\pi(\theta;N)$, has the asymptotically smallest constant risk up to $\mathrm{O}(N^{-2})$. □

Remark 2. *In Theorem 2, we choose* $\log\tilde{f}(\theta)$*, satisfying Condition (C2) among the solutions of (A1). We consider the model with a one-dimensional parameter,* θ*. There are four possibilities to the solutions of (A1):*

$$\sqrt{\mathring{g}^{\theta\theta}(\theta)}\partial_\theta\log\tilde{f}(\theta)=\begin{cases}\pm\sqrt{g^{X,\theta\theta}(\theta_{\max})g^Y_{\theta\theta}(\theta_{\max})-g^{X,\theta\theta}(\theta)g^Y_{\theta\theta}(\theta)} & \text{if} \quad \theta\le\theta_{\max},\\ \pm\sqrt{g^{X,\theta\theta}(\theta_{\max})g^Y_{\theta\theta}(\theta_{\max})-g^{X,\theta\theta}(\theta)g^Y_{\theta\theta}(\theta)} & \text{if} \quad \theta\ge\theta_{\max},\end{cases}$$

where the double-sign corresponds. From the concavity around $\theta_{\max}$ *as suggested by (C2), we choose* $\log\tilde{f}(\theta)$ *as the solution of the following equation:*

$$\sqrt{\mathring{g}^{\theta\theta}(\theta)}\partial_\theta\log\tilde{f}(\theta)=\begin{cases}\sqrt{g^{X,\theta\theta}(\theta_{\max})g^Y_{\theta\theta}(\theta_{\max})-g^{X,\theta\theta}(\theta)g^Y_{\theta\theta}(\theta)} & \text{if} \quad \theta\le\theta_{\max},\\ -\sqrt{g^{X,\theta\theta}(\theta_{\max})g^Y_{\theta\theta}(\theta_{\max})-g^{X,\theta\theta}(\theta)g^Y_{\theta\theta}(\theta)} & \text{if} \quad \theta\ge\theta_{\max}.\end{cases} \tag{6}$$

Integrating both sides of Equation (6), the unique function, $\log\tilde{f}(\theta)$*, is obtained.*

Remark 3. *Compare the Kullback–Leibler risk based on the asymptotically constant-risk prior,* $\pi(\theta;N)$*, with that based on the prior,* $\lambda(\theta)$*, independent of the sample size, N. From Theorem 1 and Theorem 2, the Kullback–Leibler risk based on the asymptotically constant-risk prior,* $\pi(\theta;N)$*, is given as:*

$$\begin{aligned}R(\theta,\hat{q}_\pi(y|x^{(N)})) &= \frac{1}{2N}g^{X,ij}(\theta_{\max})g^Y_{ij}(\theta_{\max})\\ &\quad+\frac{1}{N\sqrt{N}}\mathring{g}^{ij}(\theta_{\max})\partial_{ij}\log\tilde{f}(\theta_{\max})+\mathrm{O}(N^{-2}).\end{aligned} \tag{7}$$

In contrast, the Kullback–Leibler risk based on the prior, $\lambda(\theta)$*, is given as:*

$$R(\theta,\hat{q}_\lambda(y|x^{(N)})) = \frac{1}{2N}g^{X,ij}(\theta)g^Y_{ij}(\theta)+\mathrm{O}(N^{-2}). \tag{8}$$

The N^{-1}*-order term in (8) is under the* N^{-1}*-order term in (7); although the* $N^{-3/2}$*-order term in (8) does not exist, the* $N^{-3/2}$*-order term in (7) is negative. Thus, the maximum of the risk based on the asymptotically constant-risk prior,* $\pi(\theta;N)$*, is smaller than that of the risk based on the prior,* $\lambda(\theta)$*. This result is consistent with the minimaxity of selecting the prior that constructs the predictive density with the smallest maximum of the risk.*

4. Subminimax Estimator Problem Based on the Mean Squared Error

In this section, we refer to the subminimax estimator problem based on the mean squared error, from the viewpoint of the prediction where the distributions of data and target variables are different and have a common unknown parameter. First, we give a brief review of subminimax estimator problem through the binomial example.

Example 1. *Let us consider the binomial estimation based on the mean squared error,* $R_{\mathrm{MSE}}(\theta, \hat{\theta})$*. For any finite sample size, N, the Bayes estimator,* $\hat{\theta}_\pi$*, based on the Beta prior,* $\pi(\theta; N) \propto \theta^{\sqrt{N}/2-1}(1-\theta)^{\sqrt{N}/2-1}$*, is minimax under the mean squared error. The mean squared error of the minimax Bayes estimator,* $\hat{\theta}_\pi$*, is given by:*

$$R_{\mathrm{MSE}}(\theta, \hat{\theta}_\pi) = \frac{N}{4(\sqrt{N}+N)^2} = \frac{1}{4N} - \frac{1}{2N\sqrt{N}} + \mathrm{O}(N^{-2}). \tag{9}$$

In contrast, the mean squared error of the maximum likelihood estimator, $\hat{\theta}_{\mathrm{MLE}}$*, is given by:*

$$R_{\mathrm{MSE}}(\theta, \hat{\theta}_{\mathrm{MLE}}) = \frac{\theta(1-\theta)}{N}.$$

We compare the two estimators, $\hat{\theta}_\pi$ *and* $\hat{\theta}_{\mathrm{MLE}}$*. In the comparison of the* N^{-1}*-order terms of the mean squared errors, it seems that the maximum likelihood estimator,* $\hat{\theta}_{\mathrm{MLE}}$*, dominates the minimax Bayes estimator,* $\hat{\theta}_\pi$*. In other words, the* N^{-1}*-order term of* $R_{\mathrm{MSE}}(\theta, \hat{\theta}_{\mathrm{MLE}})$ *is not greater than that of* $R_{\mathrm{MSE}}(\theta, \hat{\theta}_\pi)$ *for every* $\theta \in \Theta$*, and the equality holds when* $\theta = 1/2$*. This seeming paradox is known as the subminimax estimator problem; see [14,17,18] for details. See also [15] for the conditions that such problems do not occur in estimation.*

However, this paradox does not mean the inferiority of the minimax Bayes estimator. This is because, although the mean squared error of the minimax Bayes estimator, $\hat{\theta}_\pi$*, has the negative* $N^{-3/2}$*-order term, the mean squared error of the maximum likelihood estimator,* $\hat{\theta}_{\mathrm{MLE}}$*, does not have the* $N^{-3/2}$*-order term. Hence, in comparison to the mean squared errors up to* $\mathrm{O}(N^{-2})$*, the maximum of the mean squared error,* $R_{\mathrm{MSE}}(\theta, \hat{\theta}_\pi)$*, is below the maximum of the mean squared error,* $R_{\mathrm{MSE}}(\theta, \hat{\theta}_{\mathrm{MLE}})$*.*

Next, we construct the asymptotically constant-risk prior in the estimation based on the mean squared error when the subminimax estimator problem occurs, from the viewpoint of the prediction. We consider the priors, $\pi(\theta; N)$, satisfying (C1). From Lemma A3 in the Appendix, the mean squared error of the Bayes estimator, $\hat{\theta}_\pi$, is equal to the Kullback–Leibler risk of the $\hat{\theta}_\pi$-plugin predictive density, $q(y|\hat{\theta}_\pi)$, by assuming that the target variable, y, is a d-dimensional Gaussian random variable with the mean vector, θ, and unit variance. Note that $g^Y_{ij}(\theta) = 1$, $\overset{\mathrm{m}}{\Gamma}{}^Y_{ij,k} = 0$ and $\overset{\mathrm{e}}{\Gamma}{}^Y_{ij,k} = 0$ for $i, j, k = 1, \cdots, d$. Thus, if $g^Y_{ij}(\theta) g^{X,ij}(\theta) = \Sigma_{i=1}^d g^{X,ii}(\theta)$ has a unique maximum point, we obtain the asymptotically constant-risk prior, $\pi(\theta; N)$, up to $\mathrm{O}(N^{-2})$ from Lemma A2 in the Appendix and Theorem 2.

Finally, we compare the mean squared error of the asymptotically constant-risk Bayes estimator, $\hat{\theta}_\pi$, with that of the maximum likelihood estimator, $\hat{\theta}_{\mathrm{MLE}}$. The mean squared error of the asymptotically constant-risk Bayes estimator, $\hat{\theta}_\pi$, is given as:

$$R_{\mathrm{MSE}}(\theta, \hat{\theta}_\pi) = \frac{1}{N}\sum_{i=1}^{d} g^{X,ii}(\theta_{\max}) + \frac{2}{N\sqrt{N}} \Sigma_{k=1}^d g^{X,ik}(\theta_{\max}) g^{X,jk}(\theta_{\max}) \partial_{ij} \log \tilde{f}(\theta_{\max}) + \mathrm{O}(N^{-2}).$$

In contrast, the mean squared error of the maximum likelihood estimator, $\hat{\theta}_{\mathrm{MLE}}$, is given as:

$$R_{\mathrm{MSE}}(\theta, \hat{\theta}_{\mathrm{MLE}}) = \frac{1}{N} \Sigma_{i=1}^d g^{X,ii}(\theta) + \mathrm{O}(N^{-2}).$$

See [16,19].

Thus, the maximum of the mean squared error of the asymptotically constant-risk Bayes estimator is smaller than that of estimators by the improvement of order $N^{-3/2}$ in proportion to the Hessian of the scalar function, $\log \tilde{f}(\theta)$, at $\theta_{\max}$. In the prediction where the trace, $g^{X,ij}(\theta) g^Y_{ij}(\theta)$, has a unique maximum point, the same improvement holds (Remark 3).

Example 2. *Using the above results, we consider the binomial estimation based on the mean squared error from the viewpoint of the prediction. The geometrical quantities to be used are given by:*

$$g^X_{\theta\theta}(\theta) = \frac{1}{\theta(1-\theta)}, \qquad g^Y_{\theta\theta}(\theta) = 1,$$
$$\overset{\mathrm{m}}{\Gamma}{}^{X}_{\theta\theta,\theta}(\theta) = 0, \qquad \overset{\mathrm{m}}{\Gamma}{}^{X}_{\theta\theta,\theta}(\theta) = 0,$$
$$\overset{\mathrm{e}}{\Gamma}{}^{X}_{\theta\theta,\theta}(\theta)(\theta) = -\frac{1-2\theta}{\theta^2(1-\theta)^2}, \qquad \overset{\mathrm{e}}{\Gamma}{}^{Y}_{\theta\theta,\theta}(\theta) = 0,$$
$$T^X_{\theta\theta\theta}(\theta) = \frac{1-2\theta}{\theta^2(1-\theta)^2}, \qquad \text{and} \quad T^Y_{\theta\theta\theta}(\theta) = 0,$$

respectively. Since $\overset{\mathrm{m}}{\blacksquare}{}^{X,\theta}_{\theta\theta}$, $\overset{\mathrm{m}}{\blacksquare}{}^{Y,\theta}_{\theta\theta}$ *and* $T^Y_{\theta\theta\theta}$ *vanish, the asymptotically constant-risk prior in the estimation is identical to the asymptotically constant-risk prior in the prediction; compare Theorem 1 with the expansion of* $g^{Y,ij}(\theta)\mathrm{E}_{X(N)}[(\hat{\theta}^i_\pi - \theta^i)(\hat{\theta}^j_\pi - \theta^j)]$ *in Lemma A2 in the Appendix.*

In this example, Equation (3) is given by:

$$\theta^2(1-\theta)^2\{\partial_\theta \log \tilde{f}(\theta)\}^2 = \sqrt{\frac{1}{4} - \theta(1-\theta)},$$

and the solution, $\log \tilde{f}(\theta)$, *is* $(1/2)\log\{\theta(1-\theta)\}$. *Here, the second-order derivative of the function,* $\log \tilde{f}(\theta)$, *is given by:*

$$\partial_{\theta\theta} \log \tilde{f}(\theta) = -\frac{1-2\theta+2\theta^2}{2\theta^2(1-\theta)^2}.$$

From this, Equation (4) is given by:

$$\frac{1}{2}\theta(1-\theta)(1-2\theta)\partial_\theta \log \tilde{h}(\theta) + \theta^2 - \theta = -\frac{1}{4},$$

and the solution, $\log \tilde{h}(\theta)$, *is* $(1/2)\log\{\theta(1-\theta)\}$. *Hence, the asymptotically constant-risk prior,* $\pi(\theta; N)$, *is a Beta prior with the parameters,* $\alpha = \sqrt{N}/2$ *and* $\beta = \sqrt{N}/2$. *Note that the asymptotically constant-risk prior coincides with the exact minimax prior. Since* $g^{X,\theta\theta}(\theta_{\max}) = 1/2$ *and* $g^{X,\theta\theta}(\theta_{\max})\partial_{\theta\theta}\log \tilde{f}(\theta_{\max}) = -1$, *the mean squared error of the asymptotically constant-risk Bayes estimator,* $\hat{\theta}_\pi$, *agrees with (9) up to* $\mathrm{O}(N^{-2})$.

5. Application to the Prediction of the Binary Regression Model under the Covariate Shift

In this section, we construct asymptotically constant-risk priors in the prediction based on the binary regression model under the covariate shift; see [10].

We consider that we predict a binary response variable, y, based on the binary response variables, $x^{(N)}$. We assume that the target variable, y, and the data, $x^{(N)}$, follow the logistic regression models with the same parameter, β, given by:

$$\log \frac{\Pi_x}{1-\Pi_x} = \alpha + z\beta$$

and:

$$\log \frac{\Pi_y}{1-\Pi_y} = \tilde{\alpha} + \tilde{z}\beta,$$

where Π_x is the success probability of the data and Π_y is the success probability of the target variable. Let α and $\tilde{\alpha}$ denote known constant terms, and let β denote the common unknown parameter. Further, we assume that the covariates, z and $\tilde{z}$, are different.

Using the parameter $\theta = \Pi_x$, we convert this predictive setting to binomial prediction where the data, x, and the target variable, y, are distributed according to:

$$p(x|\theta) := \begin{cases} \theta & \text{if } x = 1, \\ 1-\theta & \text{if } x = 0, \end{cases}$$

and:

$$q(y|\theta) := \begin{cases} \mathrm{e}^{\tilde{\alpha}-\tilde{z}z^{-1}\alpha}\theta^{\tilde{z}z^{-1}} / \left\{(1-\theta)^{\tilde{z}z^{-1}} + \mathrm{e}^{\tilde{\alpha}-\tilde{z}z^{-1}\alpha}\theta^{\tilde{z}z^{-1}}\right\} & \text{if } y = 1, \\ (1-\theta)^{\tilde{z}z^{-1}} / \left\{(1-\theta)^{\tilde{z}z^{-1}} + \mathrm{e}^{\tilde{\alpha}-\tilde{z}z^{-1}\alpha}\theta^{\tilde{z}z^{-1}}\right\} & \text{if } y = 0, \end{cases}$$

respectively. We obtain two Fisher information for x and y as:

$$g^X_{\theta\theta}(\theta) = \frac{1}{\theta(1-\theta)}$$

and:

$$g^Y_{\theta\theta}(\theta) = \left(\frac{\tilde{z}}{z}\right)^2 \mathrm{e}^{-\tilde{\alpha}+\tilde{z}z^{-1}\alpha} \frac{(1-\theta)^{\tilde{z}z^{-1}-2}\theta^{\tilde{z}z^{-1}-2}}{\left\{\theta^{\tilde{z}z^{-1}} + \mathrm{e}^{-\tilde{\alpha}+\tilde{z}z^{-1}\alpha}(1-\theta)^{\tilde{z}z^{-1}}\right\}^2},$$

respectively.

For simplicity, we consider the setting where $z = 1$, $\tilde{z} = 2$ and $\alpha = \tilde{\alpha} = 0$. The geometrical quantities for the model, $\{p(x|\theta) : \theta \in \Theta\}$, are given by:

$$g^X_{\theta\theta}(\theta) = \frac{1}{\theta(1-\theta)}, \qquad \overset{\mathrm{m}}{\Gamma}{}^X_{\theta\theta,\theta}(\theta) = 0,$$

$$\overset{\mathrm{e}}{\Gamma}{}^X_{\theta\theta,\theta}(\theta)\,(\theta) = -\frac{1-2\theta}{\theta^2(1-\theta)^2}, \qquad \text{and} \quad T^X_{\theta\theta\theta}(\theta) = \frac{1-2\theta}{\theta^2(1-\theta)^2},$$

respectively. In the same manner, the geometrical quantities for the model, $\{q(y|\theta) : \theta \in \Theta\}$, are given by:

$$g^Y_{\theta\theta}(\theta) = \frac{4}{\{(1-\theta)^2+\theta^2\}^2}, \qquad \overset{\mathrm{m}}{\Gamma}{}^X_{\theta\theta,\theta}(\theta) = 4\frac{(1-2\theta)(1+2\theta-2\theta^2)}{\theta(1-\theta)\{(1-\theta)^2+\theta^2\}^3},$$

$$\overset{\mathrm{e}}{\Gamma}{}^Y_{\theta\theta,\theta}(\theta) = -4\frac{1-2\theta}{\theta(1-\theta)\{(1-\theta)^2+\theta^2\}^2}, \qquad \text{and} \quad T^Y_{\theta\theta\theta}(\theta) = 8\frac{1-2\theta}{\theta(1-\theta)\{(1-\theta)^2+\theta^2\}^3},$$

respectively.

Using these quantities, Equation (3) is given by:

$$4\frac{\theta^2(1-\theta)^2}{\{\theta^2+(1-\theta)^2\}^2}(\partial_\theta \log \tilde{f}(\theta))^2 = 4 - 4\frac{\theta(1-\theta)}{\{\theta^2+(1-\theta)^2\}^2}.$$

By noting that the maximum point of $g^{X,\theta\theta}(\theta)g^{Y}_{\theta\theta}(\theta)$ is $1/2$, the solution, $\log \tilde{f}(\theta)$, of this equation is given by:

$$\begin{aligned}\log \tilde{f}(\theta) &= 2\sqrt{1-\theta+\theta^2} + \log\{\theta(1-\theta)\} \\ &\quad - \log(2-\theta+2\sqrt{1-\theta+\theta^2}) - \log(1+\theta+2\sqrt{1-\theta+\theta^2}).\end{aligned}$$

Using this solution, we obtain the solution of Equation (4) given by:

$$\begin{aligned}\log \tilde{h}(\theta) &= \frac{1}{6}\left[-\frac{1}{1-\theta} - \frac{1}{\theta} - 12\theta(1-\theta) - 12\sqrt{3}\sqrt{1-\theta+\theta^2}\right. \\ &\quad +(3-6\sqrt{3})\{\log\theta + \log(1-\theta)\} - 3\log(1-\theta+\theta^2) + 10\log\{(1-\theta)^2+\theta^2\} \\ &\quad -6\log(\sqrt{3}+2\sqrt{1-\theta+\theta^2}) + 6\sqrt{3}\log\{1+(1-\theta)+2\sqrt{1-\theta+\theta^2}\} \\ &\quad \left. +6\sqrt{3}\log\{1+\theta+2\sqrt{1-\theta+\theta^2}\}\right].\end{aligned}$$

The asymptotically constant-risk priors for the different sample sizes are shown in Figure 1. The prior weight is found to be more concentrated to $1/2$ as the sample size, N, grows.

In this example, we obtain the Kullback–Leibler risk of the Bayesian predictive density based on the asymptotically constant-risk prior, $\pi(\theta; N)$, as:

$$R(\theta, \hat{q}_\pi(y|x^{(N)})) = \frac{2}{N} - \frac{4\sqrt{3}}{N\sqrt{N}} + O(N^{-2}).$$

We compare this value with the Bayes risk calculated using the Monte Carlo simulation; see Figure 2. As the sample size, N, grows, the difference appears negligible. Further, we compare this value with the risk itself calculated by the Monte Carlo simulation; see Figure 3. As the sample size, N, grows, the risk becomes more constant.

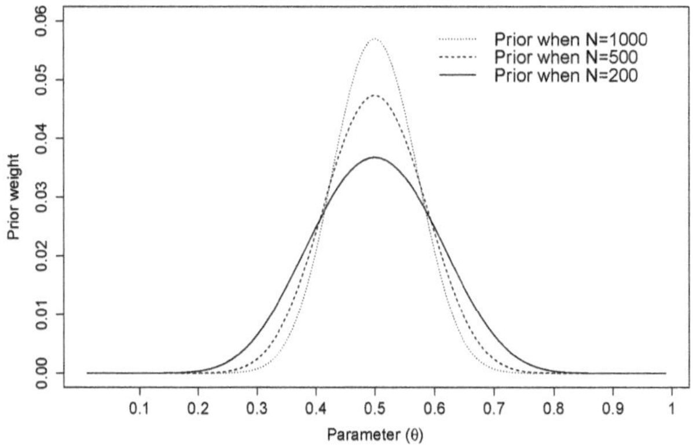

Figure 1. Asymptotically constant-risk prior in the prediction where the data are distributed according to the binomial distribution, $\mathrm{Bin}(N, \theta)$, and the target variable is distributed according to the binomial distribution, $\mathrm{Bin}(1, \theta^2/(\theta^2+(1-\theta)^2))$.

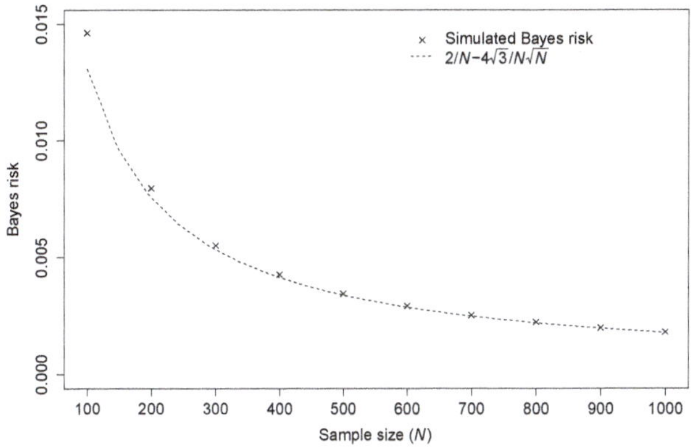

Figure 2. Bayes risk based on the asymptotically constant-risk prior in the prediction where the data are distributed according to the binomial distribution, $\mathrm{Bin}(N, \theta)$, and the target variable is distributed according to the binomial distribution, $\mathrm{Bin}(1, \theta^2/(\theta^2 + (1-\theta)^2))$.

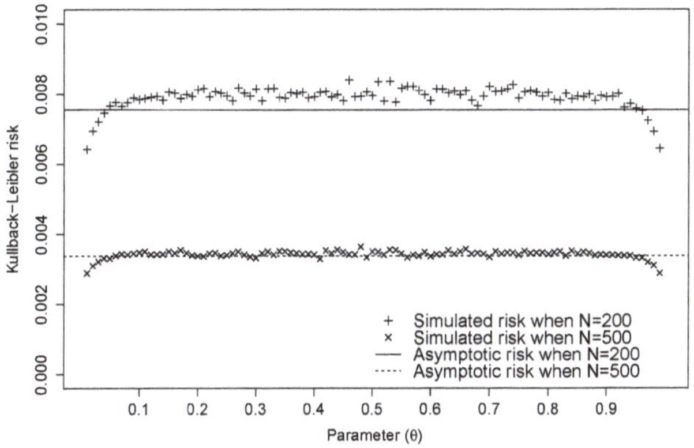

Figure 3. Comparison of the Kullback–Leibler risk calculated using the Monte Carlo simulations and the asymptotic risk, $2/N - (4\sqrt{3})/(N\sqrt{N})$, in the prediction where the data are distributed according to the binomial distribution, $\mathrm{Bin}(N, \theta)$, and the target variable is distributed according to the binomial distribution, $\mathrm{Bin}(1, \theta^2/(\theta^2 + (1-\theta)^2))$.

6. Discussion and Conclusions

We have considered the setting where the quantity, $g^{X,ij}(\theta)g^{Y}_{ij}(\theta)$—the trace of the product of the inverse Fisher information matrix, $g^{X,ij}(\theta)$, and the Fisher information matrix, $g^{Y}_{ij}(\theta)$—has a unique maximum point, and we have investigated the asymptotically constant-risk prior in the sense that the asymptotic risk is constant up to $\mathrm{O}(N^{-2})$.

In Section 3, we have considered the prior depending on the sample size, N, and constructed the asymptotically constant-risk prior using Equations (3) and (4). In Section 4, we have clarified the relationship between the subminimax estimator problem based on the mean squared error and the prediction where the distributions of data and target variables are different. In Section 5, we have constructed the asymptotically constant-risk prior in the prediction based on the logistic regression model under the covariate shift.

We have assumed that the trace, $g^{X,ij}(\theta)g^{Y}_{ij}(\theta)$, is finite. However, the trace may diverge in the non-compact parameter space; for example, it diverges under the predictive setting, where the distribution, $q(y|\theta)$, of the target variable is the Poisson distribution and the data distribution, $p(x|\theta)$, is the exponential distribution, with Θ equivalent to $\mathbb{R}$. Therefore, for our future work, in such a setting, we should adopt criteria other than minimaxity.

Acknowledgments: The authors thank the referees for their helpful comments. This research was partially supported by a Grant-in-Aid for Scientific Research (23650144, 26280005).

Author Contributions: Both authors contributed to the research and writing of this paper. Both authors read and approved the final manuscript.

Conflicts of Interest: The authors declare no conflict of interest.

Appendix A

We prove Theorem 1. First, we introduce some lemmas for the proof. For the expansion, we follow the following six steps (the first five steps are arranged in the form of lemmas): the first is to expand the MAPestimator; the second is to calculate their bias and mean squared error; the third is to expand the Kullback–Leibler risk using $\hat{\theta}_\pi$-plugin predictive density, $q(y|\hat{\theta}_\pi)$; the fourth is to expand the Bayesian predictive density based on the prior $\pi(\theta;N)$; the fifth is to expand the Bayesian estimator minimizing the Bayes risk; and the last is to prove Theorem 1 using these lemmas.

We use some additional notations for the expansion. Let $\hat{\theta}_\pi$ be the maximum point of the scalar function $\log p(x^{(N)}|\theta) + \log\{\pi(\theta;N)/|g^X(\theta)|^{1/2}\}$. Let $l(\theta|x^{(N)})$ denote the log likelihood of the data, $x^{(N)}$. Let $l_{ij}(\theta|x^{(N)})$, $l_{ijk}(\theta|x^{(N)})$ and $l_{ijkl}(\theta|x^{(N)})$ be the derivatives of order 2, 3 and 4 of the log likelihood, $l(\theta|x^{(N)})$. Let $H_{ij}(\theta|x^{(N)})$ denote the quantity, $l_{ij}(\theta|x^{(N)}) + Ng^X_{ij}(\theta)$. Let $\tilde{l}_i(\theta|x^{(N)})$ and $\tilde{H}_{ij}(\theta|x^{(N)})$ denote $(1/\sqrt{N})l_i(\theta|x^{(N)})$ and $(1/\sqrt{N})H_{ij}(\theta|x^{(N)})$, respectively. In addition, the brackets ( ) denotes the symmetrization: for any two tensors, a_{ij} and b_{ij}, $a_{i(j}b_{k)l}$ denotes $a_{i(j}b_{k)l} = (a_{ij}b_{kl} + a_{ik}b_{jl})/2$.

Lemma A1. *Let $\hat{\theta}_\pi$ be the maximum point of $\log p(x^{(N)}|\theta) + \log\{\pi(\theta;N)/|g^X(\theta)|^{1/2}\}$. Then, the i-th component of this estimator $\hat{\theta}_\pi$ is expanded as follows:*

$$\begin{aligned}
\hat{\theta}^i_\pi &= \theta^i + \frac{1}{\sqrt{N}}g^{X,ik}(\theta)\tilde{l}_k(\theta|x^{(N)}) + \frac{1}{\sqrt{N}}g^{X,ik}(\theta)\partial_k \log f(\theta) \\
&+\frac{1}{N}g^{X,ik}(\theta)\tilde{H}_{km}(\theta|x^{(N)})g^{X,mr}(\theta)\tilde{l}_r(\theta|x^N) \\
&+\frac{1}{2N}g^{X,ik}(\theta)L^X_{kmr}(\theta)g^{X,mq}(\theta)g^{X,rs}(\theta)\tilde{l}_q(\theta|x^N)\tilde{l}_s(\theta|x^{(N)}) \\
&+\frac{1}{N}g^{X,ik}(\theta)\tilde{H}_{km}(\theta|x^N)g^{X,mr}(\theta)\partial_r \log f(\theta) \\
&+\frac{1}{N}g^{X,ik}(\theta)L^X_{kmr}(\theta)g^{X,mq}(\theta)g^{X,rs}(\theta)\tilde{l}_q(\theta|x^N)\partial_s \log f(\theta) \\
&+\frac{1}{2N}g^{X,ik}(\theta)L^X_{kmr}(\theta)g^{X,mq}(\theta)g^{X,rs}(\theta)\partial_q \log f(\theta)\partial_s \log f(\theta) \\
&+\frac{1}{N}g^{X,ik}(\theta)g^{X,mq}(\theta)\partial_{km} \log f(\theta)\tilde{l}_q(\theta|x^{(N)}) \\
&+\frac{1}{N}g^{X,ik}(\theta)g^{X,mq}(\theta)\partial_{km} \log f(\theta)\partial_q \log f(\theta) \\
&+\frac{1}{N}g^{X,ik}(\theta)\partial_k \log h(\theta) + O_P(N^{-3/2}).
\end{aligned} \tag{A1}$$

Proof. By the definition of $\hat{\theta}_\pi$, we get the equation given by:

$$\partial_i \log p(x^{(N)}|\hat{\theta}_\pi) + \partial_i \log \frac{\pi(\hat{\theta}_\pi;N)}{|g^X(\hat{\theta}_\pi)|^{1/2}} = 0.$$

From our assumption that prior $\pi(\theta; N)$ has the form given by:

$$\frac{\pi(\theta;N)}{|g^X(\theta)|^{1/2}} \quad \propto \quad \exp\{\sqrt{N}\log f(\theta) + \log h(\theta)\},$$

we rewrite this equation as:

$$\partial_i \log p(x^{(N)}|\hat{\theta}_\pi) + \sqrt{N}\partial_i \log f(\hat{\theta}_\pi) + \partial_i \log h(\hat{\theta}_\pi) \quad = \quad 0.$$

By applying Taylor expansion around θ to this new equation, we derive the following expansion:

$$\begin{aligned}
&\partial_i \log p(x^{(N)}|\theta) + \{\partial_{ij} \log p(x^{(N)}|\theta)\}(\hat{\theta}^j_\pi - \theta^j) \\
&+\frac{1}{2}\{\partial_{ijk} \log p(x^{(N)}|\theta)\}(\hat{\theta}^j_\pi - \theta^j)(\hat{\theta}^k_\pi - \theta^k) + \sqrt{N}\partial_i \log f(\theta) \\
&+\sqrt{N}\{\partial_{ij} \log f(\theta)\}(\hat{\theta}^j_\pi - \theta^j) + \partial_i \log h(\theta) + o_P(1) = 0.
\end{aligned}$$

From the law of large numbers and the central limit theorem, we rewrite the above expansion as:

$$\begin{aligned}
N g^X_{ij}(\theta)(\hat{\theta}^j_\pi - \theta^j) \quad = \quad & \partial_i \log p(x^{(N)}|\theta) + \sqrt{N}\partial_i \log f(\theta) + H_{ij}(\theta|x^{(N)})(\hat{\theta}^j_\pi - \theta^j) \\
&+\frac{N}{2}L_{ijk}(\theta)(\hat{\theta}^j_\pi - \theta^j)(\hat{\theta}^k_\pi - \theta^k) + \sqrt{N}\partial_{ij} \log f(\theta)(\hat{\theta}^j_\pi - \theta^j) \\
&+\partial_i \log h(\theta) + o_P(1). \qquad \text{(A2)}
\end{aligned}$$

By substituting the deviation, $\hat{\theta}_\pi - \theta$, recursively into Expansion (A2), we obtain Expansion (A1). □

Lemma A2. *Let $\hat{\theta}_\pi$ be the maximum point of $\log p(x^{(N)}|\theta) + \log\{\pi(\theta;N)/|g^X(\theta)|^{1/2}\}$. Then, the i-th component of the bias of the estimator, $\hat{\theta}_\pi$, is given by:*

$$\begin{aligned}
\mathrm{E}_{X^{(N)}}[\hat{\theta}^i_\pi] \quad = \quad & \theta^i + \frac{1}{\sqrt{N}} g^{X,ik}\partial_k \log f(\theta) \\
&-\frac{1}{2N}\overset{\mathrm{m}}{\Gamma}{}^{X,i}(\theta) + \frac{1}{2N} g^{X,ik}(\theta) g^{X,mq}(\theta) g^{X,rs}(\theta) L^X_{kmr}(\theta)\partial_q \log f(\theta)\partial_s \log f(\theta) \\
&+\frac{1}{N} g^{X,ik}(\theta) g^{X,mq}(\theta)\partial_{km} \log f(\theta)\partial_q \log f(\theta) \\
&+\frac{1}{N} g^{X,ik}(\theta)\partial_k \log h(\theta) + \mathrm{O}(N^{-3/2}). \qquad \text{(A3)}
\end{aligned}$$

The (i,j)*-component of the mean squared error of* $\hat{\theta}_\pi$ *is given by:*

$$
\begin{aligned}
&\mathrm{E}_{X^{(N)}}[(\hat{\theta}^i_\pi - \theta^i)(\hat{\theta}^j_\pi - \theta^j)] \\
&= \frac{1}{N} g^{X,ij}(\theta) + \frac{1}{N} g^{X,ik}(\theta) g^{X,jl}(\theta) \partial_k \log f(\theta) \partial_l \log f(\theta) \\
&\quad - \frac{1}{N\sqrt{N}} g^{X,k(i}(\theta) \overset{\mathrm{m}}{\Gamma}{}^{X,j)}(\theta) \partial_k \log f(\theta) + \frac{2}{N\sqrt{N}} g^{X,k(i}(\theta) g^{X,j)l}(\theta) \partial_{kl} \log f(\theta) \\
&\quad + \frac{2}{N\sqrt{N}} g^{X,k(i}(\theta) \partial_k g^{X,j)l}(\theta) \partial_l \log f(\theta) \\
&\quad + \frac{1}{N\sqrt{N}} g^{X,k(i}(\theta) g^{X,j)l}(\theta) g^{X,nr}(\theta) g^{X,pt}(\theta) L^X_{lrt}(\theta) \partial_k \log f(\theta) \partial_n \log f(\theta) \partial_p \log f(\theta) \\
&\quad + \frac{2}{N\sqrt{N}} g^{X,k(i}(\theta) g^{X,j)l}(\theta) g^{X,nr}(\theta) \partial_{ln} \log f(\theta) \partial_r \log f(\theta) \partial_k \log f(\theta) \\
&\quad + \frac{2}{N\sqrt{N}} g^{X,k(i}(\theta) g^{X,j)l}(\theta) \partial_k \log f(\theta) \partial_l \log h(\theta) \\
&\quad + \mathrm{O}(N^{-2}),
\end{aligned} \tag{A4}
$$

where $g^{X,k(i}(\theta)\overset{\mathrm{m}}{\Gamma}{}^{X,j)}(\theta)$ *denotes* $(1/2)\{g^{X,ki}(\theta)\overset{\mathrm{m}}{\Gamma}{}^{X,j}(\theta) + g^{X,ki}(\theta)\overset{\mathrm{m}}{\Gamma}{}^{X,j}(\theta)\}$ *and* $g^{X,k(i}(\theta)\partial_k g^{X,j)l}(\theta)$ *denotes* $(1/2)\{g^{X,ki}(\theta)\partial_k g^{X,jl}(\theta) + g^{X,kj}(\theta)\partial_k g^{X,il}(\theta)\}$*. The* (i,j,k)*-component of the mean of the third power of the deviation,* $\hat{\theta}_\pi - \theta$*, is given by:*

$$
\begin{aligned}
&\mathrm{E}_{X^{(N)}}[(\hat{\theta}^i_\pi - \theta^i)(\hat{\theta}^j_\pi - \theta^j)(\hat{\theta}^k_\pi - \theta^k)] \\
&= \frac{1}{N\sqrt{N}} g^{X,is}(\theta) g^{X,jt}(\theta) g^{X,ku}(\theta) \partial_s \log f(\theta) \partial_t \log f(\theta) \partial_u \log f(\theta) \\
&\quad + \frac{3}{N\sqrt{N}} g^{X,(ij}(\theta) g^{X,k)l}(\theta) \partial_l \log f(\theta) + \mathrm{O}(N^{-2}).
\end{aligned} \tag{A5}
$$

Proof. First, using Lemma A1, we determine the i-th component of the bias of $\hat{\theta}_\pi$ given by:

$$
\begin{aligned}
&\mathrm{E}_{X^{(N)}}[\hat{\theta}^i_\pi - \theta^i] \\
&= \frac{1}{\sqrt{N}} g^{X,ik} \partial_k \log f(\theta) \\
&\quad - \frac{1}{2N} \overset{\mathrm{m}}{\Gamma}{}^{X,i}(\theta) + \frac{1}{2N} g^{X,ik}(\theta) g^{X,mq}(\theta) g^{X,rs}(\theta) L^X_{kmr}(\theta) \partial_q \log f(\theta) \partial_s \log f(\theta) \\
&\quad + \frac{1}{N} g^{X,ik}(\theta) g^{X,mq}(\theta) \partial_{km} \log f(\theta) \partial_q \log f(\theta) + \frac{1}{N} g^{X,ik}(\theta) \partial_k \log h(\theta) + \mathrm{O}(N^{-3/2}).
\end{aligned}
$$

Second, consider the following relationship:

$$
\begin{aligned}
&\mathrm{E}_{X^{(N)}}\left[\left\{\hat{\theta}^i_\pi - \theta^i - \frac{1}{\sqrt{N}} g^{X,ik}(\theta) \tilde{l}_k(\theta|x^{(N)}) - \frac{1}{\sqrt{N}} g^{X,ik}(\theta) \partial_k \log f(\theta)\right\}\right. \\
&\qquad \left. \times \left\{\hat{\theta}^j_\pi - \theta^j - \frac{1}{\sqrt{N}} g^{X,jl}(\theta) \tilde{l}_l(\theta|x^N) - \frac{1}{\sqrt{N}} g^{X,jl}(\theta) \partial_l \log f(\theta)\right\}\right] \\
&= \mathrm{E}_{X^{(N)}}[(\hat{\theta}^i_\pi - \theta^i)(\hat{\theta}^j_\pi - \theta^j)] + \frac{1}{N} g^{X,ij}(\theta) + \frac{1}{N} g^{X,ik}(\theta) g^{X,jl}(\theta) \partial_k \log f(\theta) \partial_l \log f(\theta) \\
&\quad - \frac{1}{\sqrt{N}} g^{X,ki}(\theta) \mathrm{E}_{X^{(N)}}[(\hat{\theta}^j_\pi - \theta^j) \tilde{l}_k(\theta|x^{(N)})] - \frac{1}{\sqrt{N}} g^{X,kj}(\theta) \mathrm{E}_{X^{(N)}}[(\hat{\theta}^i_\pi - \theta^i) \tilde{l}_k(\theta|x^{(N)})] \\
&\quad - \frac{1}{\sqrt{N}} g^{X,ki}(\theta) \mathrm{E}_{X^{(N)}}[(\hat{\theta}^j_\pi - \theta^j) \partial_k \log f(\theta)] - \frac{1}{\sqrt{N}} g^{X,kj}(\theta) \mathrm{E}_{X^{(N)}}[(\hat{\theta}^i_\pi - \theta^i) \partial_k \log f(\theta)].
\end{aligned} \tag{A6}
$$

By differentiating the j-th component of the bias, $\mathrm{E}_{X(N)}[\hat{\theta}^j_\pi - \theta^j]$, we obtain the equation given by:

$$\frac{1}{N}\partial_k \mathrm{E}_{X(N)}[\hat{\theta}^j_\pi - \theta^j] \quad = \quad -\frac{1}{N}\delta^j_k + \frac{1}{\sqrt{N}}\mathrm{E}_{X(N)}[(\hat{\theta}^j_\pi - \theta^j)\tilde{l}_k(\theta|x^N)], \tag{A7}$$

where δ^i_j denotes the delta function: if the upper and the lower indices agree, then the value of this function is one and otherwise zero. Equation (A7) has been used by [2,16,19]. By substituting Equations (A7) and (A3) into Relationship (A6), we obtain the (i,j)-component of the mean squared error of $\hat{\theta}_\pi$ given by:

$$\begin{aligned}
&\mathrm{E}_{X(N)}[(\hat{\theta}^i_\pi - \theta^i)(\hat{\theta}^j_\pi - \theta^j)] \\
&= \frac{1}{N}g^{X,ij}(\theta) + \frac{1}{N}g^{X,ik}(\theta)g^{X,jl}(\theta)\partial_k \log f(\theta)\partial_l \log f(\theta) \\
&\quad - \frac{1}{N\sqrt{N}}g^{X,k(i}(\theta)\overset{\mathrm{m}}{\Gamma}{}^{X,j)}(\theta)\partial_k \log f(\theta) + \frac{2}{N\sqrt{N}}g^{X,k(i}(\theta)g^{X,j)l}(\theta)\partial_{kl}\log f(\theta) \\
&\quad + \frac{2}{N\sqrt{N}}g^{X,k(i}(\theta)\partial_k g^{X,j)l}(\theta)\partial_l \log f(\theta) \\
&\quad + \frac{1}{N\sqrt{N}}g^{X,k(i}(\theta)g^{X,j)l}(\theta)g^{X,nr}(\theta)g^{X,pt}(\theta)L^X_{lrt}(\theta)\partial_k \log f(\theta)\partial_n \log f(\theta)\partial_p \log f(\theta) \\
&\quad + \frac{2}{N\sqrt{N}}g^{X,k(i}(\theta)g^{X,j)l}(\theta)g^{X,nr}(\theta)\partial_{ln}\log f(\theta)\partial_r \log f(\theta)\partial_k \log f(\theta) \\
&\quad + \frac{2}{N\sqrt{N}}g^{X,k(i}(\theta)g^{X,j)l}(\theta)\partial_k \log f(\theta)\partial_l \log h(\theta) + \mathrm{O}(N^{-2}).
\end{aligned}$$

Finally, by taking the expectation of the third power of the deviation, $\hat{\theta}^i_\pi - \theta^i$, we obtain the following expansion:

$$\begin{aligned}
&\mathrm{E}_{X(N)}[(\hat{\theta}^i_\pi - \theta^i)(\hat{\theta}^j_\pi - \theta^j)(\hat{\theta}^k_\pi - \theta^k)] \\
&= \frac{1}{N\sqrt{N}}g^{X,is}(\theta)g^{X,jt}(\theta)g^{X,ku}(\theta)\partial_s \log f(\theta)\partial_t \log f(\theta)\partial_u \log f(\theta) \\
&\quad + \frac{3}{N\sqrt{N}}g^{X,(ij}(\theta)g^{X,k)l}(\theta)\partial_l \log f(\theta) + \mathrm{O}(N^{-2}).
\end{aligned}$$

□

Lemma A3. *Let* $\hat{\theta}_\pi$ *be the maximum point of* $\log p(x^{(N)}|\theta) + \log\{\pi(\theta;N)/|g^X(\theta)|^{1/2}\}$. *The Kullback–Leibler risk of the plug-in predictive density,* $q(y^{(N)}|\hat{\theta}_\pi)$, *with the estimator,* $\hat{\theta}_\pi$, *is expanded as follows:*

$$
\begin{aligned}
&R(\theta, q(y|\hat{\theta}_\pi)) \\
&= \frac{1}{2N} g^Y_{ij}(\theta) g^{X,ij}(\theta) + \frac{1}{2N} g^{ij}(\theta)\partial_i \log f(\theta)\partial_j \log f(\theta) + \frac{1}{N\sqrt{N}} g^{ij}(\theta)\left\{ \overset{\mathrm{e}}{\nabla}_i \partial_j \log f(\theta)\right\} \\
&\quad + \frac{1}{N\sqrt{N}} g^{ij}(\theta) g^{X,kl}(\theta)\left\{ \overset{\mathrm{e}}{\nabla}_i \partial_k \log f(\theta)\right\}\partial_j \log f(\theta)\partial_l \log f(\theta) \\
&\quad - \frac{1}{3N\sqrt{N}} T^Y_{ijk}(\theta) g^{X,is}(\theta) g^{X,jt}(\theta) g^{X,ku}(\theta)\partial_s \log f(\theta)\partial_t \log f(\theta)\partial_u \log f(\theta) \\
&\quad + \frac{1}{2N\sqrt{N}} g^Y_{kl}(\theta) M^l_{ij} g^{X,is}(\theta) g^{X,jt}(\theta) g^{X,ku}(\theta)\partial_s \log f(\theta)\partial_t \log f(\theta)\partial_u \log f(\theta) \\
&\quad + \frac{1}{2N\sqrt{N}} g^{X,ij}(\theta) g^Y_{kl}(\theta) g^{X,kl}(\theta) M^m_{ij}\partial_m \log f(\theta) - \frac{1}{N\sqrt{N}} T^Y_{ijk}(\theta) g^{X,ij}(\theta) g^{X,kl}(\theta)\partial_l \log f(\theta) \\
&\quad + \frac{1}{N\sqrt{N}} g^{ij}(\theta) M^k_{ij}\partial_k \log f(\theta) + \frac{1}{N\sqrt{N}} g^{ij}(\theta)\partial_i \log f(\theta)\partial_j \log h(\theta) + \mathrm{O}(N^{-2}).
\end{aligned}
\tag{A8}
$$

Proof. By applying the Taylor expansion, the Kullback–Leibler risk, $R(\theta, q(y|\hat{\theta}_\pi))$, is expanded as:

$$
\begin{aligned}
&E_{x^{(N)}}[D(q(\cdot|\theta), q(\cdot|\hat{\theta}_\pi))] \\
&= \mathrm{E}_{X^{(N)}}\Bigg[\int q(y|\theta)\Big\{ -l_i(\theta|y)\tilde{\theta}^i_\pi - \frac{1}{2} l_{ij}(\theta|y)(\hat{\theta}^i_\pi - \theta^i)(\hat{\theta}^j_\pi - \theta^j) \\
&\qquad - \frac{1}{6} l_{ijk}(\theta|y)(\hat{\theta}^i_\pi - \theta^i)(\hat{\theta}^j_\pi - \theta^j)(\hat{\theta}^k_\pi - \theta^k) + \mathrm{O_P}(N^{-2})\Big\}\,\mathrm{d}y\Bigg] \\
&= \frac{1}{2} g^Y_{ij}(\theta)\mathrm{E}_{X^{(N)}}[(\hat{\theta}^i_\pi - \theta^i)(\hat{\theta}^j_\pi - \theta^j)] - \frac{1}{6} L^Y_{ijk}(\theta)\mathrm{E}_{X^{(N)}}[(\hat{\theta}^i_\pi - \theta^i)(\hat{\theta}^j_\pi - \theta^j)(\hat{\theta}^k_\pi - \theta^k)] + \mathrm{O}(N^{-2}) \\
&= \frac{1}{2} g^Y_{ij}(\theta)\mathrm{E}_{X^{(N)}}[(\hat{\theta}^i_\pi - \theta^i)(\hat{\theta}^j_\pi - \theta^j)] \\
&\quad + \left\{ \frac{3}{2}\overset{\mathrm{m}}{\Gamma}{}^Y_{(ij,k)}(\theta) - \frac{1}{3} T^Y_{ijk}(\theta)\right\}\mathrm{E}_{X^{(N)}}[(\hat{\theta}^i_\pi - \theta^i)(\hat{\theta}^j_\pi - \theta^j)(\hat{\theta}^k_\pi - \theta^k)] + \mathrm{O}(N^{-2}) \\
&= \frac{1}{2} g^Y_{ij}(\theta)\mathrm{E}_{X^{(N)}}[(\hat{\theta}^i_\pi - \theta^i)(\hat{\theta}^j_\pi - \theta^j)] - \frac{1}{3} T^Y_{ijk}(\theta)\mathrm{E}_{X^{(N)}}[(\hat{\theta}^i_\pi - \theta^i)(\hat{\theta}^j_\pi - \theta^j)(\hat{\theta}^k_\pi - \theta^k)] \\
&\quad + \frac{1}{2}\left\{ g^Y_{kl}(\theta)\overset{\mathrm{m}}{\Gamma}{}^{Y,l}_{ij}(\theta) - g^Y_{kl}(\theta)\overset{\mathrm{m}}{\Gamma}{}^{X,l}_{ij}(\theta)\right\}\mathrm{E}_{X^{(N)}}[(\hat{\theta}^i_\pi - \theta^i)(\hat{\theta}^j_\pi - \theta^j)(\hat{\theta}^k_\pi - \theta^k)] \\
&\quad + \frac{1}{2} g^Y_{kl}(\theta)\overset{\mathrm{m}}{\Gamma}{}^{X,l}_{ij}(\theta)\mathrm{E}_{X^{(N)}}[(\hat{\theta}^i_\pi - \theta^i)(\hat{\theta}^j_\pi - \theta^j)(\hat{\theta}^k_\pi - \theta^k] + \mathrm{O}(N^{-2}),
\end{aligned}
\tag{A9}
$$

where $\overset{\mathrm{e}}{\Gamma}{}^Y_{(ij,k)}$ denotes $(1/3)\{\overset{\mathrm{e}}{\Gamma}{}^Y_{ij,k} + \overset{\mathrm{e}}{\Gamma}{}^Y_{jk,i} + \overset{\mathrm{e}}{\Gamma}{}^Y_{ki,j}\}$.

By the definition of the predictive metric, $\mathring{g}_{ij}(\theta) = g^X_{ik}(\theta) g^{Y,kl}(\theta) g^X_{lj}(\theta)$, by Expansions (A4) and (A5) and by the relationship $L^X_{ijk}(\theta) = -\overset{\mathrm{e}}{\Gamma}{}^X_{ij,k}(\theta) - \overset{\mathrm{e}}{\blacksquare}{}^X_{jk,i}(\theta) - \overset{\mathrm{e}}{\blacksquare}{}^X_{ki,j}(\theta) - T^X_{ijk}(\theta)$, the last two terms of the above expansion (A9) are expanded as:

$$
\begin{aligned}
&\frac{1}{2} g^Y_{ij}(\theta) \mathrm{E}_{X^{(N)}}[(\hat{\theta}^i_\pi - \theta^i)(\hat{\theta}^j_\pi - \theta^j)] + \frac{1}{2} g^Y_{kl}(\theta) \overset{\mathrm{m}}{\Gamma}{}^{X,l}_{ij}(\theta) \mathrm{E}_{X^{(N)}}[(\hat{\theta}^i_\pi - \theta^i)(\hat{\theta}^j_\pi - \theta^j)(\hat{\theta}^k_\pi - \theta^k)] \\
&= \frac{1}{2N} g^Y_{ij}(\theta) g^{X,ij}(\theta) + \frac{1}{2N} \mathring{g}^{ij}(\theta) \partial_i \log f(\theta) \partial_j \log f(\theta) \\
&\quad + \frac{1}{N\sqrt{N}} \mathring{g}^{ij}(\theta) \left\{ \partial_{ij} \log f(\theta) - \overset{\mathrm{e}}{\Gamma}{}^{X,k}_{ij}(\theta) \partial_k \log f(\theta) \right\} \\
&\quad + \frac{1}{N\sqrt{N}} \mathring{g}^{ij}(\theta) g^{X,kl}(\theta) \left\{ \partial_{ik} \log f(\theta) - \overset{\mathrm{e}}{\blacksquare}{}^{X,m}_{ik} \partial_m \log f(\theta) \right\} \partial_j \log f(\theta) \partial_l \log f(\theta) \\
&\quad + \frac{1}{N\sqrt{N}} \mathring{g}^{ij}(\theta) \partial_i \log f(\theta) \partial_j \log h(\theta) + \mathrm{O}(N^{-2}).
\end{aligned}
\tag{A10}
$$

By substituting Expansion (A10) into Expansion (A9), Expansion (A8) is obtained. □

Note that Expansion (A8) is invariant up to $\mathrm{O}(N^{-2})$ under the reparametrization, so that each term of this expansion is a scalar function of θ.

Lemma A4. *Let $\hat{\theta}_\pi$ be the maximum point of $\log p(x^{(N)}|\theta) + \log\{\pi(\theta;N)/|g^X(\theta)|^{1/2}\}$. The Bayesian predictive density based on the prior, $\pi(\theta;N)$, is expanded as:*

$$
\begin{aligned}
\hat{q}_\pi(y|x^{(N)}) &= q(y|\hat{\theta}_\pi) + \frac{1}{N} g^{X,ij}(\hat{\theta}_\pi) \left\{ \partial_i \log |g^X(\hat{\theta}_\pi)|^{\frac{1}{2}} - \overset{\mathrm{e}}{\Gamma}{}^{X,k}_{ik}(\hat{\theta}_\pi) \right\} \partial_j q(y|\hat{\theta}_\pi) \\
&\quad + \frac{1}{2N} g^{X,ij}(\hat{\theta}_\pi) \left\{ \partial_{ij} q(y|\hat{\theta}_\pi) - \overset{\mathrm{m}}{\Gamma}{}^{X,k}_{ij}(\hat{\theta}_\pi) \partial_k q(y|\hat{\theta}_\pi) \right\} + \mathrm{O}_{\mathrm{P}}(N^{-3/2}).
\end{aligned}
\tag{A11}
$$

Proof. Let $\tilde{\theta}_\pi$ denote $\hat{\theta}_\pi - \theta$. First, using a Taylor expansion twice, we expand the posterior density, $\pi(\theta|x^{(N)})$, as:

$$
\begin{aligned}
\pi(\theta|x^{(N)}) &= |g^X(\hat{\theta}_\pi)|^{\frac{1}{2}} \frac{\pi(\hat{\theta}_\pi)}{|g^X(\hat{\theta}_\pi)|^{\frac{1}{2}}} p(x^{(N)}|\hat{\theta}_\pi) \exp\left[-\frac{1}{2} \{-l_{ij}(\hat{\theta}_\pi|x^{(N)})\} \tilde{\theta}^i_\pi \tilde{\theta}^j_\pi \right] \\
&\quad \times \left[1 - \{\partial_i \log |g^X(\hat{\theta}_\pi)|^{\frac{1}{2}}\} \tilde{\theta}^i_\pi + \frac{1}{2} \left\{ \frac{\partial_{ij} |g^X(\hat{\theta}_\pi)|^{\frac{1}{2}}}{|g^X(\hat{\theta}_\pi)|^{\frac{1}{2}}} \right\} \tilde{\theta}^i_\pi \tilde{\theta}^j_\pi + \mathrm{O}_{\mathrm{P}}(N^{-3/2}) \right] \\
&\quad \times \Bigg(1 + \frac{1}{2} \{\sqrt{N} \partial_{ij} \log f(\hat{\theta}_\pi)\} \tilde{\theta}^i_\pi \tilde{\theta}^j_\pi - \frac{1}{6} \{l_{ijk}(\hat{\theta}_\pi|x^{(N)})\} \tilde{\theta}^i_\pi \tilde{\theta}^j_\pi \tilde{\theta}^k_\pi + \frac{1}{2} \{\log h(\hat{\theta}_\pi)\} \tilde{\theta}^i_\pi \tilde{\theta}^j_\pi \\
&\qquad - \frac{1}{6} \{\sqrt{N} \partial_{ijk} \log f(\hat{\theta}_\pi)\} \tilde{\theta}^i_\pi \tilde{\theta}^j_\pi \tilde{\theta}^k_\pi + \frac{1}{24} l_{ijkl}(\hat{\theta}_\pi|x^{(N)}) \tilde{\theta}^i_\pi \tilde{\theta}^j_\pi \tilde{\theta}^k_\pi \tilde{\theta}^l_\pi \\
&\qquad + \frac{1}{2} \left[\frac{1}{2} \{\sqrt{N} \partial_{ij} \log f(\hat{\theta}_\pi)\} \tilde{\theta}^i_\pi \tilde{\theta}^j_\pi - \frac{1}{6} l_{ijk}(\hat{\theta}_\pi|x^N) \tilde{\theta}^i_\pi \tilde{\theta}^j_\pi \tilde{\theta}^k_\pi \right] \\
&\qquad\quad \times \left[\frac{1}{2} \{\sqrt{N} \partial_{ij} \log f(\hat{\theta}_\pi)\} \tilde{\theta}^i_\pi \tilde{\theta}^j_\pi - \frac{1}{6} l_{ijk}(\hat{\theta}_\pi|x^{(N)}) \tilde{\theta}^i_\pi \tilde{\theta}^j_\pi \tilde{\theta}^k_\pi \right] + \mathrm{O}_{\mathrm{P}}(N^{-3/2}) \Bigg) \\
&\quad \times \left\{ \int p(x^{(N)}|\theta) \frac{\pi(\theta;N)}{|g^X(\theta)|^{\frac{1}{2}}} |g^X(\theta)|^{\frac{1}{2}} \mathrm{d}\theta \right\}^{-1}.
\end{aligned}
$$

We denote the $N^{-1/2}$-order, N^{-1}-order and $N^{-3/2}$-order terms by $(N^{-1/2})a_0(\tilde{\theta}_\pi;\hat{\theta}_\pi)$, $(N^{-1})a_1(\tilde{\theta}_\pi;\hat{\theta}_\pi)$ and $(N^{-3/2})a_2(\tilde{\theta}_\pi;\hat{\theta}_\pi)$, respectively. Then, this expansion is rewritten as:

$$\begin{aligned}\pi(\theta|x^{(N)}) &= |g^X(\hat{\theta}_\pi)|^{\frac{1}{2}}\frac{\pi(\hat{\theta}_\pi)}{|g^X(\hat{\theta}_\pi)|^{\frac{1}{2}}}p(x^{(N)}|\hat{\theta}_\pi)\exp\left[-\frac{1}{2}\{-l_{ij}(\hat{\theta}_\pi|x^{(N)})\}\tilde{\theta}^i_\pi\tilde{\theta}^j_\pi\right] \\ &\quad\times\left[1+\frac{1}{\sqrt{N}}a_0(\tilde{\theta}_\pi;\hat{\theta}_\pi)\right. \\ &\quad\left.+\frac{1}{N}a_1(\tilde{\theta}_\pi;\hat{\theta}_\pi)+\frac{1}{N\sqrt{N}}a_2(\tilde{\theta}_\pi;\hat{\theta}_\pi)+\mathrm{O_P}(N^{-2})\right] \\ &\quad\times\left\{\int p(x^{(N)}|\theta)\frac{\pi(\theta;N)}{|g^X(\theta)|^{\frac{1}{2}}}|g^X(\theta)|^{\frac{1}{2}}\mathrm{d}\theta\right\}^{-1}.\end{aligned}$$

To make the expansion easier to see, the following notations are used. Let $\phi(\eta;-l_{ij}(\hat{\theta}_\pi|x^{(N)}))$ be the probability density function of the d-dimensional normal distribution with the precision matrix whose (i,j)-component is $-l_{ij}(\hat{\theta}_\pi|x^{(N)})$. Let $\eta=(\eta^1,\cdots,\eta^d)$ be a d-dimensional random vector distributed according to the normal density, $\phi(\eta;-l_{ij}(\hat{\theta}_\pi|x^{(N)}))$ The notations, $\bar{a}_0(\hat{\theta}_\pi)$, $\bar{a}_1(\hat{\theta}_\pi)$, $\bar{a}_2(\hat{\theta}_\pi)$ and $\hat{\omega}^{ij}(\hat{\theta}_\pi)$, denote the expectations of $a_0(\eta;\hat{\theta}_\pi)$, $a_1(\eta;\hat{\theta}_\pi)$, $a_2(\eta;\hat{\theta}_\pi)$ and $\eta^i\eta^j$, respectively.

Using the above notations, we get the following posterior expansion:

$$\begin{aligned}\pi(\theta|x^{(N)}) &= \phi(\hat{\theta}_\pi;-l_{ij}(\hat{\theta}_\pi|x^{(N)})) \\ &\quad\times\left[1+\frac{1}{\sqrt{N}}\{a_0(\tilde{\theta}_\pi;\hat{\theta}_\pi)-\bar{a}_0(\hat{\theta}_\pi)\}+\frac{1}{N}\{a_1(\tilde{\theta}_\pi;\hat{\theta}_\pi)-\bar{a}_1(\hat{\theta}_\pi)\}\right. \\ &\quad-\frac{1}{N}\bar{a}_0(\hat{\theta}_\pi)\{a_0(\tilde{\theta}_\pi;\hat{\theta}_\pi)-\bar{a}_0(\hat{\theta}_\pi)\}+\frac{1}{N\sqrt{N}}\{a_2(\tilde{\theta}_\pi;\hat{\theta}_\pi)-\bar{a}_2(\hat{\theta}_\pi)\} \\ &\quad-\frac{1}{N\sqrt{N}}\bar{a}_0(\hat{\theta}_\pi)\{a_1(\tilde{\theta}_\pi;\hat{\theta}_\pi)-\bar{a}_1(\hat{\theta}_\pi)\}-\frac{1}{N\sqrt{N}}\bar{a}_1(\hat{\theta}_\pi)\{a_0(\tilde{\theta}_\pi;\hat{\theta}_\pi)-\bar{a}_0(\hat{\theta}_\pi)\} \\ &\quad\left.+\frac{1}{N\sqrt{N}}\bar{a}_0^2(\hat{\theta}_\pi)\{a_1(\tilde{\theta}_\pi;\hat{\theta}_\pi)-\bar{a}_1(\hat{\theta}_\pi)\}+\mathrm{O_P}(N^{-2})\right].\end{aligned}\quad\text{(A12)}$$

Second, using (A12), the Bayesian predictive density, $\hat{q}_\pi(y|x^{(N)})$, based on the prior, $\pi(\theta;N)$, is expanded as:

$$\begin{aligned}&\hat{q}_\pi(y|x^{(N)}) \\ &=\int q(y|\hat{\theta}_\pi)\left[1-\{\partial_i\log q(y|\hat{\theta}_\pi)\}\tilde{\theta}^i_\pi+\frac{1}{2}\frac{\partial_{ij}q(y|\hat{\theta}_\pi)}{q(y|\hat{\theta}_\pi)}\tilde{\theta}^i_\pi\tilde{\theta}^j_\pi+\mathrm{o_P}(N^{-1})\right]\pi(\theta|x^N)\mathrm{d}\theta \\ &=\int q(y|\hat{\theta}_\pi)\left[1+\{\partial_i\log|g^X(\hat{\theta}_\pi)|^{\frac{1}{2}}\}\{\partial_j\log q(y|\hat{\theta}_\pi)\}\tilde{\theta}^i_\pi\tilde{\theta}^j_\pi\right. \\ &\quad+\frac{1}{6}\{\partial_{ijk}\log p(x^{(N)}|\hat{\theta}_\pi)+\sqrt{N}\partial_{ijk}\log f(\hat{\theta}_\pi)\}\{\partial_l\log q(y|\hat{\theta}_\pi)\}\tilde{\theta}^i_\pi\tilde{\theta}^j_\pi\tilde{\theta}^k_\pi\tilde{\theta}^l_\pi \\ &\quad\left.+\frac{1}{2}\frac{\partial_{ij}q(y|\hat{\theta}_\pi)}{q(y|\hat{\theta}_\pi)}\tilde{\theta}^i_\pi\tilde{\theta}^j_\pi+\mathrm{o_P}(N^{-1})\right]\phi(\tilde{\theta}_\pi;-l_{ij}(\hat{\theta}_\pi|x^N))\mathrm{d}\tilde{\theta}_\pi \\ &=q(y|\hat{\theta}_\pi)+\hat{\omega}^{ij}(\hat{\theta}_\pi)\{\partial_i\log|g^X(\hat{\theta}_\pi)|^{\frac{1}{2}}\}\partial_jq(y|\hat{\theta}_\pi)+\frac{1}{2}\hat{\omega}^{ik}(\hat{\theta}_\pi)\hat{\omega}^{jl}(\hat{\theta}_\pi)l_{ijk}(\hat{\theta}_\pi|x^N)\partial_lq(y|\hat{\theta}_\pi) \\ &\quad+\frac{1}{2}\hat{\omega}^{ij}(\hat{\theta}_\pi)\partial_{ij}q(y|\hat{\theta}_\pi)+\mathrm{O_P}(N^{-3/2}).\end{aligned}\quad\text{(A13)}$$

Here, the following two equations hold:

$$-l_{ij}(\hat{\theta}_\pi|x^{(N)}) = Ng^X_{ij}(\hat{\theta}_\pi)-\sqrt{N}\tilde{H}_{ij}(\hat{\theta}_\pi|x^N)+\mathrm{O_P}(1),\quad\text{(A14)}$$

$$l_{ijk}(\hat{\theta}_\pi|x^{(N)}) = -2N\overset{\mathrm{e}}{\Gamma}{}^{X}_{ij,k}(\hat{\theta}_\pi) - N\overset{\mathrm{m}}{\Gamma}{}^{X}_{ik,j}(\hat{\theta}_\pi) + \sqrt{N}\tilde{H}_{ijk}(\hat{\theta}|x^N). \tag{A15}$$

By combining Equation (A14) with the Sherman–Morrison–Woodbury formula, the following expansion is obtained:

$$\hat{\omega}^{ij}(\hat{\theta}_\pi) = \frac{1}{N}g^{X,ij}(\hat{\theta}_\pi) + \frac{1}{N\sqrt{N}}g^{X,ik}(\hat{\theta}_\pi)g^{X,jl}(\hat{\theta}_\pi)H_{kl}(\hat{\theta}_\pi|x^{(N)}) + \mathrm{O_P}(N^{-2}). \tag{A16}$$

By substituting Equations (A14), (A15) and (A16) into Expansion (A13), Expansion (A11) is obtained. □

Note that the integration of Expansion (A11) is one up to $\mathrm{O_P}(N^{-2})$. Further, Expansion (A11) is similar to the expansion in [2]. However, the estimator that is the center of the expansion is different, because of the dependence of the prior on the sample size.

Lemma A5. *The Bayesian estimator, $\hat{\theta}_{\mathrm{opt}}$, minimizing the Bayes risk, $\int R(\theta, q(y|\hat{\theta}))\mathrm{d}\pi(\theta; N)$, among plug-in predictive densities is given by:*

$$\begin{aligned}\hat{\theta}^i_{\mathrm{opt}} &= \hat{\theta}^i_\pi + \frac{1}{2N}g^{X,ij}(\hat{\theta}_\pi)T^X_j(\hat{\theta}_\pi) \\ &+ \frac{1}{2N}g^{X,jk}(\hat{\theta}_\pi)\left\{\overset{\mathrm{m}}{\Gamma}{}^{Y,i}_{jk}(\hat{\theta}_\pi) - \overset{\mathrm{m}}{\Gamma}{}^{X,i}_{jk}(\hat{\theta}_\pi)\right\} + \mathrm{O_P}(N^{-3/2}).\end{aligned} \tag{A17}$$

Proof. The Bayes risk, $\int R(\theta, q(y|\hat{\theta}))\mathrm{d}\pi(\theta; N)$, is decomposed as:

$$\begin{aligned}\int R(\theta, q(y|\hat{\theta}))\mathrm{d}\pi(\theta; N) &= \int \pi(\theta; N)\int p(x^{(N)}|\theta)\int q(y|\theta)\log\frac{q(y|\theta)}{\hat{q}_\pi(y|x^{(N)})}\mathrm{d}y\mathrm{d}x^{(N)}\mathrm{d}\theta \\ &+ \int \pi(\theta; N)\int p(x^{(N)}|\theta)\int q(y|\theta)\log\frac{\hat{q}_\pi(y|x^{(N)})}{q(y|\hat{\theta})}\mathrm{d}y\mathrm{d}x^{(N)}\mathrm{d}\theta.\end{aligned}$$

The first term of this decomposition is not dependent on $\hat{\theta}$. From Fubini's theorem and Lemma A4, the proof is completed. □

Using these lemmas, we prove Theorem 1. First, we find that the Kullback–Leibler risk of the plug-in predictive density with the estimator, $\hat{\theta}_{\mathrm{opt}}$, defined in Lemma A5, is given by:

$$\begin{aligned}R(\theta, q(y|\hat{\theta}_{\mathrm{opt}})) &= R(\theta, q(y|\hat{\theta}_\pi)) + \frac{1}{2N\sqrt{N}}g^{ij}(\theta)T^X_i(\theta)\partial_j\log f(\theta) \\ &+ \frac{1}{2N\sqrt{N}}g^{X,im}(\theta)g^Y_{ij}(\theta)g^{X,kl}(\theta) \\ &\times\left\{\overset{\mathrm{m}}{\Gamma}{}^{Y,j}_{kl}(\theta) - \overset{\mathrm{m}}{\Gamma}{}^{X,j}_{kl}(\theta)\right\}\partial_m\log f(\theta).\end{aligned} \tag{A18}$$

Using Expansion (A18) and Lemma A3, we expand the Kullback–Leibler risk, $R(\theta, \hat{q}_\pi(y|x^{(N)}))$. Here, the risk, $R(\theta, \hat{q}_\pi(y|x^{(N)}))$, is equal to the risk, $R(\theta, q(y|\hat{\theta}_{\mathrm{opt}}))$, up to $\mathrm{O}(N^{-2})$, because we expand the Bayesian predictive density, $\hat{q}_\pi(y|x^{(N)})$ as:

$$q(y|x^{(N)}) = q(y|\hat{\theta}_{\mathrm{opt}}) + \frac{1}{2N}g^{X,ij}(\hat{\theta}_\pi)\left\{\partial_{ij}q(y|\hat{\theta}_\pi) - \overset{\mathrm{m}}{\Gamma}{}^{Y,k}_{ij}(\hat{\theta}_\pi)\partial_k q(y|\hat{\theta}_\pi)\right\} + \mathrm{O_P}(N^{-3/2}). \tag{A19}$$

Thus, we obtain Expansion (1).

References

1. Aitchison, J. Goodness of prediction fit. *Biometrika* **1975**, *62*, 547–554.
2. Komaki, F. On asymptotic properties of predictive distributions. *Biometrika* **1996**, *83*, 299–313.
3. Hartigan, J. The maximum likelihood prior. *Ann. Stat.* **1998**, *26*, 2083–2103.
4. Bernardo, J. Reference posterior distributions for Bayesian inference. *J. R. Stat. Soc. B* **1979**, *41*, 113–147.
5. Clarke, B.; Barron, A. Jeffreys prior is asymptotically least favorable under entropy risk. *J. Stat. Plan. Inference* **1994**, *41*, 37–60.
6. Aslan, M. Asymptotically minimax Bayes predictive densities. *Ann. Stat.* **2006**, *34*, 2921–2938.
7. Komaki, F. Bayesian predictive densities based on latent information priors. *J. Stat. Plan. Inference* **2011**, *141*, 3705–3715.
8. Komaki, F. Asymptotically minimax Bayesian predictive densities for multinomial models. *Electron. J. Stat.* **2012**, *6*, 934–957.
9. Kanamori, T.; Shimodaira, H. Active learning algorithm using the maximum weighted log-likelihood estimator. *J. Stat. Plan. Inference* **2003**, *116*, 149–162.
10. Shimodaira, H. Improving predictive inference under covariate shift by weighting the log-likelihood function. *J. Stat. Plan. Inference* **2000**, *90*, 227–244.
11. Fushiki, T.; Komaki, F.; Aihara, K. On parametric bootstrapping and Bayesian prediction. *Scand. J. Stat.* **2004**, *31*, 403–416.
12. Suzuki, T.; Komaki, F. On prior selection and covariate shift of β-Bayesian prediction under α-divergence risk. *Commun. Stat. Theory* **2010**, *39*, 1655–1673.
13. Komaki, F. Asymptotic properties of Bayesian predictive densities when the distributions of data and target variables are different. *Bayesian Anal.* **2014**, submitted for publication.
14. Hodges, J.L.; Lehmann, E.L. Some problems in minimax point estimation. *Ann. Math. Stat.* **1950**, *21*, 182–197.
15. Ghosh, M.N. Uniform approximation of minimax point estimates. *Ann. Math. Stat.* **1964**, *35*, 1031–1047.
16. Amari, S. *Differential-Geometrical Methods in Statistics*; Springer: New York, NY, USA, 1985.
17. Robbins, H. Asymptotically Subminimax solutions of Compound Statistical Decision Problems. In Proceedings of the Second Berkley Symposium Mathematical Statistics and Probability, Berkeley, CA, USA, 31 July–12 August 1950; University of California Press: Oakland, CA, USA, 1950; pp. 131–148.
18. Frank, P.; Kiefer, J. Almost subminimax and biased minimax procedures. *Ann. Math. Stat.* **1951**, *22*, 465–468.
19. Efron, B. Defining curvature of a statistical problem (with applications to second order efficiency). *Ann. Stat.* **1975**, *3*, 189–1372.

entropy

Article

Information-Geometric Markov Chain Monte Carlo Methods Using Diffusions

Samuel Livingstone [1,*] and Mark Girolami [2]

[1] Department of Statistical Science, University College London, Gower Street, London WC1E 6BT, UK
[2] Department of Statistics, University of Warwick, Coventry CV4 7AL, UK; E-Mail:m.girolami@warwick.ac.uk
* E-Mail: samuel.livingstone@ucl.ac.uk; Tel.: +44-20-7679-1872.

Received: 29 March 2014; in revised form: 23 May 2014 / Accepted: 28 May 2014 /
Published: 3 June 2014

Abstract: Recent work incorporating geometric ideas in Markov chain Monte Carlo is reviewed in order to highlight these advances and their possible application in a range of domains beyond statistics. A full exposition of Markov chains and their use in Monte Carlo simulation for statistical inference and molecular dynamics is provided, with particular emphasis on methods based on Langevin diffusions. After this, geometric concepts in Markov chain Monte Carlo are introduced. A full derivation of the Langevin diffusion on a Riemannian manifold is given, together with a discussion of the appropriate Riemannian metric choice for different problems. A survey of applications is provided, and some open questions are discussed.

Keywords: information geometry; Markov chain Monte Carlo; Bayesian inference; computational statistics; machine learning; statistical mechanics; diffusions

---

1. Introduction

There are three objectives to this article. The first is to introduce geometric concepts that have recently been employed in Monte Carlo methods based on Markov chains [1] to a wider audience. The second is to clarify what a "diffusion on a manifold" is, and how this relates to a diffusion defined on Euclidean space. Finally, we review the state-of-the-art in the field and suggest avenues for further research.

The connections between some Monte Carlo methods commonly used in statistics, physics and application domains, such as econometrics, and ideas from both Riemannian and information geometry [2,3] were highlighted by Girolami and Calderhead [1] and the potential benefits demonstrated empirically. Two Markov chain Monte Carlo methods were introduced, the manifold Metropolis-adjusted Langevin algorithm and Riemannian manifold Hamiltonian Monte Carlo. Here, we focus on the former for two reasons. First, the intuition for why geometric ideas can improve standard algorithms is the same in both cases. Second, the foundations of the methods are quite different, and since the focus of the article is on using geometric ideas to improve performance, we considered a detailed description of both to be unnecessary. It should be noted, however, that impressive empirical evidence exists for using Hamiltonian methods in some scenarios (e.g., [4]). We refer interested readers to [5,6].

We take an expository approach, providing a review of some necessary preliminaries from Markov chain Monte Carlo, diffusion processes and Riemannian geometry. We assume only a minimal familiarity with measure-theoretic probability. More informed readers may prefer to skip these sections. We then provide a full derivation of the Langevin diffusion on a Riemannian manifold and offer some intuition for how to think about such a process. We conclude Section 4 by presenting the Metropolis-adjusted Langevin algorithm on a Riemannian manifold.

A key challenge in the geometric approach is which manifold to choose. We discuss this in Section 4.4 and review some candidates that have been suggested in the literature, along with the reasoning for each. Rather than provide a simulation study here, we instead reference studies where the methods we describe have been applied in Section 5. In Section 6, we discuss several open questions, which we feel could be interesting areas of further research and of interest to both theorists and practitioners.

Throughout, $\pi(\cdot)$ will refer to an n-dimensional probability distribution and $\pi(x)$ its density with respect to the Lebesgue measure.

2. Markov Chain Monte Carlo

Markov chain Monte Carlo (MCMC) is a set of methods for drawing samples from a distribution, $\pi(\cdot)$, defined on a measurable space $(\mathcal{X}, \mathcal{B})$, whose density is only known up to some proportionality constant. Although the i-th sample is dependent on the $(i-1)$-th, the Ergodic Theorem ensures that for an appropriately constructed Markov chain with invariant distribution $\pi(\cdot)$, long-run averages are consistent estimators for expectations under $\pi(\cdot)$. As a result, MCMC methods have proven useful in Bayesian statistical inference, where often, the posterior density $\pi(x|y) \propto f(y|x)\pi_0(x)$ for some parameter, x (where $f(y|x)$ denotes the likelihood for data y and $\pi_0(x)$ the prior density), is only known up to a constant [7]. Here, we briefly introduce some concepts from general state space Markov chain theory together with a short overview of MCMC methods. The exposition follows [8].

2.1. Markov Chain Preliminaries

A time-homogeneous Markov chain, $\{X_m\}_{m\in\mathbb{N}}$, is a collection of random variables, X_m, each of which is defined on a measurable space $(\mathcal{X}, \mathcal{B})$, such that:

$$\mathbb{P}[X_m \in A | X_0 = x_0, ..., X_{m-1} = x_{m-1}] = \mathbb{P}[X_m \in A | X_{m-1} = x_{m-1}], \tag{1}$$

for any $A \in \mathcal{B}$. We define the transition kernel $P(x_{m-1}, A) = \mathbb{P}[X_m \in A | X_{m-1} = x_{m-1}]$ for the chain to be a map for which $P(x, \cdot)$ defines a distribution over $(\mathcal{X}, \mathcal{B})$ for any $x \in \mathcal{X}$, and $P(\cdot, A)$ is measurable for any $A \in \mathcal{B}$. Intuitively, P defines a map from points to distributions in $\mathcal{X}$. Similarly, we define the m-step transition kernel to be:

$$P^m(x_0, A) = \mathbb{P}[X_m \in A | X_0 = x_0]. \tag{2}$$

We call a distribution $\pi(\cdot)$ invariant for $\{X_m\}_{m\in\mathbb{N}}$ if:

$$\pi(A) = \int_{\mathcal{X}} P(x, A)\pi(dx) \tag{3}$$

for all $A \in \mathcal{B}$. If $P(x, \cdot)$ admits a density, $p(x'|x)$, this can be equivalently written:

$$\pi(x') = \int_{\mathcal{X}} \pi(x)p(x'|x)dx. \tag{4}$$

The connotation of Equations (3) and (4) is that if $X_m \sim \pi(\cdot)$, then $X_{m+s} \sim \pi(\cdot)$ for any $s \in \mathbb{N}$. In this instance, we say the chain is "at stationarity". Of interest to us will be Markov chains for which there is a unique invariant distribution, which is also the limiting distribution for the chain, meaning that for any $x_0 \in \mathcal{X}$ for which $\pi(x_0) > 0$:

$$\lim_{m\to\infty} P^m(x_0, A) = \pi(A) \tag{5}$$

for any $A \in \mathcal{B}$. Certain conditions are required for Equation (5) to hold, but for all Markov chains presented here, these are satisfied (though, see [8]).

A useful condition, which is sufficient (though not necessary) for $\pi(\cdot)$ to be an invariant distribution, is reversibility, which can be shown by the relation:

$$\pi(x)p(x'|x) = \pi(x')p(x|x'). \tag{6}$$

Integrating over both sides with respect to x, we recover Equation (4). In other words, a chain is reversible if, at stationarity, the probability that $x_i \in A$ and $x_{i+1} \in B$ are equal to the probability that $x_{i+1} \in A$ and $x_i \in B$. The relation (6) will be the primary tool used to construct Markov chains with a desired invariant distribution in the next section.

2.1.1. Monte Carlo Estimates from Markov Chains

Of most interest here are estimators constructed from a Markov chain. The Ergodic Theorem states that for any chain, $\{X_m\}_{m\in\mathbb{N}}$, satisfying Equation (5) and any $g \in L^1(\pi)$, we have that:

$$\lim_{m\to\infty} \frac{1}{m} \sum_{i=1}^{m} g(X_i) = \mathbb{E}_\pi[g(X)] \tag{7}$$

with probability one [7]. This is a Markov chain analogue to the Law of large numbers.

The efficiency of estimators of the form $\hat{t}_m = \sum_i g(X_i)/m$ can be assessed through the autocorrelation between elements in the chain. We will assess the efficiency of $\hat{t}_m$ relative to estimators $\bar{t}_m = \sum_i g(Z_i)/m$, where $\{Z_i\}_{m\in\mathbb{N}}$ is a sequence of independent random variables, each having distribution $\pi(\cdot)$. Provided $\text{Var}_\pi[g(Z_i)] < \infty$, then $\text{Var}[\bar{t}_m] = \text{Var}_\pi[g(Z_i)]/m$. We now seek a similar result for estimators of the form, $\hat{t}_m$.

It follows directly from the Kipnis–Varadhan Theorem [9] that an estimator, $\hat{t}_m$, from a reversible Markov chain for which $X_0 \sim \pi(\cdot)$ satisfies:

$$\lim_{m\to\infty} \frac{\text{Var}[\hat{t}_m]}{\text{Var}[\bar{t}_m]} = 1 + 2\sum_{i=1}^{\infty} \rho^{(0,i)} = \tau, \tag{8}$$

provided that $\sum_{i=1}^{\infty} i|\rho^{(0,i)}| < \infty$, where $\rho^{(0,i)} = \text{Corr}_\pi[g(X_0), g(X_i)]$. We will refer to the constant, τ, as the autocorrelation time for the chain.

Equation (8) implies that for large enough m, $\text{Var}[\hat{t}_m] \approx \tau\text{Var}[\bar{t}_m]$. In practical applications, the sum in Equation (8) is truncated to the first $p-1$ realisations of the chain, where p is the first instance at which $|\rho^{(0,p)}| < \epsilon$ for some $\epsilon > 0$. For example, in the Convergence Diagnosis and Output Analysis for MCMC (CODA) package within the R statistical software $\epsilon = 0.05$ [10,11].

Another commonly used measure of efficiency is the effective sample size $m_{eff} = m/\tau$, which gives the number of independent samples from $\pi(\cdot)$ needed to give an equally efficient estimate for $\mathbb{E}_\pi[g(X)]$. Clearly, minimising τ is equivalent to maximising m_{eff}.

The measures arising from Equation (8) give some intuition for what sort of Markov chain gives rise to efficient estimators. However, in practice, the chain will never be at stationarity. Therefore, we also assess Markov chains according to how far away they are from this point. For this, we need to measure how close $P^m(x_0,\cdot)$ is from $\pi(\cdot)$, which requires a notion of distance between probability distributions.

Although there are several appropriate choices [12], a common option in the Markov chain literature is the total variation distance:

$$\|\mu(\cdot) - \nu(\cdot)\|_{TV} := \sup_{A\in\mathcal{B}} |\mu(A) - \nu(A)|, \tag{9}$$

which informally gives the largest possible difference between the probabilities of a single event in $\mathcal{B}$ according to $\mu(\cdot)$ and $\nu(\cdot)$. If both distributions admit densities, Equation (9) can be written (see Appendix A):

$$\|\mu(\cdot) - \nu(\cdot)\|_{TV} = \frac{1}{2}\int_{\mathcal{X}} |\mu(x) - \nu(x)|dx. \tag{10}$$

which is proportional to the L_1 distance between $\mu(x)$ and $\nu(x)$. Our metric, $\|\cdot\|_{TV} \in [0,1]$, with $\|\cdot\|_{TV} = 1$ for distributions with disjoint supports and $\|\mu(\cdot) - \nu(\cdot)\|_{TV} = 0$, implies $\mu(\cdot) \equiv \nu(\cdot)$.

Typically, for an unbounded $\mathcal{X}$, the distance $\|P^m(x_0, \cdot) - \pi(\cdot)\|_{TV}$ will depend on x_0 for any finite m. Therefore, bounds on the distance are often sought via some inequality of the form:

$$\|P^m(x_0, \cdot) - \pi(\cdot)\|_{TV} \leq MV(x_0)f(m), \tag{11}$$

for some $M < \infty$, where $V : \mathcal{X} \to [1, \infty)$ depends on x_0 and is called a drift function, and $f : \mathbb{N} \to [0, \infty)$ depends on the number of iterations, m (and is often defined, such that $f(0) = 1$).

A Markov chain is called geometrically ergodic if $f(m) = r^m$ in Equation (11) for some $0 < r < 1$. If in addition to this, V is bounded above, the chain is called uniformly ergodic. Intuitively, if either condition holds, then the distribution of X_m will converge to $\pi(\cdot)$ geometrically quickly as m grows, and in the uniform case, this rate is independent of x_0. As well as providing some (often qualitative if M and r are unknown) bounds on the convergence rate of a Markov chain, geometric ergodicity implies that a central limit theorem exists for estimators of the form, $\hat{t}_m$. For more detail on this, see [13,14].

In practice several approximate methods also exist to assess whether a chain is close enough to stationarity for long-run averages to provide suitable estimators (e.g., [15]). The MCMC practitioner also uses a variety of visual aids to judge whether an estimate from the chain will be appropriate for his or her needs.

2.2. Markov Chain Monte Carlo

Now that we have introduced Markov chains, we turn to simulating them. The objective here is to devise a method for generating a Markov chain, which has a desired limiting distribution, $\pi(\cdot)$. In addition, we would strive for the convergence rate to be as fast as possible and the effective sample size to be suitably large relative to the number of iterations. Of course, the computational cost of performing an iteration is also an important practical consideration. Ideally, any method would also require limited problem-specific alterations, so that practitioners are able to use it with as little knowledge of the inner workings as is practical.

Although other methods exist for constructing chains with a desired limiting distribution, a popular choice is the Metropolis–Hastings algorithm [7]. At iteration i, a sample is drawn from some candidate transition kernel, $Q(x_{i-1}, \cdot)$, and then either accepted or rejected (in which case, the state of the chain remains x_{i-1}). We focus here on the case where $Q(x_{i-1}, \cdot)$ admits a density, $q(x'|x_{i-1})$, for all $x_{i-1} \in \mathcal{X}$ (though, see [8]). In this case, a single step is shown below (the wedge notation $a \wedge b$ denotes the minimum of a and b). The "acceptance rate", $\alpha(x_{i-1}, x')$, governs the behaviour of the chain, so that, when it is close to one, then many proposed moves are accepted, and the current value in the chain is constantly changing. If it is on average close to zero, then many proposals are rejected, so that the chain will remain in the same place for many iterations. However, $\alpha \approx 1$ is typically not ideal, often resulting in a large autocorrelation time (see below). The challenge in practice is to find the right acceptance rate to balance these two extremes.

Algorithm 1 Metropolis–Hastings, single iteration.

Require: x_{i-1}
Draw $X' \sim Q(x_{i-1}, \cdot)$
Draw $Z \sim U[0, 1]$
Set $\alpha(x_{i-1}, x') \leftarrow 1 \wedge \frac{\pi(x')q(x_{i-1}|x')}{\pi(x_{i-1})q(x'|x_{i-1})}$
if $z < \alpha(x_{i-1}, x')$ **then**
  Set $x_i \leftarrow x'$
else
  Set $x_i \leftarrow x_{i-1}$
end if

Combining the "proposal" and "acceptance" steps, the transition kernel for the resulting Markov chain is:

$$P(x, A) = r(x)\delta_x(A) + \int_A \alpha(x, x')q(x'|x)dx', \tag{12}$$

for any $A \in \mathcal{B}$, where:

$$r(x) = 1 - \int_{\mathcal{X}} \alpha(x, x')q(x'|x)dx'$$

is the average probability that a draw from $Q(x, \cdot)$ will be rejected, and $\delta_x(A) = 1$ if $x \in A$ and zero, otherwise. A Markov chain defined in this way will have $\pi(\cdot)$ as an invariant distribution, since the chain is reversible for $\pi(\cdot)$. We note here that:

$$\begin{aligned}\pi(x_{i-1})q(x_i|x_{i-1})\alpha(x_{i-1}, x_i) &= \pi(x_{i-1})q(x_i|x_{i-1}) \wedge \pi(x_i)q(x_{i-1}|x_i) \\ &= \alpha(x_i, x_{i-1})q(x_{i-1}|x_i)\pi(x_i)\end{aligned}$$

in the case that the proposed move is accepted and that if the proposed move is rejected, then $x_i = x_{i-1}$; so the chain is reversible for $\pi(\cdot)$. It can be shown that $\pi(\cdot)$ is also the limiting distribution for the chain [7].

The convergence rate and autocorrelation time of a chain produced by the algorithm are dependent on both the choice of proposal, $Q(x_{i-1}, \cdot)$, and the target distribution, $\pi(\cdot)$. For simple forms of the latter, less consideration is required when choosing the former. A broad objective among researchers in the field is to find classes of proposal kernels that produce chains that converge and mix quickly for a large class of target distributions. We first review a simple choice before discussing one that is more sophisticated, and the will be the focus of the rest of the article.

2.3. Random Walk Proposals

An extremely simple choice for $Q(x, \cdot)$ is one for which:

$$q(x'|x) = q(\|x' - x\|) \tag{13}$$

where $\|\cdot\|$ denotes some appropriate norm on $\mathcal{X}$, meaning the proposal is symmetric. In this case, the acceptance rate reduces to:

$$\alpha(x, x') = 1 \wedge \frac{\pi(x')}{\pi(x)}. \tag{14}$$

In addition to simplifying calculations, Equation (14) strengthens the intuition for the method, since proposed moves with higher density under $\pi(\cdot)$ will always be accepted. A typical choice for $Q(x, \cdot)$ is $\mathcal{N}(x, \lambda^2\Sigma)$, where the matrix, Σ, is often chosen in an attempt to match the correlation structure of $\pi(\cdot)$ or simply taken as the identity [16]. The tuning parameter, λ, is the only other user-specific input required.

Much research has been conducted into properties of the random walk Metropolis algorithm (RWM). It has been shown that the optimal acceptance rate for proposals tends to 0.234 as the dimension, n, of the state space, $\mathcal{X}$, tends to ∞ for a wide class of targets (e.g., [17,18]). The intuition for an optimal acceptance rate is to find the right balance between the distance of proposed moves and the chances of acceptance. Increasing the former will reduce the autocorrelation in the chain if the proposal is accepted, but if it is rejected, the chain will not move at all, so autocorrelation will be high. Random walk proposals are sometimes referred to as blind (e.g., [19]), as no information about $\pi(\cdot)$ is used when generating proposals, so typically, very large moves will result in a very low chance of acceptance, while small moves will be accepted, but result in very high autocorrelation for the chain. Figure 1 demonstrates this in the simple case where $\pi(\cdot)$ is a one-dimensional $\mathcal{N}(0, 1^2)$ distribution.

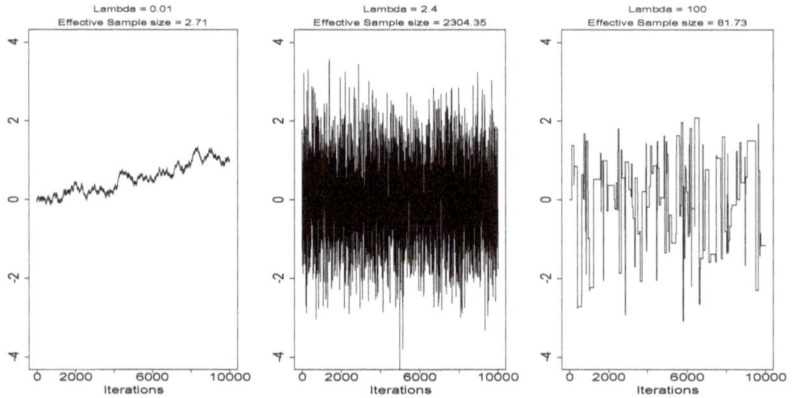

Figure 1. These traceplots show the evolution of three RWM Markov chains for which $\pi(\cdot)$ is a $\mathcal{N}(0, 1^2)$ distribution, with different choices for λ.

Several authors have also shown that for certain classes of $\pi(\cdot)$, the tuning parameter, λ, should be chosen, such that $\lambda^2 \propto n^{-1}$, so that $\alpha \nrightarrow 0$ as $n \to \infty$ [20]. Because of this, we say that algorithm efficiency "scales" $O(n^{-1})$ as the dimension n of $\pi(\cdot)$ increases.

Ergodicity results for a Markov chain constructed using the RWM algorithm also exist [21–23]. At least exponentially light tails are a necessity for $\pi(x)$ for geometric ergodicity, which means that $\pi(x)/e^{-\|x\|} \to c$ as $\|x\| \to \infty$, for some constant, c. For super-exponential tails (where $\pi(x) \to 0$ at a faster than the exponential rate), additional conditions are required [21,23]. We demonstrate with a simple example why heavy-tailed forms of $\pi(x)$ pose difficulties here (where $\pi(x) \to 0$ at a rate slower than $e^{-\|x\|}$).

Example: Take $\pi(x) \propto 1/(1+x^2)$, so that $\pi(\cdot)$ is a Cauchy distribution. Then, if $X' \sim \mathcal{N}(x, \lambda^2)$, the ratio $\pi(x')/\pi(x) = (1+x^2)/(1+(x')^2) \to 1$ as $|x| \to \infty$. Therefore, if x_0 is far away from zero, the Markov chain will dissolve into a random walk, with almost every proposal being accepted.

It should be noted that starting the chain from at or near zero can also cause problems in the above example, as the tails of the distribution may not be explored. See [7] for more detail here.

Ergodicity results for the RWM also exist for specific classes of the statistical model. Conditions for geometric ergodicity in the case of generalised linear mixed models are given in [24], while spherically constrained target densities are discussed in [25]. In [26], the authors provide necessary conditions for the geometric convergence of RWM algorithms, which are related to the existence of exponential moments for $\pi(\cdot)$ and $P(x, \cdot)$. Weaker forms of ergodicity and corresponding conditions are also discussed in the paper.

In the remainder of the article, we will primarily discuss another approach to choosing Q, which has been shown empirically [1] and, in some cases, theoretically [20] to be superior to the RWM algorithm, though it should be noted that random walk proposals are still widely used in practice and are often sufficient for more straightforward problems [16].

3. Diffusions

In MCMC, we are concerned with discrete time processes. However, often, there are benefits to first considering a continuous time process with the properties we desire. For example, some continuous time processes can be specified via a form of differential equation. In this section, we derive a choice for a Metropolis–Hastings proposal kernel based on approximations to diffusions, those continuous-time n-dimensional Markov processes $(X_t)_{t\geq 0}$ for which any sample path $t \mapsto X_t(\omega)$ is a continuous function with probability one. For any fixed t, we assume X_t is a random variable taking values on the measurable space $(\mathcal{X}, \mathcal{B})$ as before. The motivation for this section is to define a class of diffusions for which $\pi(\cdot)$ is the invariant distribution. First, we provide some preliminaries, followed by an introduction to our main object of study, the Langevin diffusion.

3.1. Preliminaries

We focus on the class of time-homogeneous Itô diffusions, whose dynamics are governed by a stochastic differential equation of the form:

$$dX_t = b(X_t)dt + \sigma(X_t)dB_t, \;\; X_0 = x_0, \tag{15}$$

where $(B_t)_{t\geq 0}$ is a standard Brownian motion and the drift vector, b, and volatility matrix, σ, are Lipschitz continuous [27]. Since $\mathbb{E}[B_{t+\triangle t} - B_t | B_t = b_t] = 0$ for any $\triangle t \geq 0$, informally, we can see that:

$$\mathbb{E}[X_{t+\triangle t} - X_t | X_t = x_t] = b(x_t)\triangle t + o(\triangle t), \tag{16}$$

implying that the drift dictates how the mean of the process changes over a small time interval, and if we define the process $(M_t)_{t\geq 0}$ through the relation:

$$M_t = X_t - \int_0^t b(X_s)ds \tag{17}$$

then we have:

$$\mathbb{E}[(M_{t+\triangle t} - M_t)(M_{t+\triangle t} - M_t)^T | M_t = m_t, X_t = x_t] = \sigma(x_t)\sigma(x_t)^T \triangle t + o(\triangle t), \tag{18}$$

giving the stochastic part of the relationship between $X_{t+\triangle t}$ and X_t for small enough $\triangle t$; see, e.g., [28].

While Equation(15) is often a suitable description of an Itô diffusion, it can also be characterised through an infinitesimal generator, $\mathcal{A}$, which describes how functions of the process are expected to evolve. We define this partial differential operator through its action on a function, $f \in C_0(\mathcal{X})$, as:

$$\mathcal{A}f(X_t) = \lim_{\triangle t \to 0} \frac{\mathbb{E}[f(X_{t+\triangle t}) | X_t = x_t] - f(x_t)}{\triangle t}, \tag{19}$$

though $\mathcal{A}$ can be associated with the drift and volatility of $(X_t)_{t\geq 0}$ by the relation:

$$\mathcal{A}f(x) = \sum_i b_i(x) \frac{\partial f}{\partial x_i}(x) + \frac{1}{2} \sum_{i,j} V_{ij}(x) \frac{\partial^2 f}{\partial x_i \partial x_j}(x), \tag{20}$$

where $V_{ij}(x)$ denotes the component in row i and column j of $\sigma(x)\sigma(x)^T$ [27].

As in the discrete case, we can describe the transition kernel of a continuous time Markov process, $P^t(x_0, \cdot)$. In the case of an Itô diffusion, $P^t(x_0, \cdot)$ admits a density, $p_t(x|x_0)$, which, in fact, varies smoothly as a function of t. The Fokker–Planck equation describes this variation in terms of the drift and volatility and is given by:

$$\frac{\partial}{\partial t} p_t(x|x_0) = -\sum_i \frac{\partial}{\partial x_i}[b_i(x) p_t(x|x_0)] + \frac{1}{2} \sum_{i,j} \frac{\partial^2}{\partial x_i \partial x_j}[V_{ij}(x) p_t(x|x_0)]. \tag{21}$$

Although, typically, the form of $P^t(x_0,\cdot)$ is unknown, the expectation and variance of $X_t \sim P^t(x_0,\cdot)$ are given by the integral equations:

$$\mathbb{E}[X_t|X_0 = x_0] = x_0 + \mathbb{E}\left[\int_0^t b(X_s)ds\right],$$
$$\mathbb{E}[(X_t - \mathbb{E}[X_t])(X_t - \mathbb{E}[X_t])^T|X_0 = x_0] = \mathbb{E}\left[\int_0^t \sigma(X_s)\sigma(X_s)^T ds\right],$$

where the second of these is a result of the Itô isometry [27]. Continuing the analogy, a natural question is whether a diffusion process has an invariant distribution, $\pi(\cdot)$, and whether:

$$\lim_{t\to\infty} P^t(x_0, A) = \pi(A) \tag{22}$$

for any $A \in \mathcal{B}$ and any $x_0 \in \mathcal{X}$, in some sense. For a large class of diffusions (which we confine ourselves to), this is, in fact, the case. Specifically, in the case of positive Harris recurrent diffusions with invariant distribution $\pi(\cdot)$, all compact sets must be small for some skeleton chain, see [29] for details. In addition, Equation (21) provides a means of finding $\pi(\cdot)$, given b and σ. Setting the left-hand side of Equation (21) to zero gives:

$$\sum_i \frac{\partial}{\partial x_i}[b_i(x)\pi(x)] = \frac{1}{2}\sum_{i,j}\frac{\partial^2}{\partial x_i \partial x_j}[V_{ij}(x)\pi(x)], \tag{23}$$

which can be solved to find $\pi(\cdot)$.

3.2. Langevin Diffusions

Given Equation (23), our goal becomes clearer: find drift and volatility terms, so that the resulting dynamics describe a diffusion, which converges to some user-defined invariant distribution, $\pi(\cdot)$. This process can then be used as a basis for choosing Q in a Metropolis–Hastings algorithm. The Langevin diffusion, first used to describe the dynamics of molecular systems [30], is such a process, given by the solution to the stochastic differential equation:

$$dX_t = \frac{1}{2}\nabla \log \pi(X_t)dt + dB_t, \quad X_0 = x_0. \tag{24}$$

Since $V_{ij}(x) = \mathbb{1}_{\{i=j\}}$, it is clear that

$$\frac{1}{2}\frac{\partial}{\partial x_i}[\log \pi(x)]\pi(x) = \frac{1}{2}\frac{\partial}{\partial x_i}\pi(x), \quad \forall i, \tag{25}$$

which is a sufficient condition for Equation (23) to hold. Therefore, for any case in which $\pi(x)$ is suitably regular, so that $\nabla \log \pi(x)$ is well-defined and the derivatives in Equation (23) exist, we can use (24) to construct a diffusion, which has invariant distribution, $\pi(\cdot)$.

Roberts and Tweedie [31] give sufficient conditions on $\pi(\cdot)$ under which a diffusion, $(X_t)_{t\geq 0}$, with dynamics given by Equation (24), will be ergodic, meaning:

$$\|P^t(x_0,\cdot) - \pi(\cdot)\|_{TV} \to 0 \tag{26}$$

as $t \to \infty$, for any $x_0 \in \mathcal{X}$.

3.3. Metropolis-Adjusted Langevin Algorithm

We can use Langevin diffusions as a basis for MCMC in many ways, but a popular variant is known as the Metropolis-adjusted Langevin algorithm (MALA), whereby $Q(x,\cdot)$ is constructed through a Euler–Maruyama discretisation of (24) and used as a candidate kernel in a Metropolis–Hastings algorithm. The resulting Q is:

$$Q(x,\cdot) \equiv \mathcal{N}\left(x + \frac{\lambda^2}{2}\nabla \log \pi(x), \lambda^2 I\right), \tag{27}$$

where λ is again a tuning parameter.

Before we discuss the theoretical properties of the approach, we first offer an intuition for the dynamics. From Equation (27), it can be seen that Langevin-type proposals comprise a deterministic shift towards a local mode of $\pi(x)$, combined with some random additive Gaussian noise, with variance λ^2 for each component. The relative weights of the deterministic and random parts are fixed, given as they are by the parameter, λ. Typically, if $\lambda^{1/2} \gg \lambda$, then the random part of the proposal will dominate and *vice versa* in the opposite case, though this also depends on the form of $\nabla \log \pi(x)$ [31].

Again, since this is a Metropolis–Hastings method, choosing λ is a balance between proposing large enough jumps and ensuring that a reasonable proportion are accepted. It has been shown that in the limit, as $n \to \infty$, the optimal acceptance rate for the algorithm is 0.574 [20] for forms of $\pi(\cdot)$, which either have independent and identically distributed components or whose components only differ by some scaling factor [20]. In these cases, as $n \to \infty$, the parameter, λ, must be $\propto n^{-1/3}$, so we say the algorithm efficiency scales $O(n^{-1/3})$. Note that these results compare favourably with the $O(n^{-1})$ scaling of the random walk algorithm.

Convergence properties of the method have also been established. Roberts and Tweedie [31] highlight some cases in which MALA is either geometrically ergodic or not. Typically, results are based on the tail behaviour of $\pi(x)$. If these tails are heavier than exponential, then the method is typically not geometrically ergodic and similarly if the tails are lighter than Gaussian. However, in the in between case, the converse is true. We again offer two simple examples for intuition here.

Example: Take $\pi(x) \propto 1/(1+x^2)$ as in the previous example. Then, $\nabla \log \pi(x) = -2x/(1+x^2)^2 \to 0$ as $|x| \to \infty$. Therefore, if x_0 is far away from zero, then the MALA will be approximately equal to the RWM algorithm and, so, will also dissolve into a random walk.

Example: Take $\pi(x) \propto e^{-x^4}$. Then, $\nabla \log \pi(x) = -4x^3$ and $X' \sim \mathcal{N}(x - 4\lambda^2 x^3, \lambda^2)$. Therefore, for any fixed λ, there exists $c > 0$, such that, for $|x_0| > c$, we have $|4\lambda^2 x^3| >> x$ and $|x - 4\lambda^2 x^3| >> \lambda$, suggesting that MALA proposals will quickly spiral further and further away from any neighbourhood of zero, and hence, nearly all will be rejected.

For cases where there is a strong correlation between elements of x or each element has a different marginal variance, the MALA can also be "pre-conditioned" in a similar way to the RWM, so that the covariance structure of proposals more accurately reflects that of $\pi(x)$ [32]. In this case, proposals take the form:

$$Q(x,\cdot) \equiv \mathcal{N}\left(x + \frac{\lambda^2}{2}\Sigma\nabla \log \pi(x), \lambda^2 \Sigma\right), \tag{28}$$

where λ is again a tuning parameter. It can be shown that provided Σ is a constant matrix, $\pi(x)$ is still the invariant distribution for the diffusion on which Equation (28) is based [33].

4. Geometric Concepts in Markov Chain Monte Carlo

Ideas from information geometry have been successfully applied to statistics from as early as [34]. More widely, other geometric ideas have also been applied, offering new insight into common problems (e.g., [35,36]). A survey is given in [37]. In this section, we suggest why some ideas from differential geometry may be beneficial for sampling methods based on Markov chains. We then review what is meant by a "diffusion on a manifold", before turning to the specific case of Equation (24). After this, we discuss what can be learned from work in information geometry in this context.

4.1. Manifolds and Markov Chains

We often make assumptions in MCMC about the properties of the space, $\mathcal{X}$, in which our Markov chains evolve. Often $\mathcal{X} = \mathbb{R}^n$ or a simple re-parametrisation would make it so. However, here, $\mathbb{R}^n = \{(a_1, ..., a_n) : a_i \in (-\infty, \infty)\ \forall i\}$. The additional assumption that is often made is that $\mathbb{R}^n$ is Euclidean, an inner product space with the induced distance metric:

$$d(x, y) = \sqrt{\sum_i (x_i - y_i)^2}. \tag{29}$$

For sampling methods based on Markov chains that explore the space locally, like the RWM and MALA, it may be advantageous to instead impose a different metric structure on the space, $\mathcal{X}$, so that some points are drawn closer together and others pushed further apart. Intuitively, one can picture distances in the space being defined, such that if the current position in the chain is far from an area of $\mathcal{X}$, which is "likely to occur" under $\pi(\cdot)$, then the distance to such a typical set could be reduced. Similarly, once this region is reached, the space could be "stretched" or "warped", so that it is explored as efficiently as possible.

While the idea is attractive, it is far from a constructive definition. We only have the pre-requisite that $(\mathcal{X}, d)$ must be a metric space. However, as Langevin dynamics use gradient information, we will require $(\mathcal{X}, d)$ to be a space on which we can do differential calculus. Riemannian manifolds are an appropriate choice, therefore, as the rules of differentiation are well understand for functions defined on them [38,39], while we are still free to define a more local notion of distance than Euclidean. In this section, we write $\mathbb{R}^n$ to denote the Euclidean vector space.

4.2. Preliminaries

We do not provide a full overview of Riemannian geometry here [38–40]. We simply note that for our purposes, we can consider an n-dimensional Riemannian manifold (henceforth, manifold) to be an n-dimensional metric space, in which distances are defined in a specific way. We also only consider manifolds for which a global coordinate chart exists, meaning that a mapping $r : \mathbb{R}^n \to M$ exists, which is both differentiable and invertible and for which the inverse is also differentiable (a diffeomorphism). Although this restricts the class of manifolds available (the sphere, for example, is not in this class), it is again suitable for our needs and avoids the practical challenges of switching between coordinate patches. The connection with $\mathbb{R}^n$ defined through r is crucial for making sense of differentiability in M. We say a function $f : M \to \mathbb{R}$ is "differentiable" if $(f \circ r) : \mathbb{R}^n \to \mathbb{R}$ is [39].

As has been stated, Equation (29) can be induced via a Euclidean inner product, which we denote $\langle \cdot, \cdot \rangle$. However, it will aid intuition to think of distances in $\mathbb{R}^n$ via curves:

$$\gamma : [0, 1] \to \mathbb{R}^n. \tag{30}$$

We could think of the distance between two points in $x, y \in \mathbb{R}^n$ as the minimum length among all curves that pass through x and y. If $\gamma(0) = x$ and $\gamma(1) = y$, the length is defined as:

$$L(\gamma) = \int_0^1 \sqrt{\langle \gamma'(t), \gamma'(t) \rangle} dt, \tag{31}$$

giving the metric:

$$d(x, y) = \inf \{L(\gamma) : \gamma(0) = x, \gamma(1) = y\}. \tag{32}$$

In $\mathbb{R}^n$, the curve with a minimum length will be a straight line, so that Equation (32) agrees with Equation (29). More generally, we call a solution to Equation (32) a geodesic [38].

In a vector space, metric properties can always be induced through an inner product (which also gives a notion of orthogonality). Such a space can be thought of as "flat", since for any two points, y and z, the straight line $ay + (1-a)z$, $a \in [0,1]$ is also contained in the space. In general, manifolds do not have vector space structure globally, but do so at the infinitesimal level. As such, we can think of them as "curved". We cannot always define an inner product, but we can still define distances through (32). We define a curve on a manifold, M, as $\gamma_M : [0,1] \to M$. At each point $\gamma_M(t) = p \in M$, the velocity vector, $\gamma'_M(t)$, lies in an n-dimensional vector space, which touches M at p. These are known as tangent spaces, denoted T_pM, which can be thought of as local linear approximations to M. We can define an inner product on each as $g_p : T_pM \to \mathbb{R}$, which allows us to define a generalisation of (31) as:

$$L(\gamma_M) = \int_0^1 \sqrt{g_p(\gamma'_M(t), \gamma'_M(t))}dt. \tag{33}$$

and provides a means to define a distance metric on the manifold as $d(x,y) = \inf\{L(\gamma_M) : \gamma_M(0) = x, \gamma_M(1) = y\}$. We emphasise the difference between this distance metric on M and g_p, which is called a Riemannian metric or metric tensor and which defines an inner product on T_pM.

Embeddings and Local Coordinates

So far, we have introduced manifolds as abstract objects. In fact, they can also be considered as objects that are embedded in some higher-dimensional Euclidean space. A simple example is any two-dimensional surface, such as the unit sphere, lying in $\mathbb{R}^3$. If a manifold is embedded in this way, then metric properties can be induced from the ambient Euclidean space.

We seek to make these ideas more concrete through an example, the graph of a function, $f(x_1, x_2)$, of two variables, x_1 and x_2. The resulting map, r, is:

$$r : \mathbb{R}^2 \to M \tag{34}$$

$$r(x_1, x_2) = (x_1, x_2, f(x_1, x_2)). \tag{35}$$

We can see that M is embedded in $\mathbb{R}^3$, but that any point can be identified using only two coordinates, x_1 and x_2. In this case, each T_pM is a plane, and therefore, a two-dimensional subspace of $\mathbb{R}^3$, so: (i) it inherits the Euclidean inner product, $\langle \cdot, \cdot \rangle$; and (ii) any vector, $v \in T_pM$, can be expressed as a linear combination of any two linearly independent basis vectors (a canonical choice is the partial derivatives $\partial r / \partial x_1 := r_1$ and r_2, evaluated at $x = r^{-1}(p) \in \mathbb{R}^2$). The resulting inner product, $g_p(v, w)$, between two vectors, $v, w \in T_pM$, can be induced from the Euclidean inner product as:

$$\begin{aligned}
\langle v, w \rangle &= \langle v_1 r_1(x) + v_2 r_2(x), w_1 r_1(x) + w_2 r_2(x) \rangle, \\
&= v_1 w_1 \langle r_1(x), r_1(x) \rangle + v_1 w_2 \langle r_1(x), r_2(x) \rangle + v_2 w_1 \langle r_2(x), r_1(x) \rangle + v_2 w_2 \langle r_2(x), r_2(x) \rangle, \\
&= v^T G(x) w,
\end{aligned}$$

where:

$$G(x) = \begin{pmatrix} \langle r_1(x), r_1(x) \rangle & \langle r_1(x), r_2(x) \rangle \\ \langle r_1(x), r_2(x) \rangle & \langle r_2(x), r_2(x) \rangle \end{pmatrix} \tag{36}$$

and we use v_i, w_i to denote the components of v and w. To write (31) using this notation, we define the curve, $x(t) \in \mathbb{R}^2$, corresponding to $\gamma_M(t) \in M$ as $x = (r^{-1} \circ \gamma_M) : [0,1] \to \mathbb{R}^2$. Equation (31) can then be written:

$$L(\gamma_M) = \int_0^1 \sqrt{x'(t)^T G(x(t)) x'(t)}dt, \tag{37}$$

which can be used in (32) as before.

The key point is that, although we have started with an object embedded in $\mathbb{R}^3$, we can compute the Riemannian metric, $g_p(v, w)$ (and, hence, distances in M), using only the two-dimensional "local" coordinates (x_1, x_2). We also need not have explicit knowledge of the mapping, r, only the components of the positive definite matrix, $G(x)$. The Nash embedding theorem [41] in essence enables us to define manifolds by the reverse process: simply choose the matrix, $G(x)$, so that we define a metric space with suitable distance properties, and some object embedded in some higher-dimensional Euclidean space will exist for which these metric properties can be induced as above. Therefore, to define our new space, we simply choose an appropriate matrix-valued map, $G(x)$ (we discuss this choice in Section 4.4). If $G(x)$ does not depend on x, then M has a vector space structure and can be thought of as "flat". Trivially, $G(x) = I$ gives Euclidean n-space.

We can also define volumes on a Riemannian manifold in local coordinates. Following standard coordinate transformation rules, we can see that for the above example, the area element, dx, in $\mathbb{R}^2$ will change according to a Jacobian $J = |(Dr)^T(Dr)|^{1/2}$, where $Dr = \partial(p_1, p_2, p_3)/\partial(x_1, x_2)$. This reduces to $J = |G(x)|^{1/2}$, which is also the case for more general manifolds [38]. We therefore define the Riemannian volume measure on a manifold, M, in local coordinates as:

$$\mathrm{Vol}_M(dx) = |G(x)|^{\frac{1}{2}} dx. \tag{38}$$

If $G(x) = I$, then this reduces to the Lebesgue measure.

4.3. Diffusions on Manifolds

By a "diffusion on a manifold" in local coordinates, we actually mean a diffusion defined on Euclidean space. For example, a realisation of Brownian motion on the surface, $S \subset \mathbb{R}^3$, defined in Figure 2 through $r(x_1, x_2) = (x_1, x_2, \sin(x_1) + 1)$ will be a sample path, which is defined on S and "looks locally" like Brownian motion in a neighbourhood of any point, $p \in S$. However, the pre-image of this sample path (through r^{-1}) will not be a realisation of a Brownian motion defined on $\mathbb{R}^2$, owing to the nonlinearity of the mapping. Therefore, to define "Brownian motion on S", we define some diffusion $(X_t)_{t\geq 0}$ that takes values in $\mathbb{R}^2$, for which the process $(r(X_t))_{t\geq 0}$ "looks locally" like a Brownian motion (and lies on S). See [42] for more intuition here.

Our goal, therefore, is to define a diffusion on Euclidean space, which, when mapped onto a manifold through r, becomes the Langevin diffusion described in (24) by the above procedure. Such a diffusion takes the form:

$$dX_t = \frac{1}{2}\tilde{\nabla} \log \tilde{\pi}(X_t)dt + d\tilde{B}_t, \tag{39}$$

where those objects marked with a tilde must be defined appropriately. The next few paragraphs are technical, and readers aiming to simply grasp the key points may wish to skip to the end of this Subsection.

We turn first to $(\tilde{B}_t)_{t\geq 0}$, which we use to denote Brownian motion on a manifold. Intuitively, we may think of a construction based on embedded manifolds, by setting $\tilde{B}_0 = p \in M$, and for each increment sampling some random vector in the tangent space T_pM, and then moving along the manifold in the prescribed direction for an infinitesimal period of time before re-sampling another velocity vector from the next tangent space [42]. In fact, we can define such a construction using Stratonovich calculus and show that the infinitesimal generator can be written using only local coordinates [28]. Here, we instead take the approach of generalising the generator directly from Euclidean space to the local coordinates of a manifold, arriving at the same result. We then deduce the stochastic differential equation describing $(\tilde{B}_t)_{t\geq 0}$ in Itô form using (20).

For a standard Brownian motion on $\mathbb{R}^n$, $\mathcal{A} = \triangle/2$, where $\triangle$ denotes the Laplace operator:

$$\triangle f = \sum_i \frac{\partial^2 f}{\partial x_i^2} = \mathrm{div}(\nabla f). \tag{40}$$

Substituting $\mathcal{A} = \triangle/2$ into (20) trivially gives $b_i(x) = 0\ \forall i$, $V_{ij}(x) = \mathbb{1}_{\{i=j\}}$, as required. The Laplacian, $\triangle f(x)$, is the divergence of the gradient vector field of some function, $f \in C^2(\mathbb{R}^n)$, and its value at $x \in \mathbb{R}^n$ can be thought of as the average value of f in some neighbourhood of x [43].

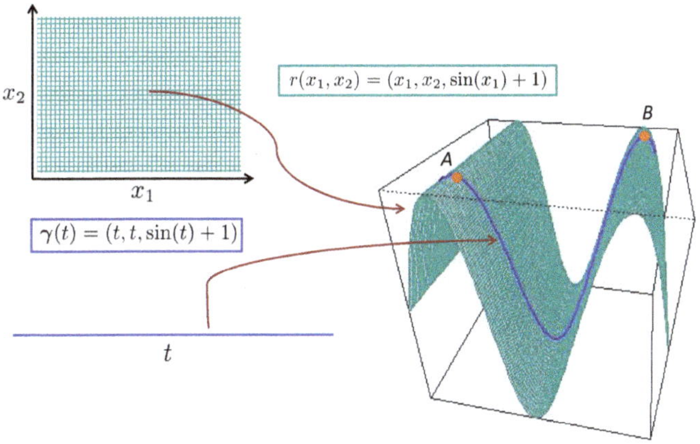

Figure 2. A two-dimensional manifold (surface) embedded in $\mathbb{R}^3$ through $r(x_1, x_2) = (x_1, x_2, \sin(x_1) + 1)$, parametrised by the local coordinates, x_1 and x_2. The distance between points A and B is given by the length of the curve $\gamma(t) = (t, t, \sin(t) + 1))$.

To define a Brownian motion on any manifold, the gradient and divergence must be generalised. We provide a full derivation in Appendix B, which shows that the gradient operator on a manifold can be written in local coordinates as $\nabla_M = G^{-1}(x)\nabla$. Combining with the operator, div_M, we can define a generalisation of the Laplace operator, known as the Laplace–Beltrami operator (e.g., [44,45]), as:

$$\triangle_{LB} f = \text{div}_M(\nabla_M f) = |G(x|^{-\frac{1}{2}} \sum_{i=1}^{n} \frac{\partial}{\partial x_i} \left(|G(x)|^{\frac{1}{2}} \sum_{j=1}^{n} \{G^{-1}(x)\}_{ij} \frac{\partial f}{\partial x_j} \right), \tag{41}$$

for some $f \in C_0^2(M)$.

The generator of a Brownian motion on M is $\triangle_{LB}/2$ [44]. Using (20), the resulting diffusion has dynamics given by:

$$d\tilde{B}_t = \Omega(X_t)dt + \sqrt{G^{-1}(X_t)}dB_t,$$
$$\Omega_i(X_t) = \frac{1}{2}|G(X_t)|^{-\frac{1}{2}} \sum_{j=1}^{n} \frac{\partial}{\partial x_j} \left(|G(X_t)|^{\frac{1}{2}} \{G^{-1}(X_t)\}_{ij} \right).$$

Those familiar with the Itô formula will not be surprised by the additional drift term, $\Omega(X_t)$. As Itô integrals do not follow the chain rule of ordinary calculus, non-linear mappings of martingales, such as $(B_t)_{t\geq 0}$, typically result in drift terms being added to the dynamics (e.g., [27]).

To define $\tilde{\nabla}$, we simply note that this is again the gradient operator on a general manifold, so $\tilde{\nabla} = G^{-1}(x)\nabla$. For the density, $\tilde{\pi}(x)$, we note that this density will now implicitly be defined with respect to the volume measure, $|G(x)|^{\frac{1}{2}}dx$, on the manifold. Therefore, to ensure the diffusion (39) has the correct invariant density with respect to the Lebesgue measure, we define:

$$\tilde{\pi}(x) = \pi(x)|G(x)|^{-\frac{1}{2}}. \tag{42}$$

Putting these three elements together, Equation (39) becomes:

$$dX_t = \frac{1}{2}G^{-1}(X_t)\nabla \log\left\{\pi(X_t)|G(X_t)|^{-\frac{1}{2}}\right\}dt + \Omega(X_t)dt + \sqrt{G^{-1}(X_t)}dB_t,$$

which, upon simplification, becomes:

$$dX_t = \frac{1}{2}G^{-1}(X_t)\nabla \log \pi(X_t)dt + \Lambda(X_t)dt + \sqrt{G^{-1}(X_t)}dB_t, \tag{43}$$

$$\Lambda_i(X_t) = \frac{1}{2}\sum_j \frac{\partial}{\partial x_j}\{G^{-1}(X_t)\}_{ij}. \tag{44}$$

It can be shown that this diffusion has invariant Lebesgue density $\pi(x)$, as required [33]. Intuitively, when a set is mapped onto the manifold, distances are changed by a factor, $\sqrt{G(x)}$. Therefore, to end up with the initial distances, they must first be changed by a factor of $\sqrt{G^{-1}(x)}$ before the mapping, which explains the volatility term in Equation (43).

The resulting Metropolis–Hastings proposal kernel for this "MALA on a manifold" was clarified in [33] and is given by:

$$Q(x, \cdot) \equiv \mathcal{N}\left(x + \frac{\lambda^2}{2}G^{-1}(x)\nabla \log \pi(x) + \lambda^2\Lambda(x), \lambda^2 G^{-1}(x)\right), \tag{45}$$

where λ^2 is a tuning parameter. The nonlinear drift term here is slightly different to that reported in [1,32], for reasons discussed in [33].

4.4. Choosing a Metric

We now turn to the question of which metric structure to put on the manifold, or equivalently, how to choose $G(x)$. In this section, we sometimes switch notation slightly, denoting the target density, $\pi(x|y)$, as some of the discussion is directed towards Bayesian inference, where $\pi(\cdot)$ is the posterior distribution for some parameter, x, after observing some data, y. The problem statement is: what is an appropriate choice of distance between points in the sample space of a given probability distribution?

A related (but distinct) question is how to define a distance between two probability distributions from the same parametric family, but with different parameters. This has been a key theme in information geometry, explored by Rao [46] and others [2] for many years. Although generic measures of distance between distributions (such as total variation) are often appropriate, based on information-theoretic principles, one can deduce that for a given parametric family, $\{p_x(y) : x \in \mathcal{X}\}$, it is in some sense natural to consider this "space of distributions" to be a manifold, where the Fisher information is the matrix, $G(x)$ (with the $\alpha = 0$ connection employed; see [2] for details).

Because of this, Girolami and Calderhead [1] proposed a variant of the Fisher metric for geometric Markov chain Monte Carlo, as:

$$G(x) = \mathbb{E}_{y|x}\left[-\frac{\partial^2}{\partial x_i \partial x_j}\log f(y|x)\right] - \frac{\partial^2}{\partial x_i \partial x_j}\log \pi_0(x), \tag{46}$$

where $\pi(x|y) \propto f(y|x)\pi_0(x)$ is the target density, f denotes the likelihood and π_0 the prior. The metric is tailored to Bayesian problems, which are a common use for MCMC, so the Fisher information is combined with the negative Hessian of the log-prior. One can also view this metric as the expected negative Hessian of the log target, since this naturally reduces to (46).

The motivation for a Hessian-style metric can also be understood from studying MCMC proposals. From (45) and by the same logic as for general pre-conditioning methods [32], the objective is to choose $G^{-1}(x)$ to match the covariance structure of $\pi(x|y)$ locally. If the target density were Gaussian with covariance matrix, Σ, then:

$$-\frac{\partial^2}{\partial x_i \partial x_j}\log \pi(x|y) = \Sigma. \tag{47}$$

In the non-Gaussian case, the negative Hessian is no longer constant, but we can imagine that it matches the correlation structure of $\pi(x|y)$ locally at least. Such ideas have been discussed in the geostatistics literature previously [47]. One problem with simply using (47) to define a metric is that unless $\pi(x|y)$ is log-concave, the negative Hessian will not be globally positive-definite, although Petra *et al.* [48] conjecture that it may be appropriate for use in some realistic scenarios and suggest some computationally efficient approximation procedures [48].

Example: Take $\pi(x) \propto 1/(1+x^2)$*, and set* $G(x) = -\partial^2 \log \pi(x)/\partial x^2$*. Then,* $G^{-1}(x) = (1+x^2)^2/(2-2x^2)$*, which is negative if* $x^2 > 1$*, so unusable as a proposal variance.*

Girolami and Calderhead [1] use the Fisher metric in part to counteract this problem. Taking expectations over the data ensures that the likelihood contribution to $G(x)$ in (46) will be positive (semi-)definite globally (e.g., [49]); so, provided a log-concave prior is chosen, then (46) should be a suitable choice for $G(x)$. Indeed, Girolami and Calderhead [1] provide several examples in which geometric MCMC methods using this Fisher metric perform better than their "non-geometric" counterparts.

Betancourt [50] also starts from the viewpoint that the Hessian (47) is an appropriate choice for $G(x)$ and defines a mapping from the set of $n \times n$ matrices to the set of positive-definite $n \times n$ matrices by taking a "smooth" absolute value of the eigenvalues of the Hessian. This is done in a way such that derivatives of $G(x)$ are still computable, inspiring the author to the name, SoftAbs metric. For a fixed value of x, the negative Hessian, $H(x)$, is first computed and, then, decomposed into $U^T DU$, where D is the diagonal matrix of eigenvalues. Each diagonal element of D is then altered by the mapping $t_\alpha : \mathbb{R} \to \mathbb{R}$, given by:

$$t_\alpha(\lambda_i) = \lambda_i \coth(\alpha \lambda_i), \tag{48}$$

where α is a tuning parameter (typically chosen to be as large as possible for which eigenvalues remain non-zero numerically). The function, t_α, acts as an absolute value function, but also uplifts eigenvalues, which are close to zero to $\approx 1/\alpha$. It should be noted that while the Fisher metric is only defined for models in which a likelihood is present and for which the expectation is tractable, the SoftAbs metric can be found for any target distribution, $\pi(\cdot)$.

Many authors (e.g., [1,48]) have noted that for many problems, the terms involving derivatives of $G(x)$ are often small, and so, it is not always worth the computational effort of evaluating them. Girolami and Calderhead [1] propose the simplified manifold, MALA, in which proposals are of the form:

$$Q(x, \cdot) \equiv \mathcal{N}\left(x + \frac{\lambda^2}{2} G^{-1}(x) \nabla \log \pi(x), \lambda^2 G^{-1}(x)\right) \tag{49}$$

Using this method means derivatives of $G(x)$ are no longer needed, so more pragmatic ways of regularising the Hessian are possible. One simple approach would be to take the absolute values of each eigenvalue, giving $G(x) = U^T|D|U$, where $H(x) = U^T DU$ is the negative Hessian and $|D|$ is a diagonal matrix with $\{|D|\}_{ii} = |\lambda_i|$ (this approach may fall into difficulties if eigenvalues are numerically zero). Another would be choosing $G(x)$ as the "nearest" positive-definite matrix to the negative Hessian, according to some distance metric on the set of $n \times n$ matrices. The problem has, in fact, been well-studied in mathematical finance, in the context of finding correlations using incomplete data sets [51], and tackled using distances induced by the Frobenius norm. Approximate solution algorithms are discussed in Higham [51]. It is not clear to us at present whether small changes to the Hessian would result in large changes to the corresponding positive definite matrix under a given distance or, indeed, whether given a distance metric on the space of matrices, there is always a well-defined unique "nearest" positive definite matrix. Below, we provide two simple examples, here showing how a "Hessian-style metric" can alleviate some of the difficulties associated with both heavy and light-tailed target densities.

Example: Take $\pi(x) \propto 1/(1+x^2)$*, and set* $G(x) = |-\partial^2 \log \pi(x)/\partial x^2|$*. Then,* $G^{-1}(x)\nabla \log \pi(x) = -x(1+x^2)/|1-x^2|$*, which no longer tends to zero as* $|x| \to \infty$*, suggesting a manifold variant of MALA with a Hessian-style metric may avoid some of the pitfalls of the standard algorithm. Note that the drift may become very large if* $|x| \approx 1$*, but since this event occurs with probability zero, we do not see it as a major cause for concern.*

Example: Take $\pi(x) \propto e^{-x^4}$*, and set* $G(x) = |-\partial^2 \log \pi(x)/\partial x^2|$*. Then,* $G^{-1}(x)\nabla \log \pi(x) = -x/3$*, which is* $O(x)$*, so alleviating the problem of spiralling proposals for light-tailed targets demonstrated by MALA in an earlier example.*

Other choices for $G(x)$ have been proposed, which are not based on the Hessian. These have the advantage that gradients need not be computed (either analytically or using computational methods). Sejdinovic *et al.* [52] propose a Metropolis–Hastings method, which can be viewed as a geometric variant of the RWM, where the choice for $G(x)$ is based on mapping samples to an appropriate feature space, and performing principal component analysis on the resulting features to choose a local covariance structure for proposals.

If we consider the RWM with Gaussian proposals to be a Euler–Maruyama discretisation of Brownian motion on a manifold, then proposals will take the form $Q(x,\cdot) \equiv \mathcal{N}(x + \lambda^2\Omega(x), \lambda^2 G^{-1}(x))$. If we assume (like in the simplified manifold MALA) that $\Omega(x) \approx 0$, then we have proposals centred at the current point in the Markov chain with a local covariance structure (the full Hastings acceptance rate must now be used as $q(x'|x) \neq q(x|x')$ in general).

As no gradient information is needed, the Sejdinovic *et al.* metric can be used in conjunction with the pseudo-marginal MCMC algorithm, so that $\pi(x|y)$ need not be known exactly. Examples from the article demonstrate the power of the approach [52].

An important property of any Riemannian metric is how it transforms under coordinate change (e.g., [2]). The Fisher information metric commonly studied in information geometry is an example of a "coordinate invariant" choice for $G(x)$. If we consider two parametrisations for a statistical model given by x and $z = t(x)$, computing the Fisher information under x and then transforming this matrix using the Jacobian for the mapping, t, will give the same result as computing the Fisher information under z. It should be noted that because of either the prior contribution in (46) or the nonlinear transformations applied in other cases, none of the metrics we have reviewed here have this property, which means that we have no principled way of understanding how $G(x)$ will relate to $G(z)$. It is intuitive, however, that using information from all of $\pi(x)$, rather than only the likelihood contribution, $f(y|x)$, would seem sensible when trying to sample from $\pi(\cdot)$.

5. Survey of Applications

Rather than conduct our own simulation study, we instead highlight some cases in the literature where geometric MCMC methods have been used with success.

Martin *et al.* [53] consider Bayesian inference for a statistical inverse problem, in which a surface explosion causes seismic waves to travel down into the ground (the subsurface medium). Often, the properties of the subsurface vary with distance from ground level or because of obstacles in the medium, in which case, a fraction of the waves will scatter off these boundaries and be reflected back up to ground level at later times. The observations here are the initial explosion and the waves, which return to the surface, together with return times. The challenge is to infer the properties of the subsurface medium from this data. The authors construct a likelihood based on the wave equation for the data and perform Bayesian inference using a variant of the manifold MALA. Figures are provided showing the local correlations present in the posterior and, therefore, highlighting the need for an algorithm that can navigate the high density region efficiently. Several methods are compared in the paper, but the variant of MALA that incorporates a local correlation structure is shown to be the most efficient, particularly as the dimension of the problem increases [53].

Calderhead and Girolami [54] dealt with two models for biological phenomena based on nonlinear dynamical systems. A model of circadian control in the *Arabidopsis thaliana* plant comprised a system of six nonlinear differential equations, with twenty two parameters to be inferred. Another model for cell signalling consisted of a system of six nonlinear differential equations with eight parameters, with inference complicated by the fact that observations of the model are not recorded directly [54]. The resulting inference was performed using RWM, MALA and geometric methods, with the results highlighting the benefits of taking the latter approach. The simplified variant of MALA on a manifold is reported to have produced the most efficient inferences overall, in terms of effective sample size per unit of computational time.

Stathopoulos and Girolami [55] considered the problem of inferring parameters in Markov jump processes. In the paper, a linear noise approximation is shown, which can make inference in such models more straightforward, enabling an approximate likelihood to be computed. Models based on chemical reaction dynamics are considered; one such from chemical kinetics contained four unknown parameters; another from gene expression consisted of seven. Inference was performed using the RWM, the simplified manifold MALA and Hamiltonian methods, with the MALA reported as most efficient according to the chosen diagnostics. The authors note that the simplified manifold method is both conceptually simple and able to account for local correlations, making it an attractive choice for inference [55].

Konukoglu *et al.* [56] designed a method for personalising a generic model for a physiological process to a specific patient, using clinical data. The personalisation took the form of patient-specific parameter inference. The authors highlight some of the difficulties of this task in general, including the complexity of the models and the relative sparsity of the datasets, which often result in a parameter identifiability issue [56]. The example discussed in the paper is the Eikonal-diffusion model describing electrical activity in cardiac tissue, which results in a likelihood for the data based on a nonlinear partial differential equation, combined with observation noise [56]. A method for inference was developed by first approximating the likelihood using a spectral representation and then using geometric MCMC methods on the resulting approximate posterior. The method was first evaluated on synthetic data and then repeated on clinical data taken from a study for ventricular tachycardia radio-frequency ablation [56].

6. Discussion

The geometric viewpoint in not necessary to understand manifold variants of the MALA. Indeed, several authors [32,33] have discussed these algorithms without considering them to be "geometric", rather simply Metropolis–Hastings methods in which proposal kernels have a position-dependent covariance structure. We do not claim that the geometric view is the only one that should be taken. Our goal is merely to point out that such position-dependent methods can often be viewed as methods defined on a manifold and that studying the structure of the manifold itself may lead to new insights on the methods. For example, taking the geometric viewpoint and noting the connection with information geometry enabled Girolami and Calderhead to adopt the Fisher metric for calculations [1]. We list here a few open questions on which the geometric viewpoint may help shed some insight.

Computationally-minded readers will have noted that using position-dependent covariance matrices adds a significant computational overhead in practice, with additional $O(n^3)$ matrix inversions required at each step of the corresponding Metropolis–Hastings algorithms. Clearly, there will be many problems for which the matrix, $G(x)$, does not change very much, and therefore, choosing a constant covariance $G^{-1}(x) = \Sigma$ may result in a more efficient algorithm overall. Geometrically, this would correspond to a manifold with scalar curvature close to zero everywhere. It may be that geometric ideas could be used to understand whether the manifold is flat enough that a constant choice of $G(x)$ is sufficient. To make sense of this truly would require a relationship between curvature, an inherently local property and more global statements about the manifold. Many results in differential geometry, beginning with the celebrated Gauss–Bonnet theorem, have previously related global and local properties in this way [57]. It is unknown to the authors whether results exist relating the curvature of a manifold to some global property, but this is an interesting avenue for further research.

A related question is when to choose the simplified manifold MALA over the full method. Problems in which the term, $\|\Lambda(x)\|$, is sufficient large to warrant calculation correspond to those for which the manifold has very high curvature in many places; so again, making some global statement related to curvature could help here.

Although there is a reasonably intuitive argument for why the Hessian is an appropriate starting point for $G(x)$, the lack of positive-definiteness may be seen as a cause for concern by some. After all, it could be argued that if the curvature is not positive-definite in a region, then how can it be a reasonable approximation to the local covariance structure. Many statistical models used to describe natural phenomena are characterised by distributions with heavy tails or multiple modes, for which this is the case. In addition, for target densities of the form $\pi(x) \propto e^{-|x|}$, the Hessian is everywhere equal to zero!The attempts to force positive-definiteness we have described will typically result in small moves being proposed in such regions of the sample space, which may not be an optimal strategy. Much work in information geometry has centred on the geometry of Hessian structures [58], and some insights from this field may help to better understand the question of choosing an appropriate metric. In addition, the field of pseudo-Riemannian geometry deals with forms of $G(x)$, which need not be positive-definite [39]; so again, understanding could be gained from here.

Some recent work in high-dimensional inference has centred on defining MCMC methods for which efficiency scales $O(1)$ with respect to the dimension, n, of $\pi(\cdot)$ [19,59]. In the case where X takes values in some infinite-dimensional function space, this can be done provided a Gaussian prior measure is defined for X. A striking result from infinite-dimensional probability spaces is that two different probability measures defined over some infinite dimensional space have a striking tendency to have disjoint supports [60]. The key challenge for MCMC is to define transition kernels for which proposed moves are inside the support for $\pi(\cdot)$. A straight-forward approach is to define proposals for which the prior is invariant, since the likelihood contribution to the posterior typically will not alter its support from that of the prior [19]. However, the posterior may still look very different from the prior, as noted in [61], so this proposal mechanism, though $O(1)$, can still result in slow exploration. Understanding the geometry of the support and defining methods that incorporate the likelihood term, but also respect this geometry, so as to ensure proposals remain in support of $\pi(\cdot)$, is an intriguing research proposition.

The methods reviewed in this paper are based on first order Langevin diffusions. Algorithms have also been developed that are based on second order Langevin diffusions, in which a stochastic differential equation governs the behaviour of the velocity of a process [62,63]. A natural extension to the work of Girolami and Calderhead [1] and Xifara *et al.* [33] would be to map such diffusions onto a manifold and derive Metropolis–Hastings proposal kernels based on the resulting dynamics. The resulting scheme would be a generalisation of [63], though the most appropriate discretisation scheme for a second order process to facilitate sampling is unclear and perhaps a question worthy of further exploration.

We have focused primarily here on the sample space $\mathcal{X} = \mathbb{R}^n$ and on defining an appropriate manifold on which to construct Markov chains. In some inference problems, however, the sample space is a pre-defined manifold, for example the set of $n \times n$ rotation matrices, commonly found in the field of directional statistics [64]. Such manifolds are often not globally mappable to Euclidean n-space. Methods have been devised for sampling from such spaces [65,66]. In order to use the methods described here for such problems, an appropriate approach for switching between coordinate patches at the relevant time would need to be devised, which could be an interesting area of further study.

Alongside these geometric problems, we can also discuss geometric MCMC methods from a statistical perspective. The last example given in the previous section hinted that the manifold MALA may cope better with target distributions with heavy tails. In fact, Latuszynski *et al.* [67] have shown that, in one dimension, the manifold MALA is geometrically ergodic for a class of targets of the form $\pi(x) \propto \exp(-|x|^\beta)$ for any choice of $\beta \neq 1$. This incorporates cases where tails are heavier than exponential and lighter than Gaussian, two scenarios under which geometric ergodicity fails for the MALA.

Finding optimal acceptance rates and scaling of λ with dimension are two other related challenges. In this case, the picture is more complex. Traditional results have been shown for Metropolis–Hastings methods in the case where target distributions are independent and identically-distributed or some other suitable symmetry and regularity in the shape of $\pi(\cdot)$. Manifold methods are, however, specifically tailored to scenarios in which this is not the case, scenarios in which there is a high correlation between components of x, which changes depending on the value of x. It is less clear how to proceed with finding relevant results that can serve as guidelines to practitioners here. Indeed, Sherlock [18] notes that a requirement for optimal acceptance rate results for the RWM to be appropriate is that the curvature of $\pi(x)$ does not change too much, yet this is the very scenario in which we would want to use a manifold method.

Acknowledgments: We thank the two reviewers for helpful comments and suggestions. Samuel Livingstone is funded by a PhD Scholarship from Xerox Research Centre Europe. Mark Girolami is funded by an Engineering and Physical Sciences Research Council Established Career Research Fellowship, EP/J016934/1, and a Royal Society Wolfson Research Merit Award.

Author Contributions: Author Contributions

The article was written by Samuel Livingstone under the guidance of Mark Girolami. All authors have read and approved the final manuscript.

Appendix

Appendix Total Variation Distance

We show how to obtain (10) from (9). Denoting two probability distributions, $\mu(\cdot)$ and $\nu(\cdot)$, and associated densities, $\mu(x)$ and $\nu(x)$, we have:

$$\|\mu(\cdot) - \nu(\cdot)\|_{TV} := \sup_{A \in \mathcal{B}} |\mu(A) - \nu(A)|.$$

Define the set $B = \{x \in \mathcal{X} : \mu(x) > \nu(x)\}$. To see that $B \in \mathcal{B}$, note that $B = \cup_{q \in \mathbb{Q}} \{x \in \mathcal{X} : \mu(x) > q\} \cap \{x \in \mathcal{X} : \nu(x) < q\}$, and the result follows from properties of $\mathcal{B}$ (e.g., [68]). Now, for any $A \in \mathcal{B}$:

$$\mu(A) - \nu(A) \leq \mu(A \cap B) - \nu(A \cap B) \leq \mu(B) - \nu(B),$$

and similarly:

$$\nu(A) - \mu(A) \leq \nu(B^c) - \mu(B^c),$$

so, the supremum will be attained either at B or B^c. However, since $\mu(\mathcal{X}) = \nu(\mathcal{X}) = 1$, then:

$$[\mu(B) - \nu(B)] - [\nu(B^c) - \mu(B^c)] = 0,$$

so that

$$|\mu(B) - \nu(B)| = |\mu(B^c) - \nu(B^c)|.$$

Using these facts gives an alternative characterisation of the total variation distance as:

$$\begin{aligned} \|\mu(\cdot) - \nu(\cdot)\|_{TV} &= \frac{1}{2}\left(|\mu(B) - \nu(B)| + |\mu(B^c) - \nu(B^c)|\right) \\ &= \frac{1}{2}\int_{\mathcal{X}} |\mu(x) - \nu(x)| dx \end{aligned}$$

as required.

Appendix Gradient and Divergence Operators on a Riemannian Manifold

The gradient of a function on $\mathbb{R}^n$ is the unique vector field, such that, for any unit vector, u:

$$\langle \nabla f(x), u \rangle = D_u \left[f(x) \right] = \lim_{h \to 0} \left\{ \frac{f(x+hu) - f(x)}{h} \right\}, \tag{A1}$$

the directional derivative of f along u at $x \in \mathbb{R}^n$.

On a manifold, the gradient operator, ∇_M, can still be defined, such that the inner product $g_p(\nabla_M f(x), u) = D_u[f(x)]$. Setting $\nabla_M = G(x)^{-1}\nabla$ gives:

$$\begin{aligned} g_p(\nabla_M f(x), u) &= (G^{-1}(x)\nabla f(x))^T G(x) u, \\ &= \langle \nabla f(x), u \rangle, \end{aligned}$$

which is equal to the directional derivative along u as required.

The divergence of some vector field, v, at a point, $x \in \mathbb{R}^n$, is the net outward flow generated by v through some small neighbourhood of x. Mathematically, the divergence of $v(x) \in \mathbb{R}^3$ is given by $\sum_i \partial v_i / \partial x_i$. On a more general manifold, the divergence is also a sum of derivatives, but here, they are covariant derivatives. A short introduction is provided in Appendix C. Here, we simply state that the covariant derivative of a vector field, v, at a point $p \in M$ is the orthogonal projection of the directional derivative onto the tangent space, T_pM. Intuitively, a vector field on a manifold is a field of vectors, each of which lie in the tangent space to a point, $p \in M$. It only makes sense therefore to discuss how vector fields change along the manifold or in the direction of vectors, which also lie in the tangent space. Although the idea seems simple, the covariant derivative has some attractive geometric properties; notably, it can be completely written in local coordinates,and, so, does not depend on knowledge of an embedding in some ambient space.

The divergence of a vector field, v, defined on a manifold, M, at the point, $p \in M$, is defined as:

$$\text{div}_M(v) = \sum_{i=1}^{n} D^c_{e_i}[v_i],$$

where e_i denotes the i-th basis vector for the tangent space, T_pM, at $p \in M$, and v_i denotes the i-th coefficient. This can be written in local coordinates (see Appendix C) as:

$$\text{div}_M(v) = |G(x)|^{-\frac{1}{2}} \sum_{i=1}^{n} \frac{\partial}{\partial x_i} \left(|G(x)|^{\frac{1}{2}} v_i \right),$$

and can be combined with ∇_M to form the Laplace–Beltrami operator (41).

Appendix Vector Fields and the Covariant Derivative

Here, we provide a short introduction to vector fields and differentiation on a smooth manifold; see [38,39]. The following geometric notation is used here: (i) vector components are indexed with a superscript, e.g., $v = (v^1, ..., v^n)$; and (ii) repeated subscripts and superscripts are summed over, e.g., $v^i \mathbf{e}_i = \sum_i v^i \mathbf{e}_i$ (known as the Einstein summation convention).

For any smooth manifold, M, the set of all tangent vectors to points on M is known as the tangent bundle and denoted TM.

A C^r vector field defined on M is a mapping that assigns to each point, $p \in M$, a tangent vector, $v(p) \in T_pM$. In addition, the components of $v(p)$ in any basis for T_pM must also be C^r [38]. We will denote the set of all vector fields on M as $\Gamma(TM)$. For some vector field, $v \in \Gamma(TM)$, at any point, $p \in M$, the vector, $v(p) \in T_pM$, can be written as a linear combination of some n basis vectors $\{\mathbf{e}_1, ..., \mathbf{e}_n\}$ as $v = v^i \mathbf{e}_i$. To understand how v will change in a particular direction along M, it only makes sense, therefore, to consider derivatives along vectors in T_pM. Two other things must be considered when defining a derivative along a manifold: (i) how the components, v^i, of each basis vector will change; and (ii) how each basis vector, $\mathbf{e}_i$, itself will change. For the usual directional derivative on $\mathbb{R}^n$, the basis vectors do not change, as the tangent space is the same at each point, but for a more general manifold, this is no longer the case: the $\mathbf{e}_i$'s are referred to as a "local" basis for each T_pM.

The covariant derivative, D^c, is defined so as to account for these shortcomings. When considering differentiation along a vector, $u^* \notin T_pM$, u^* is simply projected onto the tangent space. The derivative with respect to any $u \in T_pM$ can now be decomposed into a linear combination of derivatives of basis vectors and vector components:

$$D^c_u[v] = D^c_{u^i\mathbf{e}_i}[v^i\mathbf{e}_i], \tag{A2}$$

where the argument, p, has been dropped, but is implied for both components and local basis vectors. The operator, $D^c_u[v]$, is defined to be linear in both u and v and to satisfy the product rule [38]; so, Equation (A2) can be decomposed into:

$$D^c_u[v] = u^i\left(D^c_{\mathbf{e}_i}[v^j]\mathbf{e}_j + v^j D^c_{\mathbf{e}_i}[\mathbf{e}_j]\right). \tag{A3}$$

The operator, D^c, need, therefore, only be defined along the direction of basis vectors $\mathbf{e}_i$ and for vector component v^i and basis vector $\mathbf{e}_i$ arguments.

For components v^i, $D^c_{\mathbf{e_j}}[v^i]$ is defined as simply the partial derivative $\partial_j v^i := \partial v^i/\partial x^j$. The directional derivative of some basis vector $\mathbf{e}_i$ along some $\mathbf{e}_j$ is best understood through the example of a regular surface $\Sigma \subset \mathbb{R}^3$. Here, $D_{\mathbf{e}_j}[\mathbf{e}_i]$ will be a vector, $w \in \mathbb{R}^3$. Taking the basis for this space at the point, p, as $\{\mathbf{e}_1, \mathbf{e}_2, \hat{\mathbf{n}}\}$, where $\hat{\mathbf{n}}$ denotes the unit normal to $T_p\Sigma$, we can write $w = \alpha\mathbf{e}_1 + \beta\mathbf{e}_2 + \kappa\hat{\mathbf{n}}$. The covariant derivative, $D^c_{\mathbf{e}_j}[\mathbf{e}_i]$, is simply the projection of w onto $T_p\Sigma$, given by $w^* = \alpha\mathbf{e}_1 + \beta\mathbf{e}_2$. More generally, at some point, p, in a smooth manifold, M, the covariant derivative $D^c_{\mathbf{e}_j}[\mathbf{e}_i] = \Gamma^k_{ji}\mathbf{e}_k$ (with upper and lower indices summed over). The coefficients, Γ^k_{ji}, are known as the Christoffel symbols: Γ^k_{ji} denotes the coefficient of the k-th basis vector when taking the derivative of the i-th with respect to the j-th. If a Riemannian metric, g, is chosen for M; then, they can be expressed completely as a function of g (or in local coordinates as a function of the matrix, G). Using these definitions, Equation (A3) can be re-written as:

$$D^c_u[v] = u^i\left(\partial_i v^k + v^j\Gamma^k_{ij}\right)\mathbf{e}_k. \tag{A4}$$

The divergence of a vector field, $v \in \Gamma(TM)$, at the point, $p \in M$, is given by:

$$\mathrm{div}_M(v) = D^c_{\mathbf{e}_i}[v^i], \tag{A5}$$

where, again, repeated indices are summed over. If $M = \mathbb{R}^n$, this reduces to the usual sum of partial derivatives, $\partial_i v^i$. On a more general manifold, M, the equivalent expression is:"'

$$D^c_{\mathbf{e}_i}[v^i] = \partial_i v^i + v^i\Gamma^j_{ij}, \tag{A6}$$

where, again, repeated indices are summed. As has been previously stated, if a metric, g, and coordinate chart is chosen for M, the Christoffel symbols can be written in terms of the matrix, $G(x)$. In this case [69]:

$$\Gamma^j_{ij} = |G(x)|^{-\frac{1}{2}}\partial_i\left(|G(x)|^{\frac{1}{2}}\right), \tag{A7}$$

so Equation (A6) becomes:

$$D^c_{\mathbf{e}_i}[v^i] = |G(x)|^{-\frac{1}{2}}\partial_i\left(|G(x)|^{\frac{1}{2}}v^i\right), \tag{A8}$$

where $v = v(x)$.

Conflicts of Interest: Conflicts of Interest

The authors declare no conflict of interest.

References

1. Girolami, M.; Calderhead, B. Riemann manifold Langevin and Hamiltonian Monte Carlo methods. *J. R. Stat. Soc. Ser. B* **2011**, *73*, 123–214.
2. Amari, S.I.; Nagaoka, H. *Methods of Information Geometry*; American Mathematical Society: Providence, RI, USA, 2007; Volume 191.
3. Marriott, P.; Salmon, M. *Applications of Differential Geometry to Econometrics*; Cambridge University Press: Cambridge, UK, 2000.
4. Betancourt, M.; Girolami, M. Hamiltonian Monte Carlo for Hierarchical Models. **2013**, arXiv: 1312.0906.
5. Neal, R. MCMC using Hamiltonian Dynamics. In *Handbook of Markov Chain Monte Carlo*; Chapman and Hall/CRC: Boca Raton, FL, USA, 2011; pp. 113–162.
6. Betancourt, M.; Stein, L.C. The Geometry of Hamiltonian Monte Carlo. **2011**, arXiv: 1112.4118.
7. Robert, C.P.; Casella, G. *Monte Carlo Statistical Methods*; Springer: New York, NY, USA, 2004; Volume 319.
8. Tierney, L. Markov chains for exploring posterior distributions. *Ann. Stat.* **1994**, *22*, 1701–1728.
9. Kipnis, C.; Varadhan, S. Central limit theorem for additive functionals of reversible Markov processes and applications to simple exclusions. *Commun. Math. Phys.* **1986**, *104*, 1–19.
10. R Core Team. *R: A Language and Environment for Statistical Computing*; R Foundation for Statistical Computing: Vienna, Austria, 2012.
11. Plummer, M.; Best, N.; Cowles, K.; Vines, K. CODA: Convergence diagnosis and output analysis for MCMC. *R. News* **2006**, *6*, 7–11.
12. Gibbs, A.L.; Su, F.E. On choosing and bounding probability metrics. *Int. Stat. Rev.* **2002**, *70*, 419–435.
13. Jones, G.L.; Hobert, J.P. Honest exploration of intractable probability distributions via Markov chain Monte Carlo. *Stat. Sci.* **2001**, *16*, 312–334.
14. Jones, G.L. On the Markov chain central limit theorem. *Probab. Surv.* **2004**, *1*, 299–320.
15. Gelman, A.; Rubin, D.B. Inference from iterative simulation using multiple sequences. *Stat. Sci.* **1992**, *7*, 457–472.
16. Sherlock, C.; Fearnhead, P.; Roberts, G.O. The random walk Metropolis: Linking theory and practice through a case study. *Stat. Sci.* **2010**, *25*, 172–190.
17. Sherlock, C.; Roberts, G. Optimal scaling of the random walk Metropolis on elliptically symmetric unimodal targets. *Bernoulli* **2009**, *15*, 774–798.
18. Sherlock, C. Optimal scaling of the random walk Metropolis: General criteria for the 0.234 acceptance rule. *J. Appl. Probab.* **2013**, *50*, 1–15.
19. Beskos, A.; Kalogeropoulos, K.; Pazos, E. Advanced MCMC methods for sampling on diffusion pathspace. *Stoch. Processes Appl.* **2013**, *123*, 1415–1453.
20. Roberts, G.O.; Rosenthal, J.S. Optimal scaling for various Metropolis–Hastings algorithms. *Stat. Sci.* **2001**, *16*, 351–367.
21. Roberts, G.O.; Tweedie, R.L. Geometric convergence and central limit theorems for multidimensional Hastings and Metropolis algorithms. *Biometrika* **1996**, *83*, 95–110.
22. Mengersen, K.L.; Tweedie, R.L. Rates of convergence of the Hastings and Metropolis algorithms. *Ann. Stat.* **1996**, *24*, 101–121.
23. Jarner, S.F.; Hansen, E. Geometric ergodicity of Metropolis algorithms. *Stoch. Processes Appl.* **2000**, *85*, 341–361.
24. Christensen, O.F.; Møller, J.; Waagepetersen, R.P. Geometric ergodicity of Metropolis–Hastings algorithms for conditional simulation in generalized linear mixed models. *Methodol. Comput. Appl. Probab.* **2001**, *3*, 309–327.
25. Neal, P.; Roberts, G. Optimal scaling for random walk Metropolis on spherically constrained target densities. *Methodol. Comput. Appl. Probab.* **2008**, *10*, 277–297.
26. Jarner, S.F.; Tweedie, R.L. Necessary conditions for geometric and polynomial ergodicity of random-walk-type. *Bernoulli* **2003**, *9*, 559–578.
27. Øksendal, B. *Stochastic Differential Equations*; Springer: New York, NY, USA, 2003.

28. Rogers, L.C.G.; Williams, D. *Diffusions, Markov Processes and Martingales: Volume 2, Itô Calculus*; Cambridge University Press: Cambridge, UK, 2000; Volume 2.
29. Meyn, S.P.; Tweedie, R.L. Stability of Markovian processes III: Foster–Lyapunov criteria for continuous-time processes. *Adv. Appl. Probab.* **1993**, *25*, 518–518.
30. Coffey, W.; Kalmykov, Y.P.; Waldron, J.T. *The Langevin Equation: with Applications to Stochastic Problems in Physics, Chemistry, and Electrical Engineering*; World Scientific: Singapore, Singapore, 2004; Volume 14.
31. Roberts, G.O.; Tweedie, R.L. Exponential convergence of Langevin distributions and their discrete approximations. *Bernoulli* **1996**, *2*, 341–363.
32. Roberts, G.O.; Stramer, O. Langevin diffusions and Metropolis–Hastings algorithms. *Methodol. Comput. Appl. Probab.* **2002**, *4*, 337–357.
33. Xifara, T.; Sherlock, C.; Livingstone, S.; Byrne, S.; Girolami, M. Langevin diffusions and the Metropolis-adjusted Langevin algorithm. *Stat. Probab. Lett.* **2013**, *91*, 14–19.
34. Jeffreys, H. An invariant form for the prior probability in estimation problems. *Proc. R. Soc. Lond. Ser. A Math. Phys. Sci.* **1946**, *186*, 453–461.
35. Critchley, F.; Marriott, P.; Salmon, M. Preferred point geometry and statistical manifolds. *Ann. Stat.* **1993**, *21*, 1197–1224.
36. Marriott, P. On the local geometry of mixture models. *Biometrika* **2002**, *89*, 77–93.
37. Barndorff-Nielsen, O.; Cox, D.; Reid, N. The role of differential geometry in statistical theory. *Int. Stat. Rev.* **1986**, *54*, 83–96.
38. Boothby, W.M. *An Introduction to Differentiable Manifolds and Riemannian Geometry*; Academic Press: San Diego, CA, USA, 1986; Volume 120.
39. Lee, J.M. *Smooth Manifolds*; Springer: New York, NY, USA, 2003.
40. Do Carmo, M.P. *Riemannian Geometry*; Springer: New York, NY, USA, 1992.
41. Nash, J.F., Jr. The imbedding problem for Riemannian manifolds. In *The Essential John Nash*; Princeton University Press: Princeton, NJ, USA, 2002; p. 151.
42. Manton, J.H. A Primer on Stochastic Differential Geometry for Signal Processing. **2013**, arXiv: 1302.0430.
43. Stewart, J. *Multivariable Calculus*; Cengage Learning: Boston, MA, USA, 2011.
44. Hsu, E.P. *Stochastic Analysis on Manifolds*; American Mathematical Society: Providence, RI, USA, 2002; Volume 38.
45. Kent, J. Time-reversible diffusions. *Adv. Appl. Probab.* **1978**, *10*, 819–835.
46. Radhakrishna Rao, C. Information and accuracy attainable in the estimation of statistical parameters. *Bull. Calcutta Math. Soc.* **1945**, *37*, 81–91.
47. Christensen, O.F.; Roberts, G.O.; Sköld, M. Robust Markov chain Monte Carlo methods for spatial generalized linear mixed models. *J. Comput. Graph. Stat.* **2006**, *15*, 1–17.
48. Petra, N.; Martin, J.; Stadler, G.; Ghattas, O. A computational framework for infinite-dimensional Bayesian inverse problems: Part II. Stochastic Newton MCMC with application to ice sheet flow inverse problems. **2013**, arXiv: 1308.6221.
49. Pawitan, Y. *In All Likelihood: Statistical Modelling and Inference Using Likelihood*; Oxford University Press: Oxford, UK, 2001.
50. Betancourt, M. A General Metric for Riemannian Manifold Hamiltonian Monte Carlo. In *Geometric Science of Information*; Springer: New York, NY, USA, 2013; pp. 327–334.
51. Higham, N.J. Computing the nearest correlation matrix—a problem from finance. *IMA J. Numer. Anal.* **2002**, *22*, 329–343.
52. Sejdinovic, D.; Garcia, M.L.; Strathmann, H.; Andrieu, C.; Gretton, A. Kernel Adaptive Metropolis–Hastings. **2013**, arXiv: 1307.5302.
53. Martin, J.; Wilcox, L.C.; Burstedde, C.; Ghattas, O. A stochastic Newton MCMC method for large-scale statistical inverse problems with application to seismic inversion. *SIAM J. Sci. Comput.* **2012**, *34*, A1460–A1487.
54. Calderhead, B.; Girolami, M. Statistical analysis of nonlinear dynamical systems using differential geometric sampling methods. *Interface Focus* **2011**, *1*, 821–835.
55. Stathopoulos, V.; Girolami, M.A. Markov chain Monte Carlo inference for Markov jump processes via the linear noise approximation. *Philos. Trans. R. Soc. A* **2013**, *371*, 20110541.

56. Konukoglu, E.; Relan, J.; Cilingir, U.; Menze, B.H.; Chinchapatnam, P.; Jadidi, A.; Cochet, H.; Hocini, M.; Delingette, H.; Jaïs, P.; *et al.* Efficient probabilistic model personalization integrating uncertainty on data and parameters: Application to eikonal-diffusion models in cardiac electrophysiology. *Prog. Biophys. Mol. Biol.* **2011**, *107*, 134–146.
57. Do Carmo, M.P.; Do Carmo, M.P. *Differential Geometry of Curves and Surfaces*; Englewood Cliffs: Prentice-Hall, NJ, USA, 1976; Volume 2.
58. Shima, H. *The Geometry of Hessian Structures*; World Scientific: Singapore, Singapore, 2007; Volume 1.
59. Cotter, S.; Roberts, G.; Stuart, A.; White, D. MCMC methods for functions: Modifying old algorithms to make them faster. *Stat. Sci.* **2013**, *28*, 424–446.
60. Da Prato, G.; Zabczyk, J. *Stochastic Equations in Infinite Dimensions*; Cambridge University Press: Cambridge, UK, 2008.
61. Law, K.J. Proposals which speed up function-space MCMC. *J. Comput. Appl. Math.* **2014**, *262*, 127–138.
62. Ottobre, M.; Pillai, N.S.; Pinski, F.J.; Stuart, A.M. A Function Space HMC Algorithm With Second Order Langevin Diffusion Limit. **2013**, arXiv: 1308.0543.
63. Horowitz, A.M. A generalized guided Monte Carlo algorithm. *Phys. Lett. B* **1991**, *268*, 247–252.
64. Mardia, K.V.; Jupp, P.E. *Directional Statistics*; Wiley: New York, NY, USA, 2009; Volume 494.
65. Byrne, S.; Girolami, M. Geodesic Monte Carlo on embedded manifolds. *Scand. J. Stat.* **2013**, *40*, 825–845.
66. Diaconis, P.; Holmes, S.; Shahshahani, M. Sampling from a manifold. In *Advances in Modern Statistical Theory and Applications: A Festschrift in Honor of Morris L. Eaton*; Institute of Mathematical Statistics: Washington, DC, USA, 2013; pp. 102–125.
67. Latuszynski, K.; Roberts, G.O.; Thiery, A.; Wolny, K. Discussion on "Riemann manifold Langevin and Hamiltonian Monte Carlo methods" (by Girolami, M. and Calderhead, B.). *J. R. Stat. Soc. Ser. B* **2011**, *73*, 188–189.
68. Capinski, M.; Kopp, P.E. *Measure, Integral and Probability*; Springer: New York, NY, USA, 2004.
69. Schutz, B.F. *Geometrical Methods of Mathematical Physics*; Cambridge University Press: Cambridge, UK, 1984.

entropy

MDPI

Article

Variational Bayes for Regime-Switching Log-Normal Models

Hui Zhao and Paul Marriott *

University of Waterloo, 200 University Avenue West, Waterloo, ON N2L 3G1, Canada; E-Mail: h6zhao@uwaterloo.ca

* E-Mail: pmarriot@uwaterloo.ca; Tel.: +1-519-888-4567.

Received: 14 April 2014; in revised form: 12 June 2014 / Accepted: 7 July 2014 / Published: 14 July 2014

Abstract: The power of projection using divergence functions is a major theme in information geometry. One version of this is the variational Bayes (VB) method. This paper looks at VB in the context of other projection-based methods in information geometry. It also describes how to apply VB to the regime-switching log-normal model and how it provides a computationally fast solution to quantify the uncertainty in the model specification. The results show that the method can recover exactly the model structure, gives the reasonable point estimates and is very computationally efficient. The potential problems of the method in quantifying the parameter uncertainty are discussed.

Keywords: information geometry; variational Bayes; regime-switching log-normal model; model selection; covariance estimation

1. Introduction

While, in principle, the calculation of the posterior distribution is mathematically straightforward, in practice, the computation of many of its features, such as posterior densities, normalizing constants and posterior moments, is a major challenge in Bayesian analysis. Such computations typically involve high dimensional integrals, which often have no analytical or tractable forms. The variational Bayes (VB) method was developed to generate tractable approximations to these quantities. This method provides analytic approximations to the posterior distribution by minimizing the Kullback–Leibler (KL) divergence from the approximations to the actual posterior and has been demonstrated to be computationally very fast.

VB gains its computational advantages by making simplifying assumptions about the posterior dependence structure. For example, in the simplest form, it assumes posterior independence between selected sets of parameters. Under these assumptions, the resultant approximate posterior is either known analytically or can be computed by a simple iteration algorithm similar to the Expectation-Maximization (EM) algorithm. In this paper, we show that, as well as having advantages of computational speed, the VB algorithm does an excellent job of model selection, in particular in finding the appropriate number of regimes.

While the simplification in the dependence gives computational advantages, it also comes at a cost. For example, we also found that the posterior variance may be underestimated. In [1], we propose a novel method to compute the true posterior covariance matrix by only using the information obtained from VB approximations.

The use of projections to particular families is, of course, not new to information geometry (IG). In [2], we find the famous Pythagorean results concerning projection using α-divergences to α-families, and other important results on projections based on divergences can be found in [3] and [4] (Chapter 7).

1.1. Variational Bayes

Suppose, in a Bayesian inference problem that we use $q(\boldsymbol{\tau})$ to approximate the posterior $p(\boldsymbol{\tau}|y)$, where y is the data and $\boldsymbol{\tau} = \{\tau_1, \cdots, \tau_p\}$ the model parameter vector. The KL divergence between them is defined as,

$$\text{KL}\left[q(\boldsymbol{\tau})||p(\boldsymbol{\tau}|\mathbf{y})\right] = \int q(\boldsymbol{\tau}) \log \frac{q(\boldsymbol{\tau})}{p(\boldsymbol{\tau}|\mathbf{y})} d\boldsymbol{\tau}, \tag{1}$$

provided the integral exists. We want to balance two things, having the discrepancy between p and q small, while keeping q tractable. Hence, we want to seek $q(\boldsymbol{\tau})$, which minimizes Equation (1), while keeping $q(\tau)$ in an analytically tractable form. First, note that the evaluation of Equation (1) requires $p(\boldsymbol{\tau}|\mathbf{y})$, which may be unavailable, since in the general Bayesian problem, its normalizing constant is one of the main intractable integrals. However, we note that:

$$\begin{aligned}\text{KL}\left[q(\boldsymbol{\tau})||p(\boldsymbol{\tau}|\mathbf{y})\right] &= \int q(\boldsymbol{\tau}) \log \frac{q(\boldsymbol{\tau})}{p(\boldsymbol{\tau}|\mathbf{y})p(\mathbf{y})} d\boldsymbol{\tau} + \log p(\mathbf{y}) \\ &= -\int q(\boldsymbol{\tau}) \log \frac{p(\boldsymbol{\tau},\mathbf{y})}{q(\boldsymbol{\tau})} d\boldsymbol{\tau} + \log p(\mathbf{y}). \end{aligned} \tag{2}$$

Thus, minimizing Equation (1) is equivalent to maximizing the first term of the right-hand side of Equation (2). The key computational point is that, often, the term $p(\boldsymbol{\tau}, \mathbf{y})$ is available even when the full posterior $\frac{p(\tau,y)}{\int p(\tau,y)d\tau}$ is not.

Definition 1. *Let* $F(q) = \int q(\boldsymbol{\tau}) \log \frac{p(\boldsymbol{\tau},\mathbf{y})}{q(\boldsymbol{\tau})} d\boldsymbol{\tau}$ *and:*

$$\hat{q} = \arg\max_{q \in Q} F(q), \tag{3}$$

where Q is a predetermined set of probability density functions over the parameter space. Then $\hat{q}$ is called the variational approximation or variational posterior distribution, and functions of $\hat{q}$ (such as mean, variance, etc.), are called variational parameters.

Some of the power of Definition 1 comes when we assume that all elements of Q have tractable posteriors. In that case, all variational parameters will then also be tractable when the optimization can be achieved. A prime example of a choice for Q is the set of all densities that factorize as

$$q(\boldsymbol{\tau}) = \prod_{i=1}^{d} q_i(\tau_i).$$

This reduces the computational problem from computing a high dimensional integral to one of computing a number of one-dimensional ones. Furthermore, as we see in the example of this paper, it is often the case that the variational families are standard exponential families (since they are often 'maximum entropy models' in some sense), and the optimisation problem (3) can be solved by simple iterative methods with very fast convergence.

The core of the method builds on the basis of the principle of the variational free energy minimization in physics, which is concerned with finding the maxima and minima of a functional over a class of functions, and the method gains its name from this root. Early developments of the method can be found in machine learning, especially in applications on neural networks [5,6]. The method has been successfully applied in many different disciplines and domains, for example, in independent component analysis [7,8], graphical models [9,10], information retrieval [11] and factor analysis [12].

In the statistical literature, an early application of the variational principle can be found in the work of [13] to construct Bayes estimators. In recent years, the method has obtained more attention from both the application and theoretical perspective, for example [14–18].

1.2. Regime-Switching Models

In this paper, we illustrate the strengths and weaknesses of VB through a detailed case study. In particular, we look at a model that is used in finance, risk management and actuarial science, the so-called regime-switching log-normal model (RSLN) proposed, in this context, by [19].

Switching between different states, or regimes, is a common phenomenon in many time series, and regime-switching models, originally proposed by [20], have been used to model these switching processes. As demonstrated in [21], the maximum likelihood estimate (MLE) does not give a simple method to deal with parameter uncertainty; for details of this method, see [21]. The asymptotic normality of maximum likelihood estimators may not apply for sample sizes commonly found in practice. Hence, to understand parameter uncertainty, [21] considered the RSLN model in a Bayesian framework using the Metropolis–Hastings algorithm. Furthermore, model uncertainty, in particular selecting the correct number of regimes, is a major issue. Hence, model selection criteria have to be used to choose the "best" model. Hardy [19] found that a two-regime RSLN model maximized the Bayes information criterion (BIC) [22] for both monthly TSE 300 total return data and S&P 500 total return data; however, according to the Akaike information criterion (AIC) [23], a three-regime model was the optimal on S&P data. To account for the model uncertainty associated with the number of regimes, [24] offered a trans-dimensional model using reversible jump MCMC [25]. We note that BIC is not necessarily ideal for model selection with state space models [26], while it is still commonly used in the literature.

MCMC methods make possible the computation of all posterior quantities; however there are a number of practical issues associated with their implementation. A primary concern is determining that the generated chain has, in fact, "converged". In practice, MCMC practitioners have to resort to convergence diagnostic techniques. Furthermore, the computational cost can be a concern. Other implementational issues include the difficulty of making good initalisation choices, implementing the MCMC algorithm in one long chain or several shorter chains in parallel, *etc.* Detailed discussions can be found in [27].

One of the main contributions of this paper is to apply the variational Bayes (VB) method to the RSLN model and present a solution to quantify the uncertainty in model specification. The VB method is a technique that provides analytical approximations to the posterior quantities, and in practice, it is demonstrated to be a very much faster alternative to MCMC methods.

2. Variational Bayes and Informational Geometry

In this section, we explore the relationship between VB and IG, in particular the statistical properties of divergence-based projections onto exponential families. Here, we used the IG of [2], in particular the ± 1 dual affine parameters for exponential families. One of the most striking results from [2] is the Pythagorean property of these dual affine coordinate systems. This is illustrated in Figure 1, which shows a schematic representing a model space containing the distribution $f_0(x)$ and an exponential family $f(x;\theta)$.

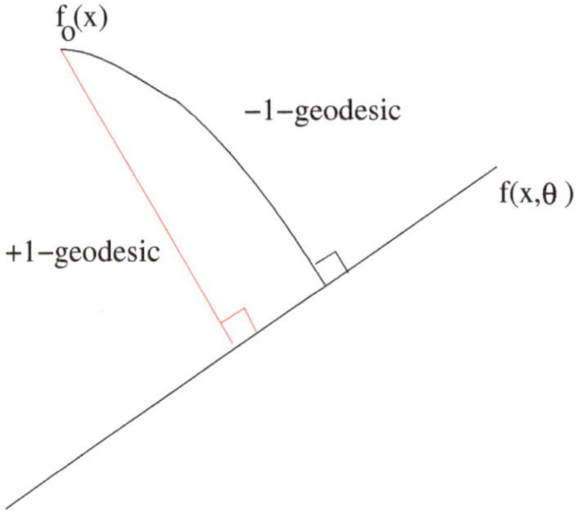

Figure 1. Projections onto an exponential family.

The Pythagorean result comes from using the KL divergence to project onto the exponential family $f(x;\theta) = \nu(x)\exp\{s(x)\theta - \psi(\theta)\}$, *i.e.*,

$$\min_{\theta} \int -\log \frac{f(x;\theta)}{f_0(x)} f_0(x)dx.$$

All distributions that project to the same point form a -1-flat space defined by all distributions $f(x)$ with the same mean, *i.e.*,

$$E_{\hat{\theta}}(s(x)) = E_{f(x)}(s(x)),$$

and further, it is Fisher orthogonal to the $+1$-flat family $f(x;\theta)$. The statistical interpretation of this concerns the behaviour of a model $f(x,\theta)$ when the data generation process does not lie in the model.

In contrast to this, we have the VB method, which uses the reverse KL divergence for the projection, *i.e.*,

$$\min_{\theta} \int \log \frac{f(x;\theta)}{f_0(x)} f(x;\theta)dx.$$

This results in a Fisher orthogonal projection, shown in Figure 1, but now using a $+1$-flat family. This does not have the property that the mean of $s(x)$ is constant, but as we shall see, it does have nice computational properties when used in the context of Bayesian analysis.

In order to investigate the information geometry of VB, we consider two examples. The first, in Section 3.1, is selected to maximally illustrate the underlying geometric issues and to get some understanding of the quality of the VB approximation. The second, in Section 3.2, shows an important real-world application from actuarial science and is illustrated with simulated and real data.

3. Applications of Variational Bayes

3.1. Geometric Foundation

We consider the simplest model that shows dependence. Let X_1, X_2 be two binary random variables, with distribution $\pi := (\pi_{00}, \pi_{10}, \pi_{01}, \pi_{11})$, where $P(X_1 = i, X_2 = j) = \pi_{ij}$, $i, j \in \{0,1\}$. Further, let the marginal distributions be denoted by $\pi_1 = P(X_1 = 1), \pi_2 = P(X_2 = 1)$. We want to consider the geometry of the VB projection from a general distribution to the family of independent distributions. This represents the way that VB gains its computational advantages by simplifying the posterior dependence structure.

The model space is illustrated in Figure 2, where π is represented by a point in the three simplex, and the independence surface, where $\pi_{00}\pi_{11} = \pi_{10}\pi_{01}$, is also shown.

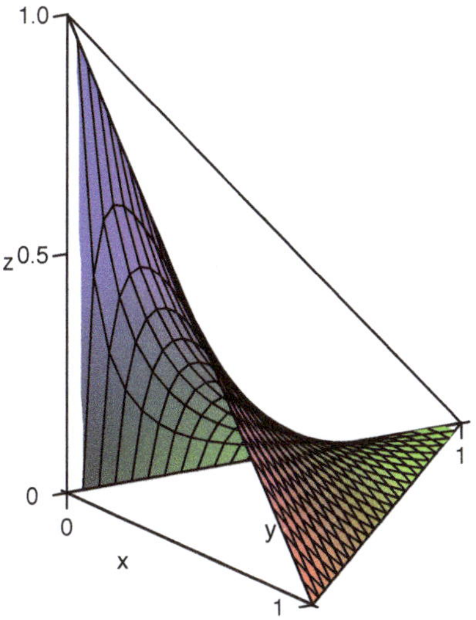

Figure 2. Space of distributions with independence surface: marginal probabilityand dependence.

Both the interior of the simplex and independence surface are exponential families, and it is convenient to use the natural parameters for the interior of the simplex:

$$\xi_1 = \log \frac{\pi_{10}}{\pi_{00}}, \xi_2 = \log \frac{\pi_{01}}{\pi_{00}}, \xi_3 = \log \frac{\pi_{11}\pi_{00}}{\pi_{10}\pi_{01}}$$

where the independence surface is given by $\xi_3 = 0$. The independence surface can also be parameterised by the marginal distributions π_1, π_2 or the corresponding natural parameters $\xi_i^{ind} := \log(\pi_i/(1-\pi_i))$. For any distribution, π, represented in natural parameters by (ξ_1, ξ_2, ξ_3), has its VB approximation defined implicitly by the simultaneous equations:

$$\xi_1^{ind}(\pi_1) = \xi_1 + \xi_3\pi_2, \quad (4)$$
$$\xi_2^{ind}(\pi_2) = \xi_2 + \xi_3\pi_1. \quad (5)$$

These can be solved, as is typical with VB methods, by iterating updated estimates of π_1 and π_2 across the two equations. We show this in a realistic example in the following section.

Having seen the VB solution in this simple model, we can investigate the quality of the approximation. If we were using the forward KL project, as proposed by [2], then the mean will be preserved by the approximation, while, of course, the variance structure is distorted. In the case of using the reverse KL projection, as used by VB, the mean will not be preserved, but in this example, we can investigate the distortion explicitly. Let $(\xi_1(\alpha), \xi_2(\alpha), \xi_3(\alpha))$ be a +1-geodesic, which cuts the independence surface orthogonally and is parameterised by α, where $\alpha = 0$ corresponds to the independence surface. In this example, all such geodesics can be computed explicitly. Figure 3 shows the distortion associated with the VB approximation. In the left-hand panel, we show the mean, which is the marginal probability, $P(X_1 = 1)$, for all points on the orthogonal geodesic. We see, as expected, that this is not constant, but it is locally constant at $\alpha = 0$, showing that the distortion of the mean can be small near the independence surface. The right-hand panel shows the dependence, as measured by the log-odds, for points on the geodesic. As expected, the VB does not preserve the dependence structure; indeed, it is designed to exploit the simplification of the dependence structure.

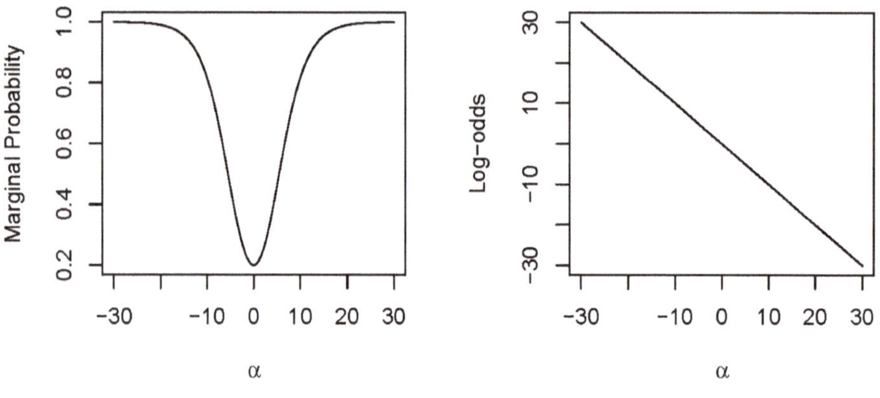

Figure 3. Distortion implied by variational Bayes (VB) approximation.

3.2. Variational Bayes for the RSLN Model

The regime-switching log-normal model [19] with a fixed finite number, K, of regimes can be described as a bivariate discrete time process with the observed data sequence $w_{1:T} = \{w_t\}_{t=1}^T$ and the unobserved regime sequence $S_{1:T} = \{S_t\}_{t=1}^T$, where $S_t \in \{1, \cdots, K\}$ and T is the number of observations. The logarithm of w_t, denoted by $y_t = \log w_t$, is assumed normally distributed, having mean μ_i and variance σ_i^2 both dependent on the hidden regime S_t. The sequence of $S_{1:T}$ is assumed to follow a first order Markov chain having transition probabilities $A = (a_{ij})$ with the probabilities $\pi = (\pi_i)_{i=1}^K$ to start the first regime.

The RSLN model is a special case of more general state-space models, which were studied in detail by [28]. In this paper, we use this model and simulated and real data to illustrate the VB method in practice. We also calibrate its performance by referring to [24], which used MCMC methods to fit the same model to the same data. Here, we are regarding the MCMC analysis as a form of "gold-standard", but with the cost of being orders-of-magnitude slower than VB in computational time.

In the Bayesian framework, we use a symmetric Dirichlet prior for π, that is $p(\pi) = \text{Dir}(\pi; \frac{C^\pi}{K}, \cdots, \frac{C^\pi}{K})$, for $C^\pi > 0$. Let a_i denote the $i - th$ row vector of A. The prior for A is chosen as $p(A) = \prod_{i=1}^K p(a_i) = \prod_{i=1}^K \text{Dir}(a_i; \frac{C^A}{K}, \cdots, \frac{C^A}{K})$, for $C^A > 0$, and the prior distribution for $\{(\mu_i, \sigma_i^2)\}_{i=1}^K$ is chosen to be normal-inverse gamma, $p(\{\mu_i, \sigma_i^2\}_{i=1}^K) = \prod_{i=1}^K \text{N}(\mu_i|\sigma_i^2; \gamma, \frac{\sigma_i^2}{\eta^2})\text{IG}(\sigma_i^2; \alpha, \beta)$. In the above setting, $C^\pi, C^A, \gamma, \eta^2, \alpha$ and β are hyper-parameters. Thus, the joint posterior distribution of $\pi, A, \{\mu_i, \sigma_i^2\}_{i=1}^K$, and $S_{1:T}$ is $P(\pi, A, \{\mu_i, \sigma_i^2\}_{i=1}^K, S_{1:T}|y_{1:T})$ and is proportional to:

$$p(S_1|\pi) \prod_{t=1}^{T-1} p(S_{t+1}|S_t; A) \prod_{t=1}^{T} p(y_t|S_t; \{\mu_i, \sigma_i^2\}_{i=1}^K) p(\pi) p(A) p(\{\mu_i, \sigma_i^2\}_{i=1}^K). \tag{6}$$

This posterior distribution and its corresponding marginal posterior distributions are analytically intractable. In VB, we seek an approximation of Equation (6), denoted by $q(\pi, A, \{\mu_i, \sigma_i^2\}_{i=1}^K, S_{1:T})$, to which we want to balance two things: having the discrepancy between Equation (6) and q small, while keeping q tractable. In general, there are two ways to choose q. The first is to specify a particular distributional family for q, for example the multivariate normal distribution. The other is to choose q with a simpler dependency structure than that of Equation (6); for example, we choose q, which factorizes as:

$$q(\pi, A, \{\mu_i, \sigma_i^2\}_{i=1}^K, S_{1:T}) = q(\pi) \prod_{i=1}^K q(a_i) \prod_{i=1}^K q(\mu_i|\sigma_i^2) q(\sigma_i^2) q(S_{1:T}). \tag{7}$$

The Kullback–Leibler (KL) divergence [29] can be used as the measure of dissimilarity between Equations (6) and (7). For succinctness, we denote $\tau = (\pi, A, \{\mu_i, \sigma_i^2\}_{i=1}^K, S_{1:T})$; thus the KL divergence is defined as:

$$\mathrm{KL}(q(\tau) \,||\, p(\tau|y)) = \int q(\tau) \log \frac{q(\tau)}{p(\tau|y)} d\tau. \tag{8}$$

Note that the evaluation of Equation (8) requires $p(\tau|y)$, which is unavailable. However, we note that:

$$\mathrm{KL}(q(\tau) \,||\, p(\tau|y)) = \log p(y) - \int q(\tau) \log \frac{p(\tau, y)}{q(\tau)} d\tau$$

Given the factorization Equation (7), this can be written as:

$$\mathrm{KL}(q(\tau) \,||\, p(\tau|y)) =$$
$$\log p(y) - \int \sum_{S_{1:T}} q(\pi)q(A) \prod_{i=1}^{K} q(\mu_i|\sigma_i^2)q(\sigma_i^2)q(S_{1:T}) \log \frac{p(\pi, A, \{\mu_i, \sigma_i^2\}_{i=1}^K, S_{1:T}, y_{1:T})}{q(\pi)q(A)\prod_{i=1}^{K} q(\mu_i|\sigma_i^2)q(\sigma_i^2)q(S_{1:T})} d\pi dA d\{\mu_i, \sigma_i^2\}_{i=1}^K$$

Consider first the $q(\pi)$ term. The right-hand side can be rearranged as:

$$\mathrm{KL}\left(q(\pi) \middle|\middle| \frac{\exp\left[\int \sum_{S_{1:T}} q(S_{1:T})q(A) \prod_{i=1}^{K} q(\mu_i|\sigma_i^2)q(\sigma_i^2) \log p(\pi, A, \{\mu_i, \sigma_i^2\}_{i=1}^K, s_{1:T}, y_{1:T}) dA d\{\mu_i, \sigma_i^2\}_{i=1}^K \right]}{Z_\pi} \right) + K_\pi, \tag{9}$$

where:

$$K_\pi = \int \sum_{S_{1:T}} q(S_{1:T})q(A) \prod_{i=1}^{K} q(\mu_i|\sigma_i^2)q(\sigma_i^2)q(S_{1:T}) \log q(A) \prod_{i=1}^{K} q(\mu_i|\sigma_i^2)q(\sigma_i^2) dA d\{\mu_i, \sigma_i^2\}_{i=1}^K - \log Z_\pi + \log p(y),$$

and Z_π is a normalizing term. The first term of Equation (9) is the only term that depends on $q(\pi)$. Thus, the minimum value of $\mathrm{KL}(q(\tau) \,||\, p(\tau|y))$ is achieved when this term equals zero. Hence, we obtained:

$$q(\pi) = \frac{\exp\left[\int \sum_{S_{1:T}} q(S_{1:T})q(A) \prod_{i=1}^{K} q(\mu_i|\sigma_i^2)q(\sigma_i^2) \log p(\pi, A, \{\mu_i, \sigma_i^2\}_{i=1}^K, s_{1:T}, y_{1:T}) dA d\{\mu_i, \sigma_i^2\}_{i=1}^K \right]}{Z_\pi} \tag{10}$$

Given the joint distribution of $p(\pi, A, \{\mu_i, \sigma_i^2\}_{i=1}^K, s_{1:T}, y_{1:T})$ in the form of Equation (6), the straightforward evaluation of Equation (10) results in:

$$q(\pi) \;\propto\; \prod_{i=1}^{K} \pi_i^{\frac{C_\pi^K}{K} + w_i^s - 1} = \mathrm{Dir}(\pi, w_1^\pi, \cdots, w_K^\pi);\; w_i^\pi = \frac{C_\pi^K}{K} + w_i^s, w_i^s = \mathrm{E}_{q(S_{1:T})}[S_{1,i}] \tag{11}$$

where $S_{1,i} = 1$, if the process is in state i at time 1, and zero otherwise.

Similarly, we can rearrange Equation (9) with respect to $\{q(a_i)\}_{i=1}^K$, $\{q(\mu_i|\sigma_i^2)\}_{i=1}^K$, $\{q(\sigma_i^2)\}_{i=1}^K$ and $q(S_{1:T})$, respectively, and using the same arguments, then we can obtain:

$$q(A) = \prod_i^k \mathrm{Dir}(a_i; w_{i1}^A, ..., w_{ik}^A);\ w_{ij}^A = \frac{C^A}{K} + v_{ij}^s, \tag{12}$$

$$q(\mu_i|\sigma_i^2) = \mathrm{N}\left(\gamma_i', \frac{\sigma_i^2}{\kappa_i}\right), \gamma_i' = \frac{\eta^2\gamma + p_i^s}{\eta^2 + q_i^s}, \kappa_i = \eta^2 + q_i^s \tag{13}$$

$$q(\sigma_i^2) = \mathrm{IG}\left(\alpha_i', \beta_i'\right), \alpha_i' = \alpha + \frac{q_i^s}{2}, \beta_i' = \beta + \frac{r_i^s}{2} + \frac{\eta^2}{2}(\gamma_i' - \gamma)^2 \tag{14}$$

$$q(S_{1:T}) = \frac{\prod_{i=1}^k \pi_i^{*S_{1,i}} \prod_{t=1}^{T-1}\prod_{i=1}^k\prod_{j=1}^k a_{ij}^{*S_{t,i}S_{t+1,j}} \prod_{t=1}^T\prod_{i=1}^k \theta^{*S_{t,i}}}{\tilde{Z}}, \tag{15}$$

where $S_{t,i} = 1$, if the process in state i at time t, and zero otherwise, and with $\pi_i^* = e^{\mathrm{E}_{q(\pi)}[\log \pi_i]}$, $a_{ij}^* = e^{\mathrm{E}_{q(A)}[\log(a_{ij})]}$, $\theta_{i,t}^* = e^{\mathrm{E}_{q(\mu_i|\sigma_i^2)q(\sigma_i^2)}[\log \phi_i(y_t)]}$, $v_{ij}^s = \sum_{t=1}^{T-1} \mathrm{E}_{q(S_{1:T})}\left[S_{t,i}S_{t+1,j}\right]$, $p_i^s = \sum_{t=1}^T \mathrm{E}_{q(S_{1:T})}[s_{t,i}]y_t$, $q_i^s = \sum_{t=1}^T \mathrm{E}_{q(S_{1:T})}[s_{t,i}]$, $r_i^s = \sum_{t=1}^T (\gamma_i' - y_t)^2 E_{q(S)}[s_{t,i}]$. Here, ψ is the digamma function, ϕ is the normal density function and the exact functional forms used in the updates are shown in Algorithm 1.

Algorithm 1 Variational Bayes algorithm for the regime-switching log-normal model (RSLN) model.

Initialize $w_i^{s(0)}, v_{ij}^{s\,(0)}, p_i^{s(0)}, q_i^{s(0)}$, and $r_i^{s(0)}$ at step 0
while $w_i^{\pi(t-1)}, w_{ij}^{A\,(t-1)}, \gamma_i'^{(t-1)}, \alpha_i'^{(t-1)}, \beta_i'^{(t-1)}, \pi_i^{*(t-1)}, a_{ij}^{*\,(t-1)}$, and $\theta_{i,t}^{*\,(t-1)}$ do not converge **do**

1. Compute $w_i^{\pi(t)}, w_{ij}^{A(t)}, \gamma_i'^{(t)}, \kappa_i^{(t)}, \alpha_i'^{(t)}$, and $\beta_i'^{(t)}$ at step t by

$$w_i^{\pi(t)} = \frac{C_\pi^K}{K} + w_i^{s(t-1)},\quad w_{ij}^{A(t)} = \frac{C_\pi^A}{K} + v_{ij}^{s\,(t-1)},\quad \gamma_i'^{(t)} = \frac{\eta^2\gamma + p_i^{s(t-1)}}{\eta^2 + q_i^{s(t-1)}},$$

$$\kappa_i^{(t)} = \eta^2 + q_i^{s(t-1)},\quad \alpha_i'^{(t)} = \alpha + \frac{q_i^{s(t-1)}}{2},\quad \beta_i'^{(t)} = \beta + \frac{r_i^{s(t-1)}}{2} + \frac{\eta^2}{2}(\gamma_i'^{(t)} - \gamma)^2$$

2. Compute $\pi_i^{*(t)}, \theta_{i,t}^{*\,(t)}$ and $a_{ij}^{*\,(t)}$ at step t by:

$$\pi_i^{*(t)} = \exp\left(\psi(w_i^{\pi(t)}) - \psi(\sum_i w_i^{\pi(t)})\right),\quad a_{ij}^{*\,(t)} = \exp\left(\psi(w_{ij}^{A\,(t)}) - \psi(\sum_{j=1} w_{ij}^{A\,(t)})\right)$$

$$\theta_{i,t}^{*\,(t)} = \exp\left(-\frac{1}{2}\log 2\pi - \frac{1}{2}(\log \beta_i'^{(t)} - \psi(\alpha_i'^{(t)})) - \frac{1}{2}\left((y_t - \gamma_i'^{(t)})^2\frac{\alpha_i'^{(t)}}{\beta_i'^{(t)}} + \frac{1}{\kappa_i^{(t)}}\right)\right)$$

3. Compute $w_i^{s(t)}, v_{ij}^{s\,(t)}, p_i^{s(t)}, q_i^{s(t)}$, and $r_i^{s(t)}$ at step t by:

$$w_i^{s(t)} = \mathrm{E}_{q^{(t)}(S_{1:T})}[S_{1,i}],\ v_{ij}^{s\,(t)} = \sum_{t=1}^{T-1}\mathrm{E}_{q^{(t)}(S_{1:T})}\left[S_{t,i}S_{t+1,j}\right],\ p_i^{s(t)} = \sum_{t=1}^{T-1}\mathrm{E}_{q^{(t)}(S_{1:T})}[s_{t,i}]y_t,$$

$$q_i^{s(t)} = \sum_{t=1}^{T-1}\mathrm{E}_{q^{(t)}(S_{1:T})}[s_{t,i}],\ r_i^{s(t)} = \sum_{t=1}^{T-1}(\gamma_i'^{(t)} - y_t)^2 E_{q^{(t)}(S)}[s_{t,i}]$$

$t \Leftarrow t+1$
end while

The VB method proceeds, as was shown with the simple Equations (4) and (5), by iterative updating the variational parameters to solve a set of simultaneous equations. In this example, the update equations for the variables $\pi, A, \{\mu_i, \sigma_i^2\}_{i=1}^K, S_{1:T}$ are given explicitly by Algorithm 1. For the initialisation, we choose symmetric values for most of the parameters and choose random values for others, as appropriate. For this example, this worked very satisfactory, although we note that for more general state space models [28], states that find good initial values can be non-trivial.

3.3. Interpretation of Results

First, all approximating distributions above turn out to lie in well-known parametric families. The only unknown quantities are the parameters of these distributions, which are often called the variational parameters.

The evaluation of parameters of $q(\pi)$, $q(A)$, $q(\mu_i|\sigma_i^2)$, and $q(\sigma_i^2)$ requires knowledge of $q(S_{1:T})$, and also, the evaluation of π_i^*, a_{ij}^* and $\theta_{i,t}^*$ requires knowledge of $q(\pi)$, $q(A)$, $q(\mu_i|\sigma_i^2)$ and $q(\sigma_i^2)$. This structure leads to an iterative updating scheme, described in Algorithm 1.

The main computational effort in Algorithm 1 is computing $\mathrm{E}_{q(S_{1:T})}[S_{t,i}]$ and $\mathrm{E}_{q(S_{1:T})}\left[S_{t,i}S_{t+1,j}\right]$, which have no simple tractable forms. We note that the distributional form of $q(S_{1:T})$ has a very similar structure as the conditional distribution of $p(S_{1:T}|Y_{1:T},\tau)$ for which the forward-backward algorithm [30] is commonly used to compute $\mathrm{E}_{p(S_{1:T}|Y_{1:T},\tau)}[S_{t,i}|Y_{1:T},\boldsymbol{\tau}]$ and $\mathrm{E}_{p(S_{1:T}|Y_{1:T},\tau)}\left[S_{t,i}S_{t+1,j}|Y_{1:T},\tau\right]$. Therefore, we also use the forward-backward algorithm to compute $\mathrm{E}_{q(S_{1:T})}[S_{t,i}]$ and $\mathrm{E}_{q(S_{1:T})}\left[S_{t,i}S_{t+1,j}\right]$.

The conditional distribution of $q(\mu_i|\sigma_i^2)$ is $\mathrm{N}\left(\mu_i|\sigma_i^2;\gamma_i',\frac{\sigma_i^2}{\kappa_i}\right)$, then the marginal distribution of μ_i is the location-scale t distribution, denoted as $t_{2\alpha_i'}(\mu_i;\gamma_i',\frac{\kappa_i}{\beta_i'/\alpha_i'})$, where the density function of $t_\nu(x;\mu,\lambda)$ is defined as $p(x|\nu,\mu,\lambda)=\frac{\Gamma(\frac{\nu+1}{2})}{\Gamma(\frac{\nu}{2})}\left(\frac{\lambda}{\pi\nu}\right)^{\frac{1}{2}}\left[1+\frac{\lambda(x-\mu)^2}{\nu}\right]^{-\frac{\nu+1}{2}}$, for $x,\mu\in(-\infty,+\infty)$ and $\nu,\lambda>0$.

4. Numerical Studies

4.1. Simulated Data

In this section, we applied the VB solutions to four sets of simulated data, which are used in [24]. Through these simulated studies, we will test the performance of VB on detecting the number of regimes and compare it with those of the BIC and the MCMC methods [24]. For this paper, we present only an initial study with a relatively small number of datasets. The results are highly promising, but more extensive studies are needed to draw comprehensive conclusions. Furthermore, see [28] for general results on VB in hidden state space models.

To estimate the number of regimes, we construct a matrix, called the relative magnitude matrix (RMM), defined as $A'=(\hat{a}_{ij}')$, where $\hat{a}_{ij}'=\frac{w_{ij}^A}{w_0^A}$, $w_0^A=\sum_{i=1}^K\sum_{j=1}^K w_{ij}^A$ and w_{ij}^A is the parameter of $q(A)$. Our model selection procedure is to fit a VB with a large number of regimes and to examine the rows and columns in the RMM. If the values of the entries in the $i-th$ row and the $i-th$ column of A' are all equal to $\frac{C^A/K}{T-1+C^A\times K}$, then we will declare the regime i nonexistent. This method is validated by the following observations. It can be shown that the parameter of v_{ij}^s in w_{ij}^A is equal to the number of times the process leaves regime i and enters regime j. Therefore, for the $i-th$ regime, the values of zero for all of v_{ji}^s and v_{ij}^s with $j=1,\cdots,K$ indicate that there is no transition process entering or leaving regime i.

Table 1 specifies the parameters for the four cases, and we generate 671 observations for each case (equal to the number of months from January 1956 to September 2011). The parameters used in Case 1 are identical to the maximum likelihood estimates for TSX monthly return data from 1956 to 1999 [19]. Case 2 only has one regime present. Case 3 is similar to Case 1, but the two regimes have the same mean. Case 4 adds a third regime. For each case, we use MLE to fit a one-regime, two-regime, three-regime and four-regime RSLN model and report the corresponding BIC and log-likelihood scores. We then misspecify the number of regimes and run a four-regime VB algorithm.

Table 1. Parameters of the simulated data.

Case	Regime 1 (μ_i, σ_i)	Regime 2 (μ_i, σ_i)	Regime 3 (μ_i, σ_i)	Transition Probability
1	(0.012, 0.035)	(−0.016, 0.078)	-	$\begin{pmatrix} 0.963 & 0.037 \\ 0.210 & 0.790 \end{pmatrix}$
2	(0.014, 0.050)	-	-	-
3	(0.000, 0.035)	(0.000, 0.078)	-	$\begin{pmatrix} 0.963 & 0.037 \\ 0.210 & 0.790 \end{pmatrix}$
4	(0.012, 0.035)	(−0.016, 0.078)	(0.04, 0.01)	$\begin{pmatrix} 0.953 & 0.037 & 0.01 \\ 0.210 & 0.780 & 0.01 \\ 0.80 & 0.190 & 0.01 \end{pmatrix}$

Table 2 shows the number of iterations that VB takes to converge in each case and the corresponding computational time (on a MacBook, 2 GHz processor). On average, VB converges after a hundred iterations and takes about one minute. On the same computer, a 10^4-iteration Reverse Jump MCMC (RJMCMC) will take about 10 h to finish. Using diagnostics, this seemed to be enough for convergence, while not being an "unfair" comparison in terms of time with VB. We can see that the computational efficiency will be a very attractive feature of the VB method. The results of the BIC with the log-likelihood (in parentheses), the relative magnitude matrices and the posterior probabilities for the models with the different number of regimes estimated by MCMC (cited from Hartman and Heaton [24]) are given in Table 3. In Case 1, the BIC favors the two-regime model. The posterior probability estimated by MCMC for the one-regime model is the largest, but there is still a large probability for the two regime model. Note that the prior specification for the number of regimes can effect these numbers and is always an issue with these forms of multidimensional MCMC. The relative magnitude matrix clearly shows that there are only two regimes whose $\hat{a}'_{ij}$ are not negligible. This implies VB removes excess transition and emission processes and discovers the exact number of hidden regimes. In Case 2 and Case 3, both VB and the BIC can select the correct number of regimes, and the posterior probability for the one-regime model estimated by MCMC is still the largest. In Case 4, VB does not detect the third regime. The transition probability to this regime is only 0.01, and the means and standard deviations of Regime 1 make the rare data from Regime 3 easily merged within the data from Regime 1. From Table 3, it is clear that for all of the cases, the log-likelihood always increases as the number of regimes increase.

Table 2. Computational efficiency of VB.

	Case 1	Case 2	Case 3	Case 4
Iterations to converge	62	182	132	94
Computational time [s]	27.161	80.842	58.510	45.044

Table 3. The estimated number of regimes by VB, BIC and MCMC.

Case	No. of Regimes	MLE BIC (Log Likelihood)	RJMCMC Posterior Probability	VB Relative Magnitude Matrix
1	1	1,108.875(1,115.384)	0.647	0.14357 0.00004 0.00004 0.03153
	2	1,158.227(1,174.499)	0.214	0.00004 0.00004 0.00004 0.00004
	3	1,156.370(1,182.405)	0.088	0.00004 0.00004 0.00004 0.00004
	4	1,153.150(1,188.948)	<0.052	0.03018 0.00004 0.00004 0.79428
2	1	1,045.448(1,051.957)	0.864	0.99944 0.00004 0.00004 0.00004
	2	1,038.360(1,054.632)	0.109	0.00004 0.00004 0.00004 0.00004
	3	1,030.733(1,056.768)	0.020	0.00004 0.00004 0.00004 0.00004
	4	1,026.882(1,062.680)	<0.006	0.00004 0.00004 0.00004 0.00004
3	1	1,110.903(1,117.411)	0.629	0.11322 0.00004 0.00004 0.02647
	2	1,139.214(1,155.486)	0.221	0.00004 0.00004 0.00004 0.00004
	3	1,131.904(1,157.719)	0.098	0.00004 0.00004 0.00004 0.00004
	4	1,121.921(1,157.940)	<0.052	0.02659 0.00004 0.00004 0.83327
4	1	1,044.819(1,051.328)	0.641	0.22643 0.00004 0.00004 0.05518
	2	1,092.610(1,108.881)	0.203	0.00004 0.00004 0.00004 0.00004
	3	1,087.435(1,113.470)	0.094	0.00004 0.00004 0.00004 0.00004
	4	1,080.240(1,116.038)	<0.06	0.05377 0.00004 0.00004 0.66417

4.2. Real Data

In this section, we apply the VB solution to the TSX monthly total return index in the period from January, 1956, to December, 1999 (528 observations in total and studied in [19,21]).

A four-regime VB is implemented first. VB converges after 100 iterations about 34.284 s (on a MacBook, 2 GHz processor). The relative magnitude matrix, given in Table 4, clearly shows that VB identifies two regimes. This matches both of the BIC and AIC-based results [19]. Based on these results, we then fit a two-regime VB, which converges after 83 iterations in about 14.241 s. Table 5 gives the marginal distributions for all of the parameters. Figure 4 presents the corresponding density functions, where we can see that all of the plots show a symmetric and bell-shaped pattern.

Table 4. Estimations of the number of regimes for TSXdata.

	January 1956–December 1999
R. M. M.	$\begin{pmatrix} 0.11496 & 0.00005 & 0.00005 & 0.02803 \\ 0.00005 & 0.00005 & 0.00005 & 0.00005 \\ 0.00005 & 0.00005 & 0.00005 & 0.00005 \\ 0.02853 & 0.00005 & 0.00005 & 0.82791 \end{pmatrix}$

Table 5. The marginal distributions of the parameters estimated by VB.

Parameter	Distribution	Mean	*s.d.*	Transition Probability
μ_1	$t_{454.61}(0.0123, 370778.19)$	0.0123	0.00165	-
σ_1^2	$IG(227.30, 0.28)$	0.00122(0.0349)	0.00008	-
μ_2	$t_{80.39}(-0.0161, 12987.55)$	−0.0161	0.00889	-
σ_2^2	$IG(40.20, 0.24)$	0.00603(0.0777)	0.00098	-
$p_{1,2}$	$Beta(15.21, 434.78)$	0.0338	0.00851	$\begin{pmatrix} 0.9662 & 0.0338 \\ 0.1969 & 0.8031 \end{pmatrix}$
$p_{2,1}$	$Beta(15.00, 61.21)$	0.1969	0.04525	

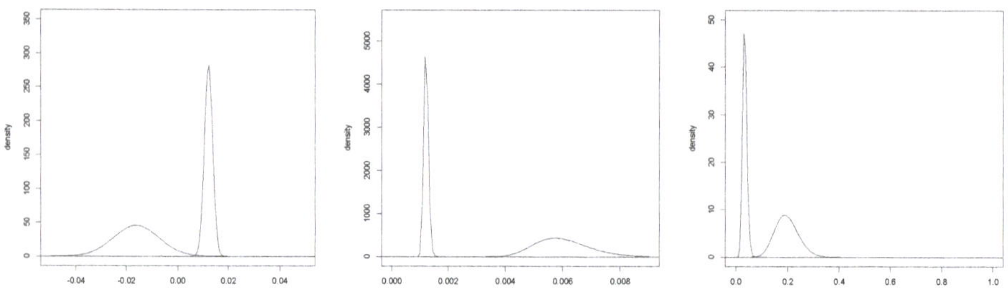

(a) (b) (c)

Figure 4. The VB marginal distributions of the parameters. (**a**) μ_2 (left) and μ_1 (right); (**b**) σ_1^2 (left) and σ_2^2 (right) ; (**c**) $p_{1,2}$ (left) and $p_{2,1}$ (right) .

Table 6 (the upper part) gives the maximum likelihood estimates (cited from [19]), mean parameters computed by the MCMC method (cited from [21]) and mean parameters computed by VB. It clearly shows that the point estimates by VB are very close to those by MLE and MCMC. The numbers in parenthesis in Table 6 are the standard deviations computed by the three methods, respectively. It is worth noting that all of the variance estimated by VB are smaller than those by the MLE or MCMC methods. In fact, some other researchers also report the underestimation of posterior variance in other VB applications, for example [31,32]. In the paper [1], we look at some diagnostics methods that can assess how well the VB approximates the true posterior, particularly with regards to its covariance structure. The methods proposed also allow us to generate simple corrections when the approximation error is large.

Table 6. Estimates and standard deviations by VB, MLE and MCMC.

	μ_1	σ_1	$p_{1,2}$	μ_2	σ_2	$p_{2,1}$
VB	0.0123(0.00165)	0.0349(0.00008)	0.0338(0.00851)	−0.0161(0.00889)	0.0777(0.00098)	0.1969(0.04525)
MLE	0.0123(0.002)	0.0347(0.001)	0.0371(0.012)	−0.0157(0.010)	0.0778(0.009)	0.2101(0.086)
MCMC	0.0122(0.002)	0.0351(0.002)	0.0334(0.012)	−0.0164(0.010)	0.0804(0.009)	0.2058(0.065)

5. Conclusions

Variational Bayes can be thought of in terms of information geometry as a projection-based approximation technique; it provides a framework to approximate posteriors. We applied this method to the regime-switching log-normal model and provide solutions to account for both model uncertainty and parameter uncertainty. The numerical results show that our method can recover exactly the number of regimes and gives reasonable point estimates. The VB method is also demonstrated to be very computationally efficient.

The application on the TSX monthly total return index data in the period from January 1956 to December 1999, confirms the similar results in the literature in finding the number of regimes.

Author Contributions

The article was written by Hui Zhao under the guidance of Paul Marriott. All authors have read and approved the final manuscript.

Conflicts of Interest

The authors declare no conflict of interest.

References

1. Zhao, H.; Marriott, P. Diagnostics for variational bayes approximations. **2013**, arXiv:1309.5117.
2. Amari, S.-I. *Differential-Geometrical Methods in Statistics*; Springer: New York, NY, USA, 1990.
3. Eguchi, S. Second order efficiency of minimum contrast estimators in a curved exponential family. *Ann. Stat.* **1983**, *11*, 793–803.
4. Kass, R.; Vos, P. *Geometrical Foundations of Asymptotic Inference*; Wiley: New York, NY, USA, 1997.
5. Hinton, G.E.; van Camp, D. Keeping neural networks simple by minimizing the description length of the weights. In Proceedings of the 6th ACM Conference on Computational Learning Theory, Santa Cruz, CA, USA, 26–28 July 1993; ACM: New York, NY, USA, 1993.
6. MacKay, D. Developments in Probabilistic Modelling with Neural Networks—Ensemble Learning. In *Neural Networks: Artifical Intelligence and Industrial Applications*; Springer: London, UK, 1995; pp. 191–198.
7. Attias, H. Independent Factor Analysis. *Neur. Comput.* **1999**, *11*, 803–851.
8. Lappalainen, H. Ensemble Learning For Independent Component Analysis. In Proceedings of the First International Workshop on Independent Component Analysis, Aussois, France, 11–15 January 1999; pp. 7–12.
9. Beal, M.; Ghahramani, Z. The variational Bayesian EM algorithm for incomplete data: With application to scoring graphical model structures. *Bayesian Stat.* **2003**, *7*, 453–463.
10. Winn, J. *Variational Message Passing and its Applications*. Ph.D. Thesis, Department of Physics, University of Cambridge, Cambridge, UK, 2003.
11. Blei, D.M.; Ng, A.Y.; Jordan, M.I.; Lafferty, J. Latent Dirichlet allocation. *J. Mach. Learn. Res.* **2003**, *3*, 993–1022.
12. Ghahramani, Z.; Beal, M.J. A Variational Inference for Bayesian Mixtures of Factor Analysers. *Adv. Neur. Inf. Process. Syst.* **2000**, *12*, 449–455.
13. Haff, L.R. The Variational Form of Certain Bayes Estimators. *Ann. Stat.* **1991**, *19*, 1163–1190.
14. Faes, C.; Ormerod, J.T.; Wand, M.P. Variational Bayesian Inference for Parametric and Nonparametric Regression With Missing Data. *J. Am. Stat. Assoc.* **2011**, *106*, 959–971.
15. McGrory, C.; Titterington, D.; Reeves, R.; Pettitt, A.N. Variational Bayes for estimating the parameters of a hidden Potts model. *Stat. Comput.* **2009**, *19*, 329–340.
16. Ormerod, J.T.; Wand, M.P. Gaussian Variational Approximate Inference for Generalized Linear Mixed Models. *J. Comput. Graph. Stat.* **2011**, *21*, 1–16.
17. Hall, P.; Humphreys, K.; Titterington, D.M. On the Adequacy of Variational Lower Bound Functions for Likelihood-Based Inference in Markovian Models with Missing Values. *J. R. Stat. Soc. Ser. B* **2002**, *64*, 549–564.
18. Wang, B.; Titterington, M. Convergence Properties of a general algorithm for calculating variational Bayesian estimates for a normal mixture model. *Bayesian Anal.* **2006**, *1*, 625–650.
19. Hardy, M.R. A Regime-Switching Model of Long-Term Stock Returns. *N. Am. Actuar. J.* **2001**, *5*, 41–53.
20. Hamilton, J.D. A New Approach to the Economic Analysis of Nonstationary Time Series and the Business Cycle. *Econometrica* **1989**, *57*, 357–384.
21. Hardy, M.R. Bayesian Risk Management for Equity-Linked Insurance. *Scand. Actuar. J.* **2002**, *2002*, 185–211.
22. Schwarz, G. Estimating the dimension of a model. *Ann. Stat.* **1978**, *6*, 461–464.
23. Akaike, H. A new look at the statistical model identification. *IEEE Trans. Autom. Control* **1974**, *19*, 716–723.
24. Hartman, B.M.; Heaton, M.J. Accounting for regime and parameter uncertainty in regime-switching models. *Insur. Math. Econ.* **2011**, *49*, 429–437.
25. Green, P.J. Reversible jump Markov chain Monte Carlo computation and Bayesian model determination. *Biometrika* **1995**, *82*, 711–732.
26. Watanabe, S. *Algebraic Geometry and Statistical Learning Theory*; Cambridge University Press: Cambridge, UK, 2009.
27. Brooks, S.P. Markov Chain Monte Carlo Method and Its Application. *J. R. Stat. Soc. Ser. D* **1998**, *47*, 69–100.

28. Ghahramani, Z.; Hinton, G.E. Variational learning for switching state-space models. *Neur. Comput.* **1998**, *12*, 831–864.
29. Kullback, S.; Leibler, R.A. On information and sufficiency. *Ann. Math. Stat* **1951**, *22*, 79–86.
30. Baum, L.E.; Petrie, T.; Soules, G.; Weiss, N. A maximization technique occurring in the statistical analysis of probabilistic functions of markov chains. *Ann. Math. Stat.* **1970**, *41*, 164–171.
31. Rue, H.; Martino, S.; Chopin, N. Approximate Bayesian inference for latent Gaussian models by using integrated nested Laplace approximations. *J. R. Stat. Soc. Ser. B* **2009**, *71*, 319–392.
32. Bishop, C.M. *Pattern Recognition and Machine Learning*; Springer: New York, NY, USA, 2006.

entropy

Article

On Clustering Histograms with k-Means by Using Mixed α-Divergences

Frank Nielsen [1,2,*], **Richard Nock** [3] **and Shun-ichi Amari** [4]

[1] Sony Computer Science Laboratories, Inc, Tokyo 141-0022, Japan
[2] École Polytechnique, 91128 Palaiseau Cedex, France
[3] NICTA and The Australian National University, Locked Bag 9013, Alexandria NSW 1435, Australia
[4] RIKEN Brain Science Institute, 2-1 Hirosawa Wako City, Saitama 351-0198, Japan; E-Mail: amari@brain.riken.jp
* E-Mail: Frank.Nielsen@acm.org; Tel.:+81-3-5448-4380.

Received: 15 May 2014; in revised form: 10 June 2014 / Accepted: 13 June 2014 /
Published: 17 June 2014

Abstract: Clustering sets of histograms has become popular thanks to the success of the generic method of bag-of-X used in text categorization and in visual categorization applications. In this paper, we investigate the use of a parametric family of distortion measures, called the α-divergences, for clustering histograms. Since it usually makes sense to deal with symmetric divergences in information retrieval systems, we symmetrize the α-divergences using the concept of mixed divergences. First, we present a novel extension of k-means clustering to mixed divergences. Second, we extend the k-means++ seeding to mixed α-divergences and report a guaranteed probabilistic bound. Finally, we describe a soft clustering technique for mixed α-divergences.

Keywords: bag-of-X; α-divergence; Jeffreys divergence; centroid; k-means clustering; k-means seeding

1. Introduction: Motivation and Background

1.1. Clustering Histograms in the Bag-of-Word Modeling Paradigm

A common task of information retrieval (IR) systems is to classify documents into categories. Given a training set of documents labeled with categories, one asks to classify new incoming documents. Text categorisation [1,2] proceeds by first defining a dictionary of words from a corpus. It then models each document by a word count yielding a word distribution histogram per document (see the University of California, Irvine, UCI, machine learning repository for such data-sets [3]). The importance of the words in the dictionary can be weighted by the term frequency-inverse document frequency [2] (tf-idf) that takes into account both the frequency of the words in a given document, but also of the frequency of the words in all documents: Namely, the tf-idf weight for a given word in a given document is the product of the frequency of that word in the document times the logarithm of the ratio of the number of documents divided by the document frequency of the word [2]. Defining a proper distance between histograms allows one to:

- Classify a new on-line document: We first calculate its word distribution histogram signature and seek for the labeled document, which has the most similar histogram to deduce its category tag.
- Find the initial set of categories: we cluster all document histograms and assign a category per cluster.

This text classification method based on the representation of the bag-of -words (BoWs) has also been instrumental in computer vision for efficient object categorization [4] and recognition in natural images [5]. This paradigm is called bag-of-features [6] (BoFs) in the general case. It first requires one to create a dictionary of "visual words" by quantizing keypoints (e.g., affine invariant descriptors of image patches) of the training database. Quantization is performed using the k-means [7–9] algorithm that partitions n data $\mathcal{X} = \{x_1, ..., x_n\}$ into k pairwise disjoint clusters $\mathcal{C}_1, ..., \mathcal{C}_k$, where each data element belongs to the closest cluster center (*i.e.*, the cluster prototype). From a given initialization, batched k-means first assigns data points to their closest centers and then updates the cluster centers and reiterates this process until convergence is met to a local minimum (not necessarily the global minimum) after a provably finite number of steps. Csurka *et al.* [4] used the squared Euclidean distance for building the visual vocabulary. Depending on the chosen features, other distances have proven useful. For example, the symmetrized Kullback–Leibler (KL) divergence was shown to perform experimentally better than the Euclidean or squared Euclidean distances for a compressed histogram of gradient descriptors [10] (CHoGs), even if it is not a metric distance, since its fails to satisfy the triangular inequality. To summarize, k-means histogram clustering with respect to the symmetrized KL (called Jeffreys divergence J) can be used to quantize both visual words and document categories. Nowadays, the seminal bag-of-word method has been generalized fruitfully to various settings using the generic bag-of-X paradigm, like the bag-of-textons [6], the bag-of-readers [11], *etc.* Bag-of-X represents each data (e.g., document, image, *etc.*) as an histogram of codeword count indices. Furthermore, the semantic space [12] paradigm has been recently explored to overcome two drawbacks of the bag-of-X paradigms: the high-dimensionality of the histograms (number of bins) and difficult human interpretation of the codewords due to the lack of semantic information. In semantic space, modeling relies on semantic multinomials that are discrete frequency histograms; see [12].

In summary, clustering histograms with respect to symmetric distances (like the symmetrized KL divergence) is playing an increasing role. It turns out that the symmetrized KL divergence belongs to a 1-parameter family of divergences, called symmetrized α-divergences, or Jeffreys α-divergence [13].

1.2. Contributions

Since divergences $D(p : q)$ are usually asymmetric distortion measures between two objects p and q, one has to often consider two kinds of centroids obtained by carrying the minimization process either on the left argument or on the right argument of the divergences; see [14]. In theory, it is enough to consider only one type of centroid, say the right centroid, since the left centroid with respect to a divergence $D(p : q)$ is equivalent to the right centroid with respect to the mirror divergence $D'(p : q) = D(q : p)$.

In this paper, we consider mixed divergences [15] that allow one to handle in a unified way the arithmetic symmetrization $S(p, q) = \frac{1}{2}(D(p : q) + D(q : p))$ of a given divergence $D(p : q)$ with both the sided divergences: $D(p : q)$ and its mirror divergence $D'(p : q)$. The mixed α-divergence is the mixed divergence obtained for the α-divergence. We term α-clustering the clustering with respect to α-divergences and mixed α-clustering the clustering w.r.t. mixed α-divergences [16]. Our main contributions are to extend the celebrated batched k-means [7–9] algorithm to mixed divergences by associating two dual centroids per cluster and to generalize the probabilistically guaranteed good seeding of k-means++ [17] to mixed α-divergences. The mixed α-seedings provide guaranteed probabilistic clustering bounds by picking up seeds from the data and do not require explicitly computing of centroids. Therefore, it follows a fast clustering technique in practice, even when cluster centers are not available in closed form. We also consider clustering histograms by explicitly building the symmetrized α-centroids and end up with a variational k-means when the centroids are not available in closed-form, Finally, we investigate soft mixed α-clustering and discuss topics related to α-clustering. Note that clustering with respect to non-symmetrized α-divergences has been recently investigated independently in [18] and proven useful in several applications.

1.3. Outline of the Paper

The paper is organized as follows: Section 2 introduces the notion of mixed divergences, presents an extension of k-means to mixed divergences and recalls some properties of α-divergences. Section 3 describes the α-seeding techniques and reports a probabilistically-guaranteed bound on the clustering quality. Section 4 investigates the various sided/symmetrized/mixed calculations of the α-centroids.

Section 5 presents the soft α-clustering with respect to α-mixed divergences. Finally, Section 6 summarises the contributions, discusses related topics and hints at further perspectives. The paper is followed by two appendices. Appendix B studies several properties of α-divergences that are used to derive the guaranteed probabilistic performance of the α-seeding. Appendix C proves that α-sided centroids are quasi-arithmetic means for the power generator functions.

2. Mixed Centroid-Based *k*-Means Clustering

2.1. Divergences, Centroids and k-Means

Consider a set $\mathcal{H}$ of n histograms $h_1, ..., h_n$, each with d bins, with all positive real-valued bins: $h_j^i > 0, \forall 1 \leq i \leq d, 1 \leq j \leq n$. A histogram h is called a frequency histogram when its bins sums up to one: $w(h) = w_h = \sum_i h^i = 1$. Otherwise, it is called a positive histogram that can eventually be normalized to a frequency histogram:

$$\tilde{h} \doteq \frac{h}{w(h)}. \tag{1}$$

The frequency histograms belong to the $(d\text{-}1)$-dimensional open probability simplex Δ_d:

$$\Delta_d \doteq \left\{ (x^1, ..., x^d) \in \mathbb{R}^d \mid \forall i, x^i > 0, \text{ and } \sum_{i=1}^{d} x^i = 1 \right\}. \tag{2}$$

That is, although frequency histograms have d bins, the constraint that those bin values should sum up to one yields d-1 degrees of freedom. In probability theory, the frequency or counting of histograms either model discrete multinomial probabilities or discrete positive measures (also called positive arrays [19]).

The celebrated k-means clustering [8,9] is one of the most famous clustering techniques that has been generalized in many ways [20,21]. In information geometry [22], a divergence $D(p:q)$ is a smooth C^3 differentiable dissimilarity measure that is not necessarily symmetric ($D(p:q) \neq D(q:p)$, hence the notation ":" instead of the classical "," reserved for metric distances), but is non-negative and satisfies the separability property: $D(p:q) = 0$ iff $p = q$. More precisely, let $\partial_i D(x:y) = \frac{\partial}{\partial x^i} D(x:y)$, $\partial_{,i} D(x:y) = \frac{\partial}{\partial y^i} D(x:y)$. Then, we require $\partial_i D(x:x) = \partial_{,i} D(x:x) = 0$ and $-\partial_i \partial_{,j} D(x:y)$ positive definite for defining a divergence. For a distance function $D(\cdot:\cdot)$, we denote by $D(x:\mathcal{H})$ the weighted average distance of x to a set a weighted histograms:

$$D(x:\mathcal{H}) \doteq \sum_{j=1}^{n} w_i D(x:h_j). \tag{3}$$

An important class of divergences on frequency histograms is the f-divergences [23–25] defined for a convex generator f (with $f(1) = f'(1) = 0$ and $f''(1) = 1$):

$$I_f(p:q) \doteq \sum_{i=1}^{d} q^i f\left(\frac{p^i}{q^i}\right).$$

Those divergences preserve information monotonicity [19] under any arbitrary transition probability (Markov morphisms). f-divergences can be extended to positive arrays [19].

The k-means algorithm on a set of weighted histograms can be tailored to any divergence as follows: First, we initialize the k cluster centers $\mathcal{C} = \{c_1, ..., c_k\}$ (say, by picking up randomly arbitrary distinct seeds). Then, we iteratively repeat until convergence the following two steps:

- Assignment: Assign each histogram h_j to its closest cluster center:

$$l(h_j) \doteq \arg\min_{l=1}^{k} D(h_j : c_l).$$

This yields a partition of the histogram set $\mathcal{H} = \cup_{l=1}^{k} \mathcal{A}_l$, where A_l denotes the set of histograms of the l-th cluster: $\mathcal{A}_l = \{h_j \, | l(h_j) = l\}$.

- Center relocation: Update the cluster centers by taking their centroids:

$$c_l \doteq \arg\min_x \sum_{h_j \in \mathcal{A}_l} w_j D(h_j : x).$$

Throughout this paper, centroid shall be understood in the broader sense of a barycenter when weights are non-uniform.

2.2. Mixed Divergences and Mixed k-Means Clustering

Since divergences are potentially asymmetric, we can define two-sided k-means or always consider a right-sided k-means, but then define another sided divergence $D'(p : q) = D(q : p)$. We can also consider the symmetrized k-means with respect to the symmetrized divergence: $S(p,q) = D(p : q) + D(q : p)$. Eventually, we may skew the symmetrization with a parameter $\lambda \in [0,1]$: $S_\lambda(p,q) = \lambda D(p : q) + (1-\lambda) D(q : p)$ (and consider other averaging schemes instead of the arithmetic mean).

In order to handle those sided and symmetrized k-means under the same framework, let us introduce the notion of mixed divergences [15] as follows:

Definition 1 (Mixed divergence).

$$M_\lambda(p : q : r) \doteq \lambda D(p : q) + (1-\lambda) D(q : r), \tag{4}$$

for $\lambda \in [0,1]$.

A mixed divergence includes the sided divergences for $\lambda \in \{0,1\}$ and the symmetrized (arithmetic mean) divergence for $\lambda = \frac{1}{2}$.

We generalize k-means clustering to mixed k-means clustering [15] by considering two centers per cluster (for the special cases of $\lambda = 0, 1$, it is enough to consider only one). Algorithm 1 sketches the generic mixed k-means algorithm. Note that a simple initialization consists of choosing randomly the k distinct seeds from the dataset with $l_i = r_i$.

Algorithm 1: Mixed divergence-based k-means clustering.

```
Input: Weighted histogram set H, divergence D(·,·), integer k > 0, real λ ∈ [0,1];
Initialize left-sided/right-sided seeds C = {(l_i, r_i)}_{i=1}^k;
repeat
    //Assignment
    for i = 1,2,...,k do
        C_i ← {h ∈ H : i = argmin_j M_λ(l_j : h : r_j)};
    // Dual-sided centroid relocation
    for i = 1,2,...,k do
        r_i ← argmin_x D(C_i : x) = Σ_{h∈C_i} w_j D(h : x);
        l_i ← argmin_x D(x : C_i) = Σ_{h∈C_i} w_j D(x : h);
until convergence;
Output: Partition of H into k clusters following C;
```

Notice that the mixed k-means clustering is different from the k-means clustering with respect to the symmetrized divergences S_λ that considers only one centroid per cluster.

2.3. Sided, Symmetrized and Mixed α-Divergences

For $\alpha \neq \pm 1$, we define the family of α-divergences [26] on positive arrays [27] as:

$$\begin{aligned} D_\alpha(p:q) &\doteq \sum_{i=1}^{d} \frac{4}{1-\alpha^2}\left(\frac{1-\alpha}{2}p^i + \frac{1+\alpha}{2}q^i - (p^i)^{\frac{1-\alpha}{2}}(q^i)^{\frac{1+\alpha}{2}}\right), \\ &= D_{-\alpha}(q:p), \alpha \in \mathbb{R}\backslash\{0,1\}, \end{aligned} \quad (5)$$

with the limit cases $D_{-1}(p:q) = \mathrm{KL}(p:q)$ and $D_1(p:q) = \mathrm{KL}(q:p)$, where KL is the extended Kullback–Leibler divergence:

$$\mathrm{KL}(p:q) \doteq \sum_{i=1}^{d} p^i \log\frac{p^i}{q^i} + q^i - p^i. \quad (6)$$

Divergence D_0 is the squared Hellinger symmetric distance (scaled by a multiplicative factor of four) extended to positive arrays:

$$D_0(p:q) = 2\int\left(\sqrt{p(x)} - \sqrt{q(x)}\right)^2 \mathrm{d}x = 4H^2(p,q), \quad (7)$$

with the Hellinger distance:

$$H(p,q) = \sqrt{\frac{1}{2}\int\left(\sqrt{p(x)} - \sqrt{q(x)}\right)^2 \mathrm{d}x}. \quad (8)$$

Note that α-divergences are defined for the full range of α values: $\alpha \in \mathbb{R}$. Observe that α-divergences of Equation (5) are homogeneous of degree one: $D_\alpha(\lambda p : \lambda q) = \lambda D_\alpha(p:q)$ for $\lambda > 0$.

When histograms p and q are both frequency histograms, we have:

$$\begin{aligned} D_\alpha(\tilde{p}:\tilde{q}) &= \frac{4}{1-\alpha^2}\left(1 - \sum_{i=1}^{d}(\tilde{p}^i)^{\frac{1-\alpha}{2}}(\tilde{q}^i)^{\frac{1+\alpha}{2}}\right), \\ &= D_{-\alpha}(\tilde{q}:\tilde{p}), \alpha \in \mathbb{R}\backslash\{0,1\}, \end{aligned} \quad (9)$$

and the extended Kullback–Leibler divergence reduces to the traditional Kullback–Leibler divergence: $\mathrm{KL}(\tilde{p}:\tilde{q}) = \sum_{i=1}^{d} \tilde{p}^i \log\frac{\tilde{p}^i}{\tilde{q}^i}$.

The Kullback–Leibler divergence between frequency histograms $\tilde{p}$ and $\tilde{q}$ ($\alpha = \pm 1$) is interpreted as the cross-entropy minus the Shannon entropy:

$$\mathrm{KL}(\tilde{p}:\tilde{q}) \doteq H^\times(\tilde{p}:\tilde{q}) - H(\tilde{p}).$$

Often, $\tilde{p}$ denotes the true model (hidden by nature), and $\tilde{q}$ is the estimated model from observations. However, in information retrieval, both $\tilde{p}$ and $\tilde{q}$ play the same symmetrical role, and we prefer to deal with a symmetric divergence.

The Pearson and Neyman χ^2 distances are obtained for $\alpha = -3$ and $\alpha = 3$, respectively:

$$D_3(\tilde{p}:\tilde{q}) = \frac{1}{2}\sum_i \frac{(\tilde{q}^i - \tilde{p}^i)^2}{\tilde{p}^i}, \quad (10)$$

$$D_{-3}(\tilde{p}:\tilde{q}) = \frac{1}{2}\sum_i \frac{(\tilde{q}^i - \tilde{p}^i)^2}{\tilde{q}^i}. \quad (11)$$

The α-divergences belong to the class of Csiszár f-divergences with the following generator:

$$f(t) = \begin{cases} \frac{4}{1-\alpha^2}\left(1 - t^{(1+\alpha)/2}\right), & \text{if } \alpha \neq \pm 1, \\ t \ln t, & \text{if } \alpha = 1, \\ -\ln t, & \text{if } \alpha = -1 \end{cases} \tag{12}$$

Remark 1. *Historically, the α-divergences have been introduced by Chernoff [28,29] in the context of hypothesis testing. In Bayesian binary hypothesis testing, we are asked to decide whether an observation belongs to one class or the other class, based on prior w_1 and w_2 and class-conditional probabilities p_1 and p_2. The average expected error of the best decision maximum a posteriori (MAP) rule is called the probability of error, denoted by P_e. When prior probabilities are identical ($w_1 = w_2 = \frac{1}{2}$), we have $P_e(p_1, p_2) = \frac{1}{2}\int \min(p_1(x), p_2(x))\mathrm{d}x$. Let $S(p,q) = \int \min(p(x), q(x))\mathrm{d}x$ denote the intersection similarity measure, with $0 < S \leq 1$ (generalizing the histogram intersection distance often used in computer vision [30]). S is bounded by the α-Chernoff affinity coefficient:*

$$S(p,q) \leq C_\beta(p,q) = \int p^\beta(x) q^{1-\beta}(x)\mathrm{d}x,$$

for all $\beta \in [0,1]$. We can convert the affinity coefficient $0 < C_\beta \leq 1$ into a divergence D_β by simply taking $D_\beta = 1 - C_\beta$. Since the absolute value of divergences does not matter, we can rescale appropriately the divergence. One nice rescaling is by multiplying by $\frac{1}{\beta(1-\beta)}$: $D_\beta = \frac{1}{\beta(1-\beta)}(1 - C_\beta)$. This lets coincide the parameterized divergence with the fundamental Kullback–Leibler divergence for the limit values $\beta \in \{0,1\}$. Last, by choosing $\beta = \frac{1-\alpha}{2}$, it yields the well-known expression of the α-divergences.

Interestingly, the α-divergences can be interpreted as a generalized α-Kullback–Leibler divergence [26] with deformed logarithms.

Next, we introduce the mixed α-divergence of a histogram x to two histograms p and q as follows:

Definition 2 (Mixed α-divergence). *The mixed α-divergence of a histogram x to two histograms p and q is defined by:*

$$\begin{aligned} M_{\lambda,\alpha}(p:x:q) &= \lambda D_\alpha(p:x) + (1-\lambda)D_\alpha(x:q), \\ &= \lambda D_{-\alpha}(x:p) + (1-\lambda)D_{-\alpha}(q:x), \\ &= M_{1-\lambda,-\alpha}(q:x:p), \end{aligned} \tag{13}$$

The α-Jeffreys symmetrized divergence is obtained for $\lambda = \frac{1}{2}$:

$$S_\alpha(p,q) = M_{\frac{1}{2},\alpha}(q:p:q) = M_{\frac{1}{2},\alpha}(p:q:p).$$

The skew symmetrized α-divergence is defined by:

$$S_{\lambda,\alpha}(p:q) = \lambda D_\alpha(p:q) + (1-\lambda)D_\alpha(q:p).$$

2.4. Notations and Hard/Soft Clusterings

Throughout the paper, superscript index i denotes the histogram bin numbers and subscript index j the histogram numbers. Index l is used to iterate on the clusters. The left-sided, right-sided and symmetrized histogram positive and frequency α-centroids are denoted by $l_\alpha, r_\alpha, s_\alpha$ and $\tilde{l}_\alpha, \tilde{r}_\alpha, \tilde{s}_\alpha$, respectively.

In this paper, we investigate the following kinds of clusterings for sets of histograms:

Hard clustering. Each histogram belongs to exactly one cluster:

- k-means with respect to mixed divergences $M_{\lambda,\alpha}$.
- k-means with respect to symmetrized divergences $S_{\lambda,\alpha}$.

- Randomized seeding for mixed/symmetrized k-means by extending k-means++ with guaranteed probabilistic bounds for α-divergences.

Soft clustering. Each histogram belongs to all clusters according to some weight distribution: the soft mixed α-clustering.

3. Coupled k-Means++ α-Seeding

It is well-known that the Lloyd k-means clustering algorithm monotonically decreases the loss function and stops after a finite number of iterations into a local optimal. Optimizing globally the k-means loss is NP-hard [17] when $d > 1$ and $k > 1$. In practice, the performance of the k-means algorithm heavily relies on the initialization. A breakthrough was obtained by the k-means++ seeding [17], which guarantees in expectation a good starting partition. We extend this scheme to the coupled α-clustering. However, we point out that although k-means++ prove popular and are often used in practice with very good results; it has been recently pointed out that "worst case" configurations exist and even in small dimensions, on which the algorithm cannot beat significantly its expected approximability with a high probability [31]. Still, the expected approximability ratio, roughly in $O(\log(k))$, is very good, as long as the number of clusters is not too large.

Algorithm 2: Mixed α-seeding; MAS($\mathcal{H}$, k, λ, α)

Input: Weighted histogram set $\mathcal{H}$, integer $k \geq 1$, real $\lambda \in [0,1]$, real $\alpha \in \mathbb{R}$;
Let $\mathcal{C} \leftarrow h_j$ with uniform probability ;
for $i = 2, 3, ..., k$ **do**
  Pick at random histogram $h \in \mathcal{H}$ with probability:

$$\pi_{\mathcal{H}}(h) \doteq \frac{w_h M_{\lambda,\alpha}(c_h : h : c_h)}{\sum_{y \in \mathcal{H}} w_y M_{\lambda,\alpha}(c_y : y : c_y)}, \tag{14}$$

  //where $(c_h, c_h) \doteq \arg\min_{(z,z)\in\mathcal{C}} M_{\lambda,\alpha}(z : h : z)$;
  $\mathcal{C} \leftarrow \mathcal{C} \cup \{(h,h)\}$;
Output: Set of initial cluster centers $\mathcal{C}$;

Algorithm 2 provides our adaptation of k-means++ seeding [15,17]. It works for all three of our sided/symmetrized and mixed clustering settings:

- Pick $\lambda = 1$ for the left-sided centroid initialization,
- Pick $\lambda = 0$ for the right-sided centroid initialization (a left-sided initialization for $-\alpha$),
- with arbitrary λ, for the λ-J_α (skew Jeffreys) centroids or mixed λ centroids. Indeed, the initialization is the same (see the MAS procedure in Algorithm 2).

Our proof follows and generalizes the proof described for the case of mixed Bregman seeding [15] (Lemma 2). In fact, our proof is more precise, as it quantifies the expected potential with respect to the optimum only, whereas in [15], the optimal potential is averaged with a dual optimal potential, which depends on the optimal centers, but may be larger than the optimum sought.

Theorem 1. *Let $C_{\lambda,\alpha}$ denote for short the cost function related to the clustering type chosen (left-, right-, skew Jeffreys or mixed) in* MAS*and $C^{opt}_{\lambda,\alpha}$ denote the optimal related clustering in k clusters, for $\lambda \in [0,1]$ and $\alpha \in (-1,1)$. Then, on average, with respect to distribution (14), the initial clustering of* MAS *satisfies:*

$$E_\pi[C_{\lambda,\alpha}] \leq 4 \begin{cases} f(\lambda)g(k)h^2(\alpha)C^{opt}_{\lambda,\alpha} & \text{if } \lambda \in (0,1) \\ g(k)z(\alpha)h^4(\alpha)C^{opt}_{\lambda,\alpha} & \text{otherwise} \end{cases}. \tag{15}$$

Here, $f(\lambda) = \max\left\{\frac{1-\lambda}{\lambda}, \frac{\lambda}{1-\lambda}\right\}$, $g(k) = 2(2+\log k)$, $z(\alpha) = \left(\frac{1+|\alpha|}{1-|\alpha|}\right)^{\frac{8|\alpha|^2}{(1-|\alpha|)^2}}$, $h(\alpha) = \max_i p_i^{|\alpha|} / \min_i p_i^{|\alpha|}$*; the min is defined on strictly positive coordinates, and* π *denotes the picking distribution of Algorithm 2.*

Remark 2. *The bound is particularly good when* λ *is close to* $1/2$*, and in particular for the* α*-Jeffreys clustering, as in these cases, the only additional penalty compared to the Euclidean case [17] is* $h^2(\alpha)$*, a penalty that relies on an optimal triangle inequality for* α*-divergences that we provide in Lemma A6 below.*

Remark 3. *This guaranteed initialization is particularly useful for* α*-Jeffreys clustering, as there is no closed form solution for the centroids (except when* $\alpha = \pm 1$*, see [32]).*

Algorithm 3: Mixed α-hard clustering: MAhC($\mathcal{H}, k, \lambda, \alpha$)

Input: Weighted histogram set $\mathcal{H}$, integer $k > 0$, real $\lambda \in [0,1]$, real $\alpha \in \mathbb{R}$;
Let $\mathcal{C} = \{(l_i, r_i)\}_{i=1}^k \leftarrow \text{MAS}(\mathcal{H}, k, \lambda, \alpha)$;
repeat
  //Assignment
  for $i = 1, 2, ..., k$ **do**
    $\mathcal{A}_i \leftarrow \{h \in \mathcal{H} : i = \arg\min_j M_{\lambda,\alpha}(l_j : h : r_j)\}$;
  // Centroid relocation
  for $i = 1, 2, ..., k$ **do**
    $r_i \leftarrow \left(\sum_{h\in\mathcal{A}_i} w_i h^{\frac{1-\alpha}{2}}\right)^{\frac{2}{1-\alpha}}$;
    $l_i \leftarrow \left(\sum_{h\in\mathcal{A}_i} w_i h^{\frac{1+\alpha}{2}}\right)^{\frac{2}{1+\alpha}}$;
until *convergence;*
Output: Partition of $\mathcal{H}$ in k clusters following $\mathcal{C}$;

Algorithm 3 presents the general hard mixed k-means clustering, which can be adapted also to left- ($\lambda = 1$) and right- ($\lambda = 0$) α-clustering.

For skew Jeffreys centers, since the centroids are not available in closed form [32], we adopt a variational approach of k-means by updating iteratively the centroid in each cluster (thus improving the overall loss function without computing the optimal centroids that would eventually require infinitely many iterations).

4. Sided, Symmetrized and Mixed α-Centroids

The k-means clustering requires assigning data elements to their closest cluster center and then updating those cluster centers by taking their centroids. This section investigates the centroid computations for the sided, symmetrized and mixed α-divergences.

Note that the mixed α-seeding presented in Section 3 does not require computing centroids and, yet, guarantees probabilistically a good clustering partition.

Since mixed α-divergences are f-divergences, we start with the generic f-centroids.

4.1. Csiszár f-Centroids

The centroids induced by f-divergences of a set of positive measures (that relaxes the normalisation constraint) have been studied by Ben-Tal *et al.* [33]. Those entropic centroids are shown to be unique, since f-divergences are convex statistical distances in both arguments. Let E_f denote the energy to minimize when considering f-divergences:

$$E_f \doteq \min_{x\in\mathcal{X}} I_f(\mathcal{H}:x) = \sum_{j=1}^{n} w_j I_f(h_j:x), \quad (16)$$

$$= \min_{x\in\mathcal{X}} \sum_{j=1}^{n} w_j \sum_{i=1}^{d} p_j^i f\left(\frac{c^i}{h_j^i}\right). \quad (17)$$

When the domain is the open probability simplex $\mathcal{X} = \Delta_d$, we get a constrained optimisation problem to solve. We transform this constrained minimisation problem (*i.e.*, $x \in \Delta_d$) into an equivalent unconstrained minimisation problem by using the Lagrange multiplier, γ:

$$\min_{x\in\mathbb{R}^d} \sum_{j=1}^{n} w_j I_f(h_j:c) + \gamma\left(\sum_{i=1}^{d} x^i - 1\right). \quad (18)$$

Taking the derivatives according to x^i, we get:

$$\forall i \in \{1,...,d\}, \sum_{j=1}^{n} w_j f'\left(\frac{x^i}{h_j^i}\right) - \gamma = 0. \quad (19)$$

We now consider this equation for α-divergences and symmetrized α-divergences, both f-divergences.

4.2. Sided Positive and Frequency α-Centroids

The positive sided α-centroids for a set of weighted histograms were reported in [34] using the representation Bregman divergence. We summarise the results in the following theorem:

Theorem 2 (Sided positive α-centroids [34]). *The left-sided l_α and right-sided r_α positive weighted α-centroid coordinates of a set of n positive histograms $h_1, ..., h_n$ are weighted α-means:*

$$r_\alpha^i = f_\alpha^{-1}\left(\sum_{j=1}^{n} w_j f_\alpha(h_j^i)\right), l_\alpha^i = r_{-\alpha}^i$$

with $f_\alpha(x) = \begin{cases} x^{\frac{1-\alpha}{2}} & \alpha \neq \pm 1, \\ \log x & \alpha = 1. \end{cases}$

Furthermore, the frequency-sided α-centroids are simply the normalized-sided α-centroids.

Theorem 3 (Sided frequency α-centroids [16]). *The coordinates of the sided frequency α-centroids of a set of n weighted frequency histograms are the normalised weighted α-means.*

Table 1 summarizes the results concerning the sided positive and frequency α-centroids.

Table 1. Positive and frequency α-centroids: the frequency α-centroids are normalized positive α-centroids, where $w(h)$ denotes the cumulative sum of the histogram bins. The arithmetic mean is obtained for $r_{-1} = l_1$ and the geometric mean for $r_1 = l_{-1}$.

	Positive centroid	Frequency centroid
Right-sided centroid	$r^i_\alpha = \begin{cases} (\sum_{j=1}^n w_j (h^i_j)^{\frac{1-\alpha}{2}})^{\frac{2}{1-\alpha}} & \alpha \neq 1 \\ r^i_1 = \prod_{j=1}^n (h^i_j)^{w_j} & \alpha = 1 \end{cases}$	$\tilde{r}^i_\alpha = \frac{r^i_\alpha}{w(\tilde{r}_\alpha)}$
Left-sided centroid	$l^i_\alpha = r^i_{-\alpha} = \begin{cases} (\sum_{j=1}^n w_j (h^i_j)^{\frac{1+\alpha}{2}})^{\frac{2}{1+\alpha}} & \alpha \neq -1 \\ l^i_{-1} = \prod_{j=1}^n (h^i_j)^{w_j} & \alpha = -1 \end{cases}$	$\tilde{l}^i_\alpha = \tilde{r}^i_{-\alpha} = \frac{r^i_{-\alpha}}{w(\tilde{r}_{-\alpha})}$

4.3. Mixed α-Centroids

The mixed α-centroids for a set of n weighted histograms is defined as the minimizer of:

$$\sum_j w_j M_{\lambda,\alpha}(l : h_j : r). \tag{20}$$

We state the theorem generalizing [15]:

Theorem 4. *The two mixed α-centroids are the left-sided and right-sided α-centroids.*

Figure 1 depicts some clustering result with our α-clustering software. We remark that the clusters found are all approximately subclusters of the "distinct" clusters that appear on the figure. When those distinct clusters are actually the optimal clusters—which is likely to be the case when they are separated by large minimal distance to other clusters—this is clearly a desirable qualitative property as long as the number of experimental clusters is not too large compared to the number of optimal clusters. We remark also that in the experiment displayed, there is no closed form solution for the cluster centers.

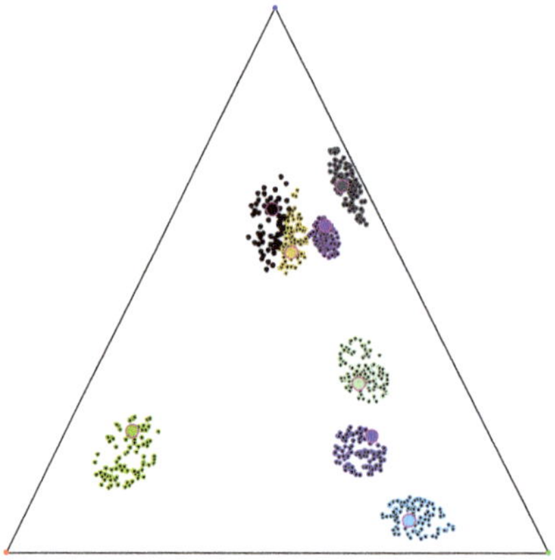

Figure 1. Snapshot of the α-clustering software. Here, $n = 800$ frequency histograms of three bins with $k = 8$, and $\alpha = 0.7$ and $\lambda = \frac{1}{2}$.

4.4. Symmetrized Jeffreys-Type α-Centroids

The Kullback–Leibler divergence can be symmetrized in various ways: Jeffreys divergence, Jensen–Shannon divergence and Chernoff information, just to mention a few. Here, we consider the following symmetrization of α-divergences extending Jeffreys J-divergence:

$$\begin{aligned} S_\alpha(p,q) &= \frac{1}{2}(D_\alpha(p:q)+D_\alpha(q:p)) = S_{-\alpha}(p,q), \\ &= M_{\frac{1}{2}}(p:q:p), \end{aligned} \tag{21}$$

For $\alpha=\pm 1$, we get half of Jeffreys divergence:

$$S_{\pm 1}(p,q) = \frac{1}{2}\sum_{i=1}^{d}(p^i-q^i)\log\frac{p^i}{q^i}$$

In particular, when p and q are frequency histograms, we have for $\alpha \neq \pm 1$:

$$J_\alpha(\tilde{p}:\tilde{q}) = \frac{8}{1-\alpha^2}\left(1+\sum_{i=1}^{d} H_{\frac{1-\alpha}{2}}(\tilde{p}^i,\tilde{q}^i)\right), \tag{22}$$

where $H_{\frac{1-\alpha}{2}}(a,b)$ a symmetric Heinz mean [35,36]:

$$H_\beta(a,b) = \frac{a^\beta b^{1-\beta}+a^{1-\beta}b^\beta}{2}.$$

Heinz means interpolate the arithmetic and geometric means and satisfies the inequality:

$$\sqrt{ab} = H_{\frac{1}{2}}(a,b) \leq H_\alpha(a,b) \leq H_0(a,b) = \frac{a+b}{2}.$$

(Another interesting property of Heinz means is the integral representation of the logarithmic mean: $L(x,y)=\frac{x-y}{\log x-\log y}=\int_0^1 H_\beta(x,y)d\beta$. This allows one to prove easily that $\sqrt{xy}\leq L(x,y)\leq\frac{x+y}{2}$.)

The J_α-divergence is a Csiszár f-divergence [24,25].

Observe that it is enough to consider $\alpha \in [0,\infty)$ and that the symmetrized α-divergence for positive and frequency histograms coincide only for $\alpha=\pm 1$.

For $\alpha=\pm 1$, $S_\alpha(p,q)$ tends to the Jeffreys divergence:

$$J(p,q) = \mathrm{KL}(p,q)+\mathrm{KL}(q,p) = \sum_{i=1}^{d}(p^i-q^i)(\log p^i - \log q^i). \tag{23}$$

The Jeffreys divergence writes mathematically the same for frequency histograms:

$$J(\tilde{p},\tilde{q}) = \mathrm{KL}(\tilde{p},\tilde{q})+\mathrm{KL}(\tilde{q},\tilde{p}) = \sum_{i=1}^{d}(\tilde{p}^i-\tilde{q}^i)(\log \tilde{p}^i - \log \tilde{q}^i). \tag{24}$$

We state the results reported in [32]:

Theorem 5 (Jeffreys positive centroid [32]). *The Jeffreys positive centroid $c=(c^1,...,c^d)$ of a set $\{h_1,...,h_n\}$ of n weighted positive histograms with d bins can be calculated component-wise exactly using the Lambert W analytic function:*

$$c^i = \frac{a^i}{W(\frac{a^i}{g^i}e)},$$

where $a^i=\sum_{j=1}^n \pi_j h_j^i$ denotes the coordinate-wise arithmetic weighted means and $g^i=\prod_{j=1}^n (h_j^i)^{\pi_j}$ the coordinate-wise geometric weighted means.

The Lambert analytic function W [37] (positive branch) is defined by $W(x)e^{W(x)} = x$ for $x \geq 0$.

Theorem 6 (Jeffreys frequency centroid [32]). *Let $\tilde{c}$ denote the Jeffreys frequency centroid and $\tilde{c}' = \frac{c}{w_c}$ the normalised Jeffreys positive centroid. Then, the approximation factor $\alpha_{\tilde{c}'} = \frac{S_1(\tilde{c}',\tilde{\mathcal{H}})}{S_1(\tilde{c},\tilde{\mathcal{H}})}$ is such that $1 \leq \alpha_{\tilde{c}'} \leq \frac{1}{w_c}$ (with $w_c \leq 1$).*

Therefore, we shall consider $\alpha \neq \pm 1$ in the remainder.

We state the following lemma generalizing the former results in [38] that were tailored to the symmetrized Kullback–Leibler divergence or the symmetrized Bregman divergence [14]:

Lemma 1 (Reduction property). *The symmetrized J_α-centroid of a set of n weighted histograms amount to computing the symmetrized α-centroid for the weighted α-mean and $-\alpha$-mean:*

$$\min J_\alpha(x, \mathcal{H}) = \min_x \left(D_\alpha(x : r_\alpha) + D_\alpha(l_\alpha : x)\right).$$

Proof. It follows that the minimization problem $\min_x S_\alpha(x, \mathcal{H}) = \sum_{j=1}^n w_j S_\alpha(x, h_j)$ reduces to the following minimization:

$$\min \sum_{i=1}^d x^i - (x^i)^{\frac{1+\alpha}{2}} \bar{h}^i_\alpha - (x^i)^{\frac{1-\alpha}{2}} \bar{h}^i_{-\alpha}. \tag{25}$$

This is equivalent to minimizing:

$$\begin{aligned}
\equiv \quad & \sum_{i=1}^d x^i - (x^i)^{\frac{1+\alpha}{2}} ((\bar{h}^i_\alpha)^{\frac{2}{1-\alpha}})^{\frac{1-\alpha}{2}} - \\
& (x^i)^{\frac{1-\alpha}{2}} ((\bar{h}^i_{-\alpha})^{\frac{2}{1+\alpha}})^{\frac{1+\alpha}{2}}, \\
\equiv \quad & \sum_{i=1}^d x^i - (x^i)^{\frac{1+\alpha}{2}} (r^i_\alpha)^{\frac{1-\alpha}{2}} - (x^i)^{\frac{1-\alpha}{2}} (l^i_\alpha)^{\frac{1+\alpha}{2}} \\
\equiv \quad & D_\alpha(x : r_\alpha) + D_\alpha(l_\alpha : x).
\end{aligned}$$

Note that $\alpha = \pm 1$, the lemma states that the minimization problem is equivalent to minimizing $\mathrm{KL}(a : x) + \mathrm{KL}(x : g)$ with respect to x, where $a = l_1$ and $g = r_1$ denote the arithmetic and geometric means, respectively. □

The lemma states that the optimization problem with n weighted histograms is equivalent to the optimization with only two equally weighted histograms.

The positive symmetrized α-centroid is equivalent to computing a representation symmetrized Bregman centroid [14,34].

The frequency symmetrized α-centroid asks to minimize the following problem:

$$\min_{\tilde{x} \in \Delta_d} \sum_j w_j S_\alpha(\tilde{x}, \tilde{h}_i).$$

Instead of seeking for $\tilde{x}$ in the probability simplex, we can optimize on the unconstrained domain $\mathbb{R}^{d-1}$ by using a reparameterization. Indeed, frequency histograms belong to the exponential families [39] (multinomials).

Exponential families also include many other continuous distributions, like the Gaussian, Beta or Dirichlet distributions. It turns out the α-divergences can be computed in closed-form for members of the same exponential family:

Lemma 2. *The α-divergence for distributions belonging to the same exponential families amounts to computing a divergence on the corresponding natural parameters:*

$$A_\alpha(p:q) = \frac{4}{1-\alpha^2}\left(1 - e^{-J_F^{\left(\frac{1-\alpha}{2}\right)}(\theta_p:\theta_q)}\right),$$

where $J_F^\beta(\theta_1:\theta_2) = \beta F(\theta_1) + (1-\beta)F(\theta_2) - F(\beta\theta_1 + (1-\beta)\theta_2)$ *is a skewed Jensen divergence defined for the log-normaliser F of the family.*

The proof follows from the fact that $\int p^\alpha(x)q^{1-\alpha}(x)\mathrm{d}x = e^{-J_F^{(\alpha)}(\theta_p:\theta_q)}$; see [40].

First, we convert a frequency histogram $\tilde{h}$ to its natural parameter θ with $\theta^i = \log\frac{\tilde{h}^i}{\tilde{h}^d}$; see [39]. The log-normaliser is a non-separable convex function $F(\theta) = \log(1 + \sum_i e^{\theta_i})$. To convert back a multinomial to a frequency histogram with d bins, we first set $\tilde{h}^d = \frac{1}{1+\sum_{l=1}^{d-1} e^{\theta^l}}$ and then retrieve the other bin values as $\tilde{h}^i = \tilde{h}^d e^{\theta^i}$.

The centroids with respect to skewed Jensen divergences has been investigated in [13,40].

Remark 4. *Note that for the special case of* $\alpha = 0$ *(squared Hellinger centroid), the sided and symmetrized centroids coincide. In that case, the coordinates* s_0^i *of the squared Hellinger centroid are:*

$$s_0^i = \left(\sum_{j=1}^n w_j\sqrt{h_j^i}\right)^2, 1 \le i \le d.$$

Remark 5. *The symmetrized positive α-centroids can be solved in special cases* ($\alpha = \pm 3$, $\alpha = \pm 1$ *corresponding to the symmetrized* χ^2 *and Jeffreys positive centroids). For frequency centroids, when dealing with binary histograms* ($d = 2$)*, we have only one degree of freedom and can solve the binary frequency centroids. Binary histograms (and mixtures thereof) are used in computer vision and pattern recognition* [41].

Remark 6. *Since α-divergences are Csiszár f-divergences and f-divergences can always be symmetrized by taking generator* $s(t) = f(t) + tf(\frac{1}{t})$*, we deduce that symmetrized α-divergences* S_α *are f-divergences for the generator:*

$$f(t) = -\log((1-\alpha) + \alpha t) - t\log\left((1-\alpha) + \frac{\alpha}{t}\right). \tag{26}$$

Hence, S_α *divergences are convex in both arguments, and the* s_α *centroids are unique.*

5. Soft Mixed α-Clustering

Algorithm 4 reports the general clustering with soft membership, which can be adapted to left ($\lambda_{\text{init}} = 1$), right ($\lambda_{\text{init}} = 0$) or mixed clustering. We have not considered a weighted histogram set in order not load the notations and because the extension is straightforward.

Again, for skew Jeffreys centers, we shall adopt a variational approach. Notice that the soft clustering approach learns all parameters, including λ (if not constrained to zero or one) and $\alpha \in \mathbb{R}$. This is not the case for Matsuyama's α-expectation maximization (EM) algorithm [42] in which α is fixed beforehand (and, thus, not learned).

Assuming we model the prior for histograms by:

$$p_{\lambda,\alpha,j}(h_i) \propto \lambda\exp -D_\alpha(l_j : h_i) + (1-\lambda)\exp -D_\alpha(h_i : r_j), \tag{27}$$

the negative log-likelihood involves the α-depending quantity:

$$\sum_{j=1}^{k}\sum_{i=1}^{m} p(j|h_i)\log p_{\lambda,\alpha,j}(h_i) \geq \sum_{j=1}^{k}\sum_{i=1}^{m} M_{\lambda,\alpha}(l_j : h_i : r_j)p(j|h_i), \tag{28}$$

because of the concavity of the logarithm function. Therefore, the maximization step for α involves finding:

$$\arg\max_{\alpha} \sum_{j=1}^{k}\sum_{i=1}^{m} M_{\lambda,\alpha}(l_j : h_i : r_j)p(j|h_i)\ . \tag{29}$$

Algorithm 4: Mixed α-soft clustering; MAsC($\mathcal{H}$, k, λ, α)

Input: Histogram set $\mathcal{H}$ with $|\mathcal{H}| = m$, integer $k > 0$, real $\lambda \leftarrow \lambda_{\text{init}} \in [0,1]$, real $\alpha \in \mathbb{R}$;
Let $\mathcal{C} = \{(l_i, r_i)\}_{i=1}^{k} \leftarrow \text{MAS}(\mathcal{H}, k, \lambda, \alpha)$;
repeat
  //Expectation
  for $i = 1, 2, ..., m$ **do**
    for $j = 1, 2, ..., k$ **do**
      $p(j|h_i) = \frac{\pi_j \exp(-M_{\lambda,\alpha}(l_j:h_i:r_j))}{\sum_{j'} \pi_{j'} \exp(-M_{\lambda,\alpha}(l_{j'}:h_i:r_{j'}))}$;
  //Maximization
  for $j = 1, 2, ..., k$ **do**
    $\pi_j \leftarrow \frac{1}{m}\sum_i p(j|h_i)$;
    $l_i \leftarrow \left(\frac{1}{\sum_i p(j|h_i)} \sum_i p(j|h_i) h_i^{\frac{1+\alpha}{2}}\right)^{\frac{2}{1+\alpha}}$;
    $r_i \leftarrow \left(\frac{1}{\sum_i p(j|h_i)} \sum_i p(j|h_i) h_i^{\frac{1-\alpha}{2}}\right)^{\frac{2}{1-\alpha}}$;
  //Alpha - Lambda
  $\alpha \leftarrow \alpha - \eta_1 \sum_{j=1}^{k}\sum_{i=1}^{m} p(j|h_i) \frac{\partial}{\partial\alpha} M_{\lambda,\alpha}(l_j : h_i : r_j)$;
  if $\lambda_{\text{init}} \neq 0, 1$ **then**
    $\lambda \leftarrow \lambda - \eta_2 \Big(\sum_{j=1}^{k}\sum_{i=1}^{m} p(j|h_i) D_\alpha(l_j : h_i) -$
    $\sum_{j=1}^{k}\sum_{i=1}^{m} p(j|h_i) D_\alpha(h_i : r_j)\Big)$;
    //for some small η_1, η_2; ensure that $\lambda \in [0,1]$.
until *convergence*;
Output: Soft clustering of $\mathcal{H}$ according to k densities $p(j|.)$ following $\mathcal{C}$;

No closed-form solution are known, so we compute the gradient update in Algorithm 4 with:

$$\frac{\partial M_{\lambda,\alpha}(l_j : h_i : r_j)}{\partial\alpha} = \lambda\frac{\partial D_\alpha(l_j : h_i)}{\partial\alpha} + (1-\lambda)\frac{\partial D_\alpha(h_i : r_j)}{\partial\alpha}\ , \tag{30}$$

$$\frac{\partial D_\alpha(p:q)}{\partial \alpha} = \frac{2}{(1-\alpha)^2}\times \left(q - \left(\frac{1-\alpha}{1+\alpha}\right)^2 p + p^{\frac{1-\alpha}{2}} q^{\frac{1+\alpha}{2}} \left(\frac{4\alpha}{1-\alpha^2} - \ln\left(\frac{q}{p}\right)\right)\right) . \quad (31)$$

The update in λ is easier as:

$$\frac{\partial M_{\lambda,\alpha}(l_j : h_i : r_j)}{\partial \lambda} = D_\alpha(l_j : h_i) - D_\alpha(h_i : r_j) . \quad (32)$$

Maximizing the likelihood in λ would imply choosing $\lambda \in \{0, 1\}$ (a hard choice for left/right centers), yet we prefer the soft update for the parameter, like for α.

6. Conclusions

The family of α-divergences plays a fundamental role in information geometry: These statistical distortion measures are the canonical divergences of dual spaces on probability distribution manifolds with constant curvature $\kappa = \frac{1-\alpha^2}{4}$ and the canonical divergences of dually flat manifolds for positive distribution manifolds [19].

In this work, we have presented three techniques for clustering (positive or frequency) histograms using k-means:

(1) Sided left or right α-centroid k-means,
(2) Symmetrized Jeffreys-type α-centroid (variational) k-means, and
(3) Coupled k-means with respect to mixed α-divergences relying on dual α-centroids.

Sided and mixed dual centroids are always available in closed-forms and are therefore highly attractive from the standpoint of implementation. Symmetrized Jeffreys centroids are in general not available in closed-form and require one to implement a variational k-means by updating incrementally the cluster centroids in order to monotonically decrease the k-means loss function. From the clustering standpoint, this appears not to be a problem when guaranteed expected approximations to the optimal clustering are enough.

Indeed, we also presented and analyzed an extension of k-means++ [17] for seeding those k-means algorithms. The mixed α-seeding initializations do not require one to calculate centroids and behaves like a discrete k-means by picking up the seeds among the data. We reported guaranteed probabilistic clustering bounds. Thus, it yields a fast hard/soft data partitioning technique with respect to mixed or symmetrized α-divergences. Recently, the advantage of clustering using α-divergences by tuning α in applications has been demonstrated in [18]. We thus expect the computationally fast mixed α-seeding with guaranteed performance to be useful in a growing number of applications.

Acknowledgments

NICTA is funded by the Australian Government as represented by the Department of Broadband, Communication and the Digital Economy and the Australian Research Council through the ICT Centre of Excellence program.

Author Contributions: Author Contributions

All authors contributed equally to the design of the research. The research was carried out by all authors. Frank Nielsen and Richard Nock wrote the paper. Frank Nielsen implemented the algorithms and performed experiments. All authors have read and approved the final manuscript.

Conflicts of Interest: Conflicts of Interest

The authors declare no conflict of interests.

Appendix Proof Sketch of Theorem 1

We give here the key results allowing one to obtain the proof of the Theorem, following the proof scheme of [15]. In order not to load the notations, weights are considered uniform. The extension to non-uniform weights is immediate as it boils down to duplicate histograms in the histogram set and does not change the approximation result.

Let $\mathcal{A} \subseteq \mathcal{H}$ be an arbitrary cluster of $\mathcal{C}_{\text{opt}}$. Let us define $U_{\mathcal{A}}$ and $\pi_{\mathcal{A}}$ as the uniform and biased distributions conditioned to $\mathcal{A}$. The key to the proof is to relate the expected potential of $\mathcal{A}$ under $U_{\mathcal{A}}$ and $\pi_{\mathcal{A}}$ to its contribution to the optimal potential.

Lemma A1. *Let $\mathcal{A} \subseteq \mathcal{H}$ be an arbitrary cluster of $\mathcal{C}_{\text{opt}}$. Then:*

$$\begin{aligned} \mathbf{E}_{c \sim U_{\mathcal{A}}}[M_{\lambda,\alpha}(\mathcal{A}, c)] &= M_{\text{opt},\lambda,\alpha}(\mathcal{A}) + M_{\text{opt},\lambda,-\alpha}(\mathcal{A}) \\ &= \mathbf{E}_{c \sim U_{\mathcal{A}}}[M_{\lambda,-\alpha}(\mathcal{A}, c)] \ , \end{aligned}$$

where $U_{\mathcal{A}}$ is the uniform distribution over $\mathcal{A}$.

Proof. α-coordinates have the property that for any subset $\mathcal{A} \subseteq \mathcal{H}$, $(1/|\mathcal{A}|)\sum_{p \in \mathcal{A}} u_\alpha(p) = u_\alpha(r_{\alpha,\mathcal{A}})$. Hence, we have:

$$\begin{aligned} & \forall c \in \mathcal{A} \ , \sum_{p \in \mathcal{A}} D_\alpha(p : c) \\ = \ & \sum_{p \in \mathcal{A}} D_{\varphi_\alpha}(u_\alpha(p) : u_\alpha(c)) \\ = \ & \sum_{p \in \mathcal{A}} D_{\varphi_\alpha}(u_\alpha(p) : u_\alpha(r_{\alpha,\mathcal{A}})) + |\mathcal{A}| D_{\varphi_\alpha}(u_\alpha(r_{\alpha,\mathcal{A}}) : u_\alpha(c)) \\ = \ & \sum_{p \in \mathcal{A}} D_\alpha(p : r_{\alpha,\mathcal{A}}) + |\mathcal{A}| D_\alpha(r_{\alpha,\mathcal{A}} : c) \ . \end{aligned} \tag{A1}$$

Because $D_\alpha(p : q) = D_{-\alpha}(q : p)$ and $l_\alpha = r_{-\alpha}$, we obtain:

$$\begin{aligned} & \forall c \in \mathcal{A} \ , \sum_{p \in \mathcal{A}} D_\alpha(c : p) \\ = \ & \sum_{p \in \mathcal{A}} D_{-\alpha}(p : c) \\ = \ & \sum_{p \in \mathcal{A}} D_{-\alpha}(p : r_{-\alpha,\mathcal{A}}) + |\mathcal{A}| D_{-\alpha}(r_{-\alpha,\mathcal{A}} : c) \\ = \ & \sum_{p \in \mathcal{A}} D_\alpha(l_{\alpha,\mathcal{A}} : p) + |\mathcal{A}| D_\alpha(c : l_{\alpha,\mathcal{A}}) \ . \end{aligned} \tag{A2}$$

It comes now from (A1) and (A2) that:

$$\begin{aligned} & \mathbf{E}_{c \sim U_{\mathcal{A}}}[M_{\lambda,\alpha}(\mathcal{A}, c)] \\ = \ & \frac{1}{|\mathcal{A}|} \sum_{c \in \mathcal{A}} \sum_{p \in \mathcal{A}} \{\lambda D_\alpha(c : p) + (1-\lambda) D_\alpha(p : c)\} \\ = \ & (1-\lambda) \sum_{p \in \mathcal{A}} D_\alpha(p : r_{\alpha,\mathcal{A}}) + (1-\lambda) \sum_{p \in \mathcal{A}} D_\alpha(r_{\alpha,\mathcal{A}} : p) \\ & + \lambda \sum_{p \in \mathcal{A}} D_\alpha(l_{\alpha,\mathcal{A}} : p) + \lambda \sum_{p \in \mathcal{A}} D_\alpha(p : l_{\alpha,\mathcal{A}}) \\ = \ & (1-\lambda) M_{\text{opt},0,\alpha}(\mathcal{A}) + \lambda M_{\text{opt},1,\alpha}(\mathcal{A}) \\ & + (1-\lambda) M_{\text{opt},0,-\alpha}(\mathcal{A}) + \lambda M_{\text{opt},1,-\alpha}(\mathcal{A}) \\ = \ & M_{\text{opt},\lambda,\alpha}(\mathcal{A}) + M_{\text{opt},\lambda,-\alpha}(\mathcal{A}) \ . \end{aligned} \tag{A3}$$

This gives the left-hand side equality of the Lemma. The right-hand side follows from the fact that $\mathbf{E}_{c\sim U_{\mathcal{A}}}[M_{\lambda,-\alpha}(\mathcal{A},c)] = M_{\mathrm{opt},1-\lambda,\alpha}(\mathcal{A}) + M_{\mathrm{opt},1-\lambda,-\alpha}(\mathcal{A})$. □

Instead of $M_{\mathrm{opt},\lambda,\alpha}(\mathcal{A}) + M_{\mathrm{opt},\lambda,-\alpha}(\mathcal{A})$, we want a term depending solely on $M_{\mathrm{opt},\lambda,\alpha}(\mathcal{A})$ as it is the "true" optimum. We now give two lemmata that shall be useful in obtaining this upper bound. The first is of independent interest, as it shows that any α-divergence is a scaled, squared Hellinger distance between geometric means of points.

Lemma A2. *For any p, q and $\alpha \neq 1$, there exists $r \in [p, q]$, such that $(1-\alpha)^2 D_\alpha(p : q) = D_0(p^{1-\alpha}r^\alpha : q^{1-\alpha}r^\alpha)$.*

Proof. By the definition of Bregman divergences, for any x, y, there exists some $z \in [x, y]$, such that:

$$\begin{aligned} D_{\varphi_\alpha}(x : y) &= \frac{1}{2}(x-y)^2 \varphi''_\alpha(z) \\ &= \frac{1}{2}(x-y)^2\left(1 + \frac{1-\alpha}{2}z\right)^{\frac{2\alpha}{1-\alpha}} , \end{aligned}$$

and since u_α is continuous and strictly increasing, for any p, q, there exists some $r \in [p, q]$, such that:

$$\begin{aligned} & D_\alpha(p : q) \\ &= D_{\varphi_\alpha}(u_\alpha(p) : u_\alpha(q)) \\ &= \frac{1}{2}(u_\alpha(p) - u_\alpha(q))^2\left(1 + \frac{1-\alpha}{2}u_\alpha(r)\right)^{\frac{2\alpha}{1-\alpha}} \\ &= \frac{2}{(1-\alpha)^2}\left(p^{\frac{1-\alpha}{2}} - q^{\frac{1-\alpha}{2}}\right)^2 r^\alpha \\ &= \frac{2}{(1-\alpha)^2}\left(p^{1-\alpha} + q^{1-\alpha} - 2(pq)^{\frac{1-\alpha}{2}}\right) r^\alpha \\ &= \frac{1}{(1-\alpha)^2} D_0(p^{1-\alpha}r^\alpha : q^{1-\alpha}r^\alpha) . \end{aligned}$$

□

Lemma A3. *Let discrete random variable x take non-negative values $x_1, x_2, ..., x_m$ with uniform probabilities. Then, for any $\beta > -1$, we have $\mathrm{var}(x^{1+\beta}/u^\beta) \leq \mathrm{var}(x)$, with $u \doteq (1+\beta)^\beta \max_i x_i$.*

Proof. First, $\forall \beta > -1$, remark that for any x, function $f(x) = x(u^\beta - x^\beta)$ is increasing for $x \leq u/(1+\beta)^\beta$. Hence, assuming that the x_is are put in non-increasing order without loss of generality, we have $f(x_i) \geq f(x_j)$, and so, $x_i(u^\beta - x_i^\beta) \geq x_j(u^\beta - x_j^\beta), \forall i \geq j$, as long as $x_i \leq u/(1+\beta)^\beta$. Choosing

$u = x_1(1+\beta)^\beta$ yields, after reordering and putting the exponent, $(x_i^{1+\beta} - x_j^{1+\beta})^2 \leq (x_i u^\beta - x_j u^\beta)^2$. Hence:

$$\begin{aligned}
&\frac{1}{m}\sum_i x_i^{2(1+\beta)} - \left(\frac{1}{m}\sum_i x_i^{(1+\beta)}\right)^2 \\
&= \frac{1}{2m^2}\sum_{i,j}(x_i^{1+\beta} - x_j^{1+\beta})^2 \\
&\leq \frac{1}{2m^2}\sum_{i,j}(x_i u^\beta - x_j u^\beta)^2 \\
&= \frac{u^{2\beta}}{2m^2}\sum_{i,j}(x_i - x_j)^2 \\
&= u^{2\beta}\left(\frac{1}{m}\sum_i x_i^2 - \left(\frac{1}{m}\sum_i x_i\right)^2\right) .
\end{aligned}$$

Dividing by $u^{2\beta}$ the leftmost and rightmost terms and using the fact that $\mathrm{var}(\lambda x) = \lambda^2 \mathrm{var}(x)$ yields the statement of the Lemma. □

We are now ready to upper bound $M_{\mathrm{opt},\lambda,-\alpha}(\mathcal{A})$ as a function of $M_{\mathrm{opt},\lambda,\alpha}(\mathcal{A})$.

Lemma A4. *For any cluster $\mathcal{A}$ of $\mathcal{C}_{\mathrm{opt}}$,*

$$M_{\mathrm{opt},\lambda,-\alpha}(\mathcal{A}) \leq M_{\mathrm{opt},\lambda,\alpha}(\mathcal{A}) \times \left\{ \begin{array}{cl} f(\lambda) & \text{if } \lambda \in (0,1) \\ z(\alpha)h^2(\alpha) & \text{otherwise} \end{array} \right. ,$$

where $z(\alpha)$, $f(\lambda)$ and $h(\alpha)$ are defined in Theorem 1.

Proof. The case $\lambda \neq 0, 1$ is fast, as we have by definition:

$$\begin{aligned}
M_{\mathrm{opt},\lambda,-\alpha}(\mathcal{A}) &= \sum_{p\in\mathcal{A}} \lambda D_{-\alpha}(l_{-\alpha,\mathcal{A}} : p) + (1-\lambda) D_{-\alpha}(p : r_{-\alpha,\mathcal{A}}) \\
&= \sum_{p\in\mathcal{A}} \lambda D_{\alpha}(p : l_{-\alpha,\mathcal{A}}) + (1-\lambda) D_{\alpha}(r_{-\alpha,\mathcal{A}} : p) \\
&= \sum_{p\in\mathcal{A}} \lambda D_{\alpha}(p : r_{\alpha,\mathcal{A}}) + (1-\lambda) D_{\alpha}(l_{\alpha,\mathcal{A}} : p) \\
&\leq \max\left\{\frac{1-\lambda}{\lambda}, \frac{\lambda}{1-\lambda}\right\} M_{\mathrm{opt},\lambda,\alpha}(\mathcal{A}) \\
&= f(\lambda) M_{\mathrm{opt},\lambda,\alpha}(\mathcal{A}) .
\end{aligned}$$

Suppose now that $\lambda = 0$ and $\alpha \geq 0$. Because $M_{\mathrm{opt},0,-\alpha}(\mathcal{A}) = \sum_{p\in\mathcal{A}} D_{-\alpha}(p : r_{-\alpha,\mathcal{A}}) = \sum_{p\in\mathcal{A}} D_{\alpha}(l_{\alpha,\mathcal{A}} : p) = M_{\mathrm{opt},1,\alpha}(\mathcal{A})$, what we wish to do is upper bound $\sum_{p\in\mathcal{A}} D_{\alpha}(l_{\alpha,\mathcal{A}} : p) = M_{\mathrm{opt},1,\alpha}(\mathcal{A})$ as a function of $\sum_{p\in\mathcal{A}} D_{\alpha}(p : r_{\alpha,\mathcal{A}}) = M_{\mathrm{opt},0,\alpha}(\mathcal{A})$. We use Lemmatas A2 and A3 in the following derivations, using $r(p)$ to refer to the r in Lemma A2, assuming $\alpha \geq 0$. We also note $\mathrm{var}_{\mathcal{A}}(f(p))$ as the variance, under the uniform distribution over $\mathcal{A}$, of discrete random variable $f(p)$, for $p \in \mathcal{A}$. We have:

$$\begin{aligned}
&\sum_{p\in\mathcal{A}} D_\alpha(l_{\alpha,\mathcal{A}} : p) \\
&= \sum_{p\in\mathcal{A}} D_{-\alpha}(p : l_{\alpha,\mathcal{A}}) \\
&= \frac{1}{(1+\alpha)^2} \sum_{p\in\mathcal{A}} r(p)^{-\alpha} D_0(p^{1+\alpha} : l_{\alpha,\mathcal{A}}^{1+\alpha}) \\
&\leq \frac{1}{(1+\alpha)^2 \min_{\mathcal{A}} p^\alpha} \sum_{p\in\mathcal{A}} D_0(p^{1+\alpha} : l_{\alpha,\mathcal{A}}^{1+\alpha}) \\
&= \frac{1}{(1+\alpha)^2 \min_{\mathcal{A}} p^\alpha} \sum_{p\in\mathcal{A}} \left(p^{1+\alpha} + l_{\alpha,\mathcal{A}}^{1+\alpha} - 2p^{\frac{1+\alpha}{2}} l_{\alpha,\mathcal{A}}^{\frac{1+\alpha}{2}}\right) \\
&= \frac{|\mathcal{A}|}{(1+\alpha)^2 \min_{\mathcal{A}} p^\alpha} \left(\frac{1}{|\mathcal{A}|}\sum_{p\in\mathcal{A}} p^{1+\alpha} - \left(\frac{1}{|\mathcal{A}|}\sum_{p\in\mathcal{A}} p^{\frac{1+\alpha}{2}}\right)^2\right) \\
&= \frac{|\mathcal{A}|\mathrm{var}_{\mathcal{A}}(p^{\frac{1+\alpha}{2}})}{(1+\alpha)^2 \min_{\mathcal{A}} p^\alpha} . \qquad \text{(A4)}
\end{aligned}$$

We have used the expression of left centroid $l_{\alpha,\mathcal{A}}^{1+\alpha}$ to simplify the expressions. Now, picking $x_i = p_i^{\frac{1-\alpha}{2}}$, $\beta = 2\alpha/(1-\alpha)$ and $u = \left(\frac{1+\alpha}{1-\alpha}\right)^{\frac{2\alpha}{1-\alpha}} \max_{\mathcal{A}} p^{\frac{1-\alpha}{2}}$ in Lemma A3 yields:

$$\begin{aligned}
&\mathrm{var}_{\mathcal{A}}(p^{\frac{1+\alpha}{2}}) \\
&= u^{2\beta}\mathrm{var}_{\mathcal{A}}(p^{\frac{1+\alpha}{2}}/u^\beta) \\
&= u^{2\beta}\mathrm{var}_{\mathcal{A}}\left(p^{\frac{1-\alpha}{2}}p^\alpha/u^\beta\right) \\
&= u^{2\beta}\mathrm{var}(x^{1+\beta}/u^\beta) \\
&\leq u^{2\beta}\mathrm{var}(x) \\
&= u^{2\beta}\mathrm{var}_{\mathcal{A}}\left(p^{\frac{1-\alpha}{2}}\right) \\
&= \left(\frac{1+\alpha}{1-\alpha}\right)^{\frac{8\alpha^2}{(1-\alpha)^2}} \max_{\mathcal{A}} p^{2\alpha}\mathrm{var}_{\mathcal{A}}\left(p^{\frac{1-\alpha}{2}}\right) . \qquad \text{(A5)}
\end{aligned}$$

Plugging this in (A4) yields:

$$\begin{aligned}
&\sum_{p\in\mathcal{A}} D_\alpha(l_{\alpha,\mathcal{A}} : p) \\
\leq\ & \left(\frac{1+\alpha}{1-\alpha}\right)^{\frac{8\alpha^2}{(1-\alpha)^2}} \frac{|\mathcal{A}| \max_{\mathcal{A}} p^{2\alpha} \mathrm{var}_{\mathcal{A}}\left(p^{\frac{1-\alpha}{2}}\right)}{(1+\alpha)^2 \min_{\mathcal{A}} p^\alpha} \\
=\ & \left(\frac{1+\alpha}{1-\alpha}\right)^{\frac{8\alpha^2}{(1-\alpha)^2}-2} \left(\frac{\max_{\mathcal{A}} p}{\min_{\mathcal{A}} p}\right)^{2\alpha} \times \frac{|\mathcal{A}| \min_{\mathcal{A}} p^\alpha \mathrm{var}_{\mathcal{A}}(p^{\frac{1-\alpha}{2}})}{(1-\alpha)^2} \\
=\ & \left(\frac{1+\alpha}{1-\alpha}\right)^{\frac{8\alpha^2}{(1-\alpha)^2}-2} \left(\frac{\max_{\mathcal{A}} p}{\min_{\mathcal{A}} p}\right)^{2\alpha} \times \frac{\min_{\mathcal{A}} p^\alpha}{(1-\alpha)^2} \sum_{p\in\mathcal{A}} D_0(p^{1-\alpha} : r^{1-\alpha}_{\alpha,\mathcal{A}}) \qquad \text{(A6)} \\
\leq\ & \left(\frac{1+\alpha}{1-\alpha}\right)^{\frac{8\alpha^2}{(1-\alpha)^2}-2} \left(\frac{\max_{\mathcal{A}} p}{\min_{\mathcal{A}} p}\right)^{2\alpha} \times \frac{1}{(1-\alpha)^2} \sum_{p\in\mathcal{A}} r(p)^\alpha D_0(p^{1-\alpha} : r^{1-\alpha}_{\alpha,\mathcal{A}}) \\
=\ & \left(\frac{1+\alpha}{1-\alpha}\right)^{\frac{8\alpha^2}{(1-\alpha)^2}-2} \left(\frac{\max_{\mathcal{A}} p}{\min_{\mathcal{A}} p}\right)^{2\alpha} \times \sum_{p\in\mathcal{A}} D_\alpha(p : r_{\alpha,\mathcal{A}}) \\
\leq\ & z(\alpha) \left(\frac{\max_{\mathcal{A}} p}{\min_{\mathcal{A}} p}\right)^{2\alpha} \times \sum_{p\in\mathcal{A}} D_\alpha(p : r_{\alpha,\mathcal{A}}) \ . \qquad \text{(A7)}
\end{aligned}$$

Here, (A6) follows the path backwards of derivations that lead to (A4). The cases $\lambda = 1$ or $\alpha < 0$ are obtained using the same chains of derivations and achieve the proof of Lemma A4. □

Lemma A4 can be directly used to refine the bound of Lemma A1 in the uniform distribution. We give the Lemma for the biased distribution, directly integrating the refinement of the bound.

Lemma A5. *Let $\mathcal{A}$ be an arbitrary cluster of C_{opt} and C an arbitrary clustering. If we add a random couple (c,c) to C, chosen from $\mathcal{A}$ with π as in Algorithm 2, then:*

$$\begin{aligned}
&\mathbf{E}_{c\sim\pi_{\mathcal{A}}}[M_{\lambda,\alpha}(\mathcal{A},c)] \\
\leq\ & 4\left\{\begin{array}{ll} f(\lambda)h^2(\alpha)M_{\mathrm{opt},\lambda,\alpha}(\mathcal{A}) & \text{if} \quad \lambda\in(0,1) \\ z(\alpha)h^4(\alpha)M_{\mathrm{opt},\lambda,\alpha}(\mathcal{A}) & \text{otherwise} \end{array}\right. , \qquad \text{(A8)}
\end{aligned}$$

where $f(\lambda)$ and $h(\alpha)$ are defined in Theorem 1.

Proof. The proof essentially follows the proof of Lemma 3 in [15]. To complete it, we need a triangle inequality involving α-divergences. We give it here.

Lemma A6. *For any p, q, r and α, we have:*

$$\sqrt{D_\alpha(p:q)} \ \leq\ \left(\frac{\max_i\{p_i,q_i,r_i\}}{\min_i\{p_i,q_i,r_i\}}\right)^{|\alpha|} \left(\sqrt{D_\alpha(p:r)} + \sqrt{D_\alpha(r:q)}\right) \qquad \text{(A9)}$$

(where the min is over strictly positive values)

Remark: take $\alpha = 0$; we find the triangle inequality for the squared Hellinger distance.

Proof. Using the proof of Lemma 2 in [15] for Bregman divergence D_{φ_α}, we get:

$$\begin{aligned}
&\sqrt{D_{\varphi_\alpha}(x:z)} \\
\leq\ & \rho(\alpha)\left(\sqrt{D_{\varphi_\alpha}(x:y)} + \sqrt{D_{\varphi_\alpha}(y:z)}\right) , \qquad \text{(A10)}
\end{aligned}$$

where:

$$\rho(\alpha) = \max_{u,v} \frac{\left(1+\frac{1-\alpha}{2}u\right)^{\frac{2\alpha}{1-\alpha}}}{\left(1+\frac{1-\alpha}{2}v\right)^{\frac{2\alpha}{1-\alpha}}} . \tag{A11}$$

Taking $x = u_\alpha(p)$, $y = u_\alpha(q)$, $z = u_\alpha(r)$ yields $\rho(\alpha) = \max_{s,t\in\{p_i,q_i,r_i\}}(s/t)^{|\alpha|}$ and the statement of Lemma A6. □

The rest of the proof of Lemma A5 follows the proof of Lemma 3 in [15]. □

We get all of the ingredients to our proof, and there remains to use Lemma 4 in [15] to achieve the proof of Theorem 1.

Appendix Properties of α-Divergences

For positive arrays p and q, the α-divergence $D_\alpha(p : q)$ can be defined as an equivalent representational Bregman divergence [19,34] $B_{\varphi_\alpha}(u_\alpha(p) : u_\alpha(q))$ over the (u_α, v_α)-structure [43] with:

$$\varphi_\alpha(x) \doteq \frac{2}{1+\alpha}\left(1+\frac{1-\alpha}{2}x\right)^{\frac{2}{1-\alpha}} , \tag{A12}$$

$$u_\alpha(p) \doteq \frac{2}{1-\alpha}\left(p^{\frac{1-\alpha}{2}}-1\right) , \tag{A13}$$

$$v_\alpha(p) \doteq \frac{2}{1+\alpha}p^{\frac{1+\alpha}{2}} , \tag{A14}$$

where we assume that $\alpha \neq \pm 1$. Otherwise, for $\alpha = \pm 1$, we compute $D_\alpha(p : q)$ by taking the sided Kullback–Leibler divergence extended to positive arrays.

In the proof of Theorem 1, we have used two properties of α-divergences of independent interest:

- any α-divergence can be explained as a scaled squared Hellinger distance between geometric means of its arguments and a point that belong to their segment (Lemma A2);
- any α-divergence satisfies a generalized triangle inequality (Lemma A6). Notice that this Lemma is optimal in the sense that for $\alpha = 0$, it is possible to recover the triangle inequality of the Hellinger distance.

The following lemma shows how to bound the mixed divergence as a function of an α-divergence.

Lemma A7. *For any positive arrays l, h, r and $\alpha \neq \pm 1$, define $\eta \doteq \lambda(1-\alpha)/(1-\alpha(2\lambda-1)) \in [0,1]$, g_η with $g_\eta^i \doteq (l^i)^\eta (r^i)^{1-\eta}$ and a_η with $a_\eta^i \doteq \eta l^i + (1-\eta)r^i$. Then, we have:*

$$M_{\lambda,\alpha}(l : h : r) \leq \frac{1-\alpha^2(2\lambda-1)^2}{1-\alpha^2} D_{\alpha(2\lambda-1)}(g_\eta : h) + \frac{2(1-\alpha(2\lambda-1))}{1-\alpha^2}\sum_i \left(a_\eta^i - g_\eta^i\right) .$$

Proof. For all index i, we have:

$$M_{\lambda,\alpha}(l^i : h^i : r^i) = \lambda D_\alpha(l^i : h^i) + (1-\lambda) D_\alpha(h^i : r^i)$$

$$= \frac{4}{1-\alpha^2}\left(\frac{\lambda(1-\alpha)}{2}l^i + \frac{(1-\lambda)(1+\alpha)}{2}r^i + \frac{1+\alpha(2\lambda-1)}{2}h^i \right. \tag{A15}$$

$$\left. -\lambda (l^i)^{\frac{1-\alpha}{2}}(h^i)^{\frac{1+\alpha}{2}} - (1-\lambda)(r^i)^{\frac{1+\alpha}{2}}(h^i)^{\frac{1-\alpha}{2}}\right) . \tag{A16}$$

The arithmetic-geometric-harmonic (AGH) inequality implies:

$$\begin{aligned}\lambda(l^i)^{\frac{1-\alpha}{2}}(h^i)^{\frac{1+\alpha}{2}} + (1-\lambda)(r^i)^{\frac{1+\alpha}{2}}(h^i)^{\frac{1-\alpha}{2}} &\geq (l^i)^{\frac{\lambda(1-\alpha)}{2}}(r^i)^{\frac{(1-\lambda)(1+\alpha)}{2}}(h^i)^{\frac{1+\alpha(2\lambda-1)}{2}} \\ &= \left((l^i)^{\frac{\lambda(1-\alpha)}{1-\alpha(2\lambda-1)}}(r^i)^{\frac{(1-\lambda)(1+\alpha)}{1-\alpha(2\lambda-1)}}\right)^{\frac{1-\alpha(2\lambda-1)}{2}}(h^i)^{\frac{1+\alpha(2\lambda-1)}{2}} \\ &= \left((l^i)^{\eta}(r^i)^{1-\eta}\right)^{\frac{1-\alpha(2\lambda-1)}{2}}(h^i)^{\frac{1+\alpha(2\lambda-1)}{2}} \\ &= (g^i_\eta)^{\frac{1-\alpha(2\lambda-1)}{2}}(h^i)^{\frac{1+\alpha(2\lambda-1)}{2}}.\end{aligned}$$

It follows that (A16) yields:

$$\begin{aligned}M_{\lambda,\alpha}(l^i : h^i : r^i) &\leq \frac{4}{1-\alpha^2}\left(\frac{1-\alpha(2\lambda-1)}{2}\left(\eta l^i + (1-\eta)r^i\right) + \right. \qquad \text{(A17)} \\ &\left. \frac{1+\alpha(2\lambda-1)}{2}h^i - (g^i_\eta)^{\frac{1-\alpha(2\lambda-1)}{2}}(h^i)^{\frac{1+\alpha(2\lambda-1)}{2}}\right) \\ &= \frac{1-\alpha^2(2\lambda-1)^2}{1-\alpha^2}D_{\alpha(2\lambda-1)}(g^i_\eta : h^i) + \frac{2(1-\alpha(2\lambda-1))}{1-\alpha^2}\left(a^i_\eta - g^i_\eta\right), \qquad \text{(A18)}\end{aligned}$$

out of which we get the statement of the Lemma. □

Appendix Sided α-Centroids

For the sake of completeness, we prove the following theorem:

Theorem A1 (Sided positive α-centroids [34]). *The left-sided l_α and right-sided r_α positive weighted α-centroid coordinates of a set of n positive histograms $h_1, ..., h_n$ are weighted α-means:*

$$r^i_\alpha = f_\alpha^{-1}\left(\sum_{j=1}^n w_j f_\alpha(h^i_j)\right), l^i_\alpha = r^i_{-\alpha}$$

with:

$$f_\alpha(x) = \begin{cases} x^{\frac{1-\alpha}{2}} & \alpha \neq \pm 1, \\ \log x & \alpha = 1. \end{cases}$$

Proof. We distinguish three cases: $\alpha \neq \pm 1$, $\alpha = -1$ and $\alpha = 1$.
First, consider the general case $\alpha \neq \pm 1$. We have to minimize:

$$R_\alpha(x, \mathcal{H}) = \frac{4}{1-\alpha^2}\sum_{j=1}^n w_j \times \sum_{i=1}^d \left(\frac{1-\alpha}{2}h^i_j + \frac{1+\alpha}{2}x^i - (h^i_j)^{\frac{1-\alpha}{2}}(x^i)^{\frac{1+\alpha}{2}}\right).$$

Removing all additive terms independent of x^i and the overall constant multiplicative factor $\frac{4}{1-\alpha^2} \neq 0$, we get the following equivalent minimisation problem:

$$R'_\alpha(x, \mathcal{H}) = \sum_{i=1}^d \frac{1+\alpha}{2}x^i - (x^i)^{\frac{1+\alpha}{2}}\underbrace{\left(\sum_{j=1}^n w_j (h^i_j)^{\frac{1-\alpha}{2}}\right)}_{\bar{h}^i_\alpha}, \qquad \text{(A19)}$$

where $\bar{h}^i_\alpha$ denote the following aggregation term:

$$\bar{h}^i_\alpha = \sum_{j=1}^{n} w_j (h^i_j)^{\frac{1-\alpha}{2}}.$$

Setting coordinate-wise the derivative to zero of Equation (A19) (*i.e.*, $\nabla_x R'(x, \mathcal{H}) = 0$), we get:

$$\frac{1+\alpha}{2} - \frac{1+\alpha}{2}(x^i)^{\frac{\alpha-1}{2}} \bar{h}^i_\alpha = 0$$

Thus, we find that the coordinates of the right-sided α-centroids are:

$$c^i_\alpha = (\bar{h}^i_\alpha)^{\frac{2}{1-\alpha}} = \left(\sum_{j=1}^{n} w_j (h^i_j)^{\frac{1-\alpha}{2}} \right)^{\frac{2}{1-\alpha}} = \hat{h}^i_\alpha.$$

We recognise the expression of a quasi-arithmetic mean for the strictly monotonous generator $f_\alpha(x)$:

$$r^i_\alpha = f^{-1}_\alpha \left(\sum_{j=1}^{n} w_j f_\alpha(h^i_j) \right), \tag{A20}$$

with:

$$f_\alpha(x) = x^{\frac{1-\alpha}{2}}, \quad f^{-1}_\alpha(x) = x^{\frac{2}{1-\alpha}}, \alpha \neq \pm 1.$$

Therefore, we conclude that the coordinates of the positive α-centroid are the weighted α-means of the histogram coordinates (for $\alpha \neq \pm 1$). Quasi-arithmetic means are also called in the literature quasi-linear means or f-means.

When $\alpha = -1$, we search for the right-sided extended Kullback–Leibler divergence centroid by minimising:

$$R_{-1}(x; \tilde{\mathcal{H}}) = \sum_{j=1}^{n} w_j \sum_{i=1}^{d} h^i_j \log \frac{h^i_j}{x^i} + x^i - h^i_j.$$

It is equivalent to minimizing:

$$R'_{-1}(x; \tilde{\mathcal{H}}) = \sum_{i=1}^{d} x^i - \underbrace{\left(\sum_{j=1}^{n} w_j h^i_j \right)}_{a} \log x^i,$$

where a denotes the arithmetic mean. Solving coordinate-wise, we get $c^i = a^i = \sum_{j=1}^{n} w_j h^i_j$.

When $\alpha = 1$, the right-sided reverse extended KL centroid is a left-sided extended KL centroid. The minimisation problem is:

$$R_1(x; \tilde{\mathcal{H}}) = \sum_{j=1}^{n} w_j \sum_{i=1}^{d} x^i \log \frac{x^i}{h^i_j} + h^i_j - x^i.$$

Since $\sum_j w_j = 1$, we solve coordinate-wise and find $\log x = \sum_j w_j \log h_j$. That is, r^i_1 is the geometric mean:

$$r^i_1 = \prod_{j=1}^{n} (h^i_j)^{w_j}.$$

Both the arithmetic mean and the geometric mean are power means in the limit case (and hence quasi-arithmetic means). Thus,

$$r^i_\alpha = f^{-1}_\alpha \left(\sum_{j=1}^{n} w_j f_\alpha(h^i_j) \right), \tag{A21}$$

with:

$$f_\alpha(x) = \begin{cases} x^{\frac{1-\alpha}{2}} & \alpha \neq \pm 1, \\ \log x & \alpha = 1. \end{cases}$$

□

References

1. Baker, L.D.; McCallum, A.K. Distributional clustering of words for text classification. In Proceedings of the 21st Annual International ACM SIGIR Conference on Research and Development in Information Retrieval, Melbourne, Australia, 24–28 August 1998; ACM: New York, NY, USA, 1998; pp. 96–103.
2. Bigi, B. Using Kullback–Leibler distance for text categorization. In Proceedings of the 25th European conference on IR research (ECIR), Pisa, Italy, 14–16 April 2003; Springer-Verlag: Berlin/Heidelberg, Germany, 2003; ECIR'03, pp. 305–319.
3. Bag of Words Data Set. Available online: http://archive.ics.uci.edu/ml/datasets/Bag+of+Words (accessed on 17 June 2014).
4. Csurka, G.; Bray, C.; Dance, C.; Fan, L. *Visual Categorization with Bags of Keypoints*; Workshop on Statistical Learning in Computer Vision (ECCV); Xerox Research Centre Europe: Meylan, France, 2004, pp. 1–22.
5. Jégou, H.; Douze, M.; Schmid, C. Improving Bag-of-Features for Large Scale Image Search. *Int. J. Comput. Vis.* **2010**, *87*, 316–336.
6. Yu, Z.; Li, A.; Au, O.; Xu, C. Bag of textons for image segmentation via soft clustering and convex shift. In Proceedings of 2012 IEEE Conference on Computer Vision and Pattern Recognition (CVPR), Providence, RI, USA, 16–21 June 2012; pp. 781–788.
7. Steinhaus, H. Sur la division des corp matériels en parties. *Bull. Acad. Polon. Sci.* **1956**, *1*, 801–804. (in French)
8. Lloyd, S.P. *Least Squares Quantization in PCM*; Technical Report RR-5497; Bell Laboratories: Murray Hill, NJ, USA, 1957.
9. Lloyd, S.P. Least squares quantization in PCM. *IEEE Trans. Inf. Theory* **1982**, *28*, 129–137.
10. Chandrasekhar, V.; Takacs, G.; Chen, D.M.; Tsai, S.S.; Reznik, Y.A.; Grzeszczuk, R.; Girod, B. Compressed histogram of gradients: A low-bitrate descriptor. *Int. J. Comput. Vis.* **2012**, *96*, 384–399.
11. Nock, R.; Nielsen, F.; Briys, E. Non-linear book manifolds: Learning from associations the dynamic geometry of digital libraries. In Proceedings of the 13th ACM/IEEE-CS Joint Conference on Digital Libraries, New York, NY, USA, 2013; pp. 313–322.
12. Kwitt, R.; Vasconcelos, N.; Rasiwasia, N.; Uhl, A.; Davis, B.C.; Häfner, M.; Wrba, F. Endoscopic image analysis in semantic space. *Med. Image Anal.* **2012**, *16*, 1415–1422.
13. Nielsen, F. A family of statistical symmetric divergences based on Jensen's inequality. **2010**, arXiv:1009.4004.
14. Nielsen, F.; Nock, R. Sided and symmetrized Bregman centroids. *IEEE Trans. Inf. Theory* **2009**, *55*, 2882–2904.
15. Nock, R.; Luosto, P.; Kivinen, J. Mixed Bregman clustering with approximation guarantees. In Proceedings of the European Conference on Machine Learning and Knowledge Discovery in Databases, Antwerp, Belgium, 15–19 September 2008; Springer-Verlag: Berlin/Heidelberg, Germany, 2008; pp. 154–169.
16. Amari, S. Integration of Stochastic Models by Minimizing α-Divergence. *Neural Comput.* **2007**, *19*, 2780–2796.
17. Arthur, D.; Vassilvitskii, S. k-means++: The advantages of careful seeding. In Proceedings of the Eighteenth Annual ACM-SIAM Symposium on Discrete Algorithms (SODA), New Orleans, LA, USA, 7–9 January 2007; Society for Industrial and Applied Mathematics: Philadelphia, PA, USA, 2007; pp. 1027–1035.
18. Olszewski, D.; Ster, B. Asymmetric clustering using the alpha-beta divergence. *Pattern Recognit.* **2014**, *47*, 2031–2041.
19. Amari, S. Alpha-divergence is unique, belonging to both f-divergence and Bregman divergence classes. *IEEE Trans. Inf. Theory* **2009**, *55*, 4925–4931.
20. Banerjee, A.; Merugu, S.; Dhillon, I.S.; Ghosh, J. Clustering with Bregman divergences. *J. Mach. Learn. Res.* **2005**, *6*, 1705–1749.
21. Teboulle, M. A unified continuous optimization framework for center-based clustering methods. *J. Mach. Learn. Res.* **2007**, *8*, 65–102.
22. Amari, S.; Nagaoka, H. *Methods of Information Geometry*; Oxford University Press: Oxford, UK, 2000.
23. Morimoto, T. Markov Processes and the H-theorem. *J. Phys. Soc. Jpn.* **1963**, *18*, 328–331.

24. Ali, S.M.; Silvey, S.D. A general class of coefficients of divergence of one distribution from another. *J. R. Stat. Soc. Ser. B* **1966**, *28*, 131–142.
25. Csiszár, I. Information-type measures of difference of probability distributions and indirect observation. *Studi. Sci. Math. Hung.* **1967**, *2*, 229–318.
26. Cichocki, A.; Cruces, S.; Amari, S. Generalized alpha-beta divergences and their application to robust nonnegative matrix factorization. *Entropy* **2011**, *13*, 134–170.
27. Zhu, H.; Rohwer, R. Measurements of generalisation based on information geometry. In *Mathematics of Neural Networks;* Operations Research/Computer Science Interfaces Series; Ellacott, S., Mason, J., Anderson, I., Eds.; Springer: New York, NY, USA, 1997; Volume 8, pp. 394–398.
28. Chernoff, H. A measure of asymptotic efficiency for tests of a hypothesis based on the sum of observations. *Ann. Math. Stat.* **1952**, *23*, 493–507.
29. Nielsen, F. An information-geometric characterization of Chernoff information. *IEEE Signal Process. Lett.* **2013**, *20*, 269–272.
30. Wu, J.; Rehg, J. Beyond the euclidean distance: creating effective visual codebooks using the histogram intersection kernel. In Proceedings of 2009 IEEE 12th International Conference on Computer Vision, Kyoto, Japan, 29 September–2 October 2009; pp. 630–637.
31. Bhattacharya, A.; Jaiswal, R.; Ailon, N. A tight lower bound instance for k-means++ in constant dimension. In *Theory and Applications of Models of Computation;* Lecture Notes in Computer Science; Gopal, T., Agrawal, M., Li, A., Cooper, S., Eds.; Springer International Publishing: New York, NY, USA, 2014; Volume 8402, pp. 7–22.
32. Nielsen, F. Jeffreys centroids: A closed-form expression for positive histograms and a guaranteed tight approximation for frequency histograms. *IEEE Signal Process. Lett.* **2013**, *20*, 657–660.
33. Ben-Tal, A.; Charnes, A.; Teboulle, M. Entropic means. *J. Math. Anal. Appl.* **1989**, *139*, 537–551.
34. Nielsen, F.; Nock, R. The dual Voronoi diagrams with respect to representational Bregman divergences. In Proceedings of International Symposium on Voronoi Diagrams (ISVD), Copenhagen, Denmark, 23–26 June 2009; pp. 71–78.
35. Heinz, E. Beiträge zur Störungstheorie der Spektralzerlegung. *Math. Anna.* **1951**, *123*, 415–438. (in German)
36. Besenyei, A. On the invariance equation for Heinz means. *Math. Inequal. Appl.* **2012**, *15*, 973–979.
37. Barry, D.A.; Culligan-Hensley, P.J.; Barry, S.J. Real values of the W-function. *ACM Trans. Math. Softw.* **1995**, *21*, 161–171.
38. Veldhuis, R.N.J. The centroid of the symmetrical Kullback–Leibler distance. *IEEE Signal Process. Lett.* **2002**, *9*, 96–99.
39. Nielsen, F.; Garcia, V. Statistical exponential families: A digest with flash cards. **2009**, arXiv.org: 0911.4863.
40. Nielsen, F.; Boltz, S. The Burbea-Rao and Bhattacharyya centroids. *IEEE Trans. Inf. Theory* **2011**, *57*, 5455–5466.
41. Romberg, S.; Lienhart, R. Bundle min-hashing for logo recognition. In Proceedings of the 3rd ACM Conference on International Conference on Multimedia Retrieval, Dallas, TX, USA, 16–19 April 2013; ACM: New York, NY, USA, 2013; pp. 113–120.
42. Matsuyama, Y. The alpha-EM algorithm: Surrogate likelihood maximization using alpha-logarithmic information measures. *IEEE Trans. Inf. Theory* **2003**, *49*, 692–706.
43. Amari, S.I. New developments of information geometry (26): Information geometry of convex programming and game theory. In *Mathematical Sciences (suurikagaku);* Number 605; The Science Company: Denver, CO, USA, 2013; pp. 65–74. (In Japanese)

entropy

MDPI

Article

New Riemannian Priors on the Univariate Normal Model

Salem Said *, Lionel Bombrun and Yannick Berthoumieu

Groupe Signal et Image, CNRS Laboratoire IMS, Institut Polytechnique de Bordeaux, Université de Bordeaux, UMR 5218, Talence, 33405, France; E-Mails: lionel.bombrun@u-bordeaux.fr (L.B.); yannick.berthoumieu@u-bordeaux.fr (Y.B.)

* E-Mail: salem.said@u-bordeaux.fr; Tel.:+33-(0)5-4000-6185.

Received: 17 April 2014; in revised form: 23 June 2014 / Accepted: 9 July 2014 / Published: 17 July 2014

Abstract: The current paper introduces new prior distributions on the univariate normal model, with the aim of applying them to the classification of univariate normal populations. These new prior distributions are entirely based on the Riemannian geometry of the univariate normal model, so that they can be thought of as "Riemannian priors". Precisely, if $\{p_\theta; \theta \in \Theta\}$ is any parametrization of the univariate normal model, the paper considers prior distributions $G(\bar{\theta}, \gamma)$ with hyperparameters $\bar{\theta} \in \Theta$ and $\gamma > 0$, whose density with respect to Riemannian volume is proportional to $\exp(-d^2(\theta, \bar{\theta})/2\gamma^2)$, where $d^2(\theta, \bar{\theta})$ is the square of Rao's Riemannian distance. The distributions $G(\bar{\theta}, \gamma)$ are termed Gaussian distributions on the univariate normal model. The motivation for considering a distribution $G(\bar{\theta}, \gamma)$ is that this distribution gives a geometric representation of a class or cluster of univariate normal populations. Indeed, $G(\bar{\theta}, \gamma)$ has a unique mode $\bar{\theta}$ (precisely, $\bar{\theta}$ is the unique Riemannian center of mass of $G(\bar{\theta}, \gamma)$, as shown in the paper), and its dispersion away from $\bar{\theta}$ is given by γ. Therefore, one thinks of members of the class represented by $G(\bar{\theta}, \gamma)$ as being centered around $\bar{\theta}$ and lying within a typical distance determined by γ. The paper defines rigorously the Gaussian distributions $G(\bar{\theta}, \gamma)$ and describes an algorithm for computing maximum likelihood estimates of their hyperparameters. Based on this algorithm and on the Laplace approximation, it describes how the distributions $G(\bar{\theta}, \gamma)$ can be used as prior distributions for Bayesian classification of large univariate normal populations. In a concrete application to texture image classification, it is shown that this leads to an improvement in performance over the use of conjugate priors.

Keywords: Fisher information; Riemannian metric; prior distribution; univariate normal distribution; image classification

1. Introduction

In this paper, a new class of prior distributions is introduced on the univariate normal model. The new prior distributions, which will be called Gaussian distributions, are based on the Riemannian geometry of the univariate normal model. The paper introduces these new distributions, uncovers some of their fundamental properties and applies them to the problem of the classification of univariate normal populations. It shows that, in the context of a real-life application to texture image classification, the use of these new prior distributions leads to improved performance in comparison with the use of more standard conjugate priors.

To motivate the introduction of the new prior distributions, considered in the following, recall some general facts on the Riemannian geometry of parametric models.

In information geometry [1], it is well known that a parametric model $\{p_\theta; \theta \in \Theta\}$, where $\Theta \subset R^p$, can be equipped with a Riemannian geometry, determined by Fisher's information matrix, say $I(\theta)$.

Indeed, assuming $I(\theta)$ is strictly positive definite, for each $\theta \in \Theta$, a Riemannian metric on Θ is defined by:

$$ds^2(\theta) = \sum_{i,j=1}^{p} I_{ij}(\theta) d\theta^i d\theta^j \tag{1}$$

The fact that the length element Equation (1) is invariant to any change of parametrization was realized by Rao [2], who was the first to propose the application of Riemannian geometry in statistics.

Once the Riemannian metric Equation (1) is introduced, the whole machinery of Riemannian geometry becomes available for application to statistical problems relevant to the parametric model $\{p_\theta; \theta \in \Theta\}$. This includes the notion of Riemannian distance between two distributions, p_θ and $p_{\theta'}$, which is known as Rao's distance, say $d(\theta, \theta')$, the notion of Riemannian volume, which is exactly the same as Jeffreys prior [3], and the notion of Riemannian gradient, which can be used in numerical optimization and coincides with the so-called natural gradient of Amari [4].

It is quite natural to apply Rao's distance to the problem of classifying populations that belong to the parametric model $\{p_\theta; \theta \in \Theta\}$. In the case where this parametric model is the univariate normal model, this approach to classification is implemented in [5]. For more general parametric models, beyond the univariate normal model, similar applications of Rao's distance to problems of image segmentation and statistical tests can be found in [6–8].

The idea of [5] is quite elegant. In general, it requires that some classes $\{\mathcal{S}_L; L = 1, \ldots, C\}$, (based on a learning sequence) have been identified with "centers" $\bar{\theta}_L \in \Theta$. Then, in order to assign a test population, given by the parameter θ_t, to a class L^*, it is proposed to choose L^*, which minimizes Rao's distance $d^2(\theta_t, \bar{\theta}_L)$, over $L = 1, \ldots, C$. In the specific context of the classification of univariate normal populations [5], this leads to the introduction of hyperbolic Voronoi diagrams.

The present paper is also concerned with the case where the parametric model $\{p_\theta; \theta \in \Theta\}$ is a univariate normal model. It starts from the idea that a class $\mathcal{S}_L$ should be identified not only with a center $\bar{\theta}_L$, as in [5], but also with a kind of "variance", say γ^2, which will be called a dispersion parameter. Accordingly, assigning a test population given by the parameter θ_t to a class L should be based on a tradeoff between the square of Rao's distance $d^2(\theta_t, \bar{\theta}_L)$ and the dispersion parameter γ^2.

Of course, this idea has a strong Bayesian flavor. It proposes to give more "confidence" to classes that have a smaller dispersion parameter. Thus, in order to implement it, in a concrete way, the paper starts by introducing prior distributions on the univariate normal model, which it calls Gaussian distributions. By definition, a Gaussian distribution $G(\bar{\theta}, \gamma^2)$ has a probability density function, with respect to Riemannian volume, given by:

$$p(\theta|\bar{\theta}, \gamma) \propto \exp\left(\frac{-d^2(\theta, \bar{\theta})}{2\gamma^2}\right) \tag{2}$$

Given this definition of a Gaussian distribution (which is developed in a detailed way, in Section 3), classification of univariate normal populations can be carried out by associating to each class $\mathcal{S}_L$ of univariate normal populations a Gaussian distribution $G(\bar{\theta}_L, \gamma_L^2)$ and by assigning any test population with parameter θ_t to the class L^*, which maximizes the likelihood $p(\theta_t|\bar{\theta}_L, \gamma_L)$, over $L = 1, \ldots, C$.

The present paper develops in a rigorous way the general approach to the classification of univariate normal populations, which has just been described. It proceeds as follows.

Section 2, which is basically self-contained, provides the concepts, regarding the Riemannian geometry of the univariate normal model, which will be used throughout the paper.

Section 3 introduces Gaussian distributions on the univariate normal model and uncovers some of their general properties. In particular, Section 3.2 of this section gives a Riemannian gradient descent algorithm for computing maximum likelihood estimates of the parameters $\bar{\theta}$ and γ of a Gaussian distribution.

Section 4 states the general approach to classification of univariate normal populations proposed in this paper. It deals with two problems: (i) given a class $\mathcal{S}$ of univariate normal populations S_i, how to fit a Gaussian distribution $G(\bar{z}, \gamma)$ to this class; and (ii) given a test univariate normal population S_t and a set of classes $\{\mathcal{S}_L, L = 1, \ldots, C\}$, how to assign S_t to a suitable class $\mathcal{S}_{L^*}$.

In the present paper, the chosen approach for resolving these two problems is marginalized likelihood estimation, in the asymptotic framework where each univariate normal population contains a large number of data points. In this asymptotic framework, the Laplace approximation plays a major role [9]. In particular, it reduces the first problem, of fitting a Gaussian distribution to a class of univariate normal populations, to the problem of maximum likelihood estimation, covered in Section 3.2.

The final result of Section 4 is the decision rule Equation (37). This generalizes the one developed in [5] and already explained above, by taking into account the dispersion parameter γ, in addition to the center $\bar{\theta}$, for each class.

In Section 5, the formalism of Section 4 is applied to texture image classification, using the VisTeX image database [10]. This database is used to compare the performance obtained using Gaussian distributions, as in Section 4, to that obtained using conjugate prior distributions. It is shown that Gaussian distributions, proposed in the current paper, lead to a significant improvement in performance.

Before going on, it should be noted that probability density functions of the form (2), on general Riemannian manifolds, were considered by Pennec in [11]. However, they were not specifically used as prior distributions, but rather as a representation of uncertainty in medical image analysis and directional or shape statistics.

2. Riemannian Geometry of the Univariate Normal Model

The current section presents in a self-contained way the results on the Riemannian geometry of the univariate normal model, which are required for the remainder of the paper. Section 2.1 recalls the fact that the univariate normal model can be reparametrized, so that its Riemannian geometry is essentially the same as that of the Poincaré upper half plane. Section 2.2 uses this fact to give analytic formulas for distance, geodesics and integration on the univariate normal model. Finally, Section 2.3 presents, in general form, the Riemannian gradient descent algorithm.

2.1. Derivation of the Fisher Metric

This paper considers the Riemannian geometry of the univariate normal model, as based on the Fisher metric (1). To be precise, the univariate normal model has a two-dimensional parameter space $\Theta = \{\theta = (\mu, \sigma) | \mu \in R\ ,\ \sigma > 0\}$, and is given by:

$$p_\theta(x) = |2\pi\sigma^2|^{-1/2} \exp\left(\frac{-(x-\mu)^2}{2\sigma^2} \right) \tag{3}$$

where each p_θ is a probability density function with respect to the Lebesgue measure on R. The Fisher information matrix, obtained from Equation (3), is the following:

$$I(\theta) = \begin{pmatrix} \frac{1}{\sigma^2} & 0 \\ 0 & \frac{2}{\sigma^2} \end{pmatrix}$$

As in [12], this expression can be made more symmetric by introducing the parametrization $z = (x, y)$, where $x = \mu/\sqrt{2}$ and $y = \sigma$. This yields the Fisher information matrix:

$$I(z) = 2 \times \begin{pmatrix} \frac{1}{y^2} & 0 \\ 0 & \frac{1}{y^2} \end{pmatrix}$$

It is suitable to drop the factor two in this expression and introduce the following Riemannian metric for the univariate normal model,

$$ds^2(z) = \frac{dx^2 + dy^2}{y^2} \tag{4}$$

This is essentially the same as the Fisher metric (up to the factor tow) and will be considered throughout the following. The resulting Rao's distance and Riemannian geometry are given in the following paragraph.

2.2. Distance, Geodesics and Volume

The Riemannian metric (4), obtained in the last paragraph, happens to be a very well-known object in differential geometry. Precisely, the parameter space $H = \{z = (x, y) | y > 0\}$ equipped with the metric (4) is known as the *Poincaré upper half plane* and is a basic model of a two-dimensional hyperbolic space [13].

Rao's distance between two points $z_1 = (x_1, y_1)$ and $z_2 = (x_2, y_2)$ in H can be expressed as follows (for results in the present paragraph, see [13], or any suitable reference on hyperbolic geometry),

$$d(z_1, z_2) = \operatorname{acosh}\left(1 + \frac{(x_1 - x_2)^2 + (y_1 - y_2)^2}{2y_1 y_2}\right) \tag{5}$$

where acosh denotes the inverse hyperbolic cosine.

Starting from z_1, in any given direction, it is possible to draw a unique geodesic ray $\gamma : R_+ \to H$. This is a curve having the property that $\gamma(0) = z_1$ and, for any $t \in R_+$, if $\gamma(t) = z_2$ then $d(z_1, z_2) = t$. In other words, the length of γ between z_1 and z_2 is equal to the distance between z_1 and z_2.

The equation of a geodesic ray starting from $z \in H$ is conveniently written down in complex notation (that is, by treating points of H as complex numbers). To begin, consider the case of $z = i$ (which stands for $x = 0$ and $y = 1$). The geodesic in the direction making an angle ψ with the y-axis is the curve,

$$\gamma_i(t) = \frac{e^{t/2}\cos(\psi/2)\, i - e^{-t/2}\sin(\psi/2)}{e^{t/2}\sin(\psi/2)\, i + e^{-t/2}\cos(\psi/2)} \tag{6}$$

In particular $\psi = 0$ gives $\gamma_i(t) = e^t i$ and $\psi = \pi$ gives $\gamma_i(t) = e^{-t} i$. If ψ is not a multiple of π, $\gamma_i(t)$ traces out a portion of a circle, which is parallel to the y-axis, in the limit $t \to \infty$. For a general starting point z, the geodesic ray in the direction making an angle ψ with the y-axis can be written:

$$\gamma_z(t, \psi) = x + y\gamma_i(t/y, \psi) \tag{7}$$

where $z = (x, y)$ and $\gamma_i(t, \psi)$ is given by Equation (6). A more detailed treatment of Rao's distance (5) and of geodesics in the Poincaré upper half plane, along with applications in image clustering, can be found in [5].

The Riemannian volume (or area, since H is of dimension 2) element corresponding to the Riemannian metric (4) is $dA(z) = dxdy/y^2$. Accordingly, the integral of a function $f : H \to R$ with respect to dA is given by:

$$\int_H f(z) dA(z) = \int_0^{+\infty} \int_{-\infty}^{+\infty} \frac{f(x, y)}{y^2} dxdy \tag{8}$$

In many cases, the analytic computation of this integral can be greatly simplified by using polar coordinates (r, ϕ) defined with respect to some "origin" $\bar{z} \in H$. Polar coordinates (r, φ) map to the point $z(r, \varphi)$ given by:

$$z(r, \varphi) = \gamma_{\bar{z}}\left(r, \frac{\pi}{2} - \varphi\right) \tag{9}$$

where the right-hand side is defined according to Equation (7). The polar coordinates (r, φ) do indeed define a global coordinate system of H, in the sense that the application that takes a complex number $re^{i\varphi}$ to the point $z(r, \varphi)$ in H is a diffeomorphism. The standard notation from differential geometry is:

$$\exp_{\bar{z}}\left(re^{i\varphi}\right) = z(r, \varphi) \tag{10}$$

In these coordinates, the Riemannian metric (4) takes on the form:

$$ds^2(z) = dr^2 + \sinh^2 r d\varphi^2 \tag{11}$$

The integral Equation (8) can be computed in polar coordinates using the formula [13],

$$\int_H f(z)dA(z) = \int_0^{2\pi}\int_0^{+\infty} (f \circ \exp_{\bar{z}})\left(re^{i\varphi}\right)\sinh(r)drd\varphi \tag{12}$$

where $\exp_{\bar{z}}$ was defined in Equation (10) and $\circ$ denotes composition. This is particularly useful when $f \circ \exp_{\bar{z}}$ does not depend on φ.

2.3. Riemannian Gradient Descent

In this paper, the problem of minimizing, or maximizing, a differentiable function $f : H \to R$ will play a central role. A popular way of handling the minimization of a differentiable function defined on a Riemannian manifold (such as H) is through Riemannian gradient descent [14].

Here, the definition of Riemannian gradient is reviewed, and a generic description of Riemannian gradient descent is provided. The Riemannian gradient of f is here defined as a mapping $\nabla f : H \to C$ with the following property:

$$\frac{1}{y^2} \times \text{Re}\left\{\nabla f(z)\, h^*\right\} = \text{Re}\left\{df(z)\, h^*\right\} \tag{13}$$

for any complex number h, where Re denotes the real part, * denotes conjugation and df is the "derivative", $df = (\partial f/\partial x) + (\partial f/\partial y)\, i$. For example, if $f(z) = y$, it follows from Equation (13) that $\nabla f(z) = y^2$.

Riemannian gradient descent consists in following the direction of $-\nabla f$ at each step, with the length of the step (in other words, the step size) being determined by the user. The generic algorithm is, up to some variations, the following:

INPUT	$\hat{z} \in H$	% Initial guess
WHILE	$\lVert\nabla f(\hat{z})\rVert > \varepsilon$	% $\varepsilon \approx 0$ machine precision
	$\hat{z} \leftarrow \exp_{\hat{z}}(-\lambda\nabla f(\hat{z}))$	% $\lambda > 0$ step size, depends on $\hat{z}$
END WHILE		
OUTPUT	$\hat{z}$	% near critical point of f

Here, in the condition for the while loop, $\lVert\nabla f(z_k)\rVert$ is the Riemannian norm of the gradient $\nabla f(z_k)$. In other words,

$$\lVert\nabla f(z_k)\rVert^2 = \frac{1}{y_k^2} \times \text{Re}\left\{\nabla f(z_k)\, \nabla f(z_k)^*\right\}$$

Just like a classical gradient descent algorithm, the above Riemannian gradient descent consists in following the direction of the negative gradient $-\nabla f(\hat{z})$, in order to define a new estimate. This is repeated as long as the gradient is sensibly nonzero, in the sense of the loop condition.

The generic algorithm described above has no guarantee of convergence. Convergence and behavior near limit points depends on the function f, on the initialization of the algorithm and on the step sizes λ. For these aspects, the reader may consult [14](Chapter 4).

3. Riemannian Prior on the Univariate Normal Model

The current section introduces new prior distributions on the univariate normal model. These may be referred to as "Riemannian priors", since they are entirely based on the Riemannian geometry of this model, and will also be called "Gaussian distributions", when viewed as probability distributions on the Poincaré half plane.

Here, Section 3.1 defines in a rigorous way Gaussian distributions on H (based on the intuitive Formula (2)). A Gaussian distribution $G(\bar{z},\gamma)$ has two parameters, $\bar{z} \in H$, called the center of mass, and $\gamma > 0$, called the dispersion parameter. Section 3.2 uses the Riemannian gradient descent algorithm Section 2.3 to provide an algorithm for computing maximum likelihood estimates of $\bar{z}$ and γ. Finally, Section 3.3 proves that $\bar{z}$ is the Riemannian center of mass or Karcher mean of the distribution $G(\bar{z},\gamma)$, (Historically, it is more correct to speak of the "Fréchet mean", since this concept was proposed by Fréchet in 1948 [15]), and that γ is uniquely related to mean square Rao's distance from $\bar{z}$.

The reader may wish to note that the results of Section 3.3 are not used in the following, so this paragraph may be skipped on a first reading.

3.1. Gaussian Distributions on H

A Gaussian distribution $G(\bar{z},\gamma)$ on H is a probability distribution with the following probability density function:

$$p(z|\bar{z},\gamma) = \frac{1}{Z(\gamma)} \exp\left(\frac{-d^2(z,\bar{z})}{2\gamma^2}\right) \tag{14}$$

Here, $\bar{z} \in H$ is called the center of mass and $\gamma > 0$ the dispersion parameter of the distribution $G(\bar{z},\gamma)$. The squared distance $d^2(z,\bar{z})$ refers to Rao's distance (5). The probability density function (14) is understood with respect to the Riemannian volume element $dA(z)$. In other words, the normalization constant $Z(\gamma)$ is given by:

$$Z(\gamma) = \int_H f(z)dA(z) \qquad f(z) = \exp\left(\frac{-d^2(z,\bar{z})}{2\gamma^2}\right)$$

Using polar coordinates, as in Equation (12), it is possible to calculate this integral explicitly. To do so, let (r,φ), whose origin is $\bar{z}$. Then, $d^2(z,\bar{z}) = r^2$ when $z = z(r,\varphi)$, as in Equation (9). It follows that:

$$\left(f \circ \exp_{\bar{z}}\right)(r,\varphi) = \exp\left(\frac{-r^2}{2\gamma^2}\right) \tag{15}$$

According to Equation (12), the integral $Z(\gamma)$ reduces to:

$$Z(\gamma) = \int_0^{2\pi} \int_0^{+\infty} \exp\left(\frac{-r^2}{2\gamma^2}\right) \sinh(r) dr d\varphi$$

which is readily calculated,

$$Z(\gamma) = 2\pi \times \sqrt{\frac{\pi}{2}}\gamma \times e^{\frac{\gamma^2}{2}} \times \operatorname{erf}\left(\frac{\gamma}{\sqrt{2}}\right) \tag{16}$$

where erf denotes the error function. Formula (16) completes the definition of the Gaussian distribution $G(\bar{z},\gamma)$. This definition is the same as suggested in [11], with the difference that, in the present work, it has been possible to compute exactly the normalization constant $Z(\gamma)$.

It is noteworthy that the normalization constant $Z(\gamma)$ depends only on γ and not on $\bar{z}$. This shows that the shape of the probability density function (14) does not depend on $\bar{z}$, which only plays the role of a location parameter. At a deeper mathematical level, this reflects the fact that H is a homogeneous Riemannian space [13].

The probability density function (14) bears a clear resemblance to the usual Gaussian (or normal) probability density function. Indeed, both are proportional to the exponential minus the "square distance", but in one case, the distance is interpreted as Euclidean distance and, in the other (that of Equation (14)) as Rao's distance.

3.2. Maximum Likelihood Estimation of $\bar{z}$ and γ

Consider the problem of computing maximum likelihood estimates of the parameters $\bar{z}$ and γ of the Gaussian distribution $G(\bar{z},\gamma)$, based on independent samples $\{z_i\}_{i=1}^N$ from this distribution. Given the expression (14) of the density $p(z|\bar{z},\gamma)$, the log-likelihood function $\ell(\bar{z},\gamma)$ can be written,

$$\ell(\bar{z},\gamma) = -N\log\{Z(\gamma)\} - \frac{1}{2\gamma^2}\sum_{i=1}^{N} d^2(z_i,\bar{z}) \tag{17}$$

Since $\bar{z}$ only appears in the second term, the maximum likelihood estimate of $\bar{z}$, say $\hat{z}$, can be computed first. It is given by the minimization problem:

$$\hat{z} = \text{argmin}_{z\in H}\ \frac{1}{2}\sum_{i=1}^{N} d^2(z_i,z) \tag{18}$$

In other words, the maximum likelihood estimate $\hat{z}$ minimizes the sum of squared Rao distances to the samples z_i. This exhibits $\hat{z}$ as the Riemannian center of mass, also called the Karcher or the Fréchet mean [16], of the samples z_i.

The notion of Riemannian center of mass is currently a widely popular one in signal and image processing, with applications ranging from blind source separation and radar signal processing [17,18] to shape and motion analysis [19,20]. The definition of Gaussian distributions, proposed in the present paper, shows how the notion of Riemannian center of mass is related to maximum likelihood estimation, thereby giving it a statistical foundation.

An original result, due to Cartan and cited in Equation [16], states that $\hat{z}$, as defined in Equation (18), exists and is unique, since H, with the Riemannian distance (4), has constant negative curvature. Here, $\hat{z}$ is computed using Riemannian gradient descent, as described in Section 2.3. The cost function f to be minimized is given by (the factor N^{-1} is conventional),

$$f(z) = \frac{1}{2N}\sum_{i=1}^{N} d^2(z_i,z) \tag{19}$$

Its Riemannian gradient $\nabla f(z)$ is easily found by noting the following fact. Let $f_i(z) = (1/2)d^2(z,z_i)$. Then, the Riemannian gradient of this function is (see [21] (page 407)),

$$\nabla f_i(z) = \log_z(z_i) \tag{20}$$

where $\log_z : H \to C$ is the inverse of $\exp_z : C \to H$. It follows from Equation (20) that,

$$\nabla f(z) = \frac{1}{N}\sum_{i=1}^{N}\log_z(z_i) \tag{21}$$

The analytic expression of $\log_z$, for any $z \in H$, will be given below (see Equation (23)).

Here, the gradient descent algorithm for computing $\hat{z}$ is described. This algorithm uses a constant step size λ, which is fixed manually.

Once the maximum likelihood estimate $\hat{z}$ has been computed, using the gradient descent algorithm, the maximum likelihood estimate of γ, say $\hat{\gamma}$, is found by solving the equation:

$$F(\gamma) = \frac{1}{N}\sum_{i=1}^{N} d^2(z_i, \hat{z}) \qquad \text{where } F(\gamma) = \gamma^3 \times \frac{d}{d\gamma}\log\{Z(\gamma)\} \tag{22}$$

The gradient descent algorithm for computing $\hat{z}$ is the following,

`INPUT`	$\{z_1, \ldots, z_N\}$	% N independent samples from $G(\bar{z}, \gamma)$
	$\hat{z} \in H$	% Initial guess
`WHILE`	$\Vert\nabla f(\hat{z})\Vert > \varepsilon$	% $\varepsilon \approx 0$ machine precision
	$\hat{z} \leftarrow \exp_{\hat{z}}(-\lambda \nabla f(\hat{z}))$	% $\nabla f(\hat{z})$ given by Equation (21)
		% step size λ is constant
`END WHILE`		
`OUTPUT`	$\hat{z}$	% near Riemannian center of mass

Application of Formula (21) requires computation of $\log_{\hat{z}}(z_i)$ for $i = 1, \ldots, N$. Fortunately, this can be done analytically as follows. In general, for $\hat{z} = (\bar{x}, \bar{y})$,

$$\log_{\hat{z}}(z) = \bar{y}\log_i\left(\frac{z - \bar{x}}{\bar{y}}\right) \tag{23}$$

where $\log_i$ is found by inverting Equation (6). Precisely,

$$\log_i(z) = re^{i\varphi} \tag{24}$$

where, for $z = (x, y)$ with $x \neq 0$,

$$r = \operatorname{acosh}\left(1 + \frac{x^2 + (y-1)^2}{2y}\right)$$

and:

$$\cos(\varphi) = \frac{x}{y\sinh(r)} \qquad \sin(\varphi) = \frac{\cosh(r) - y^{-1}}{\sinh(r)}$$

and, for $z = (0, y)$,

$$\log_i(z) = \ln(y)i$$

with ln denoting the natural logarithm.

3.3. Significance of $\bar{z}$ and γ

The parameters $\bar{z}$ and γ of a Gaussian distribution $G(\bar{z}, \gamma)$ have been called the center of mass and the dispersion parameter. In the present paragraph, it is proven that,

$$\bar{z} = \operatorname{argmin}_{z \in H} \frac{1}{2}\int_H d^2(z', z)p(z'|\bar{z}, \gamma)dA(z') \tag{25}$$

and also that:

$$F(\gamma) = \int_H d^2(z', \bar{z})p(z'|\bar{z}, \gamma)dA(z') \tag{26}$$

where $F(\gamma)$ was defined in Equation (22) and $p(z'|\bar{z}, \gamma)$ is the probability density function of $G(\bar{z}, \gamma)$, given in Equation (14).

Note that Equations (25) and (26) are asymptotic versions of Equations (18) and (22). Indeed, Equations (25) and (26) can be written:

$$\bar{z} = \text{argmin}_{z \in H} \frac{1}{2} E_{\bar{z},\gamma} d^2(z', z) \qquad F(\gamma) = E_{\bar{z},\gamma} d^2(z, \bar{z}) \tag{27}$$

where $E_{\bar{z},\gamma}$ denotes the expectation with respect to $G(\bar{z}, \gamma)$, and the expectation is carried out on the variable z' in the first formula. Now, these two formulae are the same as Equations (18) and (22), but with expectation instead of empirical mean.

Note, moreover, that Equations (25) and (26) can be interpreted as follows. If z' is distributed according to the Gaussian distribution $G(\bar{z}, \gamma)$, then Equation (25) states that $\bar{z}$ is the unique point, out of all $z \in H$, which minimizes the expectation of squared Rao's distance to z'. Moreover, Equation (26) states that the expectation of squared Rao's distance between $\bar{z}$ and z' is equal to $F(\gamma)$, so $F(\gamma)$ is the least possible expected squared Rao's distance between a point $z \in H$ and z'. This interpretation justifies calling $\bar{z}$ the center of mass of $G(\bar{z}, \gamma)$ and shows that γ is uniquely related to the expected dispersion, as measured by squared Rao's distance, away from $\bar{z}$.

In order to prove Equation (25), consider the log-likelihood function,

$$\ell(\bar{z}, \gamma; z) = -\log\{Z(\gamma)\} - \frac{1}{2\gamma^2} d^2(z, \bar{z}) \tag{28}$$

Let $f_z(\bar{z}) = (1/2) d^2(z, \bar{z})$. The score function, with respect to $\bar{z}$ is, by definition,

$$\nabla_{\bar{z}} \ell(\bar{z}, \gamma; z) = \nabla f_z(\bar{z}) \tag{29}$$

where $\nabla_{\bar{z}}$ indicates the Riemannian gradient (defined in Equation (13) of Section 2.3) is with respect to the variable $\bar{z}$. Under certain regularity conditions, which are here easily verified, the expectation of the score function is identically zero,

$$E_{\bar{z},\gamma} \nabla f_z(\bar{z}) = 0 \tag{30}$$

Let $f(z)$ be defined by:

$$f(z) = E_{\bar{z},\gamma} f_{z'}(z) = \frac{1}{2} E_{\bar{z},\gamma} d^2(z', z)$$

with the expectation carried out on the variable z'. Clearly, $f(z)$ is the expression to be minimized in Equation (25) (or in the first formula in Equation (27), which is just the same). By interchanging Riemannian gradient and expectation,

$$\nabla f(\bar{z}) = E_{\bar{z},\gamma} \nabla f_z(\bar{z}) = 0$$

where the last equality follows from Equation (30).

It has just been proved that $\bar{z}$ is a stationary point of f (a point where the gradient is zero). Theorem 2.1 in [16] states the function f has one and only one stationary point, which is moreover a global minimizer. This concludes the Proof (25).

The proof of Equation (26) follows exactly the same method, defining the score function with respect to γ and noting that its expectation is identically zero.

4. Classification of Univariate Normal Populations

The previous section studied Gaussian distributions on H, "as they stand", focusing on the fundamental issue of maximum likelihood estimation of their parameters. The present Section considers the use of Gaussian distributions as prior distributions on the univariate normal model.

The main motivation behind the introduction of Gaussian distributions is that a Gaussian distribution $G(\bar{z}, \gamma)$ can be used to give a geometric representation of a cluster or class of univariate normal populations. Recall that each point $(x, y) \in H$ is identified with a univariate normal population with mean $\mu = \sqrt{2}x$ and standard deviation $\sigma = y$. The idea is that populations belonging to the same cluster, represented by $G(\bar{z}, \gamma)$, should be viewed as centered on $\bar{z}$ and lying within a typical distance determined by γ.

In the remainder of this Section, it is shown how the maximum likelihood estimation algorithm of Section 3.2 can be used to fit the hyperparameters $\bar{z}$ and γ to data, consisting in a class $\mathcal{S} = \{S_i; i = 1, \ldots, K\}$ of univariate normal populations. This is then applied to the problem of the classification of univariate normal populations. The whole development is based on marginalized likelihood estimation, as follows.

Assume each population S_i contains N_i points, $S_i = \{s_j; j = 1, \ldots, N_i\}$, and the points s_j, in any class, are drawn from a univariate normal distribution with mean μ and standard deviation σ. The focus will be on the asymptotic case where the number N_i of points in each population S_i is large.

In order to fit the hyperparameters $\bar{z}$ and γ to the data $\mathcal{S}$, assume moreover that the distribution of $z = (x, y)$, where $(x, y) = (\mu/\sqrt{2}, \sigma)$, is a Gaussian distribution $G(\bar{z}, \gamma)$. Then, the distribution of $\mathcal{S}$ can be written in integral form:

$$p(\mathcal{S}|\bar{z}, \gamma) = \prod_{i=1}^{K} \int_H p(S_i|z) p(z|\bar{z}, \gamma) dA(z) \tag{31}$$

where $p(z|\bar{z}, \gamma)$ is the probability density of a Gaussian distribution $G(\bar{z}, \gamma)$, defined in Equation (14). Moreover, expressing $p(S_i|z)$ as a product of univariate normal distributions $p(s_j|z)$, it follows,

$$p(\mathcal{S}|\bar{z}, \gamma) = \prod_{i=1}^{K} \int_H \prod_{j=1}^{N_i} p(s_j|z) p(z|\bar{z}, \gamma) dA(z) \tag{32}$$

This expression, given the data $\mathcal{S}$, is to be maximized over $(\bar{z}, \gamma)$. Using the Laplace approximation, this task is reduced to the maximum likelihood estimation problem, addressed in Section 3.2.

The Laplace approximation will here be applied in its "basic form" [9]. That is, up to terms of order N_i^{-1}. To do so, write each of the integrals in Equation (32), using Equation (8) of Section 2.2. These integrals then take on the form:

$$\int_0^{+\infty} \int_{-\infty}^{+\infty} \prod_{j=1}^{N_i} |2\pi y^2|^{-1/2} \exp\left(\frac{-\left(s_j - \sqrt{2}\,x\right)^2}{2y^2} \right) \times p(z|\bar{z}, \gamma) \times \frac{1}{y^2} dxdy \tag{33}$$

where the univariate normal distribution $p(s_j|z)$ has been replaced by its full expression. Now, this expression can be written $p(s_j|z) = \exp\left[-N_i h(x, y)\right]$, where:

$$h(x, y) = -\frac{1}{2} \ln\left(2\pi y^2\right) - \frac{B_i^2 + V_i^2}{2y^2}$$

Here, B^2 and V_i^2 are the empirical bias and variance, within population S_i,

$$B_i = \hat{S}_i - \sqrt{2}\,x \qquad V_i^2 = N_i^{-1} \sum_{j=1}^{N_i} (\hat{S}_i - s_j)^2$$

where $\hat{S}_i$ is the empirical mean of the population $\hat{S}_i = N_i^{-1} \sum_{j=1}^{N_i} s_j$.

The expression $h(x, y)$ is maximized when $x = \hat{x}_i$ and $y = \hat{y}_i$, where $\hat{z}_i = (\hat{x}_i, \hat{y}_i)$ is the couple of maximum likelihood estimates of the parameters (x, y), based on the population S_i.

According to the Laplace approximation, the integral Equation (33) is equal to:

$$2\pi \left|\partial^2 h(\hat{x}_i, \hat{y}_i)\right|^{-1/2} \times \exp\left[-N_i h(\hat{x}_i, \hat{y}_i)\right] \times p(\hat{z}_i|\bar{z}, \gamma) \times \frac{1}{\hat{y}_i^2} + O(N_i^{-1})$$

where $\partial^2 h(\hat{x}_i, \hat{y}_i)$ is the matrix of second derivatives of h, and $|\cdot|$ denotes the determinant. Now, since h is essentially the logarithm of $p(s_j|z)$, a direct calculation shows that $\partial^2 h(\hat{x}_i, \hat{y}_i)$ is the same as the Fisher information matrix derived in Section 2.1 (where it was denoted $I(z)$). Thus, the first factor in the above expression is $2\pi\hat{y}_i^2$, and cancels out with the last factor.

Finally, the Laplace approximation of the integral Equation (33) reads:

$$2\pi \times \exp\left[-N_i h(\hat{x}_i, \hat{y}_i)\right] \times p(\hat{z}_i|\bar{z}, \gamma) + O(N_i^{-1})$$

and the resulting approximation of the distribution of $\mathcal{S}$, as given by Equation (32), can be written:

$$p(\mathcal{S}|\bar{z}, \gamma) \approx \prod_{i=1}^{K} \alpha \times p(\hat{z}_i|\bar{z}, \gamma) \tag{34}$$

where α is a constant, which does not depend either on the data or on the parameters, and $p(\hat{z}_i|\bar{z}, \gamma)$ has the expression (14).

Accepting this expression to give the distribution of the data $\mathcal{S}$, conditionally on the hyperparameters $(\bar{z}, \gamma)$, the task of estimating these hyperparameters becomes the same as the maximum likelihood estimation problem, described in Section 3.2.

In conclusion, if one assumes the populations S_i belong to a single cluster or class $\mathcal{S}$ and wishes to fit the hyperparameters $\bar{z}$ and γ of a Gaussian distribution representing this cluster, it is enough to start by computing the maximum likelihood estimates $\hat{x}_i$ and $\hat{y}_i$ for each population S_i and then to consider these as input to the maximum likelihood estimation algorithm described in Section 3.2.

The same reasoning just carried out, using the Laplace approximation, can be generalized to the problem of classification of univariate normal populations. Indeed, assume that classes $\{\mathcal{S}_L, L = 1, \ldots, C\}$, each containing some number K_L of univariate normal populations, have been identified based on some training sequence. Using the Laplace approximation and the maximum likelihood estimation approach of Section 3.2, to each one of these classes, it is possible to fit hyperparameters $(\bar{z}_L, \gamma_L)$ of a Gaussian distribution $G(\bar{z}_L, \gamma_L)$ on H.

For a test population S_t, the maximum likelihood rule, for deciding which of the classes $\mathcal{S}_L$ this test population S_t belongs to, requires finding the following maximum:

$$L^* = \text{argmax}_L\, p(S_t|\bar{z}_L, \gamma_L) \tag{35}$$

and assigning the test population S_t to the class with label L^*. If the number of points N_t in the population S_t is large, the Laplace approximation, in the same way used above, approximates the maximum in Equation (35) by:

$$L^* = \text{argmax}_L\, p(\hat{z}_t|\bar{z}_L, \gamma_L) \tag{36}$$

where $\hat{z}_t = (\hat{x}_t, \hat{y}_t)$ is the couple of maximum likelihood estimates computed based on the test population S_t and where $p(\hat{z}_t|\bar{z}_L, \gamma_L)$ is given by Equation (14). Now, writing out Equation (14), the decision rule becomes:

$$L^* = \text{argmax}_L \left(-\log\{Z(\gamma_L)\} - \frac{1}{2\gamma_L^2} d^2(\hat{z}_t, \bar{z}_L) \right) \tag{37}$$

Under the homoscedasticity assumption, that all of the γ_L are equal, this decision rule essentially becomes the same as the one proposed in [5], which requires S_t to be assigned to the "nearest" cluster, in terms of Rao's distance. Indeed, if all the γ_L are equal, then Equation (37) is the same as,

$$L^* = \operatorname{argmin}_L d^2(\hat{z}_t, \bar{z}_L) \tag{38}$$

This decision rule is expected to be less efficient that the one proposed in Equation (37), which also takes into account the uncertainty associated with each cluster, as measured by its dispersion parameter γ_L.

5. Application to Image Classification

In this section, the framework proposed in Section 4, for classification of univariate normal populations, is applied to texture image classification using Gabor filters. Several authors have found that Gabor energy features are well-suited texture descriptors. In the following, consider 24 Gabor energy sub-bands that are the result of three scales and eight orientations. Hence, each texture image can be decomposed as the collection of those 24 sub-bands. For more information concerning the implementation, the interested reader is referred to [22].

Starting from the VisTeX database of 40 images [10] (these are displayed in Figure 1), each image was divided into 16 non-overlapping subimages of 128 × 128 pixels each. A training sequence was formed by choosing randomly eight subimages out of each image. To each subimage in the training sequence, a bank of 24 Gabor filters was applied. The result of applying a Gabor filter with scale s and orientation o to a subimage i belonging to an image L is a univariate normal population $S_{i,s,o}$ of 128 × 128 points (one point for each pixel, after the filter is applied).

Figure 1. Forty images of the VisTex database.

These populations $S_{i,s,o}$ (called sub-bands) are considered independent, each one of them univariate normal with mean $\mu_{i,s,o} = \sqrt{2}x_{i,s,o}$, standard deviation $\sigma_{i,s,o} = y_{i,s,o}$ and with $z_{i,s,o} = (x_{i,s,o}, y_{i,s,o})$. The couple of maximum likelihood estimates for these parameters is denoted $\hat{z}_{i,s,o} = (\hat{x}_{i,s,o}, \hat{y}_{i,s,o})$. An image L (recall, there are 40 images) contains, in each sub-band, eight populations $S_{i,s,o}$, with which hyperparameters $\bar{z}_{L,s,o}$ and $\gamma_{L,s,o}$ are associated, by applying the maximum likelihood estimation algorithm of Section 3.2 to the inputs $\hat{z}_{i,s,o}$.

If S_t is a test subimage, then one should begin by applying the 24 Gabor filters to it, obtaining independent univariate normal populations $S_{t,s,o}$, and then compute for each population the couple of maximum likelihood estimates $\hat{z}_{t,s,o} = (\hat{x}_{t,s,o}, \hat{y}_{t,s,o})$. The decision rule Equation (37) of Section 4 requires that S_t should be assigned to the image L^*, which realizes the maximum:

$$L^* = \operatorname{argmax}_L \sum_{s,o} -\log\{Z(\gamma_{L,s,o})\} - \frac{1}{2\gamma^2_{L,s,o}} d^2(\hat{z}_{t,s,o}, \bar{z}_{L,s,o}) \tag{39}$$

When considering the homoscedasticity assumption, *i.e.*, $\gamma_{L,s,o} = \gamma_{s,o}$ for all L, this decision rule becomes:

$$L^* = \text{argmin}_L \sum_{s,o} d^2(\hat{z}_{t,s,o}, \bar{z}_{L,s,o}) \tag{40}$$

For this concrete application, to the VisTex database, it is pertinent to compare the rate of successful classification (or overall accuracy) obtained using the Riemannian prior, based on the framework of Section 4, to that obtained using a more classical conjugate prior, *i.e.*, a normal-inverse gamma distribution of the mean $\mu = \sqrt{2}x$ and the standard deviation $\sigma = y$. This conjugate prior is given by:

$$p(\mu|\sigma, \mu_p, \kappa_p) = \frac{\sqrt{\kappa_p}}{\sigma\sqrt{2\pi}} \exp\left(-\frac{\kappa_p}{2\sigma^2}(\mu - \mu_p)^2\right)$$

with an inverse gamma prior, on σ^2,

$$p(\sigma^2|\alpha, \beta) = \frac{\beta^\alpha}{\Gamma(\alpha)} \left(\sigma^2\right)^{-(\alpha+1)} \exp\left(-\frac{\beta}{\sigma^2}\right) \tag{41}$$

Using this conjugate prior, instead of a Riemannian prior, and following the procedure of applying the Laplace approximation, a different decision rule is obtained, where L^* is taken to be the maximum of the following expression:

$$\begin{aligned} &\sum_{s,o} \frac{\ln \kappa_{p_{L,s,o}}}{2} - \frac{\kappa_{p_{L,s,o}}}{2\hat{y}^2_{t,s,o}} \left(\sqrt{2}\hat{x}_{t,s,o} - \mu_{p_{L,s,o}}\right)^2 \\ &+ \alpha_{L,s,o} \ln \beta_{L,s,o} - \ln \Gamma(\alpha_{L,s,o}) - 2(\alpha_{L,s,o} + 1) \ln \hat{y}_{t,s,o} - \frac{\beta_{L,s,o}}{\hat{y}^2_t} \end{aligned} \tag{42}$$

where, as in Equation (39), $\hat{x}_{t,s,o}$ and $\hat{y}_{t,s,o}$ are the maximum likelihood estimates computed for the population $S_{t,s,o}$.

Both the Riemannian and conjugate priors have been applied to the VisTex database, with half of the database used for training and half for testing. In the course of 100 Monte Carlo runs, a significant gain of about 3% is observed with the Riemannian prior compared to the conjugate prior. This is summarized in the following table.

Prior Model	Overall Accuracy
Riemannian prior Equation (39)	71.88% ± 2.16%
Riemannian prior, homoscedasticity assumption Equation (40)	69.06% ± 1.96%
Conjugate prior Equation (42)	68.73% ± 2.92%

Recall that the overall accuracy is the ratio of the number of successfully classified subimages to the total number of subimages. The table shows that the use of a Riemannian prior, even under a homoscedasticity assumption, yields significant improvement upon the use of a conjugate prior.

6. Conclusions

Motivated by the problem of the classification of univariate normal populations, this paper introduced a new class of prior distributions on the univariate normal model. With the univariate normal model viewed as the Poincaré half plane H, these new prior distributions, called Gaussian distributions, were meant to reflect the geometric picture (in terms of Rao's distance) that a cluster or class of univariate normal populations can be represented as having a center $\bar{z} \in H$ and a "variance" or dispersion γ^2. Precisely, a Gaussian distribution $G(\bar{z}, \gamma)$ has a probability density function $p(z)$, with respect to Riemannian volume of the Poincaré half plane, which is proportional to $\exp\left(-\frac{d^2(z,\bar{z})}{2\gamma^2}\right)$.

Using Gaussian distributions as prior distributions in the problem of the classification of univariate normal populations was shown to lead to a new, more general and efficient decision rule. This decision rule was implemented in a real-world application to texture image classification, where it led to significant improvement in performance, in comparison to decision rules obtained by using conjugate priors.

The general approach proposed in this paper contains several simplifications and approximations, which could be improved upon in future work. First, it is possible to use different prior distributions, which are more geometrically rich than Gaussian distributions, to represent classes of univariate normal populations. For example, it may be helpful to replace Gaussian distributions that are "isotropic", in the sense of having a scalar dispersion parameter γ, by non-isotropic distributions, with a dispersion matrix Γ (a 2×2 symmetric positive definite matrix). Another possibility would be to represent each class of univariate normal populations by a finite mixture of Gaussian distributions, instead of representing it by a single Gaussian distribution.

These variants, which would allow classes with a more complex geometric structure to be taken into account, can be integrated in the general framework proposed in the paper, based on: (i) fitting each class to a prior distribution (Gaussian non-isotropic, mixture of Gaussians); and (ii) choosing, for a test population, the most adequate class, based on a decision rule. These two steps can be realized as above, through the Laplace approximation and maximum likelihood estimation, or through alternative techniques, based on Markov chain Monte Carlo stochastic optimization.

In addition to generalizing the approach of this paper and improving its performance, a further important objective for future work will be to extend it to other parametric models, beyond univariate normal models. Indeed, there is an increasing number of parametric models (generalized Gaussian, elliptical models, *etc.*), whose Riemannian geometry is becoming well understood and where the present approach may be helpful.

Author Contributions

Salem Said carry out the mathematical development, and specify the algorithms, appearing in Sections 2, 3 and 4. Lionel Bombrun carry out all numerical simulations, and to propose the theoretical development of Section 4. Yannick Berthoumieu devise the main idea of the paper. That is, use of Riemannian priors as geometric representation of a class or cluster of univariate normal population. All authors have read and approved the final manuscript.

Conflicts of Interest

The authors declare no conflict of interest.

References

1. Amari, S.I; Nagaoka, H. *Methods of Information Geometry*; American Mathematical Society: Providence, RI, USA, 2000.
2. Rao, C.R. Information and the accuracy attainable in the estimation of statistical parameters. *Bull. Calcutta Math. Soc.* **1945**, *37*, 81–91.
3. Kass, R.E. The geometry of asymptotic inference. *Stat. Sci.* **1989**, *4*, 188–234.
4. Amari, S.I. Natural gradient works efficiently in learning. *Neur. Comput.* **1998**, *10*, 251–276.
5. Nielsen, F; Nock, R. Hyperbolic Voronoi diagrams made easy. **2009** , arXiv:0903.3287.
6. Lenglet, C.; Rousson, M.; Deriche, R.; Fougeras, O. Statistics on the manifold of multivariate normal distributions: Theory and application to diffusion tensor MRI processing. *J. Math. Imaging Vis.* **2006**, *25*, 423–444.
7. Verdoolaege, G.; Scheunders, P. On the geometry of multivariate generalized Gaussian models. *J. Math. Imaging Vis.* **2012**, *43*, 180–193.

8. Berkane, M.; Oden, K. Geodesic estimation in elliptical distributions. *J. Multival. Anal.* **1997**, *63*, 35–46.
9. Erdélyi, A. Asymptotic Expansions; Dover Books: Mineola, New York, NY, USA, 2010.
10. MIT Vision and Modeling Group. Vision Texture. Available online: http://vismod.media.mit.edu/pub/VisTex (accessed on 10 June 2014).
11. Pennec, X. Intrinsic statistics on Riemannian manifold: Basic tools for geometric measurements. *J. Math. Imaging Vis.* **2006**, *25*, 127–154.
12. Atkinson, C.; Mitchell, A.F.S. Rao's distance measure. *Sankhya Ser. A* **1981**, *43*, 345–365.
13. Gallot, S.; Hulin, D.; Lafontaine, J. *Riemannian Geometry*; Springer-Verlag: Berlin, Germany, 2004.
14. Absil, P.A.; Mahony, R.; Sepulchre, R. Optimization Algorithms on Matrix Manifolds; Princeton University Press: Cambridge, MA, USA, 2006.
15. Fréchet, M. Les éléments aléatoires de nature quelconque dans un espace distancié. *Annales de l'I.H.P.* **1948**, *10*, 215–310. (In French)
16. Afsari, B. Riemannian L^p center of mass: Existence, Uniqueness and convexity. *Proc. Am. Math. Soc.* **2011**, *139*, 655–673.
17. Manton, J.H. A centroid (Karcher mean) approach to the joint approximate diagonalisation problem: The real symmetric case. *Digit. Sign. Process.* **2006**, *16*, 468–478.
18. Arnaudon, M.; Barbaresco, F. Riemannian medians and means with applications to RADAR signal processing. *IEEE J. Sel. Top. Sign. Process.* **2013**, *7*, 595–604.
19. Le, H. On the consistency of procrustean mean shapes. *Adv. Appl. Prob.* **1998**, *30*, 53–63.
20. Turaga, P.; Veeraraghavan, A.; Chellappa, R. Statistical Snalysis on Stiefel and Grassmann Manifolds with Applications in Computer Vision. In Proceedings of the IEEE Conference on Computer Vision and Pattern Recognition, Anchorage, AK, USA, 23–28 June 2008; doi: 10.1109/CVPR.2008.4587733.
21. Chavel, I. Riemannian geometry: A modern introduction; Cambridge University Press: Princeton, MA, USA, 2008.
22. Grigorescu, S.E.; Petkov, N.; Kruizinga, P. Comparison of texture features based on Gabor filter. *IEEE Trans. Image Process.* **2002**, *11*, 1160–1167.

entropy

MDPI

Article

Combinatorial Optimization with Information Geometry: The Newton Method

Luigi Malagò [1] and Giovanni Pistone [2,*]

[1] Dipartimento di Informatica, Università degli Studi di Milano, Via Comelico, 39/41, 20135 Milano, Italy; E-Mail: malago@di.unimi.it

[2] de Castro Statistics, Collegio Carlo Alberto, Via Real Collegio 30, 10024 Moncalieri, Italy

* E-Mail: giovanni.pistone@carloalberto.org; Tel.: +39-011-670-5033; Fax: +39-011-670-5082.

Received: 31 March 2014; in revised form: 10 July 2014 / Accepted: 11 July 2014 / Published: 28 July 2014

Abstract: We discuss the use of the Newton method in the computation of $\max(p \mapsto \mathbb{E}_p[f])$, where p belongs to a statistical exponential family on a finite state space. In a number of papers, the authors have applied first order search methods based on information geometry. Second order methods have been widely used in optimization on manifolds, e.g., matrix manifolds, but appear to be new in statistical manifolds. These methods require the computation of the Riemannian Hessian in a statistical manifold. We use a non-parametric formulation of information geometry in view of further applications in the continuous state space cases, where the construction of a proper Riemannian structure is still an open problem.

Keywords: statistical manifold; Riemannian Hessian; combinatorial optimization; Newton method

1. Introduction

In this paper, statistical exponential families [1] are thought of as differentiable manifolds along the approach called information geometry [2] or the exponential statistical manifold [3]. Specifically, our aim is to discuss optimization on statistical manifolds using the Newton method, as is suggested in ([4] (Ch. 5 and 6)); see also the monograph [5]. This method is based on classical Riemannian geometry [6], but here, we put our emphasis on coordinate-free differential geometry; see [7,8].

We mainly refer to the above-mentioned references [2,4], with one notable exception in the description of the tangent space. Our manifold will be an exponential family $\mathcal{E}_V$ of positive densities, V being a vector space of sufficient statistics. Given a one-dimensional statistical model $p(t) \in \mathcal{E}_V$, $t \in I$, we define its velocity at time t to be its Fisher score $s(t) = \frac{d}{dt} \ln p(t)$ [9]. The Fisher score $s(t)$ is a random variable with zero expectation with respect to $p(t)$, $\mathbb{E}_{p(t)}[s(t)] = 0$. Because of that, the tangent space at $p \in \mathcal{E}_V$ is a vector space of random variables with zero expectation at p. A vector field is a mapping from p to a random variable $V(p)$, such that for all $p \in \mathcal{E}$, the random variable $V(p)$ is centered at p, $\mathbb{E}_p[V(p)] = 0$. In other words, each point of the manifold has a different tangent space, and this tangent space can be used as a non-parametric model space of the manifold. In this formalism, a vector field is a mapping from densities to centered random variables, that is, it is what in statistics is called a pivot of the statistical model. To avoid confusion with the product of random variables, we do not use the standard notation for the action of a vector field on a real function. This approach is possibly unusual in differential geometry, but it is fully natural from the statistical point of view, where the Fisher score has a central place. Moreover, this approach scales nicely from the finite state space to the general state space; see the discussion in [9] and the review in [3].

A complete construction of the geometric framework based on the idea of using the Fisher scores as elements of the tangent bundle has been actually worked out. In this paper, we go on by considering a second order geometry based on the non-parametric settings.

Our main motivation for such a geometrical construction is its application to combinatorial optimization using exponential families, whose first order version was developed in [10–14]. We give here an illustration of the methods in the following toy example.

Consider the function $f(x_1, x_2) = a_0 + a_1 x_1 + a_2 x_2 + a_{12} x_1 x_2$, with $x_1, x_2 = \pm 1$, $a_0, a_1, a_2, a_{12} \in \mathbb{R}$. The function f is a real random variable on the sample space $\Omega = \{+1, -1\}^2$ with the uniform probability λ. Note that the coordinate mappings X_1, X_2 of Ω generate an orthonormal basis $1, X_1, X_2, X_1X_2$ of $L^2(\Omega, \lambda)$ and that f is the general form of a real random variable on such a space. Let $\mathcal{P}_>$ be the open simplex of positive densities on (Ω, λ), and let $\mathcal{E}_V$ be a statistical model, *i.e.*, a subset of $\mathcal{P}_>$. The relaxed mapping $F\colon \mathcal{E}_V \to \mathbb{R}$,

$$F(p) = \mathbb{E}_p[f] = a_0 + a_1 \mathbb{E}_p[X_1] + a_2 \mathbb{E}_p[X_2] + a_{12} \mathbb{E}_p[X_1X_2], \tag{1}$$

is strictly bounded by the maximum of f, $F(p) = \mathbb{E}_p[f] < \max_{x\in\Omega} f(x)$, unless f is constant. We are looking for a sequence p_n, $n \in \mathbb{N}$, such that $\mathbb{E}_{p_n}[f] \to \max_{x\in\Omega} f(x)$ as $n \to \infty$. The existence of such a sequence is a nontrivial condition for the model $\mathcal{E}$. Precisely, the closure of $\mathcal{E}_V$ must contain a density, whose support is contained in the set of maxima $\{x \in \Omega | f(x) = \max f\}$. This condition is satisfied by the independence model, $V = \text{Span}\{X_1, X_2\}$, where we can write:

$$F(\eta^1, \eta^2) = a_0 + a_1\eta^1 + a_2\eta^2 + a_{12}\eta^1\eta^2, \quad \eta^i = \mathbb{E}_p[X_i], \tag{2}$$

See Figure 1.

The gradient of Equation (2) has components $\partial_1 F = a_1 + a_{12}\eta^2$, $\partial_2 F = a_2 + a_{12}\eta^1$, and the flow along the gradient produces increasing values for F; however, the gradient flow does not converge to the maximum of F; see the dotted line in Figure 2. However, one can follow the suggestion by [15] and use a modified gradient (the "natural" gradient) flow that produces better results in our problem; see Figure 3. Full details on this example are given in Section 2.5.2.

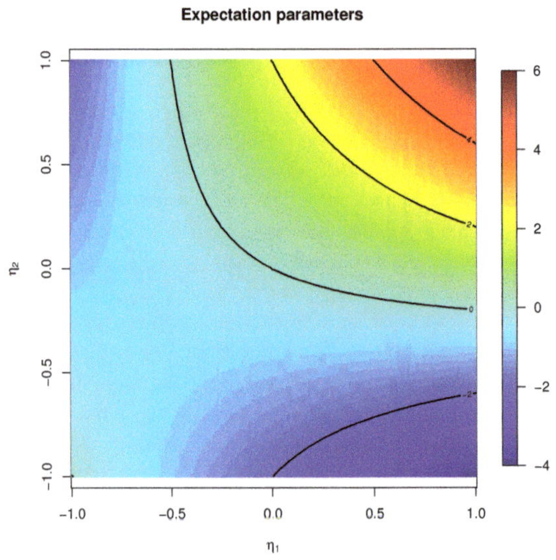

Figure 1. Relaxation of the Function (2) on the independence model. $a_1 = 1, a_2 = 2, a_{12} = 3$.

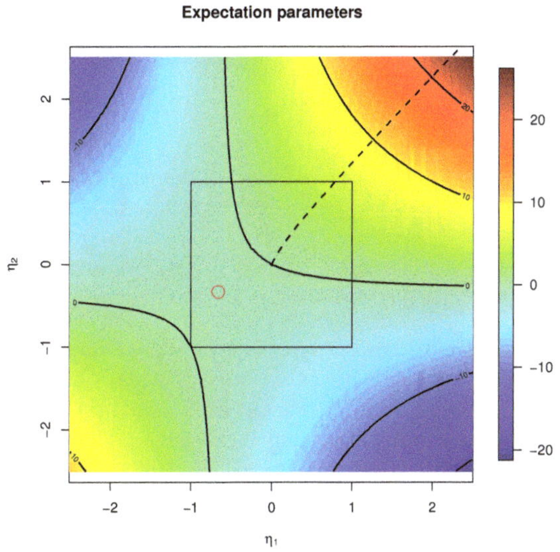

Figure 2. Gradient flow of the Function (2). The domain has been increased to include values outside the square $[-1,+1]^2$.

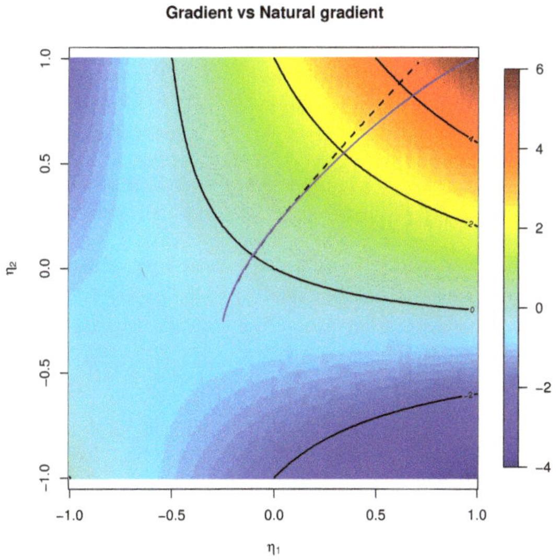

Figure 3. Gradient flow (blue line) and natural gradient flow (black line) for the Function (2), starting at $(-1/4,-1/4)$.

In combinatorial optimization, the values of the function f are assumed to be available at each point, and the curve of steepest ascent of the relaxed function is learned through a simulation procedure based on exponential statistical models.

In this paper, we introduce, in Section 2, the geometry of exponential families and its first order calculus. The second order calculus and the Hessian are discussed in Section 3. Finally, in Section 4, we apply the formalism to the discussion of the Newton method in the context of the maximization of the relaxed function.

2. Models on a Finite State Space

We consider here the exponential statistical manifold on the set of positive densities on a measure space (Ω, μ) with Ω finite and counting measure μ. The setup we describe below is not strictly required in the finite case, because in such a case, other approaches are possible, but it provides a mathematical formalism that has its own pros and that scales naturally to the infinite case.

We provide below a schematic presentation of our formalism as an introduction to this section.

- Two different exponential families can actually be the same statistical model, as the set of densities in the two exponential families are actually equal. This fact is due to both the arbitrariness of the reference density and the fact that sufficient statistics are actually a vector basis of the vector space generated by the sufficient statistics. In a non-parametric approach, we can refer directly to the vector space of centered log-densities, while the change of reference density is geometrically interpreted as a change of chart. The set of all possible such charts defines a manifold.
- We make a specific interpretation of the tangent bundle as the vector space of Fisher's scores at each density and use such tangent spaces as the space of coordinates. This produces a different tangent space/space of coordinates at each density, and different tangent spaces are mapped one onto another by a proper parallel transport, which is nothing else than the re-centering of random variables.
- If a basis is chosen, a parametrization is given, and such a parametrization is, in fact, a new chart, whose values are real vectors. In the real parametrization, the natural scalar product in each scores space is given by Fisher's information matrix.
- Riemannian gradients are defined in the usual way. It is customary in information geometry to call "natural gradient" the real coordinate presentation of the Riemannian gradient. The natural gradient is computed by applying the inverse of the Fisher information matrix to the Euclidean gradient. It seems that there are tree gradients involved, but they all represent the same object when correctly understood.
- The classical notion of expectation parameters for exponential families carries on as another chart on the statistical manifold, which gives rise to a further presentation of a geometrical object.
- While the statistical manifold is unique, there are at least three relevant connections as structures on the vector bundles of the manifold: one relating to the exponential charts, one relating to the expectation charts and one depending on the Riemannian structure.

2.1. Exponential Families As Manifolds

On the finite sample space Ω, $\#\Omega = n$, let a set of random variables $\mathcal{B} = \{X_1, \ldots, X_m\}$ be given, such that $\sum_J \alpha_j X_j$ is constant if, and only if, the α_j's are zero, or, equivalently, such that $X_0 = 1, X_1, \ldots, X_m$ are affinely independent. The condition implies, necessarily, the linear independence of $\mathcal{B}$. A common choice is to take a set of linearly independent and μ-centered random variables.

We write $\mathcal{V} = \operatorname{Span}\{X_1, \ldots, X_m\}$ and define the following exponential family of positive densities $p \in \mathcal{P}_>$:

$$\mathcal{E}_{\mathcal{V}} = \left\{ q \in \mathcal{P}_> \middle| q \propto \mathrm{e}^V p, V \in \mathcal{V} \right\}. \tag{3}$$

Given any couple $p, q \in \mathcal{E}_{\mathcal{V}}$, then there exist a unique set of parameters $\boldsymbol{\theta} = \boldsymbol{\theta}_p(q)$, such that:

$$q = \exp\left(\sum_j \theta^j \, {}^e\mathbb{U}^p X_j - \psi_p(\boldsymbol{\theta}) \right) \cdot p \tag{4}$$

where ${}^{e}\mathbb{U}^{p}$ is the centering at p, that is,

$$ {}^{e}\mathbb{U}^{p} \colon \mathcal{V} \ni U \mapsto U - \mathbb{E}_p\left[U\right] \in {}^{e}\mathbb{U}^{p}\mathcal{V}. \tag{5}$$

The linear mapping ${}^{e}\mathbb{U}^{p}$ is one-to-one on $\mathcal{V}$ and ${}^{e}\mathbb{U}^{p}X_j$, $j = 1, \ldots, m$, and is a basis of ${}^{e}\mathbb{U}^{p}\mathcal{V}$. We view each choice of a specific reference p as providing a chart centered at p on the exponential family $\mathcal{E}_{\mathcal{V}}$, namely:

$$\sigma_p \colon \exp\left(\sum_j \theta^j\, {}^{e}\mathbb{U}^{p}X_j - \psi_p(\boldsymbol{\theta})\right)\cdot p \mapsto \boldsymbol{\theta}, \tag{6}$$

If:

$$U = {}^{e}\mathbb{U}^{p}U + \mathbb{E}_p\left[U\right] = \sum_{j=1}^{m} \theta^j\, {}^{e}\mathbb{U}^{p}X_j + \mathbb{E}_p\left[U\right], \tag{7}$$

then:

$$\mathbb{E}_p\left[U\, {}^{e}\mathbb{U}^{p}X_i\right] = \sum_{j=1}^{m} \theta^j\, \mathbb{E}_p\left[{}^{e}\mathbb{U}^{p}X_i\, {}^{e}\mathbb{U}^{p}X_j\right], \tag{8}$$

so that $\boldsymbol{\theta} = I_{\mathcal{B}}^{-1}(p)\,\mathbb{E}_p\left[U\, {}^{e}\mathbb{U}^{p}\boldsymbol{X}\right]$, where:

$$I_{\mathcal{B}}(p) = \left[\mathrm{Cov}_p\left(X_i, X_j\right)\right]_{ij} = \mathbb{E}_p\left[\boldsymbol{X}\boldsymbol{X}'\right] - \mathbb{E}_p\left[\boldsymbol{X}\right]\mathbb{E}_p\left[\boldsymbol{X}'\right] \tag{9}$$

is the Fisher information matrix of the basis $\mathcal{B} = \{X_1, \ldots, X_m\}$.

The mappings:

$$\sigma_p \colon \mathcal{E}_{\mathcal{V}} \ni q \mapsto U \mapsto \boldsymbol{\theta} \in \mathbb{R}^m \tag{10}$$

where:

$$s_p \colon q \mapsto U = \log\left(\frac{q}{p}\right) - \mathbb{E}_p\left[\log\left(\frac{q}{p}\right)\right], \tag{11}$$

$$\sigma_p \colon q \mapsto \boldsymbol{\theta} = I_{\mathcal{B}}^{-1}(p)\,\mathbb{E}_p\left[U\, {}^{e}\mathbb{U}^{p}\boldsymbol{X}\right] = I_{\mathcal{B}}^{-1}(p)\,\mathbb{E}_p\left[\log\left(\frac{q}{p}\right) {}^{e}\mathbb{U}^{p}\boldsymbol{X}\right], \tag{12}$$

are global charts in the non-parametric and parametric coordinates, respectively. Notice that Equation (12) provides the regression coefficients of the least squares estimate on ${}^{e}\mathbb{U}^{p}\mathcal{V}$ of the log-likelihood.

We denote by $e_p \colon \mathbb{R}^m \to \mathcal{E}_{\mathcal{V}}$ the inverse of σ_p, *i.e.*,

$$e_p(\boldsymbol{\theta}) = \exp\left(\sum_{j=1}^{m} \theta^j\, {}^{e}\mathbb{U}^{p}X_j - \psi_p(\boldsymbol{\theta})\right)\cdot p, \tag{13}$$

so that the representation of the divergence $q \mapsto \mathrm{D}\left(p\,\|q\right)$ in the chart σ_p is ψ_p:

$$\psi_p(\boldsymbol{\theta}) = \log\left(\mathbb{E}_p\left[\mathrm{e}^{\sum_{j=1}^{m}\theta^j\, {}^{e}\mathbb{U}^{p}X_j}\right]\right) = \mathbb{E}_{\boldsymbol{\theta}}\left[\log\left(\frac{p}{e_p(\boldsymbol{\theta})}\right)\right] = \mathrm{D}\left(p\,\|e_p(\boldsymbol{\theta})\right). \tag{14}$$

The mapping $I_{\mathcal{B}} \colon p \mapsto \mathrm{Cov}_p\left(\boldsymbol{X}, \boldsymbol{X}\right) \in \mathbb{R}^{m\times m}$ is represented in the chart centered at p by:

$$I_{\mathcal{B},p}(\boldsymbol{\theta}) = I_{\mathcal{B}}(e_p(\boldsymbol{\theta})) = [\mathrm{Cov}_{e_p(\boldsymbol{\theta})}\left(X_i, X_j\right)]_{i,j} = \mathrm{Hess}\,\psi_p(\boldsymbol{\theta}), \tag{15}$$

See [1].

2.2. Change of Chart

Fix $p, \bar{p} \in \mathcal{E}_\mathcal{V}$; then, we can express p in the chart centered at $\bar{p}$,

$$p = \exp\left(\bar{U} - k_p(\bar{U})\right) \cdot \bar{p}, \quad \bar{U} \in {}^e\mathbb{U}^{\bar{p}}\mathcal{V}, \quad k_{\bar{p}}(\bar{U}) = \log\left(\mathbb{E}_{\bar{p}}\left[\mathrm{e}^{\bar{U}}\right]\right). \tag{16}$$

In coordinates $\bar{U} = \sum_{j=1}^{m}, \bar{\theta}^j\, {}^e\mathbb{U}^{\bar{p}}X_j$.

For all $q \in \mathcal{E}_\mathcal{V}$, $q = \exp\left(U - k_p(U)\right)p$, $U \in {}^e\mathbb{U}^p\mathcal{V}$, $k_p(U) = \log\left(\mathbb{E}_p\left[\mathrm{e}^U\right]\right)$, in coordinates $U = \sum_{j=1}^{m}, \theta^j\, {}^e\mathbb{U}^p X_j$, we can write:

$$\begin{aligned} q &= \exp\left(U - k_p(U)\right) \cdot p \\ &= \exp\left(U - k_p(U)\right) \exp\left(\bar{U} - k_{\bar{p}}(\bar{U})\right) \cdot \bar{p} \\ &= \exp\left(U - k_p(U) + \bar{U} - k_{\bar{p}}(\bar{U})\right) \cdot \bar{p} \\ &= \exp\left(\left((U + \bar{U}) - \mathbb{E}_{\bar{p}}[U]\right) - \left(k_p(U) - k_{\bar{p}}(\bar{U}) + \mathbb{E}_{\bar{p}}[U]\right)\right) \cdot \bar{p}, \end{aligned} \tag{17}$$

hence, the non-parametric coordinate of q in the chart centered at $\bar{p}$ is $U + \bar{U} - \mathbb{E}_{\bar{p}}[U] = {}^e\mathbb{U}^{\bar{p}}(U) + \bar{U}$.

From Equation (12):

$$\begin{aligned} \sigma_{\bar{p}}(q) &= I_\mathcal{V}^{-1}(\bar{p})\, \mathbb{E}_{\bar{p}}\left[\left({}^e\mathbb{U}^{\bar{p}}U + \bar{U}\right) {}^e\mathbb{U}^{\bar{p}}\boldsymbol{X}\right] \\ &= \boldsymbol{\theta} + \bar{\boldsymbol{\theta}} \end{aligned} \tag{18}$$

This provides the change of charts $\sigma_{\bar{p}} \circ \sigma_p^{-1} \colon \boldsymbol{\theta} \mapsto \boldsymbol{\theta} + \bar{\boldsymbol{\theta}}$. This atlas of charts defines the affine manifold $(\mathcal{E}_\mathcal{V}, (\sigma_p))$. This fact has deep consequences that we do not discuss here, e.g., our manifold is an instance of a Hessian manifold [16].

2.3. Tangent Bundle

The space of Fisher scores at p is ${}^e\mathbb{U}^p\mathcal{V}$, and it is identified with the tangent space of the manifold at p, $T_q\mathcal{E}_\mathcal{V}$; see the discussion in [3,9]. Let us check the consistency of this statement with our θ-parametrization.

Let:

$$q(\tau) = \exp\left(\sum_{j=1}^{m} \theta^j(\tau)\, {}^e\mathbb{U}^{q(0)}\boldsymbol{X} - \psi_{q(0)}(\tau)\right) \cdot q(0), \tag{19}$$

$\tau \in I$, I an open interval containing zero, a curve in $\mathcal{E}_\mathcal{V}$. In the chart centered at $q(0)$, we have from Equation (12):

$$\begin{aligned} \sigma_{q(0)}(q(\tau)) &= I_\mathcal{B}^{-1}(q(0))\, \mathbb{E}_{q(0)}\left[\log\left(\frac{q(\tau)}{q(0)}\right) {}^e\mathbb{U}^{q(0)}\boldsymbol{X}\right] \\ &= I_\mathcal{B}^{-1}(q(0))\, \mathbb{E}_{q(0)}\left[\left(\sum_{j=1}^{m} \theta^j(\tau)\, {}^e\mathbb{U}^{q(0)}X_j - \psi_{q(0)}(\boldsymbol{\theta}(\tau))\right) {}^e\mathbb{U}^{q(0)}\boldsymbol{X}\right] \\ &= I_\mathcal{B}^{-1}(q(0)) \sum_{j=1}^{m} \theta^j(\tau)\, \mathbb{E}_{q(0)}\left[{}^e\mathbb{U}^{q(0)}X_j\, {}^e\mathbb{U}^{q(0)}\boldsymbol{X}\right] \\ &= I_\mathcal{B}^{-1}(q(0))\, \mathbb{E}_{q(0)}\left[{}^e\mathbb{U}^{q(0)}\boldsymbol{X}\, {}^e\mathbb{U}^{q(0)}\boldsymbol{X}\right] \boldsymbol{\theta} \\ &= \boldsymbol{\theta}(\tau). \end{aligned} \tag{20}$$

The vector space ${}^e\mathbb{U}^p\mathcal{V}$ is represented by the coordinates in the base ${}^e\mathbb{U}^p\mathcal{B}$. The tangent bundle $T\mathcal{E}_\mathcal{V}$ as a manifold is defined by the charts $(\sigma_p, \dot{\sigma}_p)$ on the domain:

$$T\mathcal{E}_\mathcal{V} = \left\{(p, v) \middle| p \in \mathcal{E}_\mathcal{V}, v \in T_p\mathcal{E}_\mathcal{V}\right\} \tag{21}$$

with:

$$(\sigma_p, \dot\sigma_p) \colon (q, V) \mapsto \left(I_{\mathcal{B}}^{-1}(p)\, \mathbb{E}_p \left[\log\left(\frac{q}{p}\right) {}^e\mathbb{U}^p \boldsymbol{X} \right], I_{\mathcal{B}}^{-1}(p)\, \mathbb{E}_p \left[V\, {}^e\mathbb{U}^p \boldsymbol{X} \right] \right). \tag{22}$$

The dot notation $\dot\sigma_p$ for the charts on the tangent spaces is justified by the computation in Equation (23) below:

$$\begin{aligned} \frac{d}{dt}\sigma_{q(0)}(q(\tau))\Big|_{\tau=0} = I_{\mathcal{B}}^{-1}(q(0))\, \mathbb{E}_{q(0)} \left[\frac{d}{d\tau} \log(q(\tau))\Big|_{\tau=0} {}^e\mathbb{U}^{q(0)} \boldsymbol{X} \right] = \\ I_{\mathcal{B}}^{-1}(q(0))\, \mathbb{E}_{q(0)} \left[\delta q(0)\, {}^e\mathbb{U}^{q(0)} \boldsymbol{X} \right] = \dot\sigma_{q(0)}(\delta q(0)). \end{aligned} \tag{23}$$

The velocity at $\tau = 0$ is $\delta q(0) = \frac{d}{d\tau} \log(q(\tau))\big|_{\tau=0} \in T_{q(0)}\mathcal{E}_{\mathcal{V}}$ and:

$$\begin{aligned} \frac{d}{d\tau}\boldsymbol{\theta}(\tau)\Big|_{\tau=0} &= I_{\mathcal{B}}^{-1}(q(0))\, \mathbb{E}_{q(0)} \left[\frac{d}{d\tau} \log(q(\tau))\Big|_{\tau=0} {}^e\mathbb{U}^{q(0)} \boldsymbol{X} \right] \\ &= I_{\mathcal{B}}^{-1}(q(0))\, \mathbb{E}_{q(0)} \left[\delta q(0)\, {}^e\mathbb{U}^{q(0)} \boldsymbol{X} \right], \end{aligned} \tag{24}$$

which is consistent with both the definition of tangent space as set of Fisher scores and with the chart of the tangent bundle as defined in Equation (22).

The velocity at a generic τ is $\delta q(\tau) = \frac{d}{d\tau} \log(q(\tau)) \in T_{q(\tau)}\mathcal{E}_{\mathcal{V}}$ and has coordinates at p:

$$\begin{aligned} \frac{d}{d\tau}\boldsymbol{\theta}(\tau) &= I_{\mathcal{B}}^{-1}(q(0))\, \mathbb{E}_{q(0)} \left[\frac{d}{d\tau} \log(q(\tau))\, {}^e\mathbb{U}^{q(0)} \boldsymbol{X} \right] \\ &= I_{\mathcal{B}}^{-1}(q(0))\, \mathbb{E}_{q(0)} \left[\delta q(\tau)\, {}^e\mathbb{U}^{q(0)} \boldsymbol{X} \right]. \end{aligned} \tag{25}$$

If V, W are vector fields on $T\mathcal{E}_{\mathcal{V}}$, *i.e.*, $V(p), W(p) \in T_p\mathcal{E}_{\mathcal{V}} = {}^e\mathbb{U}^p\mathcal{V}$, $p \in \mathcal{E}_{\mathcal{V}}$, we define a Riemannian metric $g(V, W))$ by:

$$g(V, W)(p) = g_p(V(p), W(p)) = \mathbb{E}_p\left[V(p)W(p)\right] \tag{26}$$

In coordinates at p, $V(p) = \sum_j \dot\sigma_p^j(V)\, {}^e\mathbb{U}^p X_j$, $W(p) = \sum_j \dot\sigma_p^j(W)\, {}^e\mathbb{U}^p X_j$, so that:

$$g_p(V(p), W(p)) = \dot\sigma_p(V)' I_{\mathcal{B}}(p) \dot\sigma_p(W). \tag{27}$$

2.4. Gradients

Given a function $\phi \colon \mathcal{E}_{\mathcal{V}} \to \mathbb{R}$ let $\phi_p = \phi \circ e_p$, $e_p = \sigma_p^{-1}$, its representation in the chart centered at p:

$$\begin{array}{ccc} \mathcal{E}_{\mathcal{V}} & \xrightarrow{\ \phi\ } & \mathbb{R} \\ {\scriptstyle e_p}\uparrow & \nearrow{\scriptstyle \phi_p} & \\ \mathbb{R}^m & & \end{array} \tag{28}$$

The derivative of $\boldsymbol{\theta} \mapsto \phi_p(\boldsymbol{\theta})$ at $\boldsymbol{\theta} = \boldsymbol{0}$ along $\boldsymbol{\alpha} \in \mathbb{R}^m$ is:

$$\nabla\phi_p(\boldsymbol{0})\boldsymbol{\alpha} = \nabla\phi_p(\boldsymbol{0}) I_{\mathcal{B}}^{-1}(p) I_{\mathcal{B}}(p)\boldsymbol{\alpha} = \left(I_{\mathcal{B}}^{-1}(p) \nabla\phi_p(\boldsymbol{0})' \right)' I_{\mathcal{B}}(p)\boldsymbol{\alpha} = g_p(I_{\mathcal{B}}^{-1}(p)\nabla\phi_p(\boldsymbol{0})', \boldsymbol{\alpha}). \tag{29}$$

The mapping $\widetilde\nabla\phi \colon p \mapsto I_{\mathcal{B}}^{-1}(p)(\nabla\phi_p(\boldsymbol{0}))' \in \mathbb{R}^m$ that appears in Equation (29) is Amari's natural gradient of $\phi \colon \mathcal{E}_{\mathcal{V}}$; see [15]. It is a standard notion in Riemannian geometry; *cf.* [4] (p. 46).

More generally, the derivative of $\theta \mapsto \phi_p(\theta)$ at θ along $\alpha \in \mathbb{R}^m$ is:

$$\nabla\phi_p(\theta)\alpha = \nabla\phi_p(\theta)I_B^{-1}(e_p(\theta))I_B(e_p(\theta))\alpha =$$
$$\left(I_B^{-1}(e_p(\theta))\nabla\phi_p(\theta)'\right)' I_B(e_p(\theta))\alpha = g_{e_p(\theta)}(I_B^{-1}(e_p(\theta))\nabla\phi_p(\theta)', \alpha). \quad (30)$$

Let us compare $\nabla\phi_q(0)$ and $\nabla\phi_p(\theta)$ when $q = e_p(\theta)$. As $\phi_p = \phi \circ e_p$ and $\phi_q = \phi \circ e_q$, we have the change of charts:

$$\phi_q = \phi \circ e_q = \phi \circ e_p \circ \sigma_p \circ e_q = \phi_p \circ \sigma_p \circ e_q, \quad (31)$$

hence $\nabla\phi_q(0) = \nabla\phi_p(\sigma_p(q))J(\sigma_p \circ e_q)(0)$, where $J(\sigma_p \circ e_q)$ is the Jacobian of $\sigma_p \circ e_q$. As $\sigma_p \circ e_q(\theta) = \theta + \sigma_p(q)$, we have $J(\sigma_p \circ e_q) = \mathrm{Id}$, and in conclusion, $\nabla\phi_{e_p(\theta)}(0) = \nabla\phi_p(\theta)$. For all $p \in \mathcal{E}_V$ and $\theta \in \mathbb{R}^m$,

$$\widetilde{\nabla}\phi(e_p(\theta)) = I_B^{-1}(e_p(\theta))\nabla\phi_p(\theta). \quad (32)$$

Alternatively, for all $q, p \in \mathcal{E}_V$, $\widetilde{\nabla}\phi: \mathcal{E}_V \to \mathbb{R}^m$ is defined by:

$$\widetilde{\nabla}\phi(q) = I_B^{-1}(q)\nabla\phi_p(\sigma_p(q)). \quad (33)$$

The Riemannian gradient of $\phi: \mathcal{E}_V$ is the vector field $\nabla\phi$, such that $D_Y\phi = g(\nabla\phi, Y)$. Note that the Riemannian gradient takes values in the tangent bundle, while the natural gradient takes values in $\mathbb{R}^m$. We compute the Riemannian gradient at p as follows. If $y = \dot{\sigma}_p(Y(p))$,

$$D_Y\phi(p) = d\phi_p(0)y = g_p(\widetilde{\nabla}\phi(p), y) = \mathbb{E}_p\left[\nabla\phi(p)Y(p)\right], \quad (34)$$

hence $\widetilde{\nabla}\phi(p) = I_B^{-1}(p)\nabla\phi_p(0)'$ is the representation in the chart centered at p of the vector field $\nabla\phi: \mathcal{E}_V$. Explicitly, we have (see Equation (22)),

$$\widetilde{\nabla}\phi(p) = I_B^{-1}(p)(\nabla\phi_p(0))' = I_B^{-1}(p)\,\mathbb{E}_p\left[\nabla\phi(p)\,{}^e\mathbb{U}^p X\right], \quad (35)$$
$$\nabla\phi(p) = \sum_j (\widetilde{\nabla}\phi(p))^j\,{}^e\mathbb{U}^p X_j \quad (36)$$

The Euclidean gradient $\nabla\phi_p(\theta)$ is sometimes called the "vanilla gradient." It is equal to the covariance between the Riemannian gradient $\nabla\phi(p)$ and the basis X, $(\nabla\phi_p(0))' = \mathbb{E}_p\left[\nabla\phi(p)\,{}^e\mathbb{U}^p X\right]$.

We summarize in a display the relations between our three gradients: Euclidean $\nabla\phi_p(0)$, natural $\widetilde{\nabla}\phi(p)$ and Riemannian $\nabla\phi(p)$.

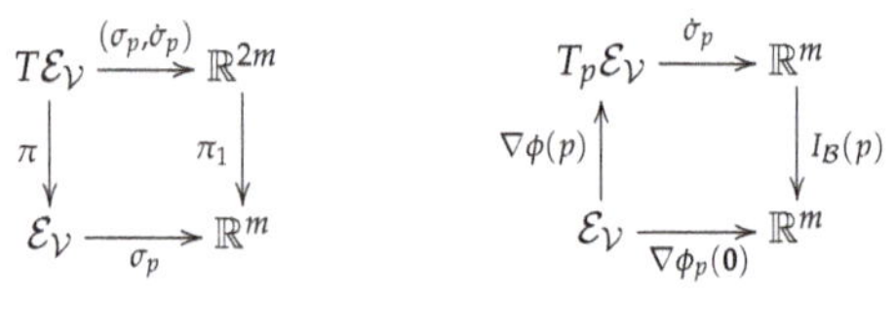

$$
\begin{array}{ccc}
T\mathcal{E}_V \xrightarrow{(\sigma_p, \dot{\sigma}_p)} \mathbb{R}^{2m} & \qquad T_p\mathcal{E}_V \xrightarrow{\dot{\sigma}_p} \mathbb{R}^m \\
\downarrow^{\pi} \qquad \downarrow^{\pi_1} & \qquad \uparrow^{\nabla\phi(p)} \qquad \downarrow^{I_B(p)} & \dot{\sigma}_p \circ \nabla\phi(p) = I_B^{-1}\nabla\phi_p(0) = \widetilde{\nabla}\phi(p) \\
\mathcal{E}_V \xrightarrow{\sigma_p} \mathbb{R}^m & \qquad \mathcal{E}_V \xrightarrow{\nabla\phi_p(0)} \mathbb{R}^m
\end{array}
$$

$$(37)$$

In the following, we shall frequently use the fact that the representation of the gradient vector field $\nabla\phi$ in a generic chart centered at p is:

$$(\nabla\phi)_p(\theta) = \dot{\sigma}_p(\nabla\phi(e_p(\theta))) = (\widetilde{\nabla}\phi)(e_p(\theta)) = I_{B,p}^{-1}(\theta)\nabla\phi_p(\theta). \quad (38)$$

It should be noted that the leftmost term $(\nabla\phi)_p(\theta)$ is the presentation of the gradient in the charts of the tangent bundle, while in the rightmost term, $\nabla\phi_p(\theta)$ denotes the Euclidean gradient of the presentation of the function ϕ in the charts of the manifold.

2.4.1. Expectation Parameters

As ψ_p is strictly convex, the gradient mapping $\boldsymbol{\theta} \mapsto (\nabla\psi_p(\boldsymbol{\theta}))'$ is a homeomorphism from the space of parameters $\mathbb{R}^m$ to the interior of the convex set generated by the image of ${}^e\mathbb{U}^p\boldsymbol{X}$; see [1]. The function $\mu_p\colon \mathcal{E}_\mathcal{V}$ defined by:

$$\mu_p(q) = \mathbb{E}_q\left[{}^e\mathbb{U}^p\boldsymbol{X}\right] = \mathbb{E}_q\left[\boldsymbol{X}\right] - \mathbb{E}_p\left[\boldsymbol{X}\right] = (\nabla\psi_p(\boldsymbol{\theta}))', \quad \boldsymbol{\theta} = \sigma_p(q) \tag{39}$$

is a chart for all $p \in \mathcal{E}_\mathcal{V}$. The value of the inverse $q = L_p(\boldsymbol{\mu})$ is characterized as the unique $q \in \mathcal{E}_\mathcal{V}$, such that $\boldsymbol{\mu} = \mathbb{E}_q\left[{}^e\mathbb{U}^p\boldsymbol{X}\right]$, *i.e.*, the maximum likelihood estimator.

Let us compute the change of chart from p to $\bar{p}$:

$$\mu_{\bar{p}} \circ \mu_p^{-1}(\boldsymbol{\eta}) = \bar{\boldsymbol{\eta}} = \boldsymbol{\eta} + \mathbb{E}_p\left[\boldsymbol{X}\right] - \mathbb{E}_{\bar{p}}\left[\boldsymbol{X}\right]. \tag{40}$$

In fact, $\boldsymbol{\mu} = \mathbb{E}_{L_p(\boldsymbol{\mu})}\left[{}^e\mathbb{U}^p\boldsymbol{X}\right]$ and $\bar{\boldsymbol{\mu}} = \mu_{\bar{p}}(L_p(\boldsymbol{\mu})) = \mathbb{E}_{L_p(\boldsymbol{\mu})}\left[{}^e\mathbb{U}^{\bar{p}}\boldsymbol{X}\right]$.

We do not discuss here the rich theory started in [2] about the duality between σ_p and μ_p. We limit ourselves to the computation of the Riemannian gradient in the expectation parameters. If $\phi\colon \mathcal{E}_\mathcal{V}$,

$$\phi_p(\boldsymbol{\theta}) = \phi \circ e_p(\boldsymbol{\theta}) = \phi \circ L_p \circ \mu_p \circ e_p(\boldsymbol{\theta}) = (\phi \circ L_p) \circ (\nabla\psi_p)(\boldsymbol{\theta}), \tag{41}$$

because $\mu_p \circ e_p(\boldsymbol{\theta}) = \mathbb{E}_{e_p(\boldsymbol{\theta})}\left[{}^e\mathbb{U}^p\boldsymbol{X}\right] = \nabla\phi_p(\boldsymbol{\theta})$, hence:

$$\nabla\phi_p(\boldsymbol{\theta}) = \nabla(\phi \circ L_p)(\nabla\psi_p(\boldsymbol{\theta}))\,\mathrm{Hess}\,\psi_p(\boldsymbol{\theta}), \tag{42}$$

$$\widetilde{\nabla}\phi(p) = I_\mathcal{V}(p)^{-1}(\nabla(\phi \circ L_p)(\mathbf{0})\,\mathrm{Hess}\,\psi_p(\mathbf{0}))' = (\nabla(\phi \circ L_p)(\mathbf{0}))', \tag{43}$$

$$\nabla\phi(p) = \nabla(\phi \circ L_p)(\mathbf{0})\,{}^e\mathbb{U}^p\boldsymbol{X}, \tag{44}$$

that is, the natural gradient $\widetilde{\nabla}\phi$ at $p = L_p(\boldsymbol{\mu})$ is equal to the Euclidean gradient of $\boldsymbol{\mu} \mapsto \phi \circ L_p(\boldsymbol{\mu})$ at $\boldsymbol{\mu} = \mathbf{0}$.

2.4.2. Vector Fields

If V is a vector field of $T\mathcal{E}_\mathcal{V}$ and $\phi\colon \mathcal{E}_\mathcal{V}$ is a real function, then we define the action of V on ϕ, $\nabla_V\phi$, to be the real function:

$$\nabla_V\phi\colon \mathcal{E}_\mathcal{V} \ni p \mapsto \nabla_V\phi(p) = \nabla\phi_p(\mathbf{0})\dot{\sigma}_p\left(V(p)\right). \tag{45}$$

We prefer to avoid the standard notation $V\phi$, because in our setting, $V(p)$ is a random variable, and the product $V(p)\phi(p)$ is otherwise defined as the ordinary product.

Let us represent $\nabla_V\phi$ in the chart centered at p:

$$(\nabla_V\phi)_p(\boldsymbol{\theta}) = \nabla_V\phi(e_p(\boldsymbol{\theta})) = \nabla\phi_{e_p(\boldsymbol{\theta})}(\mathbf{0})\dot{\sigma}_{e_p(\boldsymbol{\theta})}\left(V(e_p(\boldsymbol{\theta}))\right) = \nabla\phi_p(\boldsymbol{\theta})V_p(\boldsymbol{\theta}), \tag{46}$$

where we have used the equality $\nabla\phi_{e_p(\boldsymbol{\theta})}(\mathbf{0}) = \nabla\phi_p(\boldsymbol{\theta})$ and $V_p(\boldsymbol{\theta}) = \dot{\sigma}_{e_p(\boldsymbol{\theta})}\left(V(e_p(\boldsymbol{\theta}))\right)$.

If W is a vector field, we can compute $\nabla_W\nabla_V\phi$ at p as:

$$\begin{aligned}\nabla_W\nabla_V\phi(p) &= \nabla(\nabla_V\phi)_p(\mathbf{0})\dot{\sigma}_p(W(p)) \\ &= V_p(\mathbf{0})'\,\mathrm{Hess}\,\phi_p(\mathbf{0})W_p(\mathbf{0}) + \nabla\phi_p(\mathbf{0})JV_p(\mathbf{0})W_p(\mathbf{0}),\end{aligned} \tag{47}$$

where J denotes the Jacobian matrix.

The Lie bracket $[W, V]\phi$ (see [7] (§4.2), [8] (V, §1), [4] (Section 5.3.1)) is given by:

$$[W, V]\phi(p) = \nabla_W\nabla_V\phi(p) - \nabla_V\nabla_w\phi(p) = \nabla\phi_p(\mathbf{0})\left(JV_p(\mathbf{0})W_p(\mathbf{0}) - JW_p(\mathbf{0})V_p(\mathbf{0})\right), \tag{48}$$

because of Equation (47) and the symmetry of the Hessian.

The flow of the smooth vector field $V\colon \mathcal{E}_\mathcal{V}$ is a family of curves $\gamma(t,p)$, $p \in \mathcal{E}_\mathcal{V}$, $t \in J_p$, J_p open real interval containing zero, such that for all $p \in \mathcal{E}_\mathcal{V}$ and $t \in J_p$,

$$\gamma(0,p) = p, \tag{49}$$

$$\delta\gamma(t,p) = V(\gamma(t,p)). \tag{50}$$

As uniqueness holds in Equation (50) (see [8] (VI, §1) or [7] (§4.1)), we have semi-group property $\gamma(s+t,p) = \gamma(s,\gamma(t,p))$, and Equation (50) is equivalent to $\delta\gamma(0,p) = V(\gamma(0,p))$, $p \in \mathcal{E}_\mathcal{V}$.

If a flow of V is available, we have an interpretation of $\nabla_V\phi$ as a derivative of ϕ along $\gamma(t,p)$,

$$\left.\frac{d}{dt}\phi(\gamma(t,p))\right|_{t=0} = \nabla\phi_p(\sigma_p(\gamma(t,p)))\left.\left(\frac{d}{dt}\sigma_p(\gamma(t,p))\right)\right|_{t=0} = \nabla\phi_p(\mathbf{0})V(p) = \nabla_V\phi(p). \tag{51}$$

2.5. Examples

The following examples are intended to show how the formalism of gradients is usable in performing basic computations.

2.5.1. Expectation

Let f be any random variable, and define $F\colon \mathcal{E}_\mathcal{V}$ by $F(p) = \mathbb{E}_p[f]$. In the chart centered at p, we have:

$$F_p(\boldsymbol{\theta}) = \int f \exp\left(\sum_j \theta^j\, {}^e\mathbb{U}^p X_j - \psi_p(\boldsymbol{\theta})\right) \cdot p\, d\mu \tag{52}$$

and the Euclidean gradient:

$$\nabla F_p(\mathbf{0}) = \operatorname{Cov}_p(f, \boldsymbol{X}) \in (\mathbb{R}^m)'. \tag{53}$$

The natural gradient is:

$$\tilde{\nabla} F(p) = \operatorname{Cov}_p(\boldsymbol{X},\boldsymbol{X})^{-1} \operatorname{Cov}_p(\boldsymbol{X}, f) \in \mathbb{R}^m, \tag{54}$$

and the Riemannian gradient is:

$$\nabla F(p) = (\tilde{\nabla} F(p))'\, {}^e\mathbb{U}^p \boldsymbol{X} = \operatorname{Cov}_p(f,\boldsymbol{X}) \operatorname{Cov}_p(\boldsymbol{X},\boldsymbol{X})^{-1}\, {}^e\mathbb{U}^p\boldsymbol{X} \in T_p\mathcal{E}_\mathcal{V}. \tag{55}$$

From Equation (55), it follows that $\nabla F(p)$ is the $L^2(p)$-projection f onto ${}^e\mathbb{U}^p\mathcal{V}$, while $\tilde{\nabla} F(p)$ in Equation (54) are the coordinates of the projection. Let us consider the family of curves:

$$\gamma(t,p) = \exp\left(\sum_{j=1}^m t(\tilde{\nabla} F(p))^j\, {}^e\mathbb{U}^p X_j - \psi_p(t\tilde{\nabla} F(p))\right) \cdot p, \quad t \in \mathbb{R}. \tag{56}$$

The velocity is:

$$\delta\gamma(t,p) = \frac{d}{dt}\left(\sum_{j=1}^m t(\tilde{\nabla} F(p))^j\, {}^e\mathbb{U}^p X_j - \psi_p(t\tilde{\nabla} F(p))\right) = \nabla F(p) - \mathbb{E}_{\gamma(t,p)}[\nabla F(p)], \tag{57}$$

which is different from $\nabla F(\gamma(t,p))$, unless $f \in \mathcal{V} \oplus \mathbb{R}$. Then, γ is not, in general, the flow of ∇F, but it is a local approximation, as $\delta\gamma(0,p) = \nabla F(p)$.

These computation are the basis of model-based methods in combinatorial optimization; see [10–14].

2.5.2. Binary Independent Variables

Here, we present, in full generality, the toy example of the Introduction; see [17] for more information on the application to combinatorial optimization. Our example is a very special case of Ising exactly solvable models [18], our aim being here to explore the geometric framework.

Let $\Omega = \{+1, -1\}^m$ with counting measure μ, and let the space $\mathcal{V}$ be generated by the coordinate projections $\mathcal{B} = \{X_1, \dots, X_d\}$. Note that we use here the coding $+1, -1$ (from physics) instead of the coding $0, 1$, which is more common in combinatorial optimization. The exponential family is $\mathcal{E}_\mathcal{V} = \left\{\exp\left(\sum_{j=1}^m \theta^j X_j - \psi_\lambda(\boldsymbol{\theta})\right) \cdot 2^{-m}\right\}$, $\lambda(x) = 2^{-m}$ for $x \in \Omega$ being the uniform density. The independence of the sufficient statistics X_j under all distributions in $\mathcal{E}_\mathcal{V}$ implies:

$$\psi_\lambda(\boldsymbol{\theta}) = \sum_{j=1}^m \psi(\theta^j), \quad \psi(\theta) = \log\left(\cosh(\theta)\right). \tag{58}$$

We have:

$$\begin{aligned}
\nabla \psi_\lambda(\boldsymbol{\theta}) &= [\tanh(\theta^j) \colon j = 1, \dots, d] \\
&= \eta_\lambda(\boldsymbol{\theta}), && (59)\\
\operatorname{Hess} \psi_\lambda(\boldsymbol{\theta}) &= \operatorname{diag}\left(\cosh^{-2}(\theta^j) \colon j = 1, \dots, d\right) \\
&= \operatorname{diag}\left(\mathrm{e}^{-2\psi(\theta^j)} \colon j = 1, \dots, d\right) \\
&= I_{\mathcal{B},\lambda}(\boldsymbol{\theta}), && (60)\\
I_{\mathcal{B},\lambda}(\boldsymbol{\theta})^{-1} &= \operatorname{diag}\left(\cosh^2(\theta^j) \colon j = 1, \dots, d\right) \\
&= \operatorname{diag}\left(\mathrm{e}^{2\psi(\theta^j)} \colon j = 1, \dots, d\right). && (61)
\end{aligned}$$

The quadratic function $f(\boldsymbol{X}) = a_0 + \sum_j a_j X_j + \sum_{\{i,j\}} a_{i,j} X_i X_j$ has expected value at $p = e_\lambda(\boldsymbol{\theta})$, *i.e.*, relaxed value, equal to:

$$F(p) = F_\lambda(\boldsymbol{\theta}) = \mathbb{E}_{\boldsymbol{\theta}}\left[f(\boldsymbol{X})\right] = a_0 + \sum_j a_j \tanh(\theta^j) + \sum_{\{i,j\}} a_{i,j} \tanh(\theta^i)\tanh(\theta^j), \tag{62}$$

and covariance with $X_k \in \mathcal{B}$ equal to:

$$\begin{aligned}
\operatorname{Cov}_{\boldsymbol{\theta}}\left(f(\boldsymbol{X}), X_k\right) &= \sum_j a_j \operatorname{Cov}_{\boldsymbol{\theta}}\left(X_j, X_k\right) + \sum_{\{i,j\}} a_{i,j} \operatorname{Cov}_{\boldsymbol{\theta}}\left(X_i X_j, X_k\right) \\
&= a_k \operatorname{Var}_{\boldsymbol{\theta}}\left(X_k\right) + \sum_{i \neq k} a_{i,k} \mathbb{E}_{\boldsymbol{\theta}}\left[X_i\right] \operatorname{Var}_{\boldsymbol{\theta}}\left(X_k\right) \\
&= \cosh^{-2}(\theta^k)\left(a_k + \sum_{i \neq k} a_{i,k} \tanh(\theta^i)\right). && (63)
\end{aligned}$$

In the computation, we have used the independence and the special algebra of ± 1, which implies $X_i^2 = 1$, so that $\operatorname{Cov}_\theta\left(X_i X_j, X_k\right) = 0$ if $i, j \neq k$, otherwise $\operatorname{Cov}_\theta\left(X_i X_k, X_k\right) = \mathbb{E}_\theta\left[X_i\right] - \mathbb{E}_\theta\left[X_i\right]\mathbb{E}_\theta\left[X_k\right]^2$; see [13].

The Euclidean gradient, the natural gradient and the Riemannian gradient are, respectively,

$$\nabla F_\lambda(\boldsymbol{\theta}) = \left[\cosh^{-2}(\theta^j)\left(a_j + \sum_{i \neq j} a_{i,j}\tanh(\theta^i)\right) : j = 1,\ldots,d\right], \tag{64}$$

$$\widetilde{\nabla} F(e_\lambda(\boldsymbol{\theta})) = \left[a_j + \sum_{i \neq j} a_{i,j}\tanh(\theta^i) \colon j = 1,\ldots,d\right], \tag{65}$$

$$\nabla F(e_\lambda(\boldsymbol{\theta})) = \sum_{j=1}^{m}\left(a_j + \sum_{i \neq j} a_{i,j}\,\mathbb{E}_\theta\left[X_i\right]\right)\left(X_j - \mathbb{E}_\theta\left[X_j\right]\right). \tag{66}$$

The (natural) gradient flow equations are:

$$\dot{\theta}^j(t) = a_j + \sum_{i \neq j} a_{i,j}\tanh(\theta^i(t)), \quad j = 1,\ldots,d. \tag{67}$$

Equations (64)–(66) are usable in practice if the a_j's and the $a_{i,j}$'s are estimable. Otherwise, one can use Equation (63) and the following forms of the gradients:

$$\nabla F_\lambda(\boldsymbol{\theta}) = \left[\mathrm{Cov}_{\boldsymbol{\theta}}\left(X_j, f(\boldsymbol{X})\right) : j = 1,\ldots,d\right], \tag{68}$$

$$\widetilde{\nabla} F(e_\lambda(\boldsymbol{\theta})) = \left[\cosh^2(\theta^j)\,\mathrm{Cov}_{\boldsymbol{\theta}}\left(f(\boldsymbol{X}), X_j\right) : j = 1,\ldots,d\right], \tag{69}$$

in which case, the gradient flow equations are:

$$\dot{\theta}^j(t) = \cosh^2(\theta^j)\,\mathrm{Cov}_{\boldsymbol{\theta}}\left(f(\boldsymbol{X}), X_j\right), \quad j = 1,\ldots,d. \tag{70}$$

Let us study the relaxed function in the expectation parameters $\eta^j = \eta^j(\boldsymbol{\theta})$, $j = 1,\ldots,d$,

$$F_\lambda(\boldsymbol{\eta}) = a_0 + \sum_j a_j\eta^j + \sum_{\{i,j\}} a_{i,j}\eta^i\eta^j, \quad \boldsymbol{\eta} \in]-1,+1[^m. \tag{71}$$

The Euclidean gradient with respect to $\boldsymbol{\eta}$ has components:

$$\partial_j F_\lambda(\boldsymbol{\eta}) = a_j + \sum_{i \neq j} a_{i,j}\eta^i, \tag{72}$$

which are equal to the components of the natural gradient; see Section 2.4.1. As:

$$\dot{\eta}^j(t) = \frac{d}{dt}\tanh(\theta^j(t)) = \cosh^{-2}(\theta^j(t))\dot{\theta}^j(t) = \left(1 - \eta^j(t)^2\right)\dot{\theta}^j(t), \quad j = 1,\ldots,m, \tag{73}$$

the gradient flow expressed in the $\boldsymbol{\eta}$-parameters has equations:

$$\dot{\eta}^j(t) = \left(1 - \eta^j(t)^2\right)\left(a_j + \sum_{i \neq j} a_{i,j}\eta^i(t)\right), \quad j = 1,\ldots,d. \tag{74}$$

Alternatively, in vector form,

$$\dot{\boldsymbol{\eta}}(t) = \mathrm{diag}\left(1 - \eta^j(t)^2 \colon j = 1,\ldots,d\right)\left(\boldsymbol{a} + A\boldsymbol{\eta}(t)\right), \tag{75}$$

where $\boldsymbol{a} = [a_j \colon j = 1,\ldots,d]^t$ and $A_{i,j} = 0$ if $i = j$, $A_{ij} = a_{i,j}$. The matrix A is symmetric with zero diagonal, and it has the meaning of the adjacency matrix of the (weighted) interaction graph. We do not know a closed-form solution of Equation (74). An example of a numerical solution is shown in Figure 3.

2.5.3. Escort Probabilities

For a given $a > 0$, consider the function $C^{(a)} \colon \mathcal{E}_\mathcal{V}$ defined by $C^{(a)}(p) = \int p^a \, d\mu$. We have:

$$C_p^{(a)}(\boldsymbol{\theta}) = \int \exp\left(a \sum_{j=1}^m \theta^j \, {}^e\mathbb{U}^p X_j - a\psi_p(\boldsymbol{\theta})\right) p^a \, d\mu \tag{76}$$

and:

$$dC_p^{(a)}(\mathbf{0})\boldsymbol{\alpha} = \int a \left(\sum_{j=1}^m \alpha^j \, {}^e\mathbb{U}^p X_j\right) p^a \, d\mu = \sum_{j=1}^m \alpha^j \int a \, {}^e\mathbb{U}^p X_j p^a \, d\mu = \sum_{j=1}^m \alpha^j \operatorname{Cov}_p\left(X_j, ap^{a-1}\right), \tag{77}$$

that is, the Euclidean gradient is $\nabla C_p^{(a)}(\mathbf{0}) = \operatorname{Cov}_p\left(ap^{a-1}, \boldsymbol{X}\right)$ (row vector). The natural gradient is computed from Equation (35) as:

$$\widetilde{\nabla} C^{(a)}(p) = I_{\mathcal{B}}^{-1}(p)(\nabla C_p^{(a)}(\mathbf{0}))' = \operatorname{Cov}_p\left(\boldsymbol{X}, \boldsymbol{X}\right)^{-1} \operatorname{Cov}_p\left(\boldsymbol{X}, ap^{a-1}\right), \tag{78}$$

while the Riemannian gradient follows from Equation (36):

$$\nabla C^{(a)}(p) = \operatorname{Cov}_p\left(ap^{a-1}, \boldsymbol{X}\right) \operatorname{Cov}_p\left(\boldsymbol{X}, \boldsymbol{X}\right)^{-1} {}^e\mathbb{U}^p \boldsymbol{X}. \tag{79}$$

Note that the Riemannian gradient is the orthogonal projection of the random variable ap^{a-1} onto the tangent space $T_p\mathcal{E}_\mathcal{V} = {}^e\mathbb{U}^p\mathcal{V}$.

The probability density $p^a / C(p)$ is called the escort density in the literature on non-extensive statistical mechanics; see, e.g., [19] (Section 7.4).

We compute now the tangent mapping of $\mathcal{E}_\mathcal{V} \ni p \mapsto p^a / C^{(a)}(a) \in \mathcal{P}_>$. Let us extend the basis $X_1, \ldots, X_m$ to a basis $X_1, \ldots, X_n$, $n \geq m$, whose exponential family is full, *i.e.*, equal to $\mathcal{P}_>$. The non-parametric coordinate of $q = \left(\exp\left(\sum_{j=1}^m \theta^j \, {}^e\mathbb{U}^p X_j - \psi_p(\boldsymbol{\theta})\right) p\right)^a / C_p^{(a)}(\boldsymbol{\theta})$ in the chart centered at $\bar{p} = p^a / C_p^{(a)}(\mathbf{0})$ is the $\bar{p}$-centering of the random variable:

$$\begin{aligned} \log\left(\frac{q}{\bar{p}}\right) &= \log\left(\frac{\left(\exp\left(\sum_{j=1}^m \theta^j \, {}^e\mathbb{U}^p X_j - \psi_p(\boldsymbol{\theta})\right) p\right)^a / C_p^{(a)}(\boldsymbol{\theta})}{p^a / C_p^{(a)}(\mathbf{0})}\right) \\ &= a \sum_{j=1}^m \theta^j \, {}^e\mathbb{U}^p X_j - a\psi_p(\boldsymbol{\theta}) + \ln C_(^{(a)}\mathbf{0}) - \ln C_p^{(a)}(\boldsymbol{\theta}), \end{aligned} \tag{80}$$

that is,

$$v = a \sum_{j=1}^m \theta^j \, {}^e\mathbb{U}^{\bar{p}} X_j. \tag{81}$$

The coordinates of v in the basis ${}^e\mathbb{U}^{\bar{p}} X_1, \ldots, {}^e\mathbb{U}^{\bar{p}} X_n$ are $(a\theta^1, \ldots, a\theta^m, 0, \ldots, 0)$, and the Jacobian of $\boldsymbol{\theta} \mapsto (a\boldsymbol{\theta}, \mathbf{0}_{n-m})$ is the $m \times n$ matrix $\left[aI_m | 0_{m\times(n-m)}\right]$.

2.5.4. Polarization Measure

The polarization measure has been introduced in Economics by [20]. Here, we consider the qualitative version of [21]. If π is a distribution of a finite set, the probability that in three independent samples from π there are exactly two equal is $3\sum_j \pi_j^2(1-\pi_j)$. If $p \in \mathcal{E}_\mathcal{V}$, define:

$$G(p) = \int p^2(1-p)\,d\mu = C^{(2)}(p) - C^{(3)}(p), \tag{82}$$

where $C^{(2)}$ and $C^{(3)}$ are defined as in Example 2.5.3.

From Equation (78), we find the natural gradient:

$$\widetilde{\nabla} G(p) = \operatorname{Cov}_p(\boldsymbol{X},\boldsymbol{X})^{-1} \operatorname{Cov}_p\left(\boldsymbol{X}, 2p - 3p^2\right). \tag{83}$$

Note that $\widetilde{\nabla} G($

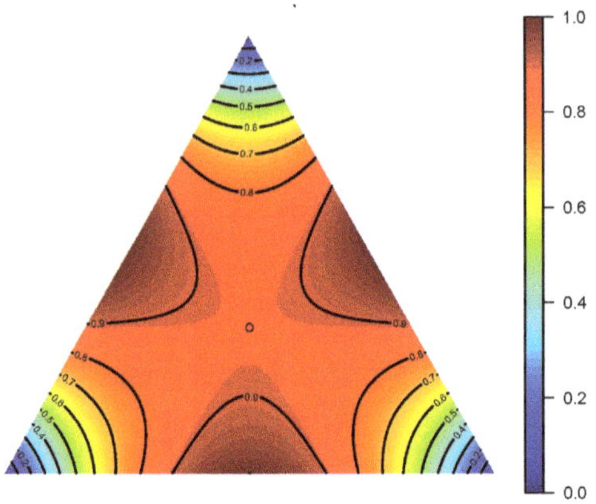

Figure 4. Normalized polarization.

3. Second Order Calculus

In this section, we turn to considering second order calculus, in particular Hessians, in order to prepare the discussion of the Newton method for the relaxed optimization of Section 4.

3.1. Metric Derivative (Levi–Civita connection)

Let $V, W \colon \mathcal{E}_\mathcal{V}$ be vector fields, that is, $V(p), W(p) \in T_p\mathcal{E}_\mathcal{V} = {}^e\mathbb{U}^p\mathcal{V}$, $p \in \mathcal{E}_\mathcal{V}$. Consider the real function $R = g(V,W) \colon \mathcal{E}_\mathcal{V} \to \mathbb{R}$, whose value at $p \in \mathcal{E}_\mathcal{V}$ is $R(p) = g_p(V(p), W(p)) = \mathbb{E}_p\left[V(p)W(p)\right]$. Assuming smoothness, we want to compute the derivative of R along the vector field $Y \colon \mathcal{E}_\mathcal{V}$, that is, $(D_Y R)(p) = dR_p(\mathbf{0})\boldsymbol{\alpha}$, with $\boldsymbol{\alpha} = \dot{\sigma}_p(Y(p))$. The expression of R in the chart centered at p is, according to Equation (27),

$$\boldsymbol{\theta} \mapsto R_p(\boldsymbol{\theta}) = \dot{\sigma}_p(V(e_p(\boldsymbol{\theta})))' I_\mathcal{B}(e_p(\boldsymbol{\theta}))\dot{\sigma}_p(W(e_p(\boldsymbol{\theta}))) = V_p(\boldsymbol{\theta})' I_{\mathcal{B},p}(\boldsymbol{\theta}) W_p(\boldsymbol{\theta}), \tag{84}$$

where V_p and W_p are the presentation in the chart of the vector fields V and W, respectively.

The i-th component $\partial_i R_p(\boldsymbol{\theta})$ of the Euclidean gradient $\nabla R_p(\boldsymbol{\theta})$ is:

$$\begin{aligned}\partial_i R_p(\boldsymbol{\theta}) &= \partial_i \left(V_p(\boldsymbol{\theta})' I_{\mathcal{B},p}(\boldsymbol{\theta}) W_p(\boldsymbol{\theta})\right) = \\ &\partial_i V_p(\boldsymbol{\theta})' I_{\mathcal{B},p}(\boldsymbol{\theta}) W_p(\boldsymbol{\theta}) + V_p(\boldsymbol{\theta})' \partial_i I_{\mathcal{B},p}(\boldsymbol{\theta}) W_p(\boldsymbol{\theta}) + V_p(\boldsymbol{\theta})' I_{\mathcal{B},p}(\boldsymbol{\theta}) \partial_i W_p(\boldsymbol{\theta}) = \\ &\left(\partial_i V_p(\boldsymbol{\theta}) + \frac{1}{2} I_{\mathcal{B},p}^{-1}(\boldsymbol{\theta}) \partial_i I_{\mathcal{B},p}(\boldsymbol{\theta}) V_p(\boldsymbol{\theta})\right)' I_{\mathcal{B},p}(\boldsymbol{\theta}) W_p(\boldsymbol{\theta}) + \\ &V_p(\boldsymbol{\theta})' I_{\mathcal{B},p}(\boldsymbol{\theta}) \left(\partial_i W_p(\boldsymbol{\theta}) + \frac{1}{2} I_{\mathcal{B},p}^{-1}(\boldsymbol{\theta}) \partial_i I_{\mathcal{B},p}(\boldsymbol{\theta}) W_p(\boldsymbol{\theta})\right), \end{aligned} \quad (85)$$

so that the derivative at $\boldsymbol{\theta}$ along $\boldsymbol{\alpha} = \dot{\sigma}_{e_p(\boldsymbol{\theta})}(Y(e_p(\boldsymbol{\theta})))$ is:

$$\begin{aligned} dR_p(\boldsymbol{\theta})\boldsymbol{\alpha} = &\left(dV_p(\boldsymbol{\theta})\boldsymbol{\alpha} + \frac{1}{2} I_{\mathcal{B},p}^{-1}(\boldsymbol{\theta}) \left(dI_{\mathcal{B},p}(\boldsymbol{\theta})\boldsymbol{\alpha}\right) V_p(\boldsymbol{\theta})\right)' I_{\mathcal{B},p}(\boldsymbol{\theta}) W_p(\boldsymbol{\theta}) + \\ &V_p(\boldsymbol{\theta})' I_{\mathcal{B},p}(\boldsymbol{\theta}) \left(dW_p(\boldsymbol{\theta})\boldsymbol{\alpha} + \frac{1}{2} I_{\mathcal{B},p}^{-1}(\boldsymbol{\theta}) \left(dI_{\mathcal{B},p}(\boldsymbol{\theta})\boldsymbol{\alpha}\right) W_p(\boldsymbol{\theta})\right). \end{aligned} \quad (86)$$

Proposition 1. *If we define $D_Y V$ to be the vector field on $\mathcal{E}_\mathcal{V}$, whose value at $q = e_p(\boldsymbol{\theta})$ has coordinates centered at p given by:*

$$\dot{\sigma}_p(D_Y V(q)) = dV_p(\boldsymbol{\theta})\boldsymbol{\alpha} + \frac{1}{2} I_{\mathcal{B}}^{-1}(p) \left(dI_{\mathcal{B},p}(\boldsymbol{\theta})\boldsymbol{\alpha}\right) V_p(\boldsymbol{\theta}), \quad \boldsymbol{\alpha} = \dot{\sigma}_p(Y(q)), \quad (87)$$

then:

$$D_Y g(V, W) = g(D_Y V, W) + g(V, D_Y W), \quad (88)$$

i.e., *Equation (87) is a metric covariant derivative; see [6] (Ch. 2 §3), [8] (VIII §4), [4] (§5.3.2).*

The metric derivative Equation (87) could be computed from the flow of the vector field Y. Let $(t, p) \mapsto \gamma(t, p)$ be the flow of the vector field V, *i.e.*, $\delta\gamma(t, p) = V(\gamma(t, p))$ and $\gamma(0, p) = p$. Using Equation (23), we have:

$$\begin{aligned} \left.\frac{d}{dt}\dot{\sigma}(V(\gamma(t,p)))\right|_{t=0} &= \left.\frac{d}{dt} V_p(\sigma_p(\gamma(t,p)))\right|_{t=0} \\ &= \left. dV_p(\sigma_p(\gamma(t,p))) \frac{d}{dt}\sigma_p(\gamma(t,p))\right|_{t=0} \\ &= dV_p(\mathbf{0})\dot{\sigma}_p(\delta\gamma(0,p)) = dV_p(\mathbf{0})\dot{\sigma}_p(Y(p)), \end{aligned} \quad (89)$$

and:

$$\left.\frac{d}{dt} I_\mathcal{V}(\gamma(t,p))\right|_{t=0} = \left.\frac{d}{dt} I_{\mathcal{B},p}(\sigma_p\gamma(t,p))\right|_{t=0} = dI_{\mathcal{B},p}(\mathbf{0})\dot{\sigma}_p(\delta\gamma(0,p)) = dI_{\mathcal{B},p}(\mathbf{0})\dot{\sigma}_p(Y(p))V_p(\mathbf{0}), \quad (90)$$

so that:

$$\dot{\sigma}(D_Y V(p)) = \left.\frac{d}{dt}\dot{\sigma} V(\gamma(t,p))\right|_{t=0} + \frac{1}{2} I_\mathcal{V}^{-1}(p) \left.\frac{d}{dt} I_\mathcal{V}(\gamma(t,p))\right|_{t=0}. \quad (91)$$

Let us check the symmetry of the metric covariant derivative to show that it is actually the unique Riemannian or Levi–Civita affine connection; see [6] (Th. 3.6).

The Lie bracket of the vector fields V and W is the vector field $[V, W]$, whose coordinates are:

$$[V, W]_p(\boldsymbol{\theta}) = dV_p(\mathbf{0})\dot{\sigma}_p(W(p)) - dW_p(\mathbf{0})\dot{\sigma}_p(V(p)). \quad (92)$$

As the ij entry of $\partial_k I_{\mathcal{B},p}(\mathbf{0})$ is $\partial_k\partial_i\partial_j\psi_p(\mathbf{0})$, then the symmetry $(dI_{\mathcal{B},p}(\mathbf{0})\boldsymbol{\alpha})\boldsymbol{\beta} = (dI_{\mathcal{B},p}(\mathbf{0})\boldsymbol{\beta})\boldsymbol{\alpha}$ holds, and we have:

$$\begin{aligned}\dot{\sigma}_p\left(D_W V(p) - D_V W(p)\right) = \\ & dV_p(\mathbf{0})\dot{\sigma}_p(W(p)) + \frac{1}{2}I_{\mathcal{B}}^{-1}(p)\left(dI_{\mathcal{B},p}(\mathbf{0})\dot{\sigma}_p(W(p))\right)V_p(\mathbf{0}) \\ & - dW_p(\mathbf{0})\dot{\sigma}_p(V(p)) - \frac{1}{2}I_{\mathcal{B}}^{-1}(p)\left(dI_{\mathcal{B},p}(\mathbf{0})\dot{\sigma}_p(V(p))\right)W_p(\mathbf{0}) \\ & = \dot{\sigma}[V,W](p). \end{aligned} \tag{93}$$

The term $\Gamma^k(p) = \frac{1}{2}I_p^{-1}(\mathbf{0})\partial_k dI_{\mathcal{B},p}(\mathbf{0})$ of Equation (87) is sometimes referred to as the Christoffel matrix, but we do not use this terminology in this paper. As:

$$I_{\mathcal{B},p}(\boldsymbol{\theta}) = I_{\mathcal{B}}(e_p(\boldsymbol{\theta})) = \left[\mathrm{Cov}_{e_p(\boldsymbol{\theta})}\left(X_i, X_j\right)\right]_{i,j=1,\ldots,m} = \left[\partial_i\partial_j\psi_p(\boldsymbol{\theta})\right]_{i,j=i,\ldots,m}, \tag{94}$$

we have $\partial_k I_{\mathcal{B}}(e_p(\boldsymbol{\theta})) = [\partial_i\partial_j\partial_k\psi_p(\boldsymbol{\theta})]_{i,j=i,\ldots,m} = \left[\mathrm{Cov}_{e_p(\boldsymbol{\theta})}\left(X_i, X_j, X_k\right)\right]_{i,j=i,\ldots,m}$ and:

$$\Gamma^k(p) = \frac{1}{2}\left[\mathrm{Cov}_p\left(X_i, X_j\right)\right]_{i,j=i,\ldots,m}^{-1}\left[\mathrm{Cov}_p\left(X_i, X_j, X_k\right)\right]_{i,j=i,\ldots,m} \tag{95}$$

.

If V, W are vector fields of $T\mathcal{E}\mathcal{V}$, we have:

$$\begin{aligned}\Gamma(p,V,W) &= \frac{1}{2}I_{\mathcal{B}}^{-1}(p)\,\mathrm{Cov}_p\left(\mathbf{X}, V, W\right) \\ &= \frac{1}{2}I_{\mathcal{B}}^{-1}(p)\,\mathbb{E}_p\left[{}^e\mathbb{U}^p\mathbf{X}VW\right],\end{aligned} \tag{96}$$

which is the projection of $V(p)W(p)/2$ on ${}^e\mathbb{U}^p\mathcal{V}$.

Notice also that:

$$(dI_p^{-1}(\mathbf{0})\boldsymbol{\alpha})I_{\mathcal{B},p}(\mathbf{0}) = -I_p^{-1}(\mathbf{0})(dI_{\mathcal{B},p}(\mathbf{0})\boldsymbol{\alpha})I_p^{-1}(\mathbf{0})I_{\mathcal{B},p}(\mathbf{0})\boldsymbol{y} = -I_p^{-1}(\mathbf{0})\left(dI_{\mathcal{B},p}(\mathbf{0})\boldsymbol{\alpha}\right). \tag{97}$$

3.2. Acceleration

Let $p(t)$, $t \in I$, be a smooth curve in $\mathcal{E}_{\mathcal{V}}$. Then, the velocity $\delta p(t) = \frac{d}{dt}\log(p(t))$ is a vector field $V(p(t)) = \delta p(t)$, defined on the support $p(I)$ of the curve. As the curve is the flow of the velocity field, we can compute the metric derivative of the velocity along the the velocity itself $D_{\delta p}\delta p$ from Equation (91) with $V(p(0)) = \delta p(0)$; we can use Equation (91) to get:

$$\begin{aligned}\dot{\sigma}_p(D_{\delta p}\delta p)(p(0)) = \frac{d}{dt}\dot{\sigma}_{p(0)}\left(\delta(p(t))\right)\Big|_{t=0} + \frac{1}{2}I_{\mathcal{B}}^{-1}(p(0))\,\frac{d}{dt}I_{\mathcal{B}}(p(t))\Big|_{t=0} = \\ \frac{d^2}{dt^2}\sigma_{p(0)}(p(t))\Big|_{t=0} + \frac{1}{2}I_{\mathcal{B}}^{-1}(p(0))\,\frac{d}{dt}I_{\mathcal{B}}(p(t))\Big|_{t=0}.\end{aligned} \tag{98}$$

which can be defined to be the Riemannian acceleration of the curve at $t = 0$.

Let us write $\boldsymbol{\theta}(t) = \sigma_p(p(t))$, $p = p(0)$ and:

$$p(t) = \exp\left(\sum_{j=1}^{m}\theta^j(t)\,{}^e\mathbb{U}^p X_j - \psi_p(\boldsymbol{\theta}(t))\right)\cdot p, \tag{99}$$

so that $\dot{\sigma}_p(\delta p)(0) = \dot{\boldsymbol{\theta}}(0)$ and $\left.\frac{d^2}{dt^2}\sigma_p(p(t))\right|_{t=0} = \ddot{\boldsymbol{\theta}}(0)$. We have:

$$\left.\frac{d}{dt} I_{\mathcal{B}}(p(t))\right|_{t=0} = \left.\frac{d}{dt} I_{\mathcal{B},p}(\boldsymbol{\theta}(t))\right|_{t=0} = \left.\frac{d}{dt} \operatorname{Hess} \psi_p(\boldsymbol{\theta}(t))\right|_{t=0} = \operatorname{Cov}_p(\boldsymbol{X}, \boldsymbol{X}, \sum_{j=1}^{m} \dot{\theta}^j(t) X_j) \tag{100}$$

so that the acceleration at p has coordinates:

$$\ddot{\boldsymbol{\theta}}(0) + \frac{1}{2} \sum_{i,j=1}^{m} \dot{\theta}^i(0)\dot{\theta}^j(0) \operatorname{Cov}_p(\boldsymbol{X}, \boldsymbol{X})^{-1} \operatorname{Cov}_p(\boldsymbol{X}, X_i, X_j) =$$
$$\ddot{\boldsymbol{\theta}}(0) + \frac{1}{2} \operatorname{Cov}_p(\boldsymbol{X}, \boldsymbol{X})^{-1} \operatorname{Cov}_p(\boldsymbol{X}, \sum_{i}^{m} \dot{\theta}^i(0) X_i, \sum_{j=1}^{m} \dot{\theta}^j(0) X_j). \tag{101}$$

A geodesic is a curve whose acceleration is zero at each point. The exponential map is the mapping $\operatorname{Exp}\colon T\mathcal{E}_{\mathcal{V}} \to \mathcal{E}_{\mathcal{V}}$ defined by:

$$(p, U) \mapsto \operatorname{Exp}_p U = p(1), \tag{102}$$

where $t \mapsto p(t)$ is the geodesic, such that $p(0) = p$ and $\delta p(0) = U$, for all U, such that the geodesic exists for $t = 1$.

The exponential map is a particular retraction, that is, a family of mappings R_p, $p \in \mathcal{E}$, from the tangent space at p to the manifold; here $R\colon T_p\mathcal{E} \to \mathcal{E}$, such that $R_p(0) = p$ and $dR_p(0) = \operatorname{Id}$; see [4] (§5.4). It should be noted that exponential manifolds have natural retractions other than Exp, a notable one being the exponential family itself. A retraction provides a crucial step in a gradient search algorithms by mapping a direction of increase of the objective function to a new trial point.

3.2.1. Example: Binary Independent 2.5.2 Continued.

Let us consider the binary independent model of Section 2.5.2. We have

$$I_{\mathcal{B}}(e_\lambda(\boldsymbol{\theta})) = I_{\mathcal{B},\lambda}(\boldsymbol{\theta}) = \operatorname{diag}\left(\cosh^{-2}(\theta^j) \colon j = 1, \dots, d\right), \tag{103}$$

it follows that

$$\begin{aligned} \partial_k I_{\mathcal{B},\lambda}(\boldsymbol{\theta}) &= \partial_k \operatorname{diag}\left(\cosh^{-2}(\theta^j) \colon j = 1, \dots, d\right) \\ &= -2\cosh^{-3}(\theta^k)\sinh(\theta^k) E^{kk}, \end{aligned} \tag{104}$$

where E^{kk} is the $d \times d$ matrix with entry one at (k, k), zero otherwise. The k-th Christoffel's matrix in the second term in the definition of the metric derivative (aka Levi–Civita connection) is:

$$\Gamma^k_{\mathcal{B}}(e_\lambda(\boldsymbol{\theta})) = \Gamma^k_\lambda(\boldsymbol{\theta}) = \frac{1}{2} I^{-1}_{\mathcal{B},\lambda}(\boldsymbol{\theta}) \partial_k I_{\mathcal{B},\lambda}(\boldsymbol{\theta}) = -\tanh(\theta^k) E^{kk}. \tag{105}$$

In terms of the moments, we have $I_{\mathcal{B},\lambda}(\boldsymbol{\theta}) = \operatorname{Cov}_{\boldsymbol{\theta}}(\boldsymbol{X}, \boldsymbol{X}') = \operatorname{Hess} \psi_\lambda(\boldsymbol{\theta})$. As $\partial_k\partial_i\partial_j\psi_\lambda(\boldsymbol{\theta}) = \operatorname{Cov}_{\boldsymbol{\theta}}(X_k, X_i, X_j)$, we that can write:

$$\begin{aligned} \partial_k I_{\mathcal{B},\lambda}(\boldsymbol{\theta}) &= \partial_k \operatorname{diag}\left(\operatorname{Var}_{\boldsymbol{\theta}}(X_j) \colon j = 1, \dots, d\right) \\ &= \operatorname{Cov}_{\boldsymbol{\theta}}(X_k, X_k, X_k) E^{kk} \end{aligned} \tag{106}$$

and:

$$\begin{aligned} \Gamma^k_\lambda(\boldsymbol{\theta}) &= \frac{1}{2} \operatorname{Cov}_{\boldsymbol{\theta}}(X_k, X_k)^{-1} \operatorname{Cov}_{\boldsymbol{\theta}}(X_k, X_k, X_k) E^{kk} \\ &= \frac{1}{2}(1 - (\eta^k)^2)^{-1}(-2\eta^k + 2(\eta^k)^3) E^{kk} = -\eta^k E^{kk}. \end{aligned} \tag{107}$$

The equations for the geodesics starting from $\boldsymbol{\theta}(0)$ with velocity $\dot{\boldsymbol{\theta}}(0) = \boldsymbol{u}$ are:

$$\ddot{\theta}^k(t) + \sum_{ij=1}^{m} \Gamma_{ij}^k(\boldsymbol{\theta}(t))\dot{\theta}^i(t)\dot{\theta}^j(t) = \ddot{\theta}^k(t) - \tanh(\theta^k(t))(\dot{\theta}^k(t))^2 = 0, \quad k = 1, \ldots, d. \tag{108}$$

The ordinary differential equation:

$$\ddot{\theta} - \tanh(\theta)\dot{\theta}^2 = 0 \tag{109}$$

has the closed form solution:

$$\theta(t) = \mathrm{gd}^{-1}\left(\mathrm{gd}(\theta(0)) + \frac{\dot{\theta}(0)}{\cosh(\theta(0))}t\right) = \tanh^{-1}\left(\sin\left(\mathrm{gd}(\theta(0)) + \frac{\dot{\theta}(0)}{\cosh(\theta(0))}t\right)\right) \tag{110}$$

for all t, such that:

$$-\pi/2 < \mathrm{gd}(\theta(0)) + \frac{\dot{\theta}(0)}{\cosh(\theta(0))}t < \pi/2, \tag{111}$$

where $\mathrm{gd}\colon \mathbb{R} \to]-\pi/2, +\pi/2[$ is the Gudermannian function, that is, $\mathrm{gd}'(x) = 1/\cosh x$, $\mathrm{gd}(0) = 0$; in closed form, $\mathrm{gd}(x) = \arcsin(\tanh(x))$. In fact, if θ is a solution of Equation (109), then:

$$\frac{d}{dt}\mathrm{gd}(\theta(t)) = \frac{\dot{\theta}(t)}{\cosh(\theta(t))} \tag{112}$$

$$\begin{aligned}\frac{d^2}{dt^2}\mathrm{gd}(\theta(t)) &= -\frac{\sinh(\theta(t))(\dot{\theta}(t))^2}{\cosh^2(\theta(t))} + \frac{\ddot{\theta}(t)}{\cosh(\theta(t))} \\ &= \frac{1}{\cosh(\theta(t))}\left(\ddot{\theta}(t) - \tanh(\theta(t))(\dot{\theta}(t))^2\right) = 0,\end{aligned} \tag{113}$$

so that $t \mapsto \mathrm{gd}(\theta(t))$ coincides (where it is defined) with an affine function characterized by the initial conditions.

In particular, at $t = 1$, the geodesic Equation (110) defines the Riemannian exponential $\mathrm{Exp}\colon T\mathcal{E}_{\mathcal{V}} \to \mathcal{E}_{\mathcal{V}}$. If $(p, U) \in T\mathcal{E}_{\mathcal{V}}$, that is, $p \in \mathcal{E}_{\mathcal{V}}$ and $U \in T_p\mathcal{E}_{\mathcal{V}}$, then $\sigma_\lambda(p) = \boldsymbol{\theta}(\mathbf{0})$ and $U = \sum u_j {}^e\mathbb{U}^p X_j$, $\dot{\sigma}_\lambda(U) = \boldsymbol{u}$. If:

$$-\pi/2 < \mathrm{gd}(\theta^j) + \frac{u_j}{\cosh(\theta^j)} < \pi/2, \tag{114}$$

then we can take $\dot{\boldsymbol{\theta}}(0) = \boldsymbol{u}$ and $t = 1$, so that:

$$\begin{aligned}\mathrm{Exp}_p\colon U \overset{\dot{\sigma}_\lambda}{\longmapsto} \boldsymbol{u} \mapsto \left[\mathrm{gd}^{-1}\left(\mathrm{gd}(\theta^j) + \frac{u_j}{\cosh(\theta^j)}\right) : j = 1, \ldots, d\right] \overset{e_\lambda}{\longmapsto} \\ \prod_{j=1}^{m} \exp\left(\mathrm{gd}^{-1}\left(\mathrm{gd}(\theta^j) + \frac{u_j}{\cosh(\theta^j)}\right) X_j - \psi\left(\mathrm{gd}^{-1}\left(\mathrm{gd}(\theta^j) + \frac{u_j}{\cosh(\theta^j)}\right)\right)\right) 2^{-m}.\end{aligned} \tag{115}$$

We have:

$$\exp\left(\mathrm{gd}^{-1}(v)\right) = \exp\left(\tanh^{-1}(\sin(v))\right) = \sqrt{\frac{1 + \sin v}{1 - \sin v}} \tag{116}$$

and:

$$\psi\left(\mathrm{gd}^{-1}(v)\right) = +\log\left(\mathrm{gd}^{-1}(\sin v)\right) = \log\left(\frac{1}{\cos v}\right), \tag{117}$$

hence $u \mapsto \operatorname{Exp}_p\left(\sum_{j=1}^d u_j {}^e\mathbb{U}^p X_j\right)$ is given for:

$$u \in \bigtimes_{j=1}^{d} \left]\cosh(\theta^j)(-\pi/2 - \operatorname{gd}(\theta^j)), \cosh(\theta^j)(\pi/2 - \operatorname{gd}(\theta^j))\right[, \tag{118}$$

by:

$$\operatorname{Exp}_{\boldsymbol{\theta}}(\boldsymbol{u}) = \prod_{j=1}^{m} \cos\left(\operatorname{gd}(\theta^j) + \frac{u_j}{\cosh(\theta^j)}\right)\left(\frac{1+\sin\left(\operatorname{gd}(\theta^j) + \frac{u_j}{\cosh(\theta^j)}\right)}{1-\sin\left(\operatorname{gd}(\theta^j) + \frac{u_j}{\cosh(\theta^j)}\right)}\right)^{\frac{X_j}{2}} = \prod_{j=1}^{m}\left(1+\sin\left(\operatorname{gd}(\theta^j) + \frac{u_j}{\cosh(\theta^j)}\right) X_j\right) 2^{-m} \in \mathcal{E}_{\mathcal{V}}. \tag{119}$$

The expectation parameters are:

$$\eta^i(t) = \mathbb{E}_{\boldsymbol{\theta}=\mathbf{0}}\left[X_i \prod_{j=1}^{m}\left(1+\sin\left(\operatorname{gd}(\theta^j) + \frac{tu_j}{\cosh(\theta^j)}\right) X_j\right)\right] = \sin\left(\operatorname{gd}(\theta^j) + \frac{tu_j}{\cosh(\theta^j)}\right), \tag{120}$$

and:

$$\operatorname{gd}(\theta^j) = \arcsin(\eta^j), \quad \cosh(\theta^j) = \frac{1}{(1-(\eta^j)^2)^{\frac{1}{2}}}, \tag{121}$$

so that the exponential in terms of the expectation parameters is:

$$\operatorname{Exp}_{\boldsymbol{\eta}}(\boldsymbol{u}) = \left(\sin\left(\arcsin \eta^j + \left(1-(\eta^j)^2\right)^{\frac{1}{2}} u_j\right) : j = 1, \dots, m\right). \tag{122}$$

The inverse of the Riemannian exponential provides a notion of translation between two elements of the exponential model, which is a particular parametrization of the model:

$$\overrightarrow{\boldsymbol{\eta}_1\boldsymbol{\eta}_2} = \operatorname{Exp}_{\boldsymbol{\eta}_1}^{-1} \boldsymbol{\eta}_2 = \left[\left((1-(\eta_i^j)^2\right)^{-\frac{1}{2}}\left(\arcsin \eta_2^j - \arcsin \eta_1^j\right) : j = 1, \dots, m\right] \tag{123}$$

In particular, at $\boldsymbol{\theta} = \mathbf{0}$, we have the geodesic:

$$t \mapsto \prod_{j=1}^{d}(1+\sin(tu_j)X_j)\, 2^{-m}, \quad |t| < \frac{\pi}{2\max|u_j|} \tag{124}$$

See in Figure 5 some geodesic curves.

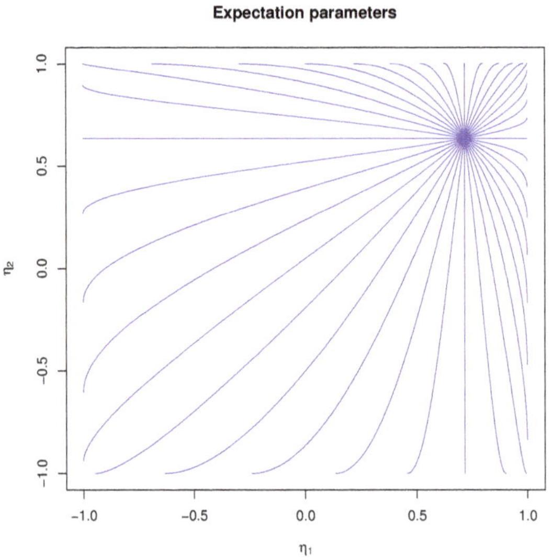

Figure 5. Geodesics from $\boldsymbol{\eta} = (0.75, 0.75)$.

3.3. *Riemannian Hessian*

Let $\phi\colon \mathcal{E}_\mathcal{V} \to \mathbb{R}$ with Riemannian gradient $\nabla\phi(p) = \sum_i (\widetilde{\nabla}\phi)_i(p)\, {}^e\mathbb{U}^p X_i$, $\widetilde{\nabla}\phi(p) = I_\mathcal{B}^{-1}(p)\nabla\phi_p(\mathbf{0})$. The Riemannian Hessian of ϕ is the metric derivative of the gradient $\nabla\phi$ along the vector field Y, that is, $\operatorname{Hess}_Y \phi = D_Y \nabla\phi$; see [6] (Ch. 6, Ex. 11), [4] (§5.5). in the following, we denote by the symbol Hess, without a subscript, the ordinary Hessian matrix.

From Equation (87), we have the coordinates of $\operatorname{Hess}_Y \phi(p)$. Given a generic tangent vector $\boldsymbol{\alpha}$, we compute from Equation (38):

$$\begin{aligned} d(\nabla\phi)_p(\boldsymbol{\theta})\boldsymbol{\alpha}\Big|_{\boldsymbol{\theta}=0} &= d\left(I_{\mathcal{B},p}^{-1}(\boldsymbol{\theta})\nabla\phi_p(\boldsymbol{\theta})\right)\boldsymbol{\alpha}\Big|_{\boldsymbol{\theta}=\mathbf{0}} \\ &= (dI_{\mathcal{B},p}^{-1}(\mathbf{0})\boldsymbol{\alpha})\nabla\phi_p(\mathbf{0}) + I_{\mathcal{B},p}^{-1}(\mathbf{0})\operatorname{Hess}\phi_p(\mathbf{0})\boldsymbol{\alpha} \\ &= -I_\mathcal{B}^{-1}(p)(dI_{\mathcal{B},p}(\mathbf{0})\boldsymbol{\alpha})\widetilde{\nabla}\phi(p) + I_\mathcal{B}^{-1}(p)\operatorname{Hess}\phi_p(\mathbf{0})\boldsymbol{\alpha} \end{aligned} \tag{125}$$

and, upon substitution of $(\nabla\phi)_p$ to V_p in Equation (87),

$$\begin{aligned} \dot{\sigma}_p(\operatorname{Hess}_Y \phi(p)) &= d(\nabla\phi)_p(\mathbf{0})\boldsymbol{\alpha} + \frac{1}{2}I_\mathcal{B}^{-1}(p)\left(dI_{\mathcal{B},p}(\mathbf{0})\boldsymbol{\alpha}\right)(\nabla\phi)_p(\mathbf{0}), \quad \boldsymbol{\alpha} = S_p(Y(p)) \\ &= -I_\mathcal{B}^{-1}(p)(dI_{\mathcal{B},p}(\mathbf{0})\boldsymbol{\alpha})\widetilde{\nabla}\phi(p) + I_\mathcal{B}^{-1}(p)\operatorname{Hess}\phi_p(\mathbf{0}) + \frac{1}{2}I_\mathcal{B}^{-1}(p)\left(dI_{\mathcal{B},p}(\mathbf{0})\boldsymbol{\alpha}\right)\widetilde{\nabla}\phi(p) \\ &= I_\mathcal{B}^{-1}(p)\operatorname{Hess}\phi_p(\mathbf{0})\boldsymbol{\alpha} - \frac{1}{2}I_\mathcal{B}^{-1}(p)\left(dI_{\mathcal{B},p}(\mathbf{0})\boldsymbol{\alpha}\right)\widetilde{\nabla}\phi(p) \\ &= I_\mathcal{B}^{-1}(p)\left(\operatorname{Hess}\phi_p(\mathbf{0})\boldsymbol{\alpha} - \frac{1}{2}\left(dI_{\mathcal{B},p}(\mathbf{0})\boldsymbol{\alpha}\right)\widetilde{\nabla}\phi(p)\right) \end{aligned} \tag{126}$$

$\operatorname{Hess}_Y \phi$ is characterized by knowing the value of $g(\operatorname{Hess}_Y \phi, X)\colon \mathcal{E}_\mathcal{V}$ for all vector fields X. We have from Equation (126), with $\boldsymbol{\alpha} = \dot{\sigma}_p(Y(p))$ and $\boldsymbol{\beta} = \dot{\sigma}_p(X(p))$,

$$g_p(\operatorname{Hess}_{Y(p)} \phi(p), X(p)) = \boldsymbol{\beta}'\operatorname{Hess}\phi_p(\mathbf{0})\boldsymbol{\alpha} - \frac{1}{2}\boldsymbol{\beta}'\left(dI_{\mathcal{B},p}(\mathbf{0})\boldsymbol{\alpha}\right)\widetilde{\nabla}\phi(p). \tag{127}$$

This is the presentation of the Riemannian Hessian as a bi-linear form on $T\mathcal{E}_{\mathcal{V}}$; see the comments in [4] (Prop. 5.5.2-3). Note that the Riemannian Hessian is positive definite if:

$$\boldsymbol{\alpha}' \operatorname{Hess} \phi_p(\mathbf{0})\boldsymbol{\alpha} \geq \frac{1}{2}\boldsymbol{\alpha}' \left(dI_{\mathcal{B},p}(\mathbf{0})\boldsymbol{\alpha}\right) \widetilde{\nabla}\phi(p), \quad \boldsymbol{\alpha} \in \mathbb{R}^m. \tag{128}$$

4. Application to Combinatorial Optimization

We conclude our paper by showing how the geometric method applies to the problem of finding the maximum of the expected value of a function.

4.1. Hessian of a Relaxed Function

Here is a key example of vector field. Let f be any bounded random variable, and define the relaxed function to be $\phi(p) = \mathbb{E}_p[f]$, $p \in \mathcal{P}_>$. Define $F(p)$ to be the projection of f, as an element of $L^2(p)$, onto $T_p\mathcal{E}_{\mathcal{V}} = {}^e\mathbb{U}^p\mathcal{V}$, *i.e.*, $F(p)$ is the element of ${}^e\mathbb{U}^p\mathcal{V}$, such that:

$$\mathbb{E}_p\left[(f - F(p))v\right] = 0, \quad v \in {}^e\mathbb{U}^p\mathcal{V} \tag{129}$$

In the basis ${}^e\mathbb{U}^p\mathcal{B}$, we have $F(p) = \sum_i \hat{f}_{p,i}\, {}^e\mathbb{U}^p X_i$ and:

$$\operatorname{Cov}_p(f, X_j) = \sum_i \hat{f}_{p,i}\, \mathbb{E}_p\left[{}^e\mathbb{U}^p X_i\, {}^e\mathbb{U}^p X_j\right], \quad j = 1, \ldots, m, \tag{130}$$

so that $\hat{f}_p = I_{\mathcal{B}}^{-1}(p) \operatorname{Cov}_p(\boldsymbol{X}, f)$ and

$$F(p) = \hat{f}_p'\, {}^e\mathbb{U}^p\boldsymbol{X} = \operatorname{Cov}_p(f, \boldsymbol{X})\, I_{\mathcal{B}}^{-1}(p)\, {}^e\mathbb{U}^p\boldsymbol{X}. \tag{131}$$

Let us compute the gradient of the relaxed function $\phi = \mathbb{E}_{\cdot}[f] : \mathcal{E}_{\mathcal{V}}$. We have $\phi_p(\boldsymbol{\theta}) = \mathbb{E}_{e_p(\boldsymbol{\theta})}[f]$, and from the properties of exponential families, the Euclidean gradient is $\nabla\phi_p(\mathbf{0}) = \operatorname{Cov}_p(f, \boldsymbol{X})$. It follows that the natural gradient is:

$$\widetilde{\nabla}\phi_p(\mathbf{0}) = I_{\mathcal{B}}^{-1}(p) \operatorname{Cov}_p(\boldsymbol{X}, f) = \hat{f}, \tag{132}$$

and the Riemannian gradient is $\nabla\phi(p) = F(p)$.

From the properties of exponential families, we have:

$$\operatorname{Hess}\phi_p(\mathbf{0}) = \operatorname{Cov}_p(\boldsymbol{X}, \boldsymbol{X}, f),$$

so that, in this case, Equation (127), when written in terms of the moments, is:

$$\boldsymbol{\beta}' \operatorname{Cov}_p(\boldsymbol{X}, \boldsymbol{X}, f)\, \boldsymbol{\alpha} - \frac{1}{2}\boldsymbol{\beta}' \operatorname{Cov}_p(\boldsymbol{X}, \boldsymbol{X}, \boldsymbol{\alpha}\cdot\boldsymbol{X}) \operatorname{Cov}_p(\boldsymbol{X}, \boldsymbol{X})^{-1} \operatorname{Cov}_p(\boldsymbol{X}, f). \tag{133}$$

4.1.1. Example: Binary Independent 2.5.2 and 3.2.1 Continued

We list below the computation of the Hessian in the case of two binary independent variables. Computations were done with Sage [22], which allows both the reduction $x_i^2 = 1$ in the ring of polynomials and the simplifications in the symbolic ring of parameters.

$$\operatorname{Cov}_\eta(\boldsymbol{X}, f) = \begin{pmatrix} -(\eta_1^2 - 1)a_1 - (\eta_1^2\eta_2 - \eta_2)a_{12} \\ -(\eta_2^2 - 1)a_2 - (\eta_1\eta_2^2 - \eta_1)a_{12} \end{pmatrix} = \begin{pmatrix} -(\eta_1 - 1)(\eta_1 + 1)(a_{12}\eta_2 + a_1) \\ -(\eta_2 - 1)(\eta_2 + 1)(a_{12}\eta_1 + a_2) \end{pmatrix} \tag{134}$$

$$\operatorname{Cov}_\eta(\boldsymbol{X}, \boldsymbol{X}) = \begin{pmatrix} -\eta_1^2 + 1 & 0 \\ 0 & -\eta_2^2 + 1 \end{pmatrix} = \begin{pmatrix} -(\eta_1 - 1)(\eta_1 + 1) & 0 \\ 0 & -(\eta_2 - 1)(\eta_2 + 1) \end{pmatrix} \tag{135}$$

$$\mathrm{Cov}_\eta(\boldsymbol{X},\boldsymbol{X})^{-1}\,\mathrm{Cov}_\eta(\boldsymbol{X},f) = \begin{pmatrix} a_{12}\eta_2 + a_1 \\ a_{12}\eta_1 + a_2 \end{pmatrix} = \nabla F(\boldsymbol{\eta}) \tag{136}$$

$$\begin{aligned}&\mathrm{Cov}_\eta(\boldsymbol{X},\boldsymbol{X},f) = \\ &\begin{pmatrix} 2\,(\eta_1^3-\eta_1)a_1 + 2\,(\eta_1^3\eta_2-\eta_1\eta_2)a_{12} & (\eta_1^2\eta_2^2-\eta_1^2-\eta_2^2+1)a_{12} \\ (\eta_1^2\eta_2^2-\eta_1^2-\eta_2^2+1)a_{12} & 2\,(\eta_1\eta_2^3-\eta_1\eta_2)a_{12} + 2\,(\eta_2^3-\eta_2)a_2 \end{pmatrix} = \\ &\begin{pmatrix} 2\,(\eta_1-1)(\eta_1+1)(a_{12}\eta_2+a_1)\eta_1 & (\eta_2-1)(\eta_2+1)(\eta_1-1)(\eta_1+1)a_{12} \\ (\eta_2-1)(\eta_2+1)(\eta_1-1)(\eta_1+1)a_{12} & 2\,(\eta_2-1)(\eta_2+1)(a_{12}\eta_1+a_2)\eta_2 \end{pmatrix}\end{aligned} \tag{137}$$

$$\mathrm{Cov}_\eta(\boldsymbol{X},\boldsymbol{X})^{-1}\,\mathrm{Cov}_\eta(\boldsymbol{X},\boldsymbol{X},f) = \begin{pmatrix} -2\,(a_{12}\eta_2+a_1)\eta_1 & -a_{12}\eta_2^2+a_{12} \\ -a_{12}\eta_1^2+a_{12} & -2\,(a_{12}\eta_1+a_2)\eta_2 \end{pmatrix} \tag{138}$$

$$\begin{aligned}&\mathrm{Cov}_\eta(\boldsymbol{X},\boldsymbol{X},\nabla F(\boldsymbol{\eta})) = \\ &\begin{pmatrix} 2\,(a_{12}\eta_2+a_1)(\eta_1+1)(\eta_1-1)\eta_1 & 0 \\ 0 & 2\,(a_{12}\eta_1+a_2)(\eta_2+1)(\eta_2-1)\eta_2 \end{pmatrix}\end{aligned} \tag{139}$$

$$\begin{aligned}&\mathrm{Cov}_\eta(\boldsymbol{X},\boldsymbol{X})^{-1}\,\mathrm{Cov}_\eta(\boldsymbol{X},\boldsymbol{X},\nabla F(\boldsymbol{\eta})) = \\ &\begin{pmatrix} -2\,(a_{12}\eta_2+a_1)\eta_1 & 0 \\ 0 & -2\,(a_{12}\eta_1+a_2)\eta_2 \end{pmatrix}\end{aligned} \tag{140}$$

The Riemannian Hessian as a matrix in the basis of the tangent space is:

$$\begin{aligned}&\mathrm{Hess}\,F(\boldsymbol{\eta}) = \mathrm{Cov}_\eta(\boldsymbol{X},\boldsymbol{X})^{-1}\left(\mathrm{Cov}_\eta(\boldsymbol{X},\boldsymbol{X},f) - \frac{1}{2}\,\mathrm{Cov}_\eta(\boldsymbol{X},\boldsymbol{X},\nabla F(\boldsymbol{\eta}))\right) = \\ &\begin{pmatrix} -(a_{12}\eta_2+a_1)\eta_1 & -a_{12}(\eta_2+1)(\eta_2-1) \\ -a_{12}(\eta_1+1)(\eta_1-1) & -(a_{12}\eta_1+a_2)\eta_2 \end{pmatrix}\end{aligned} \tag{141}$$

As a check, let us compute the Riemannian Hessian as a natural Hessian in the Riemannian parameters, $\mathrm{Hess}\,\phi \circ \mathrm{Exp}_p(\boldsymbol{u})\Big|_{u=0}$; see [4] (Prop. 5.5.4). We have:

$$\begin{aligned}&F \circ \mathrm{Exp}_\eta(\boldsymbol{u}) = \\ &a_{12}\sin\left(\sqrt{-\eta_1^2+1}u_1 + \arcsin(\eta_1)\right)\sin\left(\sqrt{-\eta_2^2+1}u_2 + \arcsin(\eta_2)\right) + \\ &a_1\sin\left(\sqrt{-\eta_1^2+1}u_1 + \arcsin(\eta_1)\right) + a_2\sin\left(\sqrt{-\eta_2^2+1}u_2 + \arcsin(\eta_2)\right)\end{aligned} \tag{142}$$

and:

$$\begin{aligned}&\mathrm{Hess}\,F \circ \mathrm{Exp}_\eta(\boldsymbol{u})\Big|_{u=0} = \\ &\begin{pmatrix} (\eta_1^2-1)a_{12}\eta_1\eta_2 + (\eta_1^2-1)a_1\eta_1 & (\eta_1^2-1)(\eta_2^2-1)a_{12} \\ (\eta_1^2-1)(\eta_2^2-1)a_{12} & (\eta_2^2-1)a_{12}\eta_1\eta_2 + (\eta_2^2-1)a_2\eta_2 \end{pmatrix} = \\ &\begin{pmatrix} (a_{12}\eta_2+a_1)(\eta_1+1)(\eta_1-1)\eta_1 & a_{12}(\eta_1+1)(\eta_1-1)(\eta_2+1)(\eta_2-1) \\ a_{12}(\eta_1+1)(\eta_1-1)(\eta_2+1)(\eta_2-1) & (a_{12}\eta_1+a_2)(\eta_2+1)(\eta_2-1)\eta_2 \end{pmatrix}.\end{aligned} \tag{143}$$

Note the presence of the factor $\mathrm{Cov}_\eta(\boldsymbol{X}, \boldsymbol{X})$.

4.2. Newton Method

The Newton method is an iterative method that generates a sequence of points p_t, with $t = 0, 1, \ldots$, that converges towards a stationary point $\hat{p}$ of a $F(p) = \mathbb{E}_p[f]$, $p \in \mathcal{E}_\mathcal{V}$, that is, a critical point of the vector field $p \mapsto \nabla F(p)$, $\nabla F(\hat{p}) = 0$. Here, we follow [4] (Ch. 5–6), and in particular Algorithm 5 on Page 113.

Let ∇F be a gradient field. We reproduce in our case the basic derivation of the Newton method in the following. Note that, in this section, we use the notation Hess $\bullet[\boldsymbol{\alpha}]$ to denote $\mathrm{Hess}_{\boldsymbol{\alpha}} \bullet$. Using the definition of metric derivative, we have for a geodesic curve $[0,1] \ni t \mapsto p(t) \in \mathcal{E}_\mathcal{V}$ connecting $p = p(0)$ to $\hat{p} = p(1)$ that:

$$\frac{d}{dt} g_{p(t)}\left(\nabla F(p(t)), \delta p(t)\right) = g_{p(t)}\left(\mathrm{Hess}\, F(p(t))[\delta p(t)], \delta p(t)\right) \tag{144}$$

hence the increment from p to $\hat{p}$ is:

$$g_{\hat{p}}\left(\nabla F(\hat{p}), \delta p(1)\right) - g_p\left(\nabla F(p), \delta p(0)\right) = \int_0^1 g_{p(t)}\left(\mathrm{Hess}\, F(p(t))[\delta p(t)], \delta p(t)\right)\, dt. \tag{145}$$

Now, we assume that $\nabla F(\hat{p}) = 0$ and that in Equation (145), the integral is approximated by the initial value of the integrand, that is to say, the Hessian is approximately constant on the geodesic from p to $\hat{p}$; we obtain:

$$-g_p\left(\nabla F(p), \delta p(0)\right) = g_p\left(\mathrm{Hess}\, F(p)[\delta p(0)], \delta p(0)\right) + \epsilon. \tag{146}$$

If we can solve the Newton equation:

$$\mathrm{Hess}\, F(p(t))[\boldsymbol{u}] = -\nabla F(p) \tag{147}$$

then $\boldsymbol{u}$ is approximately equal to the initial velocity of the geodesic connecting p to $\hat{p}$, that is, $\hat{p} = \mathrm{Exp}_p(\boldsymbol{u})$.

The particular structure of the exponential manifold suggests at least two natural retractions that could be used to move from $\boldsymbol{u}$ to $\hat{p}$. Namely, we have the Riemannian exponential $(\boldsymbol{\theta}_t, \boldsymbol{\theta}_{t+1}) \mapsto \mathrm{Exp}_{\boldsymbol{\theta}_t}(\boldsymbol{\theta}_{t+1} - \boldsymbol{\theta}_t)$ and the e-retraction coming from the exponential family itself and defined by $(\boldsymbol{\theta}_t, \boldsymbol{\theta}_{t+1}) \mapsto e_{\boldsymbol{\theta}_t}(\boldsymbol{\theta}_{t+1} - \boldsymbol{\theta}_t)$, with $\boldsymbol{\theta}_{t+1} - \boldsymbol{\theta}_t = \boldsymbol{u}_t$.

In the $\boldsymbol{\theta}$ parameters, with the e-retraction, the Newton method generates a sequence $(\boldsymbol{\theta}_t)$ according to the following updating rule:

$$\boldsymbol{\theta}_{t+1} = \boldsymbol{\theta}_t - \lambda\, \mathrm{Hess}\, F(\boldsymbol{\theta}_t)^{-1} \widetilde{\nabla} F(\boldsymbol{\theta}_t) \tag{148}$$

where $\lambda > 0$ is an extra parameter intended to control the step size and, in turn, the convergence to $\hat{\theta}$; see [5].

We can rewrite Equation (148) in terms of covariances as:

$$\boldsymbol{\theta}_{t+1} = \boldsymbol{\theta}_t - \lambda \left(\mathrm{Cov}_{\boldsymbol{\theta}_t}(\boldsymbol{X}, \boldsymbol{X}, f) - \frac{1}{2} \mathrm{Cov}_{\boldsymbol{\theta}_t}(\boldsymbol{X}, \boldsymbol{X}, \widetilde{\nabla} F(\boldsymbol{\theta}_t)) \right)^{-1} \widetilde{\nabla} F(\boldsymbol{\theta}_t). \tag{149}$$

4.3. Example: Binary Independent

In the $\boldsymbol{\eta}$ parameters, the Newton step is:

$$\boldsymbol{u} = -\,\mathrm{Hess}\, F(\boldsymbol{\eta})^{-1} \nabla F(\boldsymbol{\eta}) = \begin{pmatrix} \dfrac{a_{12}^2\eta_1 + a_{12}a_2 + (a_1a_{12}\eta_1 + a_1a_2)\eta_2}{a_{12}^2\eta_1^2 + \left(a_{12}a_2\eta_1 + a_{12}^2\right)\eta_2^2 - a_{12}^2 + \left(a_1a_{12}\eta_1^2 + a_1a_2\eta_1\right)\eta_2} \\ \dfrac{a_1a_2\eta_1 + a_1a_{12} + \left(a_{12}a_2\eta_1 + a_{12}^2\right)\eta_2}{a_{12}^2\eta_1^2 + \left(a_{12}a_2\eta_1 + a_{12}^2\right)\eta_2^2 - a_{12}^2 + \left(a_1a_{12}\eta_1^2 + a_1a_2\eta_1\right)\eta_2} \end{pmatrix} \tag{150}$$

and the new $\boldsymbol{\eta}$ in the Riemannian retraction is:

$$\operatorname{Exp}_{\boldsymbol{\eta}}(\boldsymbol{u}) = \begin{pmatrix} \sin\left(\frac{\left(a_{12}^2\eta_1+a_{12}a_2+(a_1a_{12}\eta_1+a_1a_2)\eta_2\right)\sqrt{-\eta_1^2+1}}{a_{12}^2\eta_1^2+\left(a_{12}a_2\eta_1+a_{12}^2\right)\eta_2^2-a_{12}^2+\left(a_1a_{12}\eta_1^2+a_1a_2\eta_1\right)\eta_2}+\arcsin(\eta_1)\right) \\ \sin\left(\frac{\left(a_1a_2\eta_1+a_1a_{12}+\left(a_{12}a_2\eta_1+a_{12}^2\right)\eta_2\right)\sqrt{-\eta_2^2+1}}{a_{12}^2\eta_1^2+\left(a_{12}a_2\eta_1+a_{12}^2\right)\eta_2^2-a_{12}^2+\left(a_1a_{12}\eta_1^2+a_1a_2\eta_1\right)\eta_2}+\arcsin(\eta_2)\right). \end{pmatrix} \tag{151}$$

In Figure 6, we represented the vector field associated with the Newton step in the $\boldsymbol{\eta}$ parameters, with $\lambda = 0.05$, using the Riemannian retraction, for the case $a_1 = 1$, $a_2 = 2$ and $a_{12} = 3$, with:

$$\operatorname{Exp}_{\boldsymbol{\eta}}(\boldsymbol{u}) = \begin{pmatrix} \sin\left(\lambda\frac{\sqrt{-\eta_1^2+1}((3\,\eta_1+2)\eta_2+9\,\eta_1+6)}{3\,(2\,\eta_1+3)\eta_2^2+9\,\eta_1^2+\left(3\,\eta_1^2+2\,\eta_1\right)\eta_2-9}+\arcsin(\eta_1)\right) \\ \sin\left(\lambda\frac{(3\,(2\,\eta_1+3)\eta_2+2\,\eta_1+3)\sqrt{-\eta_2^2+1}}{3\,(2\,\eta_1+3)\eta_2^2+9\,\eta_1^2+\left(3\,\eta_1^2+2\,\eta_1\right)\eta_2-9}+\arcsin(\eta_2)\right) \end{pmatrix}. \tag{152}$$

The red dotted lines represented in the figure identify the basins of attraction of the vector field and correspond to the solutions of the explicit equation in $\boldsymbol{\eta}$ for which the Newton step $\boldsymbol{u}$ is not defined. This vector field can be compared to that in Figure 7, associated with the Newton step for $F(\boldsymbol{\eta})$ using the Euclidean geometry. In the Euclidean geometry, $F(\boldsymbol{\eta})$ is a quadratic function with one saddle point, so that from any $\boldsymbol{\eta}$, the Newton step points in the direction of the critical point. This makes the Newton step unsuitable for an optimization algorithm. On the other side, in the Riemannian geometry, the vertices of the polytope are critical points for $F(\boldsymbol{\eta})$, and they determine the presence of multiple basins of attraction, as expected.

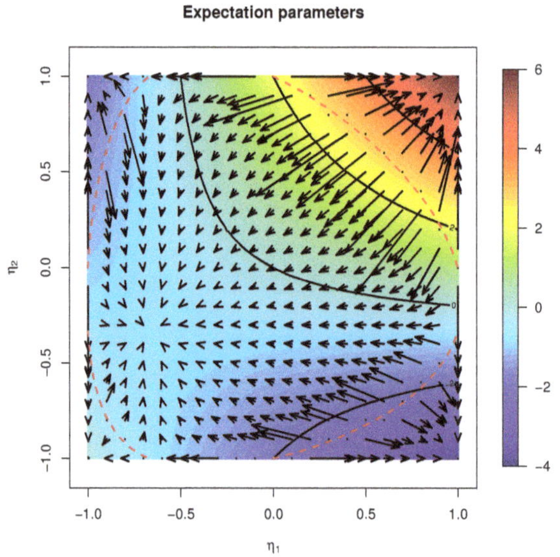

Figure 6. The Newton step in the $\boldsymbol{\eta}$ parameters, Riemannian retraction, $\lambda = 0.05$. The red dotted lines identify the different basins of attraction and correspond to the points for which the Newton step is not defined; *cf.* Equation (150). The instability close to the critical lines is represented by the longer arrows.

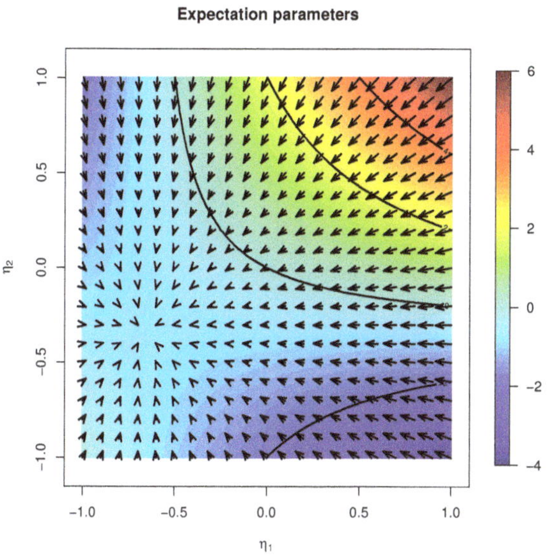

Figure 7. The Newton step in the η parameters, Euclidean geometry, $\lambda = 0.05$.

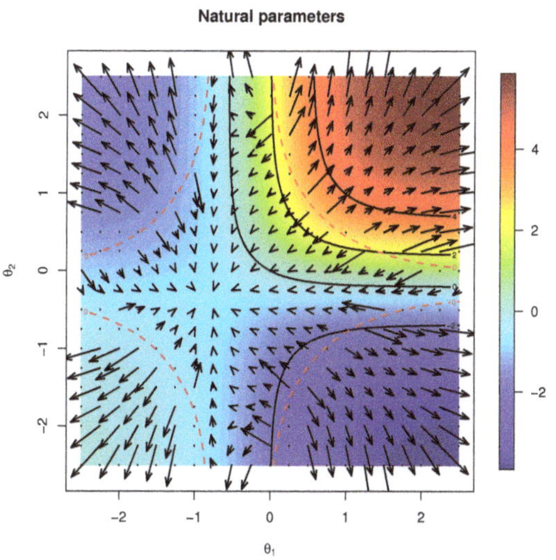

Figure 8. The Newton step in the θ parameters, exponential retraction, $\lambda = 0.015$. The red dotted lines identify the different basins of attraction and correspond to the points for which the Newton step is not defined. The instability along the critical lines, which identifies the basins of attraction, is not represented.

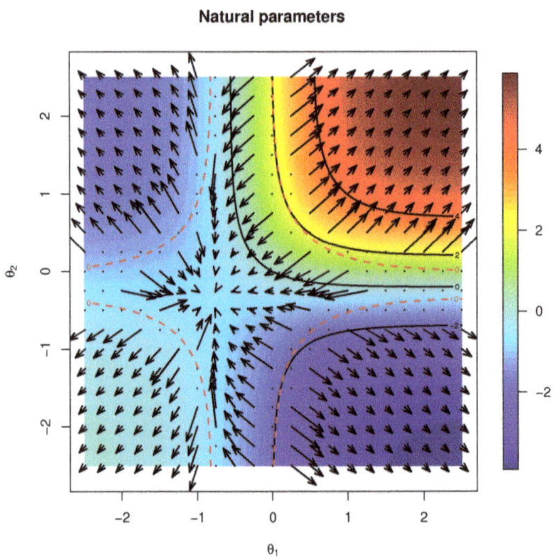

Figure 9. The Newton step in the $\boldsymbol{\theta}$ parameters, Euclidean geometry, $\lambda = 0.15$. The red dotted lines identify the different basins of attraction and correspond to the points for which the Newton step is not defined. The instability along the critical lines, which identifies the basins of attraction, is not represented.

Figure 8 shows the Newton step in the $\boldsymbol{\theta}$ parameters based on the e-retraction of Equation (149), while Figure 9 represents the Newton step evaluated with respect to the Euclidean geometry. A comparison of the two vector fields shows that, differently from the $\boldsymbol{\eta}$ parameters, the number of basins of attraction is the same in the two geometries; however, the scale of the vectors is different. In particular, notice how on the plateau, for diverging $\boldsymbol{\theta}$, the Newton step in the Euclidean geometry vanishes, while in the Riemannian geometry, it gets larger. This behavior suggests better convergence properties for an optimization algorithm based on the Newton step evaluated using the proper Riemannian geometry. In the $\boldsymbol{\theta}$ parameters, the boundaries of the basins of attraction represented by the red dotted lines have been computed numerically and correspond to the values of $\boldsymbol{\theta}$ for which the update step is not defined.

Finally, notice that in both the $\boldsymbol{\eta}$ and $\boldsymbol{\theta}$ parameters, the step is not always in the direction of descent for the function, a common behavior of the Newton method, which converges to the critical points.

5. Discussion and Conclusions

In this paper, we introduced second-order calculus over a statistical manifold, following the approach described in [4], which has been adapted to the special case of exponential statistical models [2,3]. By defining the Riemannian Hessian and using the notion of retraction, we developed the proper machinery necessary for the definition of the updating rule of the Newton method for the optimization of a function defined over an exponential family.

The examples discussed in the paper show that by taking into account the proper Riemannian geometry of a statistical exponential family, the vector fields associated with the Newton step in the different parametrizations change profoundly. Not only new basins of attraction associated with local and global minima appear, as for the expectation parameters, but also the magnitude of the Newton step is affected, as over the plateau in the natural parameters. Such differences are expected to have a strong impact on the performance of an optimization algorithm based on the Newton step, from both the point of view of achievable convergence and the speed of convergence to the optimum.

The Newton method is a popular second order optimization technique based on the computation of the Hessian of the function to be optimized and is well known for its super-linear convergence properties. However, the use of the Newton method poses a number of issues in practice.

First of all, as the examples in Figures 6 and 8 show, the Newton step does not always point in the direction of the natural gradient, and the algorithm may not converge to a (local) optimum of the function. Such behavior is not unexpected; indeed the Newton method tends to converge to critical points of the function to be optimized, which include local minima, local maxima and saddle points. In order to obtain a direction of ascent for the function to be optimized, the Hessian must be negative-definite, *i.e.*, its eigenvalues must be strictly negative, which is not guaranteed in the general case. Another important remark is related to the computational complexity associated with the evaluation of the Hessian, compared to the (natural) gradient. Indeed, to obtain the Newton step d, Christoffel matrices have to be evaluated, together with the third order covariances between sufficient statistics and the function, and the Hessian has to be inverted. Finally, notice that when the Hessian is close to being non-invertible, numerical problems may arise in the computation of the Newton step, and the algorithm may become unstable and diverge.

In the literature, different methods have been proposed to overcome these issues. Among them, we mention quasi-Newton methods, where the update vector is obtained using a modified Hessian, which has been made negative-definite, for instance, by adding a proper correction matrix.

This paper represents the first step in the design of an algorithm based on the Newton method for the optimization over a statistical model. The authors are working on the computational aspects related to the implementation of the method, and a new paper with experimental results is in progress.

Acknowledgments: Luigi Malagò was supported by the Xerox University Affairs Committee Award and by de Castro Statistics, Collegio Carlo Alberto, Moncalieri. Giovanni Pistone is supported by de Castro Statistics, Collegio Carlo Alberto, Moncalieri, and is a member of GNAMPA–INdAM, Roma.

Author Contributions

All authors contributed to the design of the research. The research was carried out by all authors. The study of the Hessian and of the Newton method in statistical manifolds was originally suggested by Luigi Malagò. The manuscript was written by Luigi Malagò and Giovanni Pistone. All authors have read and approved the final manuscript.

Conflicts of Interest: Conflicts of Interest

The authors declare no conflict of interest.

References

1. Brown, L.D. *Fundamentals of Statistical Exponential Families with Applications in Statistical Decision Theory*; Number 9 in IMS Lecture Notes. Monograph Series; Institute of Mathematical Statistics: Hayward, CA, USA, 1986; p. 283.
2. Amari, S.; Nagaoka, H. *Methods of Information Geometry*; American Mathematical Society: Providence, RI, USA, 2000; p. 206.
3. Pistone, G. Nonparametric Information Geometry. In *Geometric Science of Information*, Proceedings of the First International Conference, GSI 2013, Paris, France, 28–30 August 2013; Nielsen, F., Barbaresco, F., Eds.; Lecture Notes in Computer Science, Volume 8085; Springer: Berlin/Heidelberg, Germany, 2013; pp. 5–36.
4. Absil, P.A.; Mahony, R.; Sepulchre, R. *Optimization Algorithms on Matrix Manifolds*; Princeton University Press: Princeton, NJ, USA, 2008; pp. xvi+224.
5. Nocedal, J.; Wright, S.J. *Numerical Optimization*, 2nd ed.; Springer Series in Operations Research and Financial Engineering; Springer: New York, NY, USA, 2006; pp. xxii+664.
6. Do Carmo, M.P. *Riemannian geometry*; Mathematics: Theory & Applications; Birkhäuser Boston Inc.: Boston, MA, USA, 1992; pp. xiv+300.
7. Abraham, R.; Marsden, J.E.; Ratiu, T. *Manifolds, Tensor Analysis, and Applications*, 2nd ed.; Applied Mathematical Sciences, Volume 75; Springer: New York, NY, USA, 1988; pp. x+654.

8. Lang, S. *Differential and Riemannian Manifolds*, 3rd ed.; Graduate Texts in Mathematics; Springer: New York, NY, USA, 1995; pp. xiv+364.
9. Pistone, G. Algebraic varieties *vs.* differentiable manifolds in statistical models. In *Algebraic and Geometric Methods in Statistics*; Gibilisco, P., Riccomagno, E., Rogantin, M.P., Wynn, H.P., Eds.; Cambridge University Press: Cambridge, UK, 2010.
10. Malagò, L.; Matteucci, M.; Dal Seno, B. An information geometry perspective on estimation of distribution algorithms: Boundary analysis. In Proceedings of the 2008 GECCO Conference Companion On Genetic and Evolutionary Computation (GECCO '08); ACM: New York, NY, USA, 2008; pp. 2081–2088.
11. Malagò, L.; Matteucci, M.; Pistone, G. Stochastic Relaxation as a Unifying Approach in 0/1 Programming. In Proceedings of the NIPS 2009 Workshop on Discrete Optimization in Machine Learning: Submodularity, Sparsity & Polyhedra (DISCML), Whistler Resort & Spa, BC, Canada, 11–12 December 2009.
12. Malagò, L.; Matteucci, M.; Pistone, G. Stochastic Natural Gradient Descent by Estimation of Empirical Covariances. In Proceedings of the IEEE Congress on Evolutionary Computation (CEC), New Orleans, LA, USA, 5–8 June 2011; pp. 949–956.
13. Malagò, L.; Matteucci, M.; Pistone, G. Towards the geometry of estimation of distribution algorithms based on the exponential family. In Proceedings of the 11th Workshop on Foundations of Genetic Algorithms (FOGA '11), Schwarzenberg, Austria, 5–8 January 2011 ; ACM: New York, NY, USA, 2011; pp. 230–242.
14. Malagò, L.; Matteucci, M.; Pistone, G. Natural gradient, fitness modelling and model selection: A unifying perspective. In Proceedings of the IEEE Congress on Evolutionary Computation (CEC), Cancun, Mexico, 20–23 June 2013; pp. 486–493.
15. Amari, S.I. Natural gradient works efficiently in learning. *Neural Comput.* **1998**, *10*, 251–276.
16. Shima, H. *The Geometry of Hessian Structures*; World Scientific Publishing Co. Pte. Ltd.: Hackensack, NJ, USA, 2007; pp. xiv+246.
17. Malagò, L. On the Geometry of Optimization Based on the Exponential Family Relaxation. Ph.D. Thesis, Politecnico di Milano, Milano, Italy, 2012.
18. Gallavotti, G. *Statistical Mechanics: A Short Treatise*; Texts and Monographs in Physics; Springer: Berlin, Germany, 1999; pp. xiv+339.
19. Naudts, J. Generalised exponential families and associated entropy functions. *Entropy* **2008**, *10*, 131–149.
20. Esteban, J.; Ray, D. On the Measurement of Polarization. *Econometrica* **1994**, *62*, 819–851.
21. Montalvo, J.; Reynal-Querol, M. Ethnic polarization, potential conflict, and civil wars. *Am. Econ. Rev.* **2005**, 796–816.
22. Stein, W. *et al.* Sage Mathematics Software (Version 6.0). The Sage Development Team, 2013. Available online: http://www.sagemath.org (accessed on 27 March 2014).

entropy

MDPI

Article

Information Geometric Complexity of a Trivariate Gaussian Statistical Model

Domenico Felice [1,2,*], Carlo Cafaro [3] and Stefano Mancini [1,2]

[1] School of Science and Technology, University of Camerino, I-62032 Camerino, Italy; E-Mail: stefano.mancini@unicam.it

[2] INFN-Sezione di Perugia, Via A. Pascoli, I-06123 Perugia, Italy

[3] Department of Mathematics, Clarkson University, Potsdam, 13699 NY, USA; E-Mail: carlocafaro2000@yahoo.it

* E-Mail: domenico.felice@unicam.it

Received: 1 April 2014; in revised form: 21 May 2014 / Accepted: 22 May 2014 / Published: 26 May 2014

Abstract: We evaluate the information geometric complexity of entropic motion on low-dimensional Gaussian statistical manifolds in order to quantify how difficult it is to make macroscopic predictions about systems in the presence of limited information. Specifically, we observe that the complexity of such entropic inferences not only depends on the amount of available pieces of information but also on the manner in which such pieces are correlated. Finally, we uncover that, for certain correlational structures, the impossibility of reaching the most favorable configuration from an entropic inference viewpoint seems to lead to an information geometric analog of the well-known frustration effect that occurs in statistical physics.

Keywords: probability theory; Riemannian geometry; complexity

1. Introduction

One of the main efforts in physics is modeling and predicting natural phenomena using relevant information about the system under consideration. Theoretical physics has had a general measure of the uncertainty associated with the behavior of a probabilistic process for more than 100 years: the Shannon entropy [1]. The Shannon information theory was applied to dynamical systems and became successful in describing their unpredictability [2].

Along a similar avenue we may set Entropic Dynamics [3] which makes use of inductive inference (Maximum Entropy Methods [4]) and Information Geometry [5]. This is clearly remarkable given that microscopic dynamics can be far removed from the phenomena of interest, such as in complex biological or ecological systems. Extension of ED to temporally-complex dynamical systems on curved statistical manifolds led to relevant measures of chaoticity [6]. In particular, an information geometric approach to chaos (IGAC) has been pursued studying chaos in informational geodesic flows describing physical, biological or chemical systems. It is the information geometric analogue of conventional geometrodynamical approaches [7] where the classical configuration space is being replaced by a statistical manifold with the additional possibility of considering chaotic dynamics arising from non conformally flat metrics. Within this framework, it seems natural to consider as a complexity measure the (time average) statistical volume explored by geodesic flows, namely an Information Geometry Complexity (IGC).

This quantity might help uncover connections between microscopic dynamics and experimentally observable macroscopic dynamics which is a fundamental issue in physics [8]. An interesting manifestation of such a relationship appears in the study of the effects of microscopic external noise (noise imposed on the microscopic variables of the system) on the observed collective motion (macroscopic variables) of a globally coupled map [9]. These effects are quantified in terms of the complexity of the collective motion. Furthermore, it turns out that noise at a microscopic level reduces the complexity of the macroscopic motion, which in turn is characterized by the number of effective degrees of freedom of the system.

The investigation of the macroscopic behavior of complex systems in terms of the underlying statistical structure of its microscopic degrees of freedom also reveals effects due to the presence of microcorrelations [10]. In this article we first show which macro-states should be considered in a Gaussian statistical model in order to have a reduction in time of the Information Geometry Complexity. Then, dealing with correlated bivariate and trivariate Gaussian statistical models, the ratio between the IGC in the presence and in the absence of microcorrelations is explicitly computed, finding an intriguing, even though non yet deep understood, connection with the phenomenon of geometric frustration [11].

The layout of the article is as follows. In Section 2 we introduce a general statistical model discussing its geometry and describing both its dynamics and information geometry complexity. In Section 3, Gaussian statistical models (up to a trivariate model) are considered. There, we compute the asymptotic temporal behaviors of their IGCs. Finally, in Section 4 we draw our conclusions by outlining our findings and proposing possible further investigations.

2. Statistical Models and Information Geometry Complexity

Given n real-valued random variables $X_1, \ldots, X_n$ defined on the sample space Ω with joint probability density $p : \mathbb{R}^n \to \mathbb{R}$ satisfying the conditions

$$p(x) \geq 0 \ (\forall x \in \mathbb{R}^n) \quad \text{and} \quad \int_{\mathbb{R}^n} dx\, p(x) = 1, \tag{1}$$

let us consider a family $\mathcal{P}$ of such distributions and suppose that they can be parametrized using m real-valued variables $(\theta^1, \ldots, \theta^m)$ so that

$$\mathcal{P} = \{p_\theta = p(x|\theta) | \theta = (\theta^1, \ldots, \theta^m) \in \Theta\}, \tag{2}$$

where $\Theta \subseteq \mathbb{R}^m$ is the parameter space and the mapping $\theta \to p_\theta$ is injective. In such a way, $\mathcal{P}$ is an m-dimensional statistical model on $\mathbb{R}^n$.

The mapping $\varphi : \mathcal{P} \to \mathbb{R}^m$ defined by $\varphi(p_\theta) = \theta$ allows us to consider $\varphi = [\theta^i]$ as a coordinate system for $\mathcal{P}$. Assuming parametrizations which are C^∞, we can turn $\mathcal{P}$ into a C^∞ differentiable manifold (thus, $\mathcal{P}$ is called statistical manifold) [5].

The values $x_1, \ldots, x_n$ taken by the random variables define the *micro-state* of the system, while the values $\theta^1, \ldots, \theta^m$ taken by parameters define the *macro-state* of the system.

Let $\mathcal{P} = \{p_\theta | \theta \in \Theta\}$ be an m-dimensional statistical model. Given a point θ, the Fisher information matrix of $\mathcal{P}$ in θ is the $m \times m$ matrix $G(\theta) = [g_{ij}]$, where the (i, j) entry is defined by

$$g_{ij}(\theta) := \int_{\mathbb{R}^n} dx p(x|\theta) \partial_i \log p(x|\theta) \partial_j \log p(x|\theta), \tag{3}$$

with ∂_i standing for $\frac{\partial}{\partial \theta^i}$. The matrix $G(\theta)$ is symmetric, positive semidefinite and determines a Riemannian metric on the parameter space Θ [5]. Hence, it is possible to define a Riemannian statistical manifold $\mathcal{M} := (\Theta, g)$, where $g = g_{ij} d\theta^i \otimes d\theta^j$ $(i, j = 1, \ldots, m)$ is the metric whose components g_{ij} are given by Equation (3) (throughout the paper we use the Einstein sum convention).

Given the Riemannian manifold $\mathcal{M} = (\Theta, g)$, it is well known that there exists only one linear connection ∇(the Levi–Civita connection) on $\mathcal{M}$ that is compatible with the metric g and symmetric [12]. We remark that the manifold $\mathcal{M}$ has one chart, being Θ an open set of $\mathbb{R}^m$, and the Levi-Civita connection is uniquely defined by means of the Christoffel coefficients

$$\Gamma^k_{ij} = \frac{1}{2} g^{kl} \Big(\frac{\partial g_{lj}}{\partial \theta^i} + \frac{\partial g_{il}}{\partial \theta^j} - \frac{\partial g_{ij}}{\partial \theta^l} \Big), \quad (i, j, k = 1, \ldots, m) \tag{4}$$

where g^{kl} is the (k,l) entry of the inverse of the Fisher matrix $G(\theta)$.

The idea of curvature is the fundamental tool to understand the geometry of the manifold $\mathcal{M} = (\Theta, g)$. Actually, it is the basic geometric invariant and the intrinsic way to obtain it is by means of geodesics. It is well-known, that given any point $\theta \in \mathcal{M}$ and any vector v tangent to $\mathcal{M}$ at θ, there is a unique geodesic starting at θ with initial tangent vector v. Indeed, within the considered coordinate system, the geodesics are solutions of the following nonlinear second order coupled ordinary differential equations [12]

$$\frac{d^2\theta^k}{d\tau^2} + \Gamma^k_{ij}\frac{d\theta^i}{d\tau}\frac{d\theta^j}{d\tau} = 0, \tag{5}$$

with τ denoting the time.

The recipe to compute some curvatures at a point $\theta \in \mathcal{M}$ is the following: first, select a 2-dimensional subspace Π of the tangent space to $\mathcal{M}$ at θ; second, follow the geodesics through θ whose initial tangent vectors lie in Π and consider the 2-dimensional submanifolds S_Π swiped out by them inheriting a Riemannian metric from $\mathcal{M}$; finally, compute the Gaussian curvature of S_Π at θ, which can be obtained from its Riemannian metric as stated in the *Theorema Egregium* [13]. The number $K(\Pi)$ found in such manner is called the *sectional curvature* of $\mathcal{M}$ at θ associated with the plane Π. In terms of local coordinates, to compute the sectional curvature we need the curvature tensor,

$$R^h_{ijk} = \frac{\partial \Gamma^h_{jk}}{\partial \theta^i} - \frac{\partial \Gamma^h_{ik}}{\partial \theta^j} + \Gamma^l_{jk}\Gamma^h_{il} - \Gamma^l_{ik}\Gamma^h_{jl}. \tag{6}$$

For any basis (ξ, η) for a 2-plane $\Pi \subset T_\theta\mathcal{M}$, the sectional curvature at $\theta \in \mathcal{M}$ is given by [12]

$$K(\xi,\eta) = \frac{R(\xi,\eta,\eta,\xi)}{|\xi|^2|\eta|^2 - \langle \xi,\eta\rangle}, \tag{7}$$

where R is the Riemann curvature tensor which is written in coordinates as $R = R_{ijkl}d\theta^i \otimes d\theta^j \otimes d\theta^k \otimes d\theta^l$ with $R_{ijkl} = g_{lh}R^h_{ijk}$ and $\langle \cdot, \cdot \rangle$ is the inner product defined by the metric g.

The sectional curvature is directly related to the topology of the manifold; along this direction the *Cartan-Hadamard* Theorem [13] is enlightening by stating that any complete, simply connected n-dimensional manifold with non positive sectional curvature is diffeomorphic to $\mathbb{R}^n$.

We can consider upon the statistical manifold $\mathcal{M} = (\Theta, g)$ the macro-variables θ as accessible information and then derive the information dynamical Equation (5) from a standard principle of least action of Jacobi type [3]. The geodesic Equations (5) describe a reversible dynamics whose solution is the trajectory between an initial and a final macrostate θ^{initial} and θ^{final}, respectively. The trajectory can be equally traversed in both directions [10]. Actually, an equation relating instability with geometry exists and it makes hope that some global information about the average degree of instability (chaos) of the dynamics is encoded in global properties of the statistical manifolds [7]. The fact that this might happen is proved by the special case of constant-curvature manifolds, for which the Jacobi-Levi-Civita equation simplifies to [7]

$$\frac{d^2J^i}{d\tau^2} + KJ^i = 0, \tag{8}$$

where K is the constant sectional curvature of the manifold (see Equation (7)) and J is the geodesic deviation vector field. On a positively curved manifold, the norm of the separating vector J does not grow, whereas on a negatively curved manifold, the norm of J grows exponentially in time, and if the manifold is compact, so that its geodesic are sooner or later obliged to fold, this provide an example of chaotic geodesic motion [14].

Taking into consideration these facts, we single out as suitable indicator of dynamical (temporal) complexity, the information geometric complexity defined as the average dynamical statistical volume [15]

$$\widetilde{vol}\left[\mathcal{D}_{\Theta}^{(\text{geodesic})}(\tau)\right] := \frac{1}{\tau}\int_0^{\tau} d\tau' vol\left[\mathcal{D}_{\Theta}^{(\text{geodesic})}(\tau')\right], \tag{9}$$

where

$$vol\left[\mathcal{D}_{\Theta}^{(\text{geodesic})}(\tau')\right] := \int_{\mathcal{D}_{\Theta}^{(\text{geodesic})}(\tau')} \sqrt{\det(G(\theta))}\, d\theta, \tag{10}$$

with $G(\theta)$ the information matrix whose components are given by Equation (3). The integration space $\mathcal{D}_{\Theta}^{(\text{geodesic})}(\tau')$ is defined as follows

$$\mathcal{D}_{\Theta}^{(\text{geodesic})}(\tau') := \{\theta = (\theta^1, \ldots, \theta^m) : \theta^k(0) \leq \theta^k \leq \theta^k(\tau')\}, \tag{11}$$

where $\theta^k \equiv \theta^k(s)$ with $0 \leq s \leq \tau'$ such that $\theta^k(s)$ satisfies (5). The quantity $vol\left[\mathcal{D}_{\Theta}^{(\text{geodesic})}(\tau')\right]$ is the volume of the effective parameter space explored by the system at time τ'. The temporal average has been introduced in order to average out the possibly very complex fine details of the entropic dynamical description of the system's complexity dynamics.

Relevant properties, concerning complexity of geodesic paths on curved statistical manifolds, of the quantity (10) compared to the Jacobi vector field are discussed in [16].

3. The Gaussian Statistical Model

In the following we devote our attention to a Gaussian statistical model $\mathcal{P}$ whose element are multivariate normal joint distributions for n real-valued variables $X_1, \ldots, X_n$ given by

$$p(x|\theta) = \frac{1}{\sqrt{(2\pi)^n \det C}} \exp\left[-\frac{1}{2}(x-\mu)^t C^{-1}(x-\mu)\right], \tag{12}$$

where $\mu = \big(\mathbb{E}(X_1), \ldots, \mathbb{E}(X_n)\big)$ is the n-dimensional mean vector and C denotes the $n \times n$ covariance matrix with entries $c_{ij} = \mathbb{E}(X_i X_j) - \mathbb{E}(X_i)\mathbb{E}(X_j)$, $i, j = 1, \ldots, n$. Since μ is a n-dimensional real vector and C is a $n \times n$ symmetric matrix, the parameters involved in this model should be $n + \frac{n(n+1)}{2}$. Moreover C is a symmetric, positive definite matrix, hence we have the parameter space given by

$$\Theta := \{(\mu, C) | \mu \in \mathbb{R}^n,\ C \in \mathbb{R}^{n \times n}, C > 0\}. \tag{13}$$

Hereafter we consider the statistical model given by Equation (12) when the covariance matrix C has only variances $\sigma_i^2 = \mathbb{E}(X_i^2) - (\mathbb{E}(X_i))^2$ as parameters. In fact we assume that the non diagonal entry (i, j) of the covariance matrix C equals $\rho\sigma_i\sigma_j$ with $\rho \in \mathbb{R}$ quantifying the degree of correlation.

We may further notice that the function $f_{ij}(x) := \partial_i \log p(x|\theta) \partial_j \log p(x|\theta)$, when $p(x|\theta)$ is given by Equation (12), is a polynomial in the variables x_i ($i = 1, \ldots, n$) whose degree is not grater than four. Indeed, we have that

$$\partial_i \log p(x|\theta) = \frac{1}{p(x|\theta)}\partial_i p(x|\theta) = \partial_i \frac{1}{\sqrt{(2\pi)^n \det C}} + \partial_i \left[-\frac{1}{2}(x-\mu)^t C^{-1}(x-\mu)\right], \tag{14}$$

and, therefore, the differentiation does not affect variables x_i. With this in mind, in order to compute the integral in (3), we can use the following formula [17]

$$\frac{1}{\sqrt{(2\pi)^n \det C}} \int dx f_{ij}(x) \exp\left[-\frac{1}{2}(x-\mu)^t C^{-1}(x-\mu)\right] = \exp\left[\frac{1}{2}\sum_{h,k=1}^{n} c_{hk}\frac{\partial}{\partial x_h}\frac{\partial}{\partial x_k}\right] f_{ij}|_{x=\mu}, \tag{15}$$

where the exponential denotes the power series over its argument (the differential operator).

3.1. The monovariate Gaussian Statistical Model

We now start to apply the concepts of the previous section to a Gaussian statistical model of Equation (12) for $n = 1$. In this case, the dimension of the statistical Riemannian manifold $\mathcal{M} = (\Theta, g)$ is at most two. Indeed, to describe elements of the statistical model $\mathcal{P}$ given by Equation (12), we basically need the mean $\mu = \mathbb{E}(X)$ and variance $\sigma^2 = \mathbb{E}(X-\mu)^2$. We deal separately with the cases when the monovariate model has only μ as macro-variable (Case 1), when σ is the unique macro-variable (Case 2), and finally when both μ and σ are macro-variables (Case 3).

3.1.1. Case 1

Consider the monovariate model with only μ as macro-variable by setting $\sigma = 1$. In this case the manifold $\mathcal{M}$ is trivially the real *flat* straight line, since $\mu \in (-\infty, +\infty)$. Indeed, the integral in (3) is equal to 1 when the distribution $p(x|\theta)$ reads as $p(x|\mu) = \frac{\exp\left[-\frac{1}{2}(x-\mu)^2\right]}{\sqrt{2\pi}}$; so the metric is $g = d\mu^2$. Furthermore, from Equations (4) and (5) the information dynamics is described by the geodesic $\mu(\tau) = A_1\tau + A_2$, where $A_1, A_2 \in \mathbb{R}$. Hence, the volume of Equation (10) results $vol\left[\mathcal{D}_{\Theta}^{(\text{geodesic})}(\tau')\right] = \int d\mu = A_1\tau + A_2$; since this quantity must be positive we assume $A_1, A_2 > 0$. Finally, the asymptotic behavior of the IGC (9) is

$$\widetilde{vol}\left[\mathcal{D}_{\Theta}^{(\text{geodesic})}(\tau)\right] \approx \left(\frac{A_1}{2}\right)\tau. \tag{16}$$

This shows that the complexity linearly increases in time meaning that acquiring information about μ and updating it, is not enough to increase our knowledge about the micro state of the system.

3.1.2. Case 2

Consider now the monovariate Gaussian statistical model of Equation(12) when $\mu = \mathbb{E}(X) = 0$ and the macro-variable is only σ. In this case the probability distribution function reads $p(x|\sigma) = \frac{\exp\left[-\frac{x^2}{2\sigma^2}\right]}{\sqrt{2\pi}\sigma}$ while the Fisher–Rao metric becomes $g = \frac{2}{\sigma^2}d\sigma^2$. Emphasizing that also in this case the manifold is flat as well, we derive the information dynamics by means of Equations (4) and (5) and we obtain the geodesic $\sigma(\tau) = A_1 \exp\left[A_2\tau\right]$. The volume in Equation (10) then results

$$vol\left[\mathcal{D}_{\Theta}^{(\text{geodesic})}(\tau')\right] = \int \frac{\sqrt{2}}{\sigma}d\sigma = \sqrt{2}\log\left[A_1 \exp\left[A_2\tau\right]\right]. \tag{17}$$

Again, to have positive volume we have to assume $A_1, A_2 > 0$. Finally, the (asymptotic) IGC (9) becomes

$$\widetilde{vol}\left[\mathcal{D}_{\Theta}^{(\text{geodesic})}(\tau)\right] \approx \left(\frac{\sqrt{2}A_2}{2}\right)\tau. \tag{18}$$

This shows that also in this case the complexity linearly increases in time meaning that acquiring information about σ and updating it, is not enough to increase our knowledge about the micro-state of the system.

3.1.3. Case 3

The take home message of the previous cases is that we have to account for both mean μ and variance σ as macro-variables to look for possible non increasing complexity. Hence, consider the probability distribution function is given by,

$$p(x_1, x_2|\mu, \sigma) = \frac{\exp\left[-\frac{1}{2}\frac{(x-\mu)^2}{\sigma^2}\right]}{\sigma\sqrt{2\pi}}. \tag{19}$$

The dimension of the Riemannian manifold $\mathcal{M} = (\Theta, g)$ is two, where the parameter space Θ is given by $\Theta = \{(\mu, \sigma) | \mu \in (-\infty, +\infty), \sigma > 0\}$ and the Fisher–Rao metric reads as $g = \frac{1}{\sigma^2} d\mu^2 + \frac{2}{\sigma^2} d\sigma^2$. Here, the sectional curvature given by Equation (7) is a negative function and despite the fact that is not constant, we expect a decreasing behavior in time of the IGC. Thanks to Equation (4), we find that the only non negative Christoffel coefficients are $\Gamma^1_{12} = -\frac{1}{\sigma}$, $\Gamma^2_{11} = \frac{1}{2\sigma}$ and $\Gamma^2_{22} = -\frac{1}{\sigma}$. Substituting them into Equation (5) we derive the following geodesic equations

$$\begin{cases} \frac{d^2\mu(\tau)}{d\tau^2} - \frac{2}{\sigma}\frac{d\sigma}{d\tau}\frac{d\mu}{d\tau} = 0, \\ \frac{d^2\sigma(\tau)}{d\tau^2} - \frac{1}{\sigma}\left(\frac{d\sigma}{d\tau}\right)^2 + \frac{1}{2\sigma}\left(\frac{d\mu}{d\tau}\right)^2 = 0. \end{cases} \tag{20}$$

The integration of the above coupled differential equations is non-trivial. We follow the method described in [10] and arrive at

$$\sigma(\tau) = \frac{2\sigma_0 \exp\left[\frac{\sigma_0|A_1|}{\sqrt{2}}\tau\right]}{1 + \exp\left[\frac{2\sigma_0|A_1|}{\sqrt{2}}\tau\right]}, \qquad \mu(\tau) = -\frac{2\sigma_0\sqrt{2}A_1}{|A_1|\left(1 + \exp\left[\frac{2\sigma_0|A_1|}{\sqrt{2}}\tau\right]\right)}, \tag{21}$$

where σ_0 and A_1 are real constants. Then, using (21), the volume of Equation (10) results

$$vol\left[\mathcal{D}_{\Theta}^{(\text{geodesic})}(\tau')\right] = \int \frac{\sqrt{2}}{\sigma^2} d\sigma d\mu = \frac{\sqrt{2}A_1}{|A_1|} \exp\left[-\frac{\sigma_0|A_1|}{\sqrt{2}}\tau\right]. \tag{22}$$

Since the last quantity must be positive, we assume $A_1 > 0$. Finally, employing the above expression into Equation (9) we arrive at

$$\widetilde{vol}\left[\mathcal{D}_{\Theta}^{(\text{geodesic})}(\tau)\right] \approx \left(\frac{2}{\sigma_0 A_1}\right)\frac{1}{\tau}. \tag{23}$$

We can now see a reduction in time of the complexity meaning that acquiring information about both μ and σ and updating them allows us to increase our knowledge about the micro state of the system.

Hence, comparing Equations (16), (18) and (23) we conclude that the entropic inferences on a Gaussian distributed micro-variable is carried out in a more efficient manner when both its mean and the variance in the form of information constraints are available. Macroscopic predictions when only one of these pieces of information are available are more complex.

3.2. Bivariate Gaussian Statistical Model

Consider now the Gaussian statistical model $\mathcal{P}$ of the Equation (12) when $n = 2$. In this case the dimension of the Riemannian manifold $\mathcal{M} = (\Theta, g)$ is at most four. From the analysis of the monovariate Gaussian model in Section 3.1 we have understood that both mean and variance should be considered. Hence the minimal assumption is to consider $\mathbb{E}(X_1) = \mathbb{E}(X_2) = \mu$ and $\mathbb{E}(X_1 - \mu)^2 = \mathbb{E}(X_2 - \mu)^2 = \sigma^2$. Furthermore, in this case we have also to take into account the possible presence of (micro) correlations, which appear at the level of macro-states as off-diagonal terms in the covariance matrix. In short, this implies considering the following probability distribution function

$$p(x_1, x_2|\mu, \sigma) = \frac{\exp\left[-\frac{1}{2\sigma^2(1-\rho^2)}\left((x_1-\mu)^2 - 2\rho(x_1-\mu)(x_2-\mu) + (x_2-\mu)^2\right)\right]}{2\pi\sigma^2\sqrt{1-\rho^2}}, \tag{24}$$

where $\rho \in (-1, 1)$.

Thanks to Equation (15) we compute the Fisher-Information matrix G and find $g = g_{11}d\mu^2 + g_{22}d\sigma^2$ with,

$$g_{11} = \frac{2}{\sigma^2(\rho+1)}; \; g_{22} = \frac{4}{\sigma^2}. \tag{25}$$

The only non trivial Christoffel coefficients (4) are $\Gamma^1_{12} = -\frac{1}{\sigma}$, $\Gamma^2_{11} = \frac{1}{2\sigma(\rho+1)}$ and $\Gamma^2_{22} = -\frac{1}{\sigma}$. In this case as well, the sectional curvature (Equation (7)) of the manifold $\mathcal{M}$ is a negative function and so we may expect a decreasing asymptotic behavior for the IGC. From Equation (5) it follows that the geodesic equations are,

$$\begin{cases} \frac{d^2\mu(\tau)}{d\tau^2} - \frac{2}{\sigma}\frac{d\sigma}{d\tau}\frac{d\mu}{d\tau} = 0 \\ \\ \frac{d^2\sigma(\tau)}{d\tau^2} - \frac{1}{\sigma}\left(\frac{d\sigma}{d\tau}\right)^2 + \frac{1}{2(1+\rho)\sigma}\left(\frac{d\mu}{d\tau}\right)^2 = 0, \end{cases} \tag{26}$$

whose solutions are,

$$\sigma(\tau) = \frac{2\sigma_0 \exp\left[\frac{\sigma_0|A_1|}{\sqrt{2(1+\rho)}}\tau\right]}{1+\exp\left[\frac{2\sigma_0|A_1|}{\sqrt{2(1+\rho)}}\tau\right]}, \quad \mu(\tau) = -\frac{2\sigma_0\sqrt{2(1+\rho)}A_1}{|A_1|\left(1+\exp\left[\frac{2\sigma_0|A_1|}{\sqrt{2(1+\rho)}}\tau\right]\right)}. \tag{27}$$

Using (27) in Equation (10) gives the volume,

$$vol\left[\mathcal{D}_{\Theta}^{(\text{geodesic})}(\tau')\right] = \int \frac{2\sqrt{2}}{\sqrt{1+\rho}\,\sigma^2}d\sigma d\mu = \frac{4A_1}{|A_1|}\exp\left[-\frac{\sigma_0|A_1|}{\sqrt{2(1+\rho)}}\tau\right]. \tag{28}$$

To have it positive we have to assume $A_1 > 0$. Finally, employing (28) in (9) leads to the IGC,

$$\widetilde{\text{vol}}\left[\mathcal{D}_{\Theta}^{(\text{geodesic})}(\tau)\right] \approx \left(\frac{4\sqrt{2}}{\sigma_0 A_1}\right)\frac{\sqrt{1+\rho}}{\tau}, \tag{29}$$

with $\rho \in (-1,1)$. We may compare the asymptotic expression of the ICGs in the presence and in the absence of correlations, obtaining

$$R^{\text{strong}}_{\text{bivariate}}(\rho) := \frac{\widetilde{vol}\left[\mathcal{D}_{\Theta}^{(\text{geodesic})}(\tau)\right]}{\widetilde{vol}\left[\mathcal{D}_{\Theta}^{(\text{geodesic})}(\tau)\right]_{\rho=0}} = \sqrt{1+\rho}, \tag{30}$$

where "strong" stands for the fully connected lattice underlying the micro-variables. The ratio $R^{\text{strong}}_{\text{bivariate}}(\rho)$ results a *monotonic* increasing function of ρ.

While the temporal behavior of the IGC (29) is similar to the IGC in (23), here correlations play a fundamental role. From Equation (30), we conclude that entropic inferences on two Gaussian distributed micro-variables on a fully connected lattice is carried out in a more efficient manner when the two micro-variables are negatively correlated. Instead, when such micro-variables are positively correlated, macroscopic predictions become more complex than in the absence of correlations.

Intuitively, this is due to the fact that for anticorrelated variables, an increase in one variable implies a decrease in the other one (different directional change): variables become more distant, thus more distinguishable in the Fisher–Rao information metric sense. Similarly, for positively correlated variables, an increase or decrease in one variable always predicts the same directional change for the second variable: variables do not become more distant, thus more distinguishable in the Fisher–Rao information metric sense. This may lead us to guess that in the presence of anticorrelations, motion on curved statistical manifolds via the Maximum Entropy updating methods becomes less complex.

3.3. *Trivariate Gaussian Statistical Model*

In this section we consider a Gaussian statistical model $\mathcal{P}$ of the Equation (12) when $n = 3$. In this case as well, in order to understand the asymptotic behavior of the IGC in the presence of correlations between the micro-states, we make the minimal assumption that, given the random vector $X = (X_1, X_2, X_3)$ distributed according to a trivariate Gaussian, then $\mathbb{E}(X_1) = \mathbb{E}(X_2) = \mathbb{E}(X_3) = \mu$

and $\mathbb{E}(X_1-\mu)^2=\mathbb{E}(X_2-\mu)^2=\mathbb{E}(X_2-\mu)^2=\sigma^2$. Therefore, the space of the parameters of $\mathcal{P}$ is given by $\Theta=\{(\mu,\sigma)|\mu\in\mathbb{R},\sigma>0\}$.

The manifold $\mathcal{M}=(\Theta,g)$ changes its metric structure depending on the number of correlations between micro-variables, namely, one, two, or three . The covariance matrices corresponding to these cases read, modulo the congruence via a permutation matrix [17],

$$C_1=\sigma^2\begin{pmatrix}1&\rho&0\\\rho&1&0\\0&0&1\end{pmatrix},\qquad C_2=\sigma^2\begin{pmatrix}1&\rho&\rho\\\rho&1&0\\\rho&0&1\end{pmatrix},\qquad C_3=\sigma^2\begin{pmatrix}1&\rho&\rho\\\rho&1&\rho\\\rho&\rho&1\end{pmatrix}. \tag{31}$$

3.3.1. Case 1

First, we consider the trivariate Gaussian statistical model of Equation (12) when $C\equiv C_1$. Then proceeding like in Section 3.2 we have $g=g_{11}d\mu^2+g_{22}d\sigma^2$, where $g_{11}=\frac{3+\rho}{(1+\rho)\sigma^2}$ and $g_{22}=\frac{6}{\sigma^2}$. Also in this case we find that the sectional curvature of Equation (7) is a negative function. Hence, as we state in Section 2, we may expect a decreasing (in time) behavior of the information geometry complexity. Furthermore, we obtain the geodesics

$$\sigma(\tau)=\frac{2\sigma_0\exp\left[\sigma_0\sqrt{\mathcal{A}(\rho)}\;\tau\right]}{1+\exp\left[2\sigma_0\sqrt{\mathcal{A}(\rho)}\;\tau\right]},\ \mu(\tau)=-\frac{2\sigma_0A_1}{\sqrt{\mathcal{A}(\rho)}}\frac{1}{1+\exp\left[2\sigma_0\sqrt{\mathcal{A}(\rho)}\;\tau\right]}, \tag{32}$$

where $\mathcal{A}(\rho)=\frac{A_1^2(3+\rho)}{6(1+\rho)}$ and $A_1\in\mathbb{R}$. We remark that $\mathcal{A}(\rho)>0$ for all $\rho\in(-1,1)$. Then, the volume (10) becomes

$$vol\left[\mathcal{D}_{\Theta}^{(\text{geodesic})}(\tau')\right]=\int\sqrt{\frac{6(3-4\rho)}{(1-2\rho^2)}}\frac{1}{\sigma^2}d\sigma d\mu=\frac{6A_1}{|A_1|}\exp\left[-\sigma_0\sqrt{\mathcal{A}(\rho)}\;\tau\right], \tag{33}$$

requiring $A_1>0$ for its positivity. Finally, using (33) in (9) we arrive at the asymptotic behavior of the IGC

$$\widetilde{vol}\left[\mathcal{D}_{\Theta}^{(\text{geodesic})}(\tau)\right]\approx\left(\frac{6\sqrt{6}}{\sigma_0A_1}\right)\sqrt{\frac{1+\rho}{3+\rho}}\,\frac{1}{\tau}. \tag{34}$$

Comparing (34) in the presence and in the absence of correlations yields

$$R_{\text{trivariate}}^{\text{weak}}(\rho):=\frac{\widetilde{vol}\left[\mathcal{D}_{\Theta}^{(\text{geodesic})}(\tau)\right]}{\widetilde{vol}\left[\mathcal{D}_{\Theta}^{(\text{geodesic})}(\tau)\right]_{\rho=0}}=\sqrt{3}\sqrt{\frac{1+\rho}{3+\rho}}, \tag{35}$$

where "weak" stands for low degree of connection in the lattice underlying the micro-variables

Notice that $R_{\text{trivariate}}^{\text{weak}}(\rho)$ is a monotonic increasing function of the argument $\rho\in(-1,1)$.

3.3.2. Case 2

When the trivariate Gaussian statistical model of Equation (12) has $C\equiv C_2$, the condition $C>0$ constraints the correlation coefficient to be $\rho\in(-\frac{\sqrt{2}}{2},\frac{\sqrt{2}}{2})$. Proceeding again like in Section 3.2 we have $g=g_{11}d\mu^2+g_{22}d\sigma^2$, where $g_{11}=\frac{3-4\rho}{(1-2\rho^2)\sigma^2}$ and $g_{22}=\frac{6}{\sigma^2}$. The sectional curvature of Equation (7) is a negative function as well and so we may apply the arguments of Section 2 expecting a decreasing in time of the complexity. Furthermore, we obtain the geodesics

$$\sigma(\tau)=\frac{2\sigma_0\exp\left[\sigma_0\sqrt{\mathcal{A}(\rho)}\;\tau\right]}{1+\exp\left[2\sigma_0\sqrt{\mathcal{A}(\rho)}\;\tau\right]},\ \mu(\tau)=-\frac{2\sigma_0A_1}{\sqrt{\mathcal{A}(\rho)}}\frac{1}{1+\exp\left[2\sigma_0\sqrt{\mathcal{A}(\rho)}\;\tau\right]}, \tag{36}$$

where $\mathcal{A}(\rho) = \frac{A_1^2(3-4\rho)}{6(1-2\rho^2)}$ and $A_1 \in \mathbb{R}$. We remark that $\mathcal{A}(\rho) > 0$ for all $\rho \in (-\frac{\sqrt{2}}{2}, \frac{\sqrt{2}}{2})$. Then, the volume (10) becomes

$$vol\left[\mathcal{D}_{\Theta}^{(\text{geodesic})}(\tau')\right] = \int \sqrt{\frac{6(3-4\rho)}{(1-2\rho^2)}}\,\frac{1}{\sigma^2} d\sigma d\mu = \frac{6A_1}{|A_1|}\exp\left[-\sigma_0\sqrt{\mathcal{A}(\rho)}\,\tau\right]. \tag{37}$$

We have to set $A_1 > 0$ for the positivity of the volume (37), and using it in (9) we arrive at the asymptotic behavior of the IGC

$$\widetilde{vol}\left[\mathcal{D}_{\Theta}^{(\text{geodesic})}(\tau)\right] \approx \left(\frac{6\sqrt{6}}{\sigma_0 A_1}\right)\sqrt{\frac{1-2\rho^2}{3-4\rho}}\,\frac{1}{\tau}. \tag{38}$$

Then, comparing (38) in the presence and in the absence of correlations yields

$$R_{\text{trivariate}}^{\text{mildly weak}}(\rho) := \frac{\widetilde{vol}\left[\mathcal{D}_{\Theta}^{(\text{geodesic})}(\tau)\right]}{\widetilde{vol}\left[\mathcal{D}_{\Theta}^{(\text{geodesic})}(\tau)\right]_{\rho=0}} = \sqrt{3}\sqrt{\frac{1-2\rho^2}{3-4\rho}}, \tag{39}$$

where "mildly weak" stands for a lattice (underlying micro-variables) neither fully connected nor with minimal connection.

This is a function of the argument $\rho \in (-\frac{\sqrt{2}}{2}, \frac{\sqrt{2}}{2})$ that attains the maximum $\sqrt{\frac{3}{2}}$ at $\rho = \frac{1}{2}$, while in the extrema of the interval $(-\frac{\sqrt{2}}{2}, \frac{\sqrt{2}}{2})$ it tends to zero.

3.3.3. Case 3

Last, we consider the trivariate Gaussian statistical model of the Equation (12) when $C \equiv C_3$. In this case, the condition $C > 0$ requires the correlation coefficient to be $\rho \in (-\frac{1}{2}, 1)$. Proceeding again like in Section 3.2 we have $g = g_{11}d\mu^2 + g_{22}d\sigma^2$, where $g_{11} = \frac{3}{(1+2\rho)\sigma^2}$ and $g_{22} = \frac{6}{\sigma^2}$. We find that the sectional curvature of Equation (7) is a negative function; hence, we may expect a decreasing (in time) behavior of the complexity. It follows the geodesics

$$\sigma(\tau) = \frac{2\sigma_0 \exp\left[\sigma_0\sqrt{\mathcal{A}(\rho)}\,\tau\right]}{1+\exp\left[2\sigma_0\sqrt{\mathcal{A}(\rho)}\,\tau\right]},\ \mu(\tau) = -\frac{2\sigma_0 A_1}{\sqrt{\mathcal{A}(\rho)}}\frac{1}{1+\exp\left[2\sigma_0\sqrt{\mathcal{A}(\rho)}\,\tau\right]}, \tag{40}$$

where $\mathcal{A}(\rho) = \frac{A_1^2}{2(1+2\rho)}$ and $A_1 \in \mathbb{R}$. We note that $\mathcal{A}(\rho) > 0$ for all $\rho \in (-\frac{1}{2}, 1)$. Using (40), we compute

$$vol\left[\mathcal{D}_{\Theta}^{(\text{geodesic})}(\tau')\right] = \int \frac{3\sqrt{2}}{\sqrt{(1+2\rho)}}\,\frac{1}{\sigma^2} d\sigma d\mu = \frac{6\sqrt{2}A_1}{|A_1|}\exp\left[-\sigma_0\sqrt{\mathcal{A}(\rho)}\,\tau\right]. \tag{41}$$

Also in this case we need to assume $A_1 > 0$ to have positive volume. Finally, substituting Equation (41) into Equation (9), the asymptotic behavior of the IGC results

$$\widetilde{vol}\left[\mathcal{D}_{\Theta}^{(\text{geodesic})}(\tau)\right] \approx \left(\frac{12}{\sigma_0 A_1}\right)\sqrt{1+2\rho}\,\frac{1}{\tau}. \tag{42}$$

The comparison of (42) in the presence and in the absence of correlations yields

$$R_{\text{trivariate}}^{\text{strong}}(\rho) := \frac{\widetilde{vol}\left[\mathcal{D}_{\Theta}^{(\text{geodesic})}(\tau)\right]}{\widetilde{vol}\left[\mathcal{D}_{\Theta}^{(\text{geodesic})}(\tau)\right]_{\rho=0}} = \sqrt{1+2\rho}, \tag{43}$$

where "strong" stands for a fully connected lattice underlying the (three) micro-variables. We remark the latter ratio is a monotonically increasing function of the argument $\rho \in (-\frac{1}{2}, 1)$.

The behaviors of $R(\rho)$ of Equations (30), (35), (39) and (43) are reported in Figure 1.

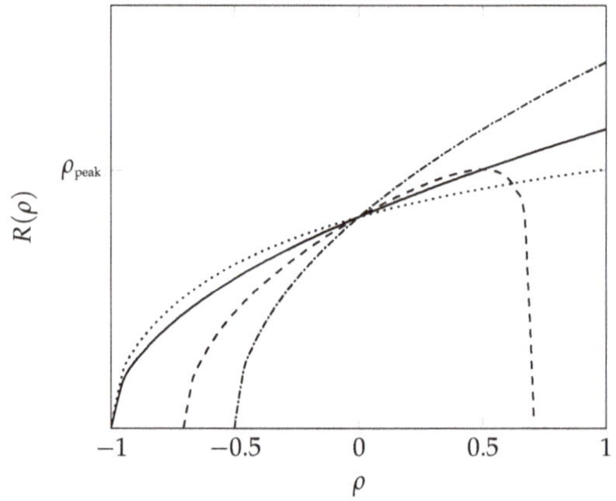

Figure 1. Ratio $R(\rho)$ of volumes *vs.* degree of correlations ρ. Solid line refers to $R^{\text{strong}}_{\text{bivariate}}(\rho)$; Dotted line refers to $R^{\text{weak}}_{\text{trivariate}}(\rho)$; Dashed line referes to $R^{\text{mildly weak}}_{\text{trivariate}}(\rho)$; Dash-dotted refers to $R^{\text{strong}}_{\text{trivariate}}(\rho)$.

The *non-monotonic* behavior of the ratio $R^{\text{mildly weak}}_{\text{trivariate}}(\rho)$ in Equation (39) corresponds to the information geometric complexities for the mildly weak connected three-dimensional lattice. Interestingly, the growth stops at a critical value $\rho_{\text{peak}} = \frac{1}{2}$ at which $R^{\text{mildly weak}}_{\text{trivariate}}(\rho_{\text{peak}}) = R^{\text{strong}}_{\text{bivariate}}(\rho_{\text{peak}})$. From Equation (30), we conclude that entropic inferences on three Gaussian distributed micro-variables on a fully connected lattice is carried out in a more efficient manner when the two micro-variables are negatively correlated. Instead, when such micro-variables are positively correlated, macroscopic predictions become more complex that in the absence of correlations. Furthermore, the ratio $R^{\text{strong}}_{\text{trivariate}}(\rho)$ of the information geometric complexities for this fully connected three-dimensional lattice increases in a *monotonic* fashion. These conclusions are similar to those presented for the bivariate case. However, there is a key-feature of the IGC to emphasize when passing from the two-dimensional to the three-dimensional manifolds associated with fully connected lattices: the effects of negative-correlations and positive-correlations are *amplified* with respect to the respective absence of correlations scenarios,

$$\frac{R^{\text{strong}}_{\text{trivariate}}(\rho)}{R^{\text{strong}}_{\text{bivariate}}(\rho)} = \sqrt{\frac{1+2\rho}{1+\rho}}, \tag{44}$$

where $\rho \in (-\frac{1}{2}, 1)$.

Specifically, carrying out entropic inferences on the higher-dimensional manifold in the presence of anti-correlations, that is for $\rho \in \left(-\frac{1}{2}, 0\right)$, is less complex than on the lower-dimensional manifold as evident form Equation (44). The vice-versa is true in the presence of positive-correlations, that is for $\rho \in (0, 1)$.

4. Concluding Remarks

In summary, we considered low dimensional Gaussian statistical models (up to a trivariate model) and have investigated their dynamical (temporal) complexity. This has been quantified by the volume of geodesics for parameters characterizing the probability distribution functions. To the best of our knowledge, there is no *dynamic* measure of complexity of geodesic paths on curved statistical manifolds that could be compared to our IGC. However, it could be worthwhile to understand the connection, if any, between our IGC and the complexity of paths of dynamic systems introduced in [20]. Specifically, according to the Alekseev-Brudno theorem in the algorithmic theory of dynamical systems [21], a way to predict each new segment of chaotic trajectory is obtained by adding information proportional to the length of this segment and independent of the full previous length of trajectory. This means that this information cannot be extracted from observation of the previous motion, even an infinitely long one! If the instability is a power law, then the required information per unit time is inversely proportional to the full previous length of the trajectory and, asymptotically, the prediction becomes possible.

For the sake of completeness, we also point out that the relevance of volumes in quantifying the *static* model complexity of statistical models was already pointed out in [22] and [23]: complexity is related to the volume of a model in the space of distributions regarded as a Riemannian manifold of distributions with a natural metric defined by the Fisher–Rao metric tensor. Finally, we would like to point out that two of the Authors have recently associated Gaussian statistical models to networks [17]. Specifically, it is assumed that random variables are located on the vertices of the network while correlations between random variables are regarded as weighted edges of the network. Within this framework, a static network complexity measure has been proposed as the volume of the corresponding statistical manifold. We emphasize that such a static measure could be, in principle, applied to time-dependent networks by accommodating time-varying weights on the edges [24]. This requires the consideration of a time-sequence of different statistical manifolds. Thus, we could follow the time-evolution of a network complexity through the time evolution of the volumes of the associated manifolds.

In this work we uncover that in order to have a reduction in time of the complexity one has to consider both mean and variance as macro-variables. This leads to different topological structures of the parameter space in (13); in particular, we have to consider at least a 2-dimensional manifold in order to have effects such as a power law decay of the complexity. Hence, the minimal hypothesis in a multivariate Gaussian model consists in considering all mean values equal and all covariances equal. In such a case, however, the complexity shows interesting features depending on the correlation among micro-variables (as summarized in Figure 1). For a trivariate model with only two correlations the information geometric complexity ratio exhibits a non monotonic behavior in ρ (correlation parameter) taking zero value at the extrema of the range of ρ. In contrast to closed configurations (bivariate and trivariate models with all micro-variables correlated each other) the complexity ratio exhibits a monotonic behavior in terms of the correlation parameter. The fact that in such a case this ratio cannot be zero at the extrema of the range of ρ is reminiscent of the geometric frustration phenomena that occurs in the presence of loops [11].

Specifically, recall that a geometrically frustrated system cannot simultaneously minimize all interactions because of geometric constraints [11,18]. For example, geometric frustration can occur in an Ising model which is an array of spins (for instance, atoms that can take states ± 1) that are magnetically coupled to each other. If one spin is, say, in the $+1$ state then it is energetically favorable for its immediate neighbors to be in the same state in the case of a ferromagnetic model. On the contrary, in antiferromagnetic systems, nearest neighbor spins want to align in opposite directions. This rule can be easily satisfied on a square. However, due to geometrical frustration, it is not possible to satisfy it on a triangle: for an antiferromagnetic triangular Ising model, any three neighboring spins are frustrated. Geometric frustration in triangular Ising models can be observed by considering spin configurations with total spin $J = \pm 1$ and analyzing the fluctuations in energy of the spin system as a function of temperature. There is no peak at all in the standard deviation of the energy in the case $J = -1$, and a monotonic behavior is recorded. This indicates that the antiferromagnetic system does not have a phase transition to a state with long-range order. Instead, in the case $J = +1$, a peak in the energy fluctuations emerges. This significant change in the behavior of energy fluctuations as a function of temperature in triangular configurations of spin systems is a signature of the presence of frustrated interactions in the system [19].

In this article, we observe a significant change in the behavior of the information geometric complexity ratios as a function of the correlation coefficient in the trivariate Gaussian statistical models. Specifically, in the fully connected trivariate case, no peak arises and a monotonic behavior in ρ of the information geometric complexity ratio is observed. In the mildly weak connected trivariate case, instead, a peak in the information geometric complexity ratio is recorded at $\rho_{\text{peak}} \geq 0$. This dramatic disparity of behavior can be ascribed to the fact that when carrying out statistical inferences with positively correlated Gaussian random variables, the maximum entropy favorable scenario is incompatible with these working hypothesis. Thus, the system appears frustrated.

These considerations lead us to conclude that we have uncovered a very interesting information geometric resemblance of the more standard geometric frustration effect in Ising spin models. However, for a conclusive claim of the existence of an information geometric analog of the frustration effect, we feel we have to further deepen our understanding. A forthcoming research project along these lines will be a detailed investigation of both arbitrary triangular and square configurations of correlated Gaussian random variables where we take into consideration both the presence of different intensities and signs of pairwise interactions ($\rho_{ij} \neq \rho_{ik}$ if $j \neq k$, $\forall i$).

Acknowledgments: Domenico Felice and Stefano Mancini acknowledge the financial support of the Future and Emerging Technologies (FET) programme within the Seventh Framework Programme for Research of the European Commission, under the FET-Open grant agreement TOPDRIM, number FP7-ICT-318121.

Author Contributions: The authors have equally contributed to the paper. All authors read and approved the final manuscript.

Conflicts of Interest: Conflicts of Interest

The authors declare no conflict of interest.

References

1. Feldman, D.F.; Crutchfield, J.P. Measures of Statistical Complexity: Why? *Phys. Lett. A* **1998**, *238*, 244–252.
2. Kolmogorov, A.N. A new metric invariant of transitive dynamical systems and of automorphism of Lebesgue spaces. *Doklady Akademii Nauk SSSR* **1958**, *119*, 861–864.
3. Caticha, A. Entropic Dynamics. *Bayesian Inference and Maximum Entropy Methods in Science and Engineering*, the 22nd International Workshop on Bayesian Inference and Maximum Entropy Methods in Science and Engineering, Moscow, Idaho, 3-7 August 2002; Fry, R.L., Ed.; American Institute of Physics: College Park, MD, USA, 2002; Volume 617, p. 302.
4. Caticha, A.; Preuss, R. Maximum entropy and Bayesian data analysis: Entropic prior distributions. *Phys. Rev. E* **2004**, *70*, 046127.
5. Amari, S.; Nagaoka, H. *Methods of Information Geometry*; Oxford University Press: New York, NY, USA, 2000.
6. Cafaro, C. Works on an information geometrodynamical approach to chaos. *Chaos Solitons Fractals* **2009**, *41*, 886–891.
7. Pettini, M. *Geometry and Topology in Hamiltonian Dynamics and Statistical Mechanics*; Springer-Verlag: Berlin/Heidelberg, Germany, 2007.
8. Lebowitz, J.L. Microscopic Dynamics and Macroscopic Laws. *Ann. N. Y. Acad. Sci.* **1981**, *373*, 220–233.
9. Shibata, T.; Chawanya, T.; Chawanya, K. Noiseless Collective Motion out of Noisy Chaos. *Phys. Rev. Lett.* **1999**, *82*, doi: http://dx.doi.org/10.1103/PhysRevLett.82.4424.
10. Ali, S.A.; Cafaro, C.; Kim, D.-H.; Mancini, S. The effect of the microscopic correlations on the information geometric complexity of Gaussian statistical models. *Physica A* **2010**, *389*, 3117–3127.
11. Sadoc, J.F.; Mosseri, R. *Geometrical Frustration*; Cambridge University Press: Cambridge, UK, 2006.
12. Lee, J.M. *Riemannian Manifolds: An Introduction to Curvature*; Springer: New York, NY, USA, 1997.
13. Do Carmo, M.P. *Riemannian Geometry*; Springer: New York, NY, USA, 1992.
14. Cafaro, C.; Ali, S.A. Jacobi fields on statistical manifolds of negative curvature. *Physica D* **2007**, *234*, 70–80.
15. Cafaro, C.; Giffin, A.; Ali, S.A.; Kim, D.-H. Reexamination of an information geometric construction of entropic indicators of complexity. *Appl. Math. Comput.* **2010**, *217*, 2944–2951.
16. Cafaro, C.; Mancini, S. Quantifying the complexity of geodesic paths on curved statistical manifolds through information geometric entropies and Jacobi fields. *Physica D* **2011**, *240*, 607–618.

17. Felice, D.; Mancini, S.; Pettini, M. Quantifying Networks Complexity from Information Geometry Viewpoint. *J. Math. Phys.* **2014**, *55*, 043505.
18. Moessner, R.; Ramirez, A.P. Geometrical Frustration. *Phys. Today* **2006**, *59*, 24–29.
19. MacKay, D.J.C. *Information Theory, Inference, and Learning Algorithms*; Cambridge University Press: Cambridge, UK, 2003.
20. Brudno, A.A. The complexity of the trajectories of a dynamical system. *Uspekhi Mat. Nauk* **1978**, *33*, 207–208.
21. Alekseev, V.M.; Yacobson, M.V. Symbolic dynamics and hyperbolic dynamic systems. *Phys. Rep.* **1981**, *75*, 290–325.
22. Myung, J.; Balasubramanian, V.; Pitt, M.A. Counting probability distributions: differential geometry and model selection. *Proc. Natl. Acad. Sci. USA* **2000**, *97*, 11170.
23. Rodriguez, C.C. The volume of bitnets. *AIP Conf. Proc.* **2004**, *735*, 555–564.
24. Motter, A.E.; Albert, R. Networks in motion. *Phys. Today* **2012**, *65*, 43–48.

entropy

MDPI

Article

Learning from Complex Systems: On the Roles of Entropy and Fisher Information in Pairwise Isotropic Gaussian Markov Random Fields

Alexandre Levada

Computing Department, Federal University of São Carlos, Rod. Washington Luiz, km 235, São Carlos, SP, Brazil; E-Mail: alexandre@dc.ufscar.br

Received: 4 December 2013; / Accepted: 30 January 2014 / Published: 18 February 2014

Abstract: Markov random field models are powerful tools for the study of complex systems. However, little is known about how the interactions between the elements of such systems are encoded, especially from an information-theoretic perspective. In this paper, our goal is to enlighten the connection between Fisher information, Shannon entropy, information geometry and the behavior of complex systems modeled by isotropic pairwise Gaussian Markov random fields. We propose analytical expressions to compute local and global versions of these measures using Besag's pseudo-likelihood function, characterizing the system's behavior through its Fisher curve , a parametric trajectory across the information space that provides a geometric representation for the study of complex systems in which temperature deviates from infinity. Computational experiments show how the proposed tools can be useful in extracting relevant information from complex patterns. The obtained results quantify and support our main conclusion, which is: in terms of information, moving towards higher entropy states (A –> B) is different from moving towards lower entropy states (B –> A), since the Fisher curves are not the same, given a natural orientation (the direction of time).

Keywords: Markov random fields; information theory; Fisher information; entropy; maximum pseudo-likelihood estimation

1. Introduction

With the increasing value of information in modern society and the massive volume of digital data that is available, there is an urgent need for developing novel methodologies for data filtering and analysis in complex systems. In this scenario, the notion of what is informative or not is a top priority. Sometimes, patterns that at first may appear to be locally irrelevant may turn out to be extremely informative in a more global perspective. In complex systems, this is a direct consequence of the intricate non-linear relationship between the pieces of data along different locations and scales.

Within this context, information theoretic measures play a fundamental role in a huge variety of applications once they represent statistical knowledge in a systematic, elegant and formal framework. Since the first works of Shannon [1], and later with many other generalizations [2–4], the concept of entropy has been adapted and successfully applied to almost every field of science, among which we can cite physics [5], mathematics [6–8], economics [9] and, fundamentally, information theory [10–12]. Similarly, the concept of Fisher information [13,14] has been shown to reveal important properties of statistical procedures, from lower bounds on estimation methods [15–17] to information geometry [18,19]. Roughly speaking, Fisher information can be thought of as the likelihood analog of entropy, which is a probability-based measure of uncertainty.

In general, classical statistical inference is focused on capturing information about location and dispersion of unknown parameters of a given family of distribution and studying how this information is related to uncertainty in estimation procedures. In typical situations, an exponential family of distributions and independence hypothesis (independent random variables) are often assumed, giving the likelihood function a series of desirable mathematical properties [15–17].

Although mathematically convenient for many problems, in complex systems modeling, independence assumption is not reasonable, because much of the information is somehow encoded in the relations between the random variables [20,21]. In order to overcome this limitation, Markov random field (MRF) models appear to be a natural generalization of the classical approach by the replacement of the independence assumption by a more realistic conditional independence assumption. Basically, in every MRF, knowledge of a finite-support neighborhood around a given variable isolates it from all the remaining variables. A further simplification consists in considering a pairwise interaction model, constraining the size of the maximum clique to be two (in other words, the model captures only binary relationships). Moreover, if the MRF model is isotropic, which means that the parameter controlling the interactions between neighboring variables is invariant to change in the directions, all the information regarding the spatial dependence structure of the system is conveyed by a single parameter, from now on denoted by β (or simply, the inverse temperature).

In this paper, we assume an isotropic pairwise Gaussian Markov random field (GMRF) model [22,23], also known as an auto-normal model or a conditional auto-regressive model [24,25]. Basically, the question that motivated this work and that we are trying to elucidate here is: What kind of information is encoded by the β parameter in such a model? We want to know how this parameter, and as a consequence, the whole spatial dependence structure of a complex system modeled by a Gaussian Markov random field, is related to both local and global information theoretic measures, more precisely the observed and expected Fisher information, as well as self-information and Shannon entropy.

In searching for answers for our fundamental question, investigations led us to an exact expression for the asymptotic variance of the maximum pseudo-likelihood (MPL) estimator of β in an isotropic pairwise GMRF model, suggesting that asymptotic efficiency is not granted. In the context of statistical data analysis, Fisher information plays a central role in providing tools and insights for modeling the interactions between complex systems and their components. The advantage of MRF models over the traditional statistical ones is that MRFs take into account the dependence between pieces of information as a function of the system's temperature, which may even be variable along time. Briefly speaking, this investigation aims to explore ways to measure and quantify distances between complex systems operating in different thermodynamical conditions. By analyzing and comparing the behavior of local patterns observed throughout the system (defined over a regular 2D lattice), it is possible to measure how informative those patterns for a given inverse temperature are, or simply β (which encodes the expected global behavior).

In summary, our idea is to describe the behavior of a complex system in terms of information as its temperature deviates from infinity (when the particles are statistically independent) to a lower bound. The obtained results suggest that, in the beginning, when the temperature is infinite and the information equilibrium prevails, the information is somehow spread along the system. However, when temperature is low and this equilibrium condition does not hold anymore, we have a more sparse representation in terms of information, since this information is concentrated in the boundaries of the regions that define a smooth global configuration. In the vast remaining of this "universe", due to this smooth constraint, the strong alignment between the particles prevails, which is exactly the expected global behavior for temperatures below a critical value (making the majority of the interaction patterns along the system uninformative).

The remainder of the paper is organized as follows: Section 2 discusses a technique for the estimation of the inverse temperature parameter, called the maximum pseudo-likelihood (MPL) approach, and provides derivations for the observed Fisher information in an isotropic pairwise GMRF model. Intuitive interpretations for the two versions of this local measure are discussed. In Section 3, we derive analytical expressions for the computation of the expected Fisher information, which allows us to assign a global information measure for a given system configuration. Similarly, in Section 4, an expression for the global entropy of a system modeled by a GMRF is shown. The results suggest a connection between maximum pseudo-likelihood and minimum entropy criteria in the estimation of the inverse temperature parameter on GMRFs. Section 5 discusses the uncertainty in the estimation of this important parameter by defining an expression for the asymptotic variance of its maximum pseudo-likelihood estimator in terms of both forms of Fisher information. In Section 6, the definition of the Fisher curve of a system as a parametric trajectory in the information space is proposed. Section 7 shows the experimental setup. Computational simulations with both Markov chain Monte Carlo algorithms and some real data were conducted, showing how the proposed tools can be used to extract relevant information from complex systems. Finally, Section 8 presents our conclusions, final remarks and possibilities for future works.

2. Fisher Information in Isotropic Pairwise GMRFs

The remarkable Hammersley–Clifford theorem [26] states the equivalence between Gibbs random fields (GRF) and Markov random fields (MRF), which implies that any MRF can be defined either in terms of a global (joint Gibbs distribution) or a local (set of local conditional density functions) model. For our purposes, we will choose the latter representation.

Definition 1. *An isotropic pairwise Gaussian Markov random field regarding a local neighborhood system, η_i, defined on a lattice $S = \{s_1, s_2, \ldots, s_n\}$ is completely characterized by a set of n local conditional density functions $p(x_i|\eta_i, \vec{\theta})$, given by:*

$$p\left(x_i|\eta_i,\vec{\theta}\right) = \frac{1}{\sqrt{2\pi}\sigma} exp\left\{-\frac{1}{2\sigma^2}\left[x_i - \mu - \beta \sum_{j\in\eta_i}(x_j - \mu)\right]^2\right\} \tag{1}$$

with $\vec{\theta} = (\mu, \sigma^2, \beta)$, where μ and σ^2 are the expected value and the variance of the random variables, and $\beta = 1/T$ is the parameter that controls the interaction between the variables (inverse temperature). Note that, for $\beta = 0$, the model degenerates to the usual Gaussian distribution. From an information geometry perspective [18,19], this means that we are constrained to a sub-manifold within the Riemannian manifold of probability distributions, where the natural Riemannian metric (tensor) is given by the Fisher information. It has been shown that the geometric structure of exponential family distributions exhibits constant curvature. However, little is known about information geometry on more general statistical models, such as GMRFs. For $\beta > 0$, some degree of correlation between the observations is expected, making the interactions grow stronger. Typical choices for η_i are the first and second order non-causal neighborhood systems, defined by the sets of four and eight nearest neighbors, respectively.

2.1. Maximum Pseudo-Likelihood Estimation

Maximum likelihood estimation is intractable in MRF parameter estimation, due to the existence of the partition function in the joint Gibbs distribution. An alternative, proposed by Besag [24], is maximum pseudo-likelihood estimation, which is based on the conditional independence principle. The pseudo-likelihood function is defined as the product of the LCDFs for all the n variables of the system, modeled as a random field.

Definition 2. *Let an isotropic pairwise GMRF be defined on a lattice $S = \{s_1, s_2, \ldots, s_n\}$ with a neighborhood system, η_i. Assuming that $\mathbf{X}^{(t)} = \{x_1^{(t)}, x_2^{(t)}, \ldots, x_n^{(t)}\}$ denotes the set corresponding to the observations at time t, the pseudo-likelihood function of the model is defined by:*

$$L\left(\vec{\theta}; \mathbf{X}^{(t)}\right) = \prod_{i=1}^{n} p(x_i|\eta_i, \vec{\theta}) \tag{2}$$

Note that the pseudo-likelihood function is a function of the parameters. For better mathematical tractability, it is usual to take the logarithm of $L(\vec{\theta};\mathbf{X}^{(t)})$. Plugging Equation (1) into Equation (2) and taking the logarithm leads to:

$$log\, L\left(\vec{\theta};\mathbf{X}^{(t)}\right) = -\frac{n}{2}log\left(2\pi\sigma^2\right) - \frac{1}{2\sigma^2}\sum_{i=1}^{n}\left[x_i - \mu - \beta\sum_{j\in\eta_i}(x_j-\mu)\right]^2 \tag{3}$$

By differentiating Equation (3) with respect to each parameter and properly solving the pseudo-likelihood equations, we obtain the following maximum pseudo-likelihood estimators for the parameters, μ, σ^2 and β:

$$\hat{\beta}_{MPL} = \frac{\sum_{i=1}^{n}\left[(x_i-\mu)\sum_{j\in\eta_i}(x_j-\mu)\right]}{\sum_{i=1}^{n}\left[\sum_{j\in\eta_i}(x_j-\mu)\right]^2} \tag{4}$$

$$\hat{\mu}_{MPL} = \frac{1}{n(1-k\beta)}\sum_{i=1}^{n}\left(x_i - \beta\sum_{j\in\eta_i}x_j\right) \tag{5}$$

$$\hat{\sigma}^2_{MPL} = \frac{1}{n}\sum_{i=1}^{n}\left[x_i-\mu-\beta\sum_{j\in\eta_i}(x_j-\mu)\right]^2 \tag{6}$$

where k denotes the cardinality of the non-causal neighborhood set η_i. Note that if $\beta = 0$, the MPL estimators of both μ and σ^2 become the widely known sample mean and sample variance.

Since the cardinality of the neighborhood system, $k = |\eta_i|$, is spatially invariant (we are assuming a regular neighborhood system) and each variable is dependent on a fixed number of neighbors on a lattice, $\hat{\beta}_{MPL}$ can be rewritten in terms of cross-covariances:

$$\hat{\beta}_{MPL} = \frac{\sum_{j\in\eta_i}\hat{\sigma}_{ij}}{\sum_{j\in\eta_i}\sum_{k\in\eta_i}\hat{\sigma}_{jk}} \tag{7}$$

where σ_{ij} denotes the sample covariance between the central variable, x_i, and $x_j \in \eta_i$. Similarly, σ_{jk} denotes the sample covariance between two variables belonging to the neighborhood system, η_i (the definition of the neighborhood system, η_i, does not include the the location, s_i).

2.2. Fisher Information of Spatial Dependence Parameters

Basically, Fisher information measures the amount of information a sample conveys about an unknown parameter. It can be thought of as the likelihood analog of entropy, which is a probability-based measure of uncertainty. Often, when we are dealing with independent and identically distributed (i.i.d) random variables, the computation of the global Fisher information presented in a random sample $\mathbf{X}^{(t)} = \{x_1^{(t)}, x_2^{(t)}, \ldots, x_n^{(t)}\}$ is quite straightforward, since each observation, x_i, $i = 1, 2, \ldots, n$, brings exactly the same amount of information (when we are dealing with independent samples, the superscript, t, is usually suppressed, since the underlying dependence structure does not change through time). However, this is not true for spatial dependence parameters in MRFs, since different configuration patterns $(x_i \cup \eta_i)$ provide distinct contributions to the local observed Fisher information, which can be used to derive a reasonable approximation to the global Fisher information [27].

2.3. The Information Equality

It is widely known from statistical inference theory that, under certain regularity conditions, information equality holds in the case of independent observations in the exponential family [15–17]. In other words, we can compute the Fisher information of a random sample regarding a parameter of interest, θ, by:

$$I\left(\theta;\mathbf{X}^{(t)}\right) = E\left[\left(\frac{\partial}{\partial\theta}logL\left(\theta;\mathbf{X}^{(t)}\right)\right)^2\right] = -E\left[\frac{\partial^2}{\partial\theta^2}logL\left(\theta;\mathbf{X}^{(t)}\right)\right] \tag{8}$$

where $L\left(\theta;\mathbf{X}^{(t)}\right)$ denotes the likelihood function at a time instant, t. In our investigations, to avoid the joint Gibbs distribution, often intractable due to the presence of the partition function (global Gibbs field), we replace the usual likelihood function by Besag's pseudo-likelihood function, and then, we work with the local model instead (local Markov field).

However, given the intrinsic spatial dependence structure of Gaussian Markov random field models, information equilibrium is not a natural condition. As we will discuss later, in general, information equality fails. Thus, in a GMRF model, we have to consider two kinds of Fisher information, from now on denoted by Type I (due to the first derivative of the pseudo-likelihood function) and Type II (due to the second derivative of the pseudo-likelihood function). Eventually, when certain conditions are satisfied, these two values of information will converge to a unique bound. Essentially, β is the parameter responsible to control whether both forms of information converge or diverge. Knowing the role of β (inverse temperature) in a GMRF model, it is expected that for $\beta = 0$ (or $T \to \infty$), information equilibrium prevails. In fact, we will see in the following sections that as β deviates from zero (and long-term correlations start to emerge), the divergence between the two kinds of information increases.

In terms of information geometry, it has been shown that the geometric structure of the exponential family of distributions is basically given by the Fisher information matrix, which is the natural Riemmanian metric (metric tensor) [18,19]. So, when the inverse temperature parameter is zero, the geometric structure of the model is a surface since the parametric space is 2D (μ and σ^2). However, as the inverse temperature parameter starts to increase, the original surface is gradually transformed to a 3D Riemmanian manifold, equipped with a novel metric tensor (the 3×3 Fisher information matrix for μ, σ^2 and β). In this context, by measuring the Fisher information regarding the inverse temperature parameter along an interval ranging from $\beta_{MIN} = A = 0$ to $\beta_{MAX} = B$, we are essentially trying to capture part of the deformation in the geometric structure of the model. In this paper, we focus on the computation of this measure. In future works we expect to derive the complete Fisher information matrix in order to completely characterize the transformations in the metric tensor.

2.4. Observed Fisher Information

In order to quantify the amount of information conveyed by a local configuration pattern in a complex system, the concept of observed Fisher information must be defined.

Definition 3. *Consider an MRF defined on a lattice $S = \{s_1, s_2, \ldots, s_n\}$ with a neighborhood system, η_i. The Type I local observed Fisher information for the observation, x_i, regarding the spatial dependence parameter, β, is defined in terms of its local conditional density function as:*

$$\phi_\beta(x_i) = \left[\frac{\partial}{\partial\beta}log\, p\left(x_i|\eta_i,\vec{\theta}\right)\right]^2 \tag{9}$$

Hence, for an isotropic pairwise GMRF model, the Type I local observed Fisher information regarding β for the observation, x_i, is given by:

$$\phi_\beta(x_i) = \frac{1}{\sigma^4}\left\{\left[x_i - \mu - \beta\sum_{j\in\eta_i}(x_j-\mu)\right]\left[\sum_{j\in\eta_i}(x_j-\mu)\right]\right\}^2$$
$$= \frac{1}{\sigma^4}\left[\sum_{j\in\eta_i}(x_i-\mu)(x_j-\mu) - \beta\sum_{j\in\eta_i}\sum_{k\in\eta_i}(x_j-\mu)(x_k-\mu)\right]^2 \quad (10)$$

Definition 4. *Consider an MRF defined on a lattice $S = \{s_1, s_2, \ldots, s_n\}$ with a neighborhood system, η_i. The Type II local observed Fisher information for the observation, x_i, regarding the spatial dependence parameter, β, is defined in terms of its local conditional density function as:*

$$\psi_\beta(x_i) = -\frac{\partial^2}{\partial\beta^2} log\ p\left(x_i|\eta_i, \vec{\theta}\right) \quad (11)$$

In case of an isotropic pairwise GMRF model, the Type II local observed Fisher information regarding β for the observation, x_i, is given by:

$$\phi_\beta(x_i) = \frac{1}{\sigma^2}\left[\sum_{j\in\eta_i}\sum_{k\in\eta_i}(x_j-\mu)(x_k-\mu)\right] \quad (12)$$

Note that $\phi_\beta(x_i)$ does not depend on x_i, only on the neighborhood system, η_i.

Therefore, we have two local measures, $\phi_\beta(x_i)$ and $\psi_\beta(x_i)$, that can be assigned to every element of a system modeled by an isotropic pairwise GMRF. In the following, we will discuss some interpretations for what is being measured with the proposed tools and how to define global versions for these measures by means of the expected Fisher information.

2.5. The Role of Fisher Information in GMRF Models

At this point, a relevant issue is the interpretation of these Fisher information measures in a complex system modeled by an isotropic pairwise GMRF. Roughly speaking, $\phi_\beta(x_i)$ is the quadratic rate of change of the logarithm of the local likelihood function at x_i, given a global value of β. As this global value of β determines what would be the expected global behavior (if β is large, a high degree of correlation among the observations is expected and if β is close to zero, the observations are independent), it is reasonable to admit that configuration patterns showing values of $\phi_\beta(x_i)$ close to zero are more likely to be observed throughout the field, once their likelihood values are high (close to the maximum local likelihood condition). In other words, these patterns are more "aligned" to what is considered to be the expected global behavior, and therefore, they convey little information about the spatial dependence structure (these samples are not informative once they are expected to exist in a system operating at that particular value of inverse temperature).

Now, let us move on to configuration patterns showing high values of $\phi_\beta(x_i)$. Those samples can be considered landmarks, because they convey a large amount of information about the global spatial dependence structure. Roughly speaking, those points are very informative once they are not expected to exist for that particular value of β (which guides the expected global behavior of the system). Therefore, Type I local observed Fisher information minimization in GMRFs can be a useful tool in producing novel configuration patterns that are more likely to exist given the chosen value of inverse temperature. Basically, $\phi_\beta(x_i)$ tells us how informative a given pattern is for that specific global behavior (represented by a single parameter in an isotropic pairwise GMRF model). In summary, this measure quantifies the degree of agreement between an observation, x_i, and the configuration defined by its neighborhood system for a given β.

As we will see later in the experiments section, typical informative patterns (those showing high values of $\phi_\beta(x_i)$) in an organized system are located at the boundaries of the regions defining homogeneous areas (since these boundary samples show an unexpected behavior for large β, which is: there is no strong agreement between x_i and its neighbors).

Let us analyze the Type II local observed Fisher information, $\psi_\beta(x_i)$. Informally speaking, this measure can be interpreted as a curvature measure, that is, how curved is the local likelihood function at x_i. Thus, patterns showing low values of $\psi_\beta(x_i)$ tend to have a nearly flat local likelihood function. This means that we are dealing with a pattern that could have been observed for a variety of β values (a large set of β values have approximately the same likelihood). An implication of this fact is that in a system dominated by this kind of patterns (patterns for which $\psi_\beta(x_i)$ is close to zero), small perturbations may cause a sharp change in β (and, therefore, in the expected global behavior). In other words, these patterns are more susceptible to changes once they do not have a "stable" configuration (it raises our uncertainty about the true value of β).

On the other hand, if the global configuration is mostly composed of patterns exhibiting large values of $\psi_\beta(x_i)$, changes on the global structure are unlikely to happen (uncertainty on β is sufficiently small). Basically, $\psi_\beta(x_i)$ measures the degree of agreement or dependence among the observations belonging to the same neighborhood system. If at a given x_i, the observations belonging to η_i are totally symmetric around the mean value, $\psi_\beta(x_i)$ would be zero. It is reasonable to expect that in this situation, as there is no information about the induced spatial dependence structure (this means that there is no contextual information available at this point). Notice that the role of $\psi_\beta(x_i)$ is not the same as $\phi_\beta(x_i)$. Actually, these two measures are almost inversely related, since if at x_i the value of $\phi_\beta(x_i)$ is high (it is a landmark or boundary pattern), then it is expected that $\psi_\beta(x_i)$ will be low (in decision boundaries or edges, the uncertainty about β is higher, causing $\psi_\beta(x_i)$ to be small). In fact, we will observe this behavior in some computational experiments conducted in future sections of the paper.

It is important to mention that these rather informal arguments define the basis for understanding the meaning of the asymptotic variance of maximum pseudo-likelihood estimators, as we will discuss ahead. In summary, $\psi_\beta(x_i)$ is a measure of how sure or confident we are about the local spatial dependence structure (at a given point, x_i), since a high average curvature is desired for predicting the system's global behavior in a reasonable manner (reducing the uncertainty of β estimation).

3. Expected Fisher Information

In order to avoid the use of approximations in the computation of the global Fisher information in an isotropic pairwise GMRF, in this section, we provide an exact expression for $\hat{\phi}_\beta$ and $\hat{\psi}_\beta$ as Type I and Type II expected Fisher information. One advantage of using the expected Fisher information instead of its global observed counterpart is the faster computing time. As we will see, instead of computing a single local measure for each observation ,$x_i \in \mathbf{X}$, and then taking the average, both Φ_β and Ψ_β expressions depend only on the covariance matrix of the configuration patterns observed along the random field.

3.1. The Type I Expected Fisher Information

Recall that the Type I expected Fisher information, from now on denoted by Φ_β, is given by:

$$\Phi_\beta = E\left[\left(\frac{\partial}{\partial\beta} log\ L\left(\vec{\theta};\mathbf{X}^{(t)}\right)\right)^2\right] \tag{13}$$

The Type II expected Fisher information, from now on denoted by Ψ_β, is given by:

$$\Psi_\beta = -E\left[\frac{\partial^2}{\partial\beta^2} log\ L\left(\vec{\theta};\mathbf{X}^{(t)}\right)\right] \tag{14}$$

We first proceed to the definition of Φ_β. Plugging Equation (3) in Equation (13), and after some algebra, we obtain the following expression, which is composed by four main terms:

$$\Phi_\beta = \frac{1}{\sigma^4} E\left\{\left[\sum_{s=1}^{n}\left(x_s - \mu - \beta \sum_{j\in\eta_s}(x_j-\mu)\right)\left(\sum_{j\in\eta_s}(x_j-\mu)\right)\right]^2\right\} \tag{15}$$

$$= \frac{1}{\sigma^4} E\left\{\sum_{s=1}^{n}\sum_{r=1}^{n}\left[x_s - \mu - \beta \sum_{j\in\eta_s}(x_j-\mu)\right]\left[x_r - \mu - \beta \sum_{k\in\eta_r}(x_k-\mu)\right] \times \left[\sum_{j\in\eta_s}(x_j-\mu)\right]\left[\sum_{k\in\eta_r}(x_k-\mu)\right]\right\}$$

$$= \frac{1}{\sigma^4} E\left\{\sum_{s=1}^{n}\sum_{r=1}^{n}\left[(x_s-\mu)(x_r-\mu) - \beta\sum_{k\in\eta_r}(x_s-\mu)(x_k-\mu) - \beta\sum_{j\in\eta_s}(x_r-\mu)(x_j-\mu) + \beta^2\sum_{j\in\eta_s}\sum_{k\in\eta_r}(x_j-\mu)(x_k-\mu)\right]\left[\sum_{j\in\eta_s}\sum_{k\in\eta_r}(x_j-\mu)(x_k-\mu)\right]\right\}$$

$$\begin{aligned}
= \frac{1}{\sigma^4}\sum_{s=1}^{n}\sum_{r=1}^{n}\Bigg\{ & \sum_{j\in\eta_s}\sum_{k\in\eta_r} E\left[(x_s-\mu)(x_r-\mu)(x_j-\mu)(x_k-\mu)\right] \\
& -\beta\sum_{j\in\eta_s}\sum_{k\in\eta_r}\sum_{l\in\eta_r} E\left[(x_s-\mu)(x_j-\mu)(x_k-\mu)(x_l-\mu)\right] \\
& -\beta\sum_{m\in\eta_s}\sum_{j\in\eta_s}\sum_{k\in\eta_r} E\left[(x_r-\mu)(x_m-\mu)(x_j-\mu)(x_k-\mu)\right] \\
& +\beta^2\sum_{m\in\eta_s}\sum_{j\in\eta_s}\sum_{k\in\eta_r}\sum_{l\in\eta_r} E\left[(x_m-\mu)(x_j-\mu)(x_k-\mu)(x_l-\mu)\right]\Bigg\}
\end{aligned}$$

Hence, the expression for Φ_β is composed by four main terms, each one of them involving a summation of higher-order cross-moments. According to Isserlis' theorem [28], for normally distributed random variables, we can compute higher order moments in terms of the covariance matrix through the following identity:

$$E[X_1X_2X_3X_4] = E[X_1X_2]\,E[X_3X_4] + E[X_1X_3]\,E[X_2X_4] + E[X_2X_3]\,E[X_1X_4] \tag{16}$$

Then, the first term of Equation (15) is reduced to:

$$\begin{aligned}
&\sum_{j\in\eta_s}\sum_{k\in\eta_r} E\left[(x_s-\mu)(x_r-\mu)(x_j-\mu)(x_k-\mu)\right] = \\
&\sum_{j\in\eta_s}\sum_{k\in\eta_r}\{E\left[(x_s-\mu)(x_r-\mu)\right]E\left[(x_j-\mu)(x_k-\mu)\right] \\
&+ E\left[(x_s-\mu)(x_j-\mu)\right]E\left[(x_r-\mu)(x_k-\mu)\right] \\
&+ E\left[(x_r-\mu)(x_j-\mu)\right]E\left[(x_s-\mu)(x_k-\mu)\right]\} = \\
&\sum_{j\in\eta_s}\sum_{k\in\eta_r}\left[\sigma_{sr}\sigma_{jk} + \sigma_{sj}\sigma_{rk} + \sigma_{rj}\sigma_{sk}\right]
\end{aligned} \tag{17}$$

where σ_{sr} denotes the covariance between variables x_s and x_r (note that in an MRF, we have $\sigma_{sr} = 0$ if $x_r \notin \eta_s$). We now proceed to the expansion of the second main term of Equation (15). Similarly, by applying Isserlis' identity, we have:

$$\sum_{j \in \eta_s} \sum_{k \in \eta_r} \sum_{l \in \eta_r} E\left[(x_s - \mu)\left(x_j - \mu\right)(x_k - \mu)(x_l - \mu)\right] = \sum_{j \in \eta_s} \sum_{k \in \eta_r} \sum_{l \in \eta_r} \left[\sigma_{sj}\sigma_{kl} + \sigma_{sk}\sigma_{jl} + \sigma_{jk}\sigma_{sl}\right] \tag{18}$$

The third term of Equation (15) can be rewritten as:

$$\begin{aligned} &\sum_{m \in \eta_s} \sum_{j \in \eta_s} \sum_{k \in \eta_r} E\left[(x_r - \mu)(x_m - \mu)\left(x_j - \mu\right)(x_k - \mu)\right] = \\ &= \sum_{m \in \eta_s} \sum_{j \in \eta_s} \sum_{k \in \eta_r} \left[\sigma_{rm}\sigma_{jk} + \sigma_{rj}\sigma_{mk} + \sigma_{mj}\sigma_{rk}\right] \end{aligned} \tag{19}$$

Finally, the fourth term of it is:

$$\begin{aligned} &\sum_{m \in \eta_s} \sum_{j \in \eta_s} \sum_{k \in \eta_r} \sum_{l \in \eta_r} E\left[(x_m - \mu)\left(x_j - \mu\right)(x_k - \mu)(x_l - \mu)\right] = \\ &= \sum_{m \in \eta_s} \sum_{j \in \eta_s} \sum_{k \in \eta_r} \sum_{l \in \eta_r} \left[\sigma_{mj}\sigma_{kl} + \sigma_{mk}\sigma_{jl} + \sigma_{ml}\sigma_{jk}\right] \end{aligned} \tag{20}$$

Therefore, by combining Expressions Equations (17)–(20), we have the complete expression for Φ_β, the Type I expected Fisher information for an isotropic pairwise GMRF model regarding the inverse temperature parameter, as:

$$\begin{aligned} \Phi_\beta = \frac{1}{\sigma^4} \sum_{s=1}^{n} \sum_{r=1}^{n} \Bigg\{ &\sum_{j \in \eta_s} \sum_{k \in \eta_r} \left[\sigma_{sr}\sigma_{jk} + \sigma_{sj}\sigma_{rk} + \sigma_{rj}\sigma_{sk}\right] \\ &- \beta \sum_{j \in \eta_s} \sum_{k \in \eta_r} \sum_{l \in \eta_r} \left[\sigma_{sj}\sigma_{kl} + \sigma_{sk}\sigma_{jl} + \sigma_{jk}\sigma_{sl}\right] \\ &- \beta \sum_{m \in \eta_s} \sum_{j \in \eta_s} \sum_{k \in \eta_r} \left[\sigma_{rm}\sigma_{jk} + \sigma_{rj}\sigma_{mk} + \sigma_{mj}\sigma_{rk}\right] \\ &+ \beta^2 \sum_{m \in \eta_s} \sum_{j \in \eta_s} \sum_{k \in \eta_r} \sum_{l \in \eta_r} \left[\sigma_{mj}\sigma_{kl} + \sigma_{mk}\sigma_{jl} + \sigma_{ml}\sigma_{jk}\right] \Bigg\} \end{aligned} \tag{21}$$

However, since we are interested in studying how the spatial correlations change as the system evolves, we need to estimate a value for Φ_β given a single global state $\mathbf{X}^{(t)} = \left\{x_1^{(t)}, x_2^{(t)}, \ldots, x_n^{(t)}\right\}$. Hence, to compute Φ_β from a single static configuration $\mathbf{X}^{(t)}$ (a photograph of the system at a given moment), we consider $n = 1$ in the previous equation, which means, among other things, that $s = r$ (which implies $\eta_s = \eta_r$) and that observations belonging to different local neighborhoods are independent from each other (as we are dealing with a pairwise interaction Markovian process, it does not make sense to model the interactions between variables that are far away from each other in the lattice).

Before proceeding, we would like to clarify some points regarding the estimation of the β parameter and the computation of the expected Fisher information in the isotropic pairwise GMRF model. Basically, there are two main possibilities: (1) the parameter is spatially-invariant, which means that we have a unique value, $\hat{\beta}^{(t)}$, for a global configuration of the system, $\mathbf{X}^{(t)}$ (this is our assumption); or (2) the parameter is spatially-variant, which means that we have a set of $\hat{\beta}_s$ values, for $s = 1, 2, \ldots, n$, each one of them estimated from $\mathbf{X}_s = \left\{x_s^{(1)}, x_s^{(2)}, \ldots, x_s^{(t)}\right\}$ (we are observing the outcomes of a random pattern along time in a fixed position of the lattice). When we are dealing with the first model (β is spatially-invariant), all possible observation patterns (samples) are extracted from the global configuration by a sliding window (with the shape of the neighborhood system) that moves through the lattice at a fixed time instant, t. In this case, we are interested in studying the spatial correlations, not the temporal ones. In other words, we would like to investigate how the the spatial structure of a GMRF model is related to Fisher information (this is exactly the scenario described above, for which $n = 1$). Our motivation here is to characterize, via information-theoretic measures, the behavior of the system as it evolves from states of minimum entropy to states of maximum entropy (and *vice versa*) by providing a geometrical tool based on the definition of the Fisher curve , which will be introduced in the following sections.

Therefore, in our case ($n = 1$), Equation (21) is further simplified for practical usage. By unifying s and r to a unique index, i, we have a final expression for Φ_β in terms of the local covariances between the random variables in a given neighborhood system (*i.e.*, for the eight nearest neighbors):

$$\Phi_\beta = \frac{1}{\sigma^4}\left\{\sum_{j\in\eta_i}\sum_{k\in\eta_i}\left[\sigma^2\sigma_{jk} + 2\sigma_{ij}\sigma_{ik}\right] - 2\beta\sum_{j\in\eta_i}\sum_{k\in\eta_i}\sum_{l\in\eta_i}\left[\sigma_{ij}\sigma_{kl} + \sigma_{ik}\sigma_{jl} + \sigma_{il}\sigma_{jk}\right]\right.$$
$$\left. + \beta^2\sum_{j\in\eta_i}\sum_{k\in\eta_i}\sum_{l\in\eta_i}\sum_{m\in\eta_i}\left[\sigma_{jk}\sigma_{lm} + \sigma_{jl}\sigma_{km} + \sigma_{jm}\sigma_{kl}\right]\right\} \tag{22}$$

Note that we have two types of covariances in the definition of Φ_β for an isotropic pairwise GMRF: (1) covariances between the central variable, x_i, and a neighboring variable, x_j, denoted by σ_{ij}, for $j \in \eta_i$; and (2) covariances between two neighboring variables, x_j and x_k, for $j, k \in \eta_i$. In the next sections, we will see how to compute the value of Ψ_β directly from the covariance matrix of the local patterns.

3.2. The Type II Expected Fisher Information

Following the same methodology of replacing the likelihood function by the pseudo-likelihood function of the GMRF model, a closed form expression for Ψ_β is developed. Plugging Equation (3) into Equation (14) leads us to:

$$\Psi_\beta = \frac{1}{\sigma^2}\sum_{i=1}^{n} E\left\{\left[\sum_{x_j\in\eta_i}(x_j - \mu)\right]^2\right\} \tag{23}$$
$$= \frac{1}{\sigma^2}\sum_{i=1}^{n} E\left[\sum_{x_j\in\eta_i}\sum_{x_k\in\eta_i}(x_j - \mu)(x_k - \mu)\right] =$$
$$= \frac{1}{\sigma^2}\sum_{i=1}^{n}\left\{\sum_{x_j\in\eta_i}\sum_{x_k\in\eta_i} E\left[(x_j - \mu)(x_k - \mu)\right]\right\} = \frac{1}{\sigma^2}\sum_{i=1}^{n}\sum_{j\in\eta_i}\sum_{k\in\eta_i}\sigma_{jk}$$

Note that unlike Φ_β, Ψ_β does not depend explicitly on β (inverse temperature). As we have seen before, Φ_β is a quadratic function of the spatial dependence parameter.

In order to simplify the notations and also to make computations easier, the expressions for Φ_β and Ψ_β can be rewritten in a matrix-vector form. Let Σ_p be the covariance matrix of the random vectors $\vec{p}_i, i = 1, 2, \ldots, n$, obtained by lexicographic ordering of the local configuration patterns $x_i \cup \eta_i$. Thus, considering a neighborhood system, η_i, of size K, we have Σ_p given by a $(K+1) \times (K+1)$ symmetric matrix (for $K+1$ odd, *i.e.*, $K = 4, 8, 12, \ldots$):

$$\Sigma_p = \begin{pmatrix} \sigma_{1,1} & \cdots & \sigma_{1,K/2} & \sigma_{1,(K/2)+1} & \sigma_{1,(K/2)+2} & \cdots & \sigma_{1,K+1} \\ \vdots & \vdots & \vdots & \vdots & \vdots & \vdots & \vdots \\ \sigma_{K/2,1} & \cdots & \sigma_{K/2,K/2} & \sigma_{K/2,(K/2)+1} & \sigma_{K/2,(K/2)+2} & \cdots & \sigma_{K/2,K+1} \\ \sigma_{(K/2)+1,1} & \cdots & \sigma_{(K/2)+1,K/2} & \sigma_{(K/2)+1,(K/2)+1} & \sigma_{(K/2)+1,(K/2)+2} & \cdots & \sigma_{(K/2)+1,K+1} \\ \sigma_{(K/2)+2,1} & \cdots & \sigma_{(K/2)+2,K/2} & \sigma_{(K/2)+2,(K/2)+1} & \sigma_{(K/2)+2,(K/2)+2} & \cdots & \sigma_{(K/2)+2,K+1} \\ \vdots & \vdots & \vdots & \vdots & \vdots & \vdots & \vdots \\ \sigma_{K+1,1} & \cdots & \sigma_{K+1,K/2} & \sigma_{K+1,(K/2)+1} & \sigma_{K+1,(K/2)+2} & \cdots & \sigma_{K+1,K+1} \end{pmatrix}$$

Let Σ_p^- be the submatrix of dimensions $K \times K$ obtained by removing the central row and central column of Σ_p (the covariances between x_i and each one of its neighbors, x_j). Then, for $K+1$ odd, we have:

$$\Sigma_p^- = \begin{pmatrix} \sigma_{1,1} & \cdots & \sigma_{1,K/2} & \sigma_{1,(K/2)+2} & \cdots & \sigma_{1,K+1} \\ \vdots & \vdots & \vdots & \vdots & \vdots & \vdots \\ \sigma_{K/2,1} & \cdots & \sigma_{K/2,K/2} & \sigma_{K/2,(K/2)+2} & \cdots & \sigma_{K/2,K+1} \\ \sigma_{(K/2)+2,1} & \cdots & \sigma_{(K/2)+2,K/2} & \sigma_{(K/2)+2,(K/2)+2} & \cdots & \sigma_{(K/2)+2,K+1} \\ \vdots & \vdots & \vdots & \vdots & \vdots & \vdots \\ \sigma_{K+1,1} & \cdots & \sigma_{K+1,K/2} & \sigma_{K+1,(K/2)+2} & \cdots & \sigma_{K+1,K+1} \end{pmatrix} \tag{24}$$

Thus, Σ_p^- is a matrix that stores only the covariances among the neighboring variables. Furthermore, let $\vec{\rho}$ be the vector of dimensions $K \times 1$ formed by all the elements of the central row of Σ_p, excluding the middle one (which is a variance actually), that is:

$$\vec{\rho} = \begin{bmatrix} \sigma_{(K/2)+1,1} & \cdots & \sigma_{(K/2)+1,K/2} & \sigma_{(K/2)+1,(K/2)+2} & \cdots & \sigma_{(K/2)+1,K+1} \end{bmatrix} \tag{25}$$

Therefore, we can rewrite Equation (23) (for $n = 1$) using Kronecker products. The following definition provides a fast way to compute Φ_β exploring these tensor products.

Definition 5. *Let an isotropic pairwise GMRF be defined on a lattice $S = \{s_1, s_2, \ldots, s_n\}$ with a neighborhood system, η_i, of size K (usual choices for K are even values: four, eight, 12, 20 or 24). Assuming that $\mathbf{X}^{(t)} = \{x_1^{(t)}, x_2^{(t)}, \ldots, x_n^{(t)}\}$ denotes the global configuration of the system at time t and $\vec{\rho}$ and Σ_p^- are defined as Equations (25) and (24), the Type I expected Fisher information, Φ_β, for this state, $\mathbf{X}^{(t)}$, is:*

$$\Phi_\beta = \frac{1}{\sigma^4}\left[\sigma^2 \left\|\Sigma_p^-\right\|_+ + 2\left\|\vec{\rho} \otimes \vec{\rho}^T\right\|_+ - 6\beta\left\|\vec{\rho}^T \otimes \Sigma_p^-\right\|_+ + 3\beta^2\left\|\Sigma_p^- \otimes \Sigma_p^-\right\|_+\right] \tag{26}$$

where $\|A\|_+$ denotes the summation of all the entries of the matrix, A (not to be confused with a matrix norm) and $\otimes$ denotes the Kronecker (tensor) product. From an information geometry perspective, the presence of tensor products indicates the intrinsic differential geometry of a manifold in the form of the Riemann curvature tensor [18]. Note that all the necessary information for computing the Fisher information is somehow encoded in the covariance matrix of the local configuration patterns, $(x_i \cup \eta_i), i = 1, 2, \ldots, n$, as would be expected in the case of Gaussian variables (second-order statistics). The same procedure is applied to the Type II expected Fisher information.

Definition 6. *Let an isotropic pairwise GMRF be defined on a lattice $S = \{s_1, s_2, \ldots, s_n\}$ with a neighborhood system, η_i, of size K (usual choices for K are four, eight, 12, 20 or 24). Assuming that $\mathbf{X}^{(t)} = \{x_1^{(t)}, x_2^{(t)}, \ldots, x_n^{(t)}\}$*

denotes the global configuration of the system at time t and Σ_p^- *is defined as Equation (24), the Type II expected Fisher information,* Ψ_β*, for this state,* $\mathbf{X^{(t)}}$*, is given by:*

$$\Psi_\beta = \frac{1}{\sigma^2}\left\|\Sigma_p^-\right\|_+ \tag{27}$$

3.3. *Information Equilibrium in GMRF Models*

From the definition of both Φ_β and Ψ_β, a natural question that raises would be: under what conditions do we have $\Phi_\beta = \Psi_\beta$ in an isotropic pairwise GMRF model? As we can see from Equations (26) and (27), the difference between Φ_β and Ψ_β, from now on denoted by $\Delta_\beta\left(\vec{\rho}, \Sigma_p^-\right)$, is simply:

$$\Delta_\beta\left(\vec{\rho}, \Sigma_p^-\right) = \frac{1}{\sigma^4}\left(2\left\|\vec{\rho}\otimes\vec{\rho}^T\right\|_+ - 6\beta\left\|\vec{\rho}^T\otimes\Sigma_p^-\right\|_+ + 3\beta^2\left\|\Sigma_p^-\otimes\Sigma_p^-\right\|_+\right) \tag{28}$$

Then, intuitively, the condition for information equality is achieved when $\Delta_\beta\left(\vec{\rho}, \Sigma_p^-\right) = 0$. As $\Delta_\beta\left(\vec{\rho}, \Sigma_p^-\right)$ is a simple quadratic function of the inverse temperature parameter, β, we can easily find that the value, β^*, for which $\Delta_\beta\left(\vec{\rho}, \Sigma_p^-\right) = 0$, is:

$$\beta^* = \frac{\left\|\vec{\rho}^T\otimes\Sigma_p^-\right\|_+}{\left\|\Sigma_p^-\otimes\Sigma_p^-\right\|_+} \pm \frac{\sqrt{3}}{3}\frac{\sqrt{3\left\|\vec{\rho}^T\otimes\Sigma_p^-\right\|_+^2 - 2\left\|\Sigma_p^-\otimes\Sigma_p^-\right\|_+\left\|\vec{\rho}\otimes\vec{\rho}^T\right\|_+}}{\left\|\Sigma_p^-\otimes\Sigma_p^-\right\|_+} \tag{29}$$

provided that $3\left\|\vec{\rho}^T\otimes\Sigma_p^-\right\|_+^2 \geq 2\left\|\Sigma_p^-\otimes\Sigma_p^-\right\|_+\left\|\vec{\rho}\otimes\vec{\rho}^T\right\|_+$ and $\left\|\Sigma_p^-\otimes\Sigma_p^-\right\|_+ \neq 0$. Note that if $\left\|\vec{\rho}\otimes\vec{\rho}^T\right\|_+ = 0$, then one solution for the above equation is $\beta^* = 0$. In other words, when $\sigma_{ij} = 0, \forall j \in \eta_i$ (no correlation between x_i and its neighbors, x_j), information equilibrium is achieved for $\beta^* = 0$, which in this case, is the maximum pseudo-likelihood estimate of β, since in this matrix-vector notation, $\hat{\beta}_{MPL}$ is given by:

$$\hat{\beta}_{MPL} = \frac{\sum_{j\in\eta_i}\hat{\sigma}_{ij}}{\sum_{j\in\eta_i}\sum_{k\in\eta_i}\hat{\sigma}_{jk}} = \frac{\|\vec{\rho}\|_+}{\|\Sigma_p^-\|_+} \tag{30}$$

In the isotropic pairwise GMRF model, if $\beta = 0$, then we have $\|\vec{\rho}\|_+ = 0$, and as a consequence, $\Phi_\beta = \Psi_\beta$. However, the opposite is not necessarily true, that is, we may observe that $\Phi_\beta = \Psi_\beta$ for a non-zero β. One example is for β^*, a solution of $\Delta_\beta\left(\vec{\rho}, \Sigma_p^-\right) = 0$.

4. Entropy in Isotropic Pairwise GMRFs

Our definition of entropy is done by repeating the same process employed to derive Φ_β and Ψ_β. Knowing that the entropy of random variable x is defined by the expected value of self-information, given by $-log\ p(x)$, it can be thought of as a probability-based counterpart to the Fisher information.

Definition 7. *Let an isotropic pairwise GMRF be defined on a lattice* $S = \{s_1, s_2, \ldots, s_n\}$ *with a neighborhood system,* η_i*. Assuming that* $\mathbf{X^{(t)}} = \{x_1^{(t)}, x_2^{(t)}, \ldots, x_n^{(t)}\}$ *denotes the global configuration of the system at time t, then the entropy,* H_β*, for this state* $\mathbf{X^{(t)}}$ *is given by:*

$$
\begin{aligned}
H_\beta &= -E\left[log\, L\left(\vec{\theta};\mathbf{X}^{(t)}\right)\right] = -E\left[log \prod_{i=1}^{n} p\left(x_i|\eta_i,\vec{\theta}\right)\right] = \\
&= \frac{n}{2}log\left(2\pi\sigma^2\right) + \frac{1}{2\sigma^2}\sum_{i=1}^{n} E\left\{\left[x_i - \mu - \beta\sum_{j\in\eta_i}\left(x_j-\mu\right)\right]^2\right\} = \\
&= \frac{n}{2}log\left(2\pi\sigma^2\right) + \frac{1}{2\sigma^2}\sum_{i=1}^{n}\left\{E\left[\left(x_i-\mu\right)^2\right] - 2\beta E\left[\sum_{j\in\eta_i}\left(x_i-\mu\right)\left(x_j-\mu\right)\right]\right. \\
&\left. + \beta^2 E\left\{\left[\sum_{j\in\eta_i}\left(x_j-\mu\right)\right]^2\right\}\right\}
\end{aligned} \tag{31}
$$

After some algebra, the expression for H_β becomes:

$$
\begin{aligned}
H_\beta &= \frac{n}{2}log\left(2\pi\sigma^2\right) + \frac{1}{2\sigma^2}\sum_{i=1}^{n}\left\{\sigma^2 - 2\beta\sum_{j\in\eta_i}\sigma_{ij} + \beta^2\sum_{j\in\eta_i}\sum_{k\in\eta_i}\sigma_{jk}\right\} = \\
&= \left[\frac{n}{2}log(2\pi\sigma^2) + \frac{n}{2}\right] - \frac{\beta}{\sigma^2}\sum_{i=1}^{n}\left[\sum_{j\in\eta_i}\sigma_{ij}\right] + \frac{\beta^2}{2\sigma^2}\sum_{i=1}^{n}\left[\sum_{j\in\eta_i}\sum_{k\in\eta_i}\sigma_{jk}\right]
\end{aligned} \tag{32}
$$

Using the same matrix-vector notation introduced in the previous sections, we can further simplify the expression for H_β (considering $n = 1$).

Definition 8. *Let an isotropic pairwise GMRF be defined on a lattice $S = \{s_1, s_2, \ldots, s_n\}$ with a neighborhood system, η_i. Assuming that $\mathbf{X}^{(\mathbf{t})} = \{x_1^{(t)}, x_2^{(t)}, \ldots, x_n^{(t)}\}$ denotes the global configuration of the system at time t and $\vec{\rho}$ and Σ_p^- are defined as Equations (25) and (24), the entropy, H_β, for this state, $\mathbf{X}^{(\mathbf{t})}$, is given by:*

$$
H_\beta = H_G - \left[\frac{\beta}{\sigma^2}\|\vec{\rho}\|_+ - \frac{\beta^2}{2\sigma^2}\left\|\Sigma_p^-\right\|_+\right] = H_G - \left[\frac{\beta}{\sigma^2}\|\vec{\rho}\|_+ - \frac{\beta^2}{2}\Psi_\beta\right] \tag{33}
$$

where H_G denotes the entropy of a Gaussian random variable with variance σ^2 and Ψ_β is the Type II expected Fisher information.

Note that Shannon entropy is a quadratic function of the spatial dependence parameter, β. Since the coefficient of the quadratic term is strictly non-negative (Ψ_β is the Type II expected Fisher information), entropy is a convex function of β. Furthermore, as expected, when $\beta = 0$ and there is no induced spatial dependence in the system, the resulting expression for H_β is the usual entropy of a Gaussian random variable, H_G. Thus, there is a value, $\hat{\beta_{MH}}$, for the inverse temperature parameter, which minimizes the entropy of the system. In fact, $\hat{\beta}_{MH}$ is given by:

$$
\begin{aligned}
\frac{\partial H_\beta}{\partial \beta} &= \frac{\beta}{\sigma^2}\left\|\Sigma_p^-\right\|_+ - \frac{1}{\sigma^2}\|\vec{\rho}\|_+ = 0 \\
\hat{\beta}_{MH} &= \frac{\|\vec{\rho}\|_+}{\|\Sigma_p^-\|_+} = \hat{\beta}_{MPL}
\end{aligned} \tag{34}
$$

showing that the maximum pseudo-likelihood and the minimum-entropy estimates are equivalent in an isotropic pairwise GMRF model. Moreover, using the derived equations, we see a relationship between Φ_β, Ψ_β and H_β:

$$\Phi_\beta - \Psi_\beta = \Delta_\beta\left(\vec{\rho}, \Sigma_p^-\right) \quad (35)$$
$$\frac{\partial^2 H_\beta}{\partial \beta^2} = \Psi_\beta$$

where the functional $\Delta_\beta\left(\vec{\rho}, \Sigma_p^-\right)$ that represents the difference between Φ_β and Ψ_β is defined by Equation (28). These equations relate the entropy and one form of Fisher information (Ψ_β) in GMRF models, showing that Ψ_β can be roughly viewed as the curvature of H_β. In this sense, in a hypothetical information equilibrium condition $\Psi_\beta = \Phi_\beta = 0$, the entropy's curvature would be null (H_β would never change). These results suggest that an increase in the value of Ψ_β, which means stability (a measure of agreement between the neighboring observations of a given point), contributes to the curve and, therefore, to inducing a change in the entropy of the system. In this context, the analysis of the Fisher information could bring us insights in predicting the entropy of a system.

5. Asymptotic Variance of MPL Estimators

It is known from the statistical inference literature that unbiasedness is a property that is not granted by maximum likelihood estimation, nor by maximum pseudo-likelihood (MPL) estimation. Actually, there is no universal method that guarantees the existence of unbiased estimators for a fixed n-sized sample. Often, in the exponential family of distributions, maximum likelihood estimators (MLEs) coincide with the UMVU (uniform minimum variance unbiased) estimators, because MLEs are functions of complete sufficient statistics. There is an important result in statistical inference that shows that if the MLE is unique, then it is a function of sufficient statistics. We could enumerate and make a huge list of several properties that make maximum likelihood estimation a reference method [15–17]. One of the most important properties concerns the asymptotic behavior of MLEs: when we make the sample size grow infinitely ($n \to \infty$), MLEs become asymptotically unbiased and efficient. Unfortunately, there is no result showing that the same occurs in maximum pseudo-likelihood estimation. The objective of this section is to propose a closed expression for the asymptotic variance of the maximum pseudo-likelihood of β in an isotropic pairwise GMRF model. Unsurprisingly, this variance is completely defined as a function of both forms of expected Fisher information, Ψ_β and Φ_β; as for general values of the inverse temperature parameter, the information equality condition fails.

5.1. *The Asymptotic Variance of the Inverse Temperature Parameter*

In mathematical statistics, asymptotic evaluations uncover several fundamental properties of inference methods, providing a powerful and general tool for studying and characterizing the behavior of estimators. In this section, our objective is to derive an expression for the asymptotic variance of the maximum pseudo-likelihood estimator of the inverse temperature parameter (β) in isotropic pairwise GMRF models. It is known from the statistical inference literature that both maximum likelihood and maximum pseudo-likelihood estimators share two important properties: consistency and asymptotic normality [29,30]. It is possible, therefore, to completely characterize their behaviors in the limiting case. In other words, the asymptotic distribution of $\hat{\beta}_{MPL}$ is normal, centered around the real parameter value (since consistency means that the estimator is asymptotically unbiased), with the asymptotic variance representing the uncertainty about how far we are from the mean (real value). From a statistical perspective, $\hat{\beta}_{MPL} \approx N\left(\beta, v_\beta\right)$, where v_β denotes the asymptotic variance of the maximum pseudo-likelihood estimator. It is known that the asymptotic covariance matrix of maximum pseudo-likelihood estimators is given by [31]:

$$C(\vec{\theta}) = H^{-1}(\vec{\theta})J(\vec{\theta})H^{-1}(\vec{\theta}) \tag{36}$$

with:

$$H(\vec{\theta}) = E_{\beta}\left[\nabla^{2} log\ L\left(\vec{\theta};\mathbf{X}^{(t)}\right)\right] \tag{37}$$

$$J(\vec{\theta}) = Var_{\beta}\left[\nabla log\ L\left(\vec{\theta};\mathbf{X}^{(t)}\right)\right] \tag{38}$$

where H and J denote, respectively, the Jacobian and Hessian matrices regarding the logarithm of the pseudo-likelihood function. Thus, considering the parameter of interest, β, we have the following definition for its asymptotic variance, v_{β} (the derivatives are taken with respect to β):

$$v_{\beta} = \frac{Var_{\beta}\left[\frac{\partial}{\partial\beta} log\ L\left(\vec{\theta};\mathbf{X}^{(t)}\right)\right]}{E_{\beta}^{2}\left[\frac{\partial^{2}}{\partial\beta^{2}} log\ L\left(\vec{\theta};\mathbf{X}^{(t)}\right)\right]} = \frac{E_{\beta}\left[\left(\frac{\partial}{\partial\beta} log\ L\left(\vec{\theta};\mathbf{X}^{(t)}\right)\right)^{2}\right] - E_{\beta}^{2}\left[\frac{\partial}{\partial\beta} log\ L\left(\vec{\theta};\mathbf{X}^{(t)}\right)\right]}{E_{\beta}^{2}\left[\frac{\partial^{2}}{\partial\beta^{2}} log\ L\left(\vec{\theta};\mathbf{X}^{(t)}\right)\right]} \tag{39}$$

However, note that the expected value of the first derivative of $log\ L\left(\vec{\theta};\mathbf{X}^{(t)}\right)$ with relation to β is zero:

$$E\left[\frac{\partial}{\partial\beta} log\ L\left(\vec{\theta};\mathbf{X}^{(t)}\right)\right] = \frac{1}{\sigma^{2}}\sum_{i=1}^{n}\left\{E\left[x_{i}-\mu\right] - \beta\sum_{j\in\eta_{i}} E\left[x_{j}-\mu\right]\right\} = 0 \tag{40}$$

Therefore, the second term of the numerator of Equation (39) vanishes and the final expression for the asymptotic variance of the inverse temperature parameter is given as the ratio between Φ_{β} and Ψ_{β}^{2}:

$$\begin{aligned} v_{\beta} = \frac{1}{\left[\sum_{j\in\eta_{i}}\sum_{k\in\eta_{i}}\sigma_{jk}\right]^{2}} & \left\{\sum_{j\in\eta_{i}}\sum_{k\in\eta_{i}}\left[\sigma^{2}\sigma_{jk} + 2\sigma_{ij}\sigma_{ik}\right] - 2\beta\sum_{j\in\eta_{i}}\sum_{k\in\eta_{i}}\sum_{l\in\eta_{i}}\left[\sigma_{ij}\sigma_{kl} + \sigma_{ik}\sigma_{jl} + \sigma_{il}\sigma_{jk}\right]\right. \\ & \left. +\beta^{2}\sum_{j\in\eta_{i}}\sum_{k\in\eta_{i}}\sum_{l\in\eta_{i}}\sum_{m\in\eta_{i}}\left[\sigma_{jk}\sigma_{lm} + \sigma_{jl}\sigma_{km} + \sigma_{jm}\sigma_{kl}\right]\right\} \end{aligned} \tag{41}$$

This derivation leads us to another definition concerning an isotropic pairwise GMRF.

Definition 9. *Let an isotropic pairwise GMRF be defined on a lattice $S = \{s_1, s_2, \ldots, s_n\}$ with a neighborhood system, η_i. Assuming that $\mathbf{X}^{(\mathbf{t})} = \{x_1^{(t)}, x_2^{(t)}, \ldots, x_n^{(t)}\}$ denotes the global configuration of the system at time t, and $\vec{\rho}$ and Σ_p^{-} are defined as Equations (25) and (24), the asymptotic variance of the maximum pseudo-likelihood estimator of the inverse temperature parameter, β, is given by (using the same matrix-vector notation from the previous sections):*

$$\begin{aligned} v_{\beta} &= \frac{\sigma^{2}\left\|\Sigma_{p}^{-}\right\|_{+} + 2\left\|\vec{\rho}\otimes\vec{\rho}^{T}\right\|_{+} - 6\beta\left\|\vec{\rho}^{T}\otimes\Sigma_{p}^{-}\right\|_{+} + 3\beta^{2}\left\|\Sigma_{p}^{-}\otimes\Sigma_{p}^{-}\right\|_{+}}{\left\|\Sigma_{p}^{-}\right\|_{+}^{2}} = \\ &= \frac{\sigma^{2}}{\left\|\Sigma_{p}^{-}\right\|_{+}} + \frac{\sigma^{4}\Delta_{\beta}\left(\vec{\rho},\Sigma_{p}^{-}\right)}{\left\|\Sigma_{p}^{-}\right\|_{+}^{2}} = \frac{1}{\Psi_{\beta}} + \frac{1}{\Psi_{\beta}^{2}}\left(\Phi_{\beta} - \Psi_{\beta}\right) \end{aligned} \tag{42}$$

Note that when information equilibrium prevails, that is $\Phi_\beta = \Psi_\beta$, the asymptotic variance is given by the inverse of the expected Fisher information. However, the interpretation of this equation indicates that the uncertainty in the estimation of the inverse temperature parameter is minimized when Ψ_β is maximized. Essentially, this means that on average, the local pseudo-likelihood functions are not flat, that is small changes on the local configuration patterns along the system cannot cause abrupt changes in the expected global behavior (the global spatial dependence structure is not susceptible to sharp changes). To reach this condition, there must be a reasonable degree of agreement between the neighboring elements throughout the system, a behavior that is usually associated to low temperature states (β is above a critical value and there is a visible induced spatial dependence structure).

6. The Fisher Curve of a System

With the definition of Φ_β, Ψ_β and H_β, we have the necessary tools to compute three important information-theoretic measures of a global configuration of the system. Our idea is that we can study the behavior of a complex system by constructing a parametric curve in this information-theoretic space as a function of the inverse temperature parameter, β. Our expectation is that the resulting trajectory provides a geometrical interpretation of how the system moves from an initial configuration, A (with a low entropy value for instance), to a desired final configuration, B (with a greater value of entropy, for instance), since the Fisher information plays an important role in providing a natural metric to the Riemannian manifolds of statistical models [18,19]. We will call the path from global State A to global State B as the Fisher curve (from A to B) of the system, denoted by $\vec{F}_A^B(\beta)$. Instead of using time as the parameter to build the curve, $\vec{F}$, we parametrize $\vec{F}$ by the inverse temperature parameter, β.

Definition 10. *Let an isotropic pairwise GMRF be defined on a lattice $S = \{s_1, s_2, \ldots, s_n\}$ with a neighborhood system, η_i, and $\mathbf{X}^{(\beta_1)}, \mathbf{X}^{(\beta_2)}, \ldots, \mathbf{X}^{(\beta_n)}$ be a sequence of outcomes (global configurations) produced by different values of β_i (inverse temperature parameters) for which $A = \beta_{MIN} = \beta_1 < \beta_2 < \cdots < \beta_n = \beta_{MAX} = B$. The system's Fisher curve from A to B is defined as the function $\vec{F} : \Re \to \Re^3$ that maps each configuration, $\mathbf{X}^{(\beta_i)}$, to a point $(\Phi_{\beta_i}, \Psi_{\beta_i}, H_{\beta_i})$ from the information space, that is:*

$$\vec{F}_A^B(\beta) = (\Phi_\beta, \Psi_\beta, H_\beta) \qquad \beta = A, \ldots, B \tag{43}$$

where Φ_β, Ψ_β and H_β denote the Type I expected Fisher information, the Type II expected Fisher information and the Shannon entropy of the global configuration, $\mathbf{X}^{(\beta)}$, defined by:

$$\Phi_\beta = \frac{1}{\sigma^4}\left[\sigma^2 \left\|\Sigma_p^-\right\|_+ + 2\left\|\vec{\rho} \otimes \vec{\rho}^T\right\|_+ - 6\beta\left\|\vec{\rho}^T \otimes \Sigma_p^-\right\|_+ + 3\beta^2\left\|\Sigma_p^- \otimes \Sigma_p^-\right\|_+\right] \tag{44}$$

$$\Psi_\beta = \frac{1}{\sigma^2}\left\|\Sigma_p^-\right\|_+ \tag{45}$$

$$H_\beta = \frac{1}{2}\left[log\left(2\pi\sigma^2 + 1\right)\right] - \left[\frac{\beta}{\sigma^2}\|\vec{\rho}\|_+ - \frac{\beta^2}{2}\Psi_\beta\right] \tag{46}$$

In the next sections, we show some computational experiments that illustrate the effectiveness of the proposed tools in measuring the information encoded in complex systems. We want to investigate what happens to the Fisher curve as the inverse temperature parameter is modified in order to control the system's global behavior. Our main conclusion, which is supported by experimental analysis, is that $\vec{F}_A^B(\beta) \neq \vec{F}_B^A(\beta)$. In other words, in terms of information, moving towards higher entropy states is not the same as moving towards lower entropy states, since the Fisher curves that represent the trajectory between the initial State A and the final State B are significantly different.

7. Computational Simulations

This section discusses some numerical experiments proposed to illustrate some applications of the derived tools in both simulations and real data. Our computational investigations were divided into two main sets of experiments:

(1) Local analysis: analysis of the local and global versions of the measures (ϕ_β, ψ_β, Φ_β, Ψ_β and H_β), considering a fixed inverse temperature parameter;
(2) Global analysis: analysis of the global versions of the measures (Φ_β, Ψ_β and H_β) along Markov chain Monte Carlo (MCMC) simulations in which the inverse temperature parameter is modified to control the expected global behavior.

7.1. Learning from Spatial Data with Local Information-Theoretic Measures

First, in order to illustrate a simple application of both forms of local observed Fisher information, ϕ_β and ψ_β, we performed an experiment using some synthetic images generated by the Metropolis–Hastings algorithm. The basic idea of this simulation process is to start at an initial configuration in which temperature is infinite (or $\beta = 0$). This basic initial condition is randomly chosen, and after a fixed number of steps, the algorithm produces a configuration that is considered to be a valid outcome of an isotropic pairwise GMRF model. Figure 1 shows an example of the initial condition and the resulting system configuration after 1,000 iterations considering a second order neighborhood system (eight nearest neighbors). The model parameters were chosen as: $\mu = 0$, $\sigma^2 = 5$ and $\beta = 0.8$.

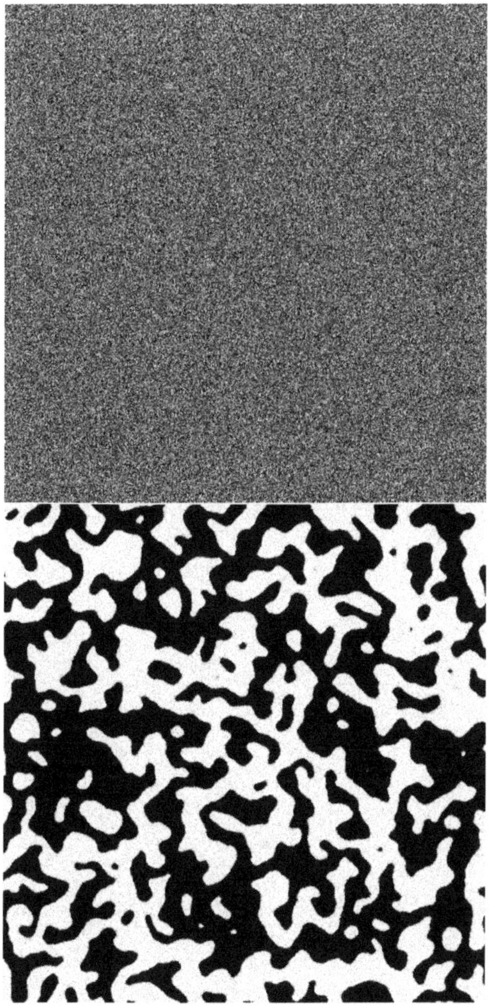

Figure 1. Example of Gaussian Markov random field (GMRF) model outputs. The values of the inverse temperature parameter, β, in the left and right configurations are zero and 0.8, respectively.

Three Fisher information maps were generated from both initial and resulting configurations. The first map was obtained by calculating the value, $\phi_\beta(x_i)$, for every point of the system, that is for $i = 1, 2, \ldots, n$. Similarly, the second one was obtained by using $\psi_\beta(x_i)$. The last information map was built by using the ratio between $\phi_\beta(x_i)$ and $\psi_\beta(x_i)$, motivated by the fact that boundaries are often composed of patterns that are not expected to be "aligned" to the global behavior (and, therefore, show high values of $\phi_\beta(x_i)$) and also are somehow unstable (show low values of $\psi_\beta(x_i)$). We will recall this measure, $L_\beta(x_i) = \phi_\beta(x_i)/\psi_\beta(x_i)$, the local L-information, once it is defined in terms of the first two derivatives of the logarithm of the local pseudo-likelihood function. Figure 2 shows the obtained information maps as images. Note that while ϕ_β has a strong response for boundaries (the edges are light), ψ_β has a weak one (so the edges are dark), evidence in favor of considering L-information in boundary detection procedures. Note also that in the initial condition, when the temperature is infinite, the informative patterns are almost uniformly spread all over the system, while the final configuration shows a more sparse representation in terms of information. Figure 3 shows the distribution of local L-information for both systems' configurations depicted in Figure 1.

Figure 2. Fisher information maps. The first row shows the information maps of the system when the temperature is infinite ($\beta = 0$). The second row shows the same maps when the temperature is low ($\beta = 0.8$). The first and second columns show information maps that were generated by computing $\phi_\beta(x_i)$ and $\psi_\beta(x_i)$ for each observation in the lattice. The column map was produced by computing the local L-information, that is the ratio between both local information measures. In terms of information, low temperature configurations are more sparse, since most local patterns are uninformative, due to the strong alignment of the particles throughout the system, which is the expected global behavior for β above a certain critical value.

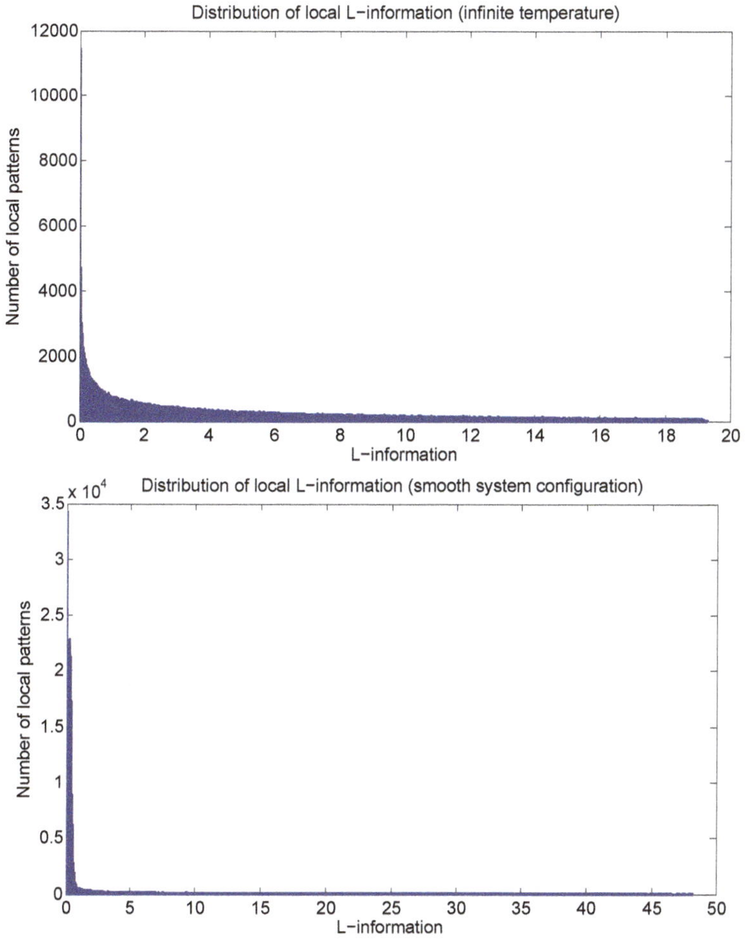

Figure 3. Distribution of local L-information. When the temperature is infinite, the information is spread along the system. For low temperature configurations, the number of local patterns with zero information content significantly increases, that is the system is more sparse in terms of Fisher information.

7.2. Analyzing Dynamical Systems with Global Information-Theoretic Measures

In order to study the behavior of a complex system that evolves from an initial State A to another State B, we use the Metropolis–Hastings algorithm, an MCMC simulation method, to generate a sequence of valid isotropic pairwise GMRF model outcomes for different values of the inverse temperature parameter, β. This process is an attempt to perform a random walk on the state space of the system, that is, in the space of all possible global configurations in order to analyze the behavior of the proposed global measures: entropy and both forms of Fisher information. The main purpose of this experiment is to observe what happens to Φ_β, Ψ_β and H_β when the system evolves from a random initial state to other global configurations. In other words, we want to investigate the Fisher curve of the system in order to characterize its behavior in terms of information. Basically, the idea is to use the Fisher curve as a kind of signature for the expected behavior of any system modeled by an isotropic pairwise GMRF, making it possible to gain insights into the understanding of large complex systems.

To simulate a system in which we can control the inverse temperature parameter, we define an updating rule for β based on fixed increments. In summary, we start with a minimum value β_{MIN} (when $\beta_{MIN} = 0$, the temperature of the system is infinite). Then, the value of β in the iteration, t, is defined as the value of β in $t-1$ plus a small increment ($\Delta\beta$), until it reaches a pre-defined upper bound, β_{MAX}. The process in then repeated with negative increments $-\Delta\beta$, until the inverse temperature reaches its minimum value, β_{MIN}, again. This process continues for a fixed number of iterations, N_{MAX}, during an MCMC simulation. As a result of this approach, a sequence of GMRF samples is produced. We use this sequence to calculate Φ_β, Ψ_β and H_β and define the Fisher curve $\vec{F}$, for $\beta = \beta_{MIN}, \ldots, \beta_{MAX}$. Figure 4 shows some of the system's configurations along an MCMC simulation. In this experiment, the parameters were defined as: $\beta_{MIN} = 0$, $\Delta\beta = 0.001$, $\beta_{MAX} = 0.15$ and $N_{MAX} = 1,000$, $\mu = 0$, $\sigma^2 = 5$ and $\eta_i = \{(i-1,j-1),(i-1,j),(i-1,j+1),(i,j-1),(i,j+1),(i+1,j-1),(i+1,j),(i+1,j+1)\}$.

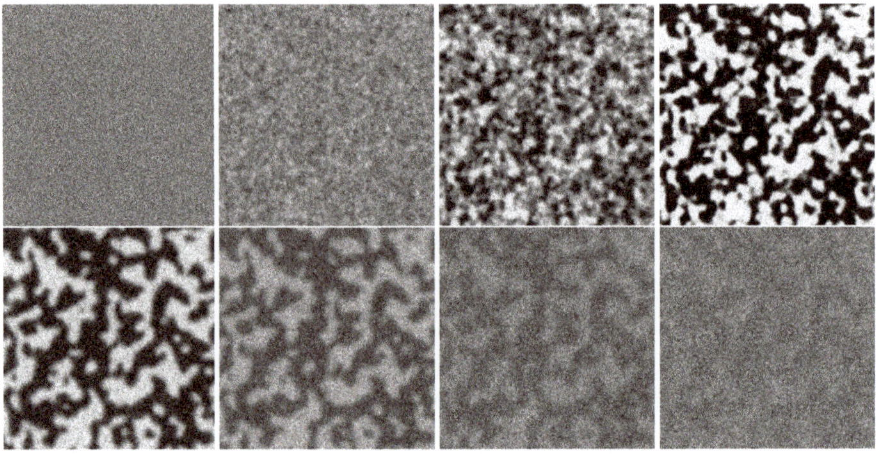

Figure 4. Global configurations along a Markov chain Monte Carlo (MCMC) simulation. Evolution of the system as the inverse temperature parameter, β, is modified to control the expected global behavior.

A plot of both forms of the expected Fisher information, Φ_β and Ψ_β, for each iteration of the MCMC simulation is shown in Figure 5. The graph produced by this experiment shows some interesting results. First of all, regarding upper and lower bounds on these measures, it is possible to note that when there is no induced spatial dependence structure ($\beta \approx 0$), we have an information equilibrium condition ($\Phi_\beta = \Psi_\beta$ and the information equality holds). In this condition, the observations are practically independent in the sense that all local configuration patterns convey approximately the same amount of information. Thus, it is hard to find and separate the two categories of patterns we know: the informative and the non-informative ones. Once they all behave in a similar manner, there is no informative pattern to highlight. Moreover, in this information equilibrium situation, Ψ_β reaches its lower bound (in this simulation, we observed that in the equilibrium $\Phi_\beta \approx \Psi_\beta \approx 8$), indicating that this condition emerges when the system is most susceptible to a change in the expected global behavior, since the uncertainty about β is maximum at this moment. In other words, modification in the behavior of a small subset of local patterns may guide the system to a totally different stable configuration in the future.

The results also show that the difference between Φ_β and Ψ_β is maximum when the system operates with large values of β, that is, when organization emerges and there is a strong dependence structure among the random variables (the global configuration shows clear visible clusters and boundaries between them). In such states, it is expected that the majority of patterns be aligned to the global behavior, which causes the appearance of few, but highly informative patterns: those connecting elements from different regions (boundaries). Besides that, the results suggest that it takes more time for the system to go from the information equilibrium state to organization than the opposite. We will see how this fact becomes clear by analyzing the Fisher curve along Markov chain Monte Carlo (MCMC) simulations. Finally, the results also suggest that both Φ_β and Ψ_β are bounded by a superior value, possibly related to the size of the neighborhood system.

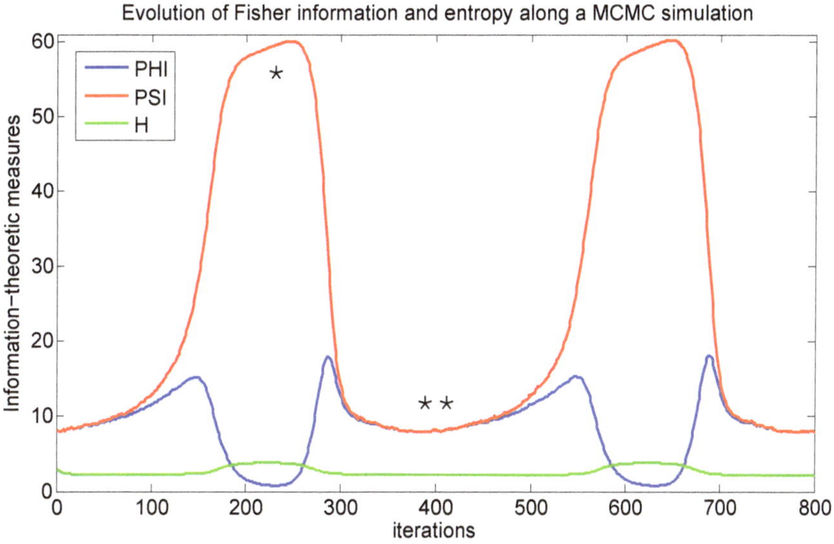

Figure 5. Evolution of Fisher information along an MCMC simulation. As the difference between Φ_β and Ψ_β is maximized (*), the uncertainty about the real inverse temperature parameter is minimized and the number of informative patterns increases. In the information equilibrium condition (**), it is hard to find informative patterns, since there is no induced spatial dependence structure.

Figure 6 shows the real parameter values used to generate the GMRF outputs (blue line), the maximum pseudo-likelihood estimative used to calculate Φ_β and Ψ_β (red line) and also a plot of the asymptotic variances (uncertainty about the inverse temperature) along the entire MCMC simulation.

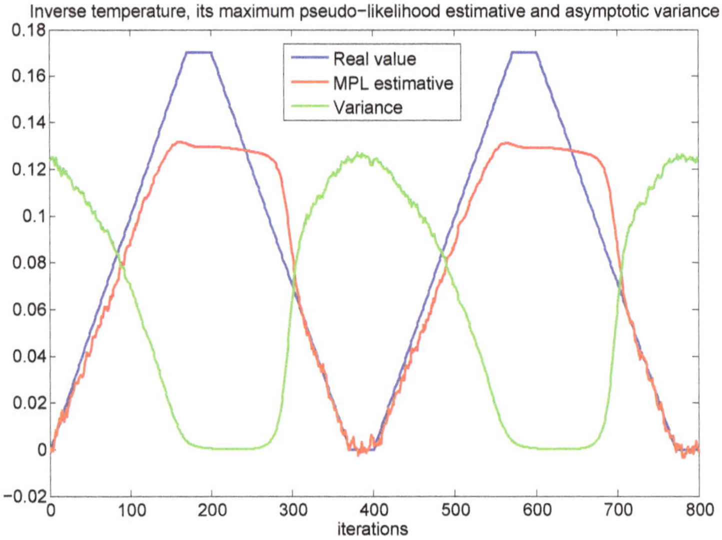

Figure 6. Real and estimated inverse temperatures along the MCMC simulation. The system's global behavior is controlled by the real inverse temperature parameter values (blue line), used to generate the GMRF outputs. The maximum pseudo-likelihood estimative is used to compute both Φ_β and Ψ_β. Note that the uncertainty about the inverse temperature increases as $\beta \to 0$ and the system approaches the information equilibrium condition.

We now proceed to the analysis of the Shannon entropy of the system along the simulation. Despite showing a behavior similar to Ψ_β, the range of values for entropy is significantly smaller. In this simulation, we observed that $0 \leq H_\beta \leq 4.5$, $0 \leq \Phi_\beta \leq 18$ and $8 \leq \Psi_\beta \leq 61$. An interesting point is that knowledge of Φ_β and Ψ_β allows us to infer the entropy of the system. For example, looking at Figures 5 and 7, we can see that Φ_β and Ψ_β start to diverge a little bit earlier ($t \approx 80$), then the entropy in a GMRF model begins to grow ($t \approx 120$). Therefore, in an isotropic pairwise GMRF model, if the system is close to the information equilibrium condition, then H_β is low, since there is little variability in the observed configuration patterns. When the difference between Φ_β and Ψ_β is large, H_β increases.

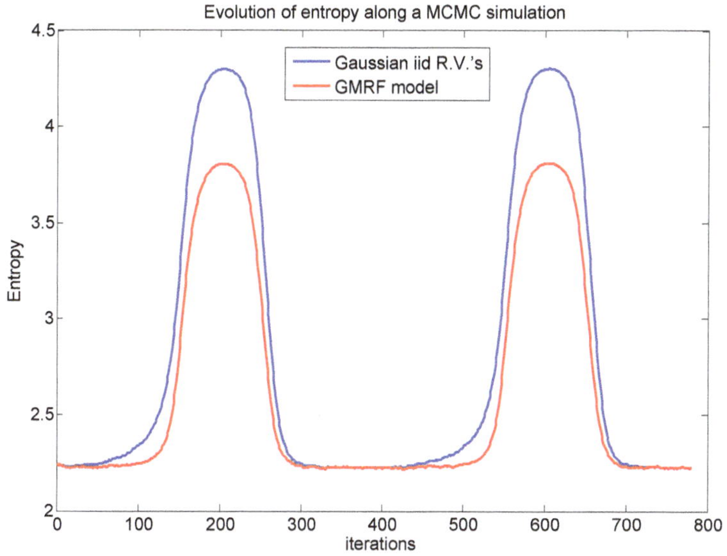

Figure 7. Evolution of Shannon entropy along an MCMC simulation. H_β start to grow when the system leaves the equilibrium condition, where the entropy in the isotropic pairwise GMRF model is identical to the entropy of a simple Gaussian random variable (since $\beta \to 0$).

Another interesting global information-theoretic measure is L-information, from now on denoted by L_β, since it conveys all the information about the likelihood function (in a GMRF model, only the first two derivatives of $L(\vec{\theta}; \mathbf{X}^{(t)})$ are not null). L_β is defined as the ratio between the two forms of expected Fisher information, Φ_β and Ψ_β. A nice property about this measure is that $0 \leq L_\beta \leq 1$. With this single measurement, it is possible to gain insights about the global system behavior. Figure 8 shows that a value close to one indicates a system approximating the information equilibrium condition, while a value close to zero indicates a system close to the maximum entropy condition (a stable configuration with boundaries and informative patterns).

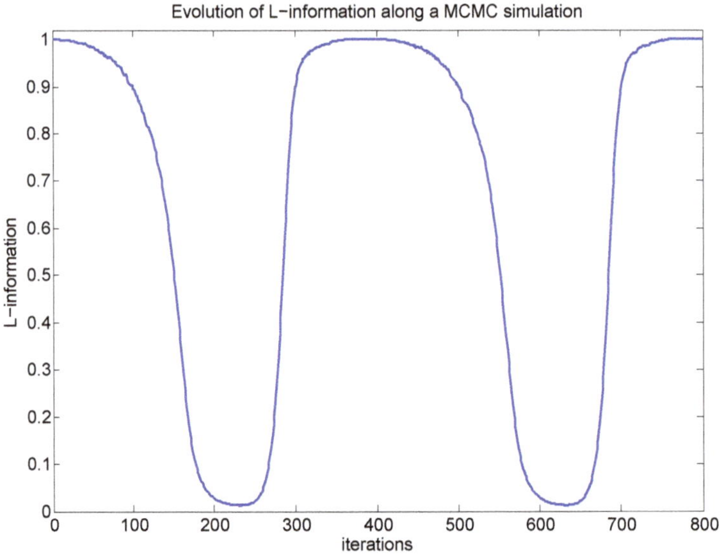

Figure 8. Evolution of L-information along an MCMC simulation. When L_β approaches one, the system tends to the information equilibrium condition. For values close to zero, the system tends to the maximum entropy condition.

To investigate the intrinsic non-linear connection between Φ_β, Ψ_β and H_β in a complex system modeled by an isotropic pairwise GMRF model, we now analyze its Fisher curves. The first curve, which is a planar one, is defined as $\vec{F}(\beta) = (\Phi_\beta, \Psi_\beta)$, for $A = \beta_{min}$ to $B = \beta_{max}$ and shows how Fisher information changes when the inverse temperature of the system is modified to control the global behavior. Figure 9 shows the results. In the first image, the blue piece of the curve is the path from A to B, that is, $\vec{F}(\beta)_A^B$, and the red piece is the inverse path (from B to A), that is, $\vec{F}(\beta)_B^A$. We must emphasize that $\vec{F}(\beta)_A^B$ is the trajectory from a lower entropy global configuration to a higher entropy global configuration. On the other hand, when the system moves from B to A, we are moving towards entropy minimization. To make this clear, the second image of Figure 9 illustrates the same Fisher curve as before, but now in three dimensions, that is, $\vec{F}(\beta) = (\Phi_\beta, \Psi_\beta, H_\beta)$. For comparison purposes, Figure 10 shows the Fisher curves for another MCMC simulation with different parameter settings. Note that the shape of the curves are quite similar to those in Figure 9.

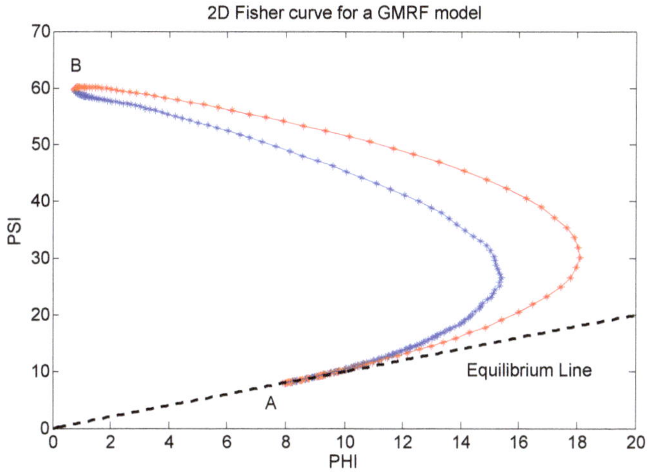

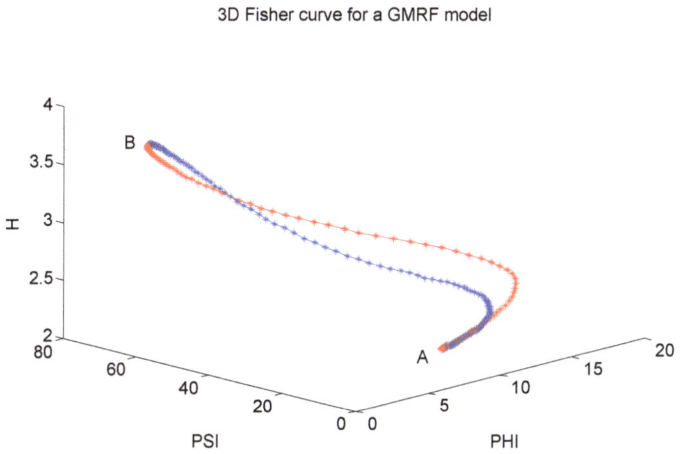

Figure 9. 2D and 3D Fisher curves of a complex system along an MCMC simulation. The graph shows a parametric curve obtained by varying the β parameter from β_{MIN} to β_{MAX} and back. Note that, from a differential geometry perspective, as the divergence between Φ_β and Ψ_β increases, the torsion of the parametric curve becomes evident (the curve leaves the plane of constant entropy).

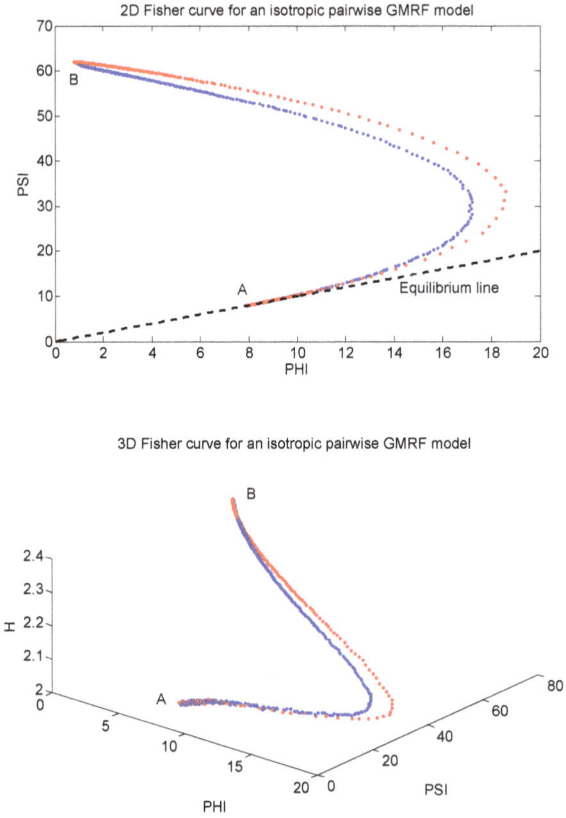

Figure 10. 2D and 3D Fisher curves along another MCMC simulation. The graph shows a parametric curve obtained by varying the β parameter from β_{MIN} to β_{MAX} and back. Note that, from a geometrical perspective, the properties of these curves are essentially the same as the ones from the previous simulation.

We can see that the majority of points along the Fisher curve is concentrated around two regions of high curvature: (A) around the information equilibrium condition (an absence of short-term and long-term correlations, since $\beta = 0$); and (B) around the maximum entropy value, where the divergence between the information values are maximum (self-organization emerges, since β is greater than a critical value, β_c). The points that lie in the middle of the path connecting these two regions represent the system undergoing a phase transition. Its properties change rapidly and in an asymmetric way, since $\vec{F}(\beta)_A^B \neq \vec{F}(\beta)_B^A$ for a given natural orientation.

By now, some observations can be highlighted. First, the natural orientation of the Fisher curve defines the direction of time. The natural A–B path (increase in entropy) is given by the blue curve and the natural B–A path (decrease in entropy) is given by the red curve. In other words, the only possible way to walk from A to B (increase H_β) by the red path or to walk from B to A (decrease H_β) by the blue path would be moving back in time (by running the recorded simulation backwards).Eventually, we believe that a possible explanation for this fact could be that the deformation process that takes the original geometric structure (with constant curvature) of the usual Gaussian model (A) to the novel geometric structure of the isotropic pairwise GMRF model (B) is not reversible. In other words, the way the model is "curved" is not simply the reversal of the "flattering" process (when it is restored to its constant curvature form). Thus, even the basic notion of time seems to be deeply connected with the relationship between entropy and Fisher information in a complex system: in the natural orientation (forward in time), it seems that the divergence between Φ_β and Ψ_β is the cause of an increase in the entropy, and the decrease of entropy is the cause of the convergence of Φ_β and Ψ_β. During the experimental analysis, we repeated the MCMC simulations with different parameter settings, and the observed behavior for Fisher information and entropy was the same. Figure 11 shows the graphs of Φ_β, Ψ_β and H_β for another recorded MCMC simulation. The results indicate that in the natural orientation (in the direction of time), an increase in Ψ_β seems to be a trigger for an increase in the entropy and a decrease in the entropy seems to be a trigger for a decrease in Ψ_β. Roughly speaking, Ψ_β "pushes H_β up" and H_β "pushes Ψ_β down".

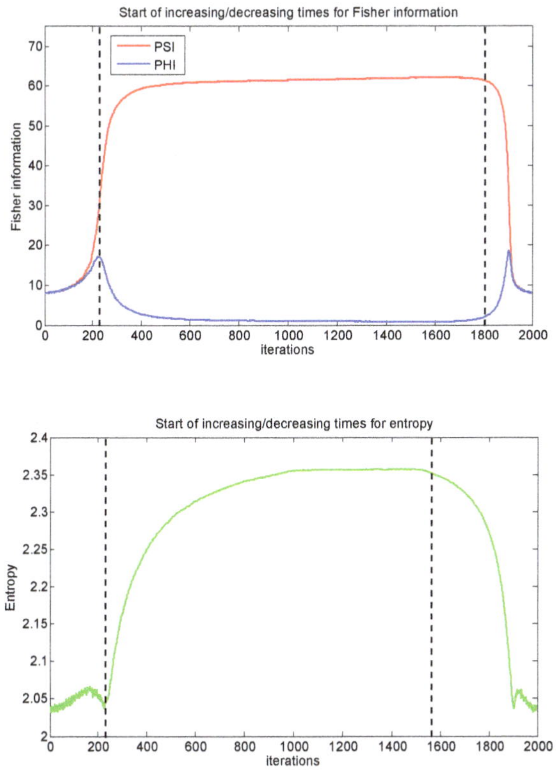

Figure 11. Relations between entropy and Fisher information. When a system modeled by an isotropic pairwise GMRF evolves in the natural orientation (forward in time), two rules that relate Fisher information and entropy can be observed: (1) an increase in Ψ_β is the cause of an increase in H_β (the increase in H_β is a consequence of the increase in Ψ_β); (2) a decrease in H_β is the cause of a decrease in Ψ_β (the decrease in Ψ_β is a consequence of the decrease in H_β). In other words, when moving towards higher entropy states, changes in Fisher information precedes changes in entropy (Ψ_β "pushes H_β up"). When moving towards lower entropy states, changes in entropy precedes changes in Fisher information (H_β "pushes Ψ_β down").

In summary, the central idea discussed here is that while entropy provides a measure of order/disorder of the system at a given configuration, $\mathbf{X}^{(t)}$, Fisher information links these thermodynamical states through a path (Fisher curve). Thus, Fisher information is a powerful mathematical tool in the study of complex and dynamical systems, since it establishes how these different thermodynamical states are related along the evolution of the inverse temperature. Instead of knowing whether the entropy, H_β, is increasing or decreasing, with Fisher information, it is possible to know how and why this change is happening.

7.2.1. The Effect of Induced Perturbations in the System

To test whether a system can recover part of its original configuration after a perturbation is induced, we conducted another computational experiment. During a stable simulation, two kinds of perturbations were induced in the system: (1) the value of the inverse temperature parameter was set to zero for the next consecutive two iterations; (2) the value of the inverse temperature parameter was set to the equilibrium value, β^* (the solution of Equation 28), for the next consecutive two iterations. We should mention that in both cases, the original value of β is recovered after these two iterations are completed.

When the system is disturbed by setting β to zero, the simulations indicate that the system is not successful in recovering components from its previous stable configuration (note that Φ_β and Ψ_β clearly touch one another in the graph). When the same perturbation is induced, but using the smallest of the two β^* values (minimum solution of Equation 28), after a short period of turbulence, the system can recover parts (components, clusters) of its previous stable state. This behavior suggests that this softer perturbation is not enough to remove all the information encoded within the spatial dependence structure of system, preserving some of the long-term correlations in data (stronger bonds), slightly remodeling the large clusters presented in the system. Figures 12 and 13 illustrate the results.

7.3. Considerations and Final Remarks

The goal of this section is to summarize the main results obtained in this paper, focusing on the interpretation of the Fisher curve of a system modeled by a GMRF. First, our system is initialized with a random configuration, simulating that in the moment of its creation, the temperature is infinite ($\beta = 0$). We observe two important things at this moment: (1) there is a perfect symmetry in information, since the equilibrium condition prevails, that is, $\Phi_\beta = \Psi_\beta$; (2) the entropy of the system is minimal. By a mere convention, we name this initial state of minimal entropy, A.

By reducing the global temperature (β increases), this "universe" is deviating from this initial condition. As the system is drifted apart from the initial condition, we clearly see a break in the symmetry of information (Φ_β diverges from Ψ_β), which apparently is the cause for an increase in the system's entropy, since this symmetry break seems to precede an increase in the entropy, H. This is a fundamental symmetry break, since other forms of ruptures that will happen in the future and will give rise to several properties of the system, including the basic notion of time as an irreversible process, follow from this first one. During this first stage of evolution, the system evolves to the condition of maximum entropy, named B.

Hence, after the break in the information equilibrium condition, there is a significant increase in the entropy as the system continues to evolve. This stage lasts while the temperature of the system is further reduced or kept established. When the temperature starts to increase (β decreases), another form of symmetry break takes place. By moving towards the initial condition (A) from B, changes in the entropy seem to precede changes in Fisher information (when moving from A to B, we observe exactly the opposite). Moreover, the variations in entropy and Fisher information towards A are not symmetric with the variations observed when we move towards B, a direct consequence of that first fundamental break of the information equilibrium condition. By continuing this process of increasing the temperature of the system until infinity (β is approaching zero), we take our system to a configuration that is equivalent to the initial condition, that is, where information equilibrium prevails.

This fundamental symmetry break becomes evident when we look at the Fisher curve of the system. We clearly see that the path from the state of minimum entropy, A, and the state of maximum entropy, B, defined by the curve, $\vec{F}_A^B$ (the blue trajectory in Figure 9), is not the same as the path from B to A, defined by the curve, $\vec{F}_B^A$ (the red trajectory in Figure 9). An implication of this behavior is that if the system is moving along the arrow of time, then we are moving through the Fisher curve in the clockwise orientation. Thus, the only way to go from A to B along $\vec{F}_B^A$ (the red path) is going back in time.

Therefore, if that first fundamental symmetry break did not exist, or even if it had happened, but all the posterior evolution of Φ_β, Ψ_β and H_β were absolutely symmetric (*i.e.*, the variations in these measures were exactly the same when moving from A to B and when moving from B to A), what we

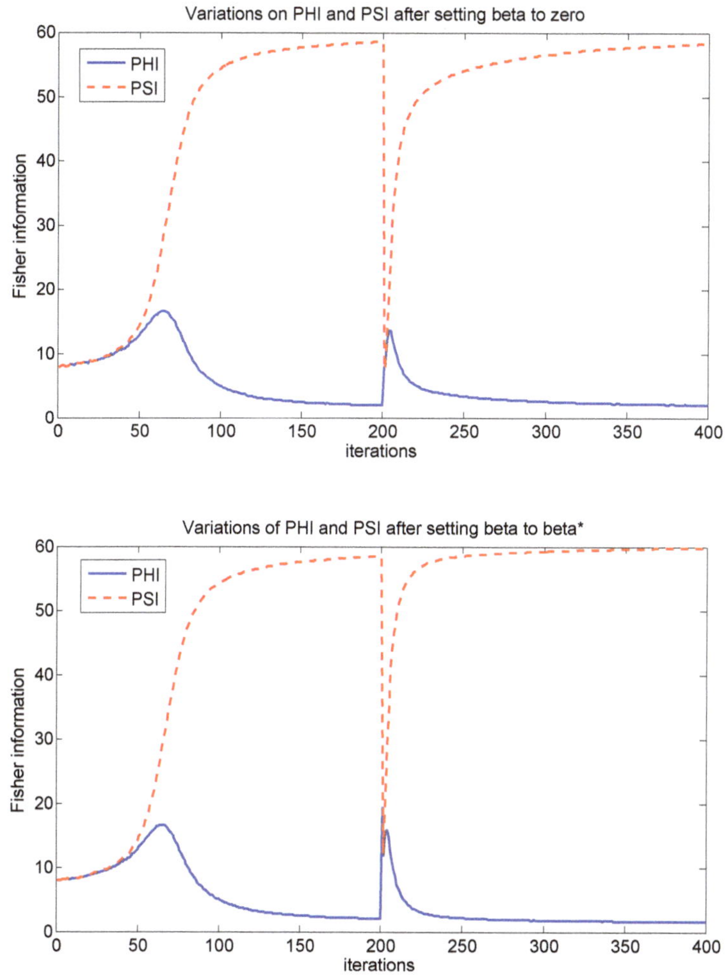

Figure 12. Disturbing the system to induce changes. Variation on Φ_β and Ψ_β after the system is disturbed by an abrupt change in the value of β. In the first image, the inverse temperature is set to zero. Note that Φ_β and Ψ_β touch one another, indicating that no residual information is kept, as if the simulation had been restarted from a random configuration. In the second image, the inverse temperature is set to the equilibrium value, β^*. The results suggest that this kind of perturbation is not enough to remove all the information within the spatial dependence structure, allowing the system to recover a significant part of its original configuration after a short stabilization period.

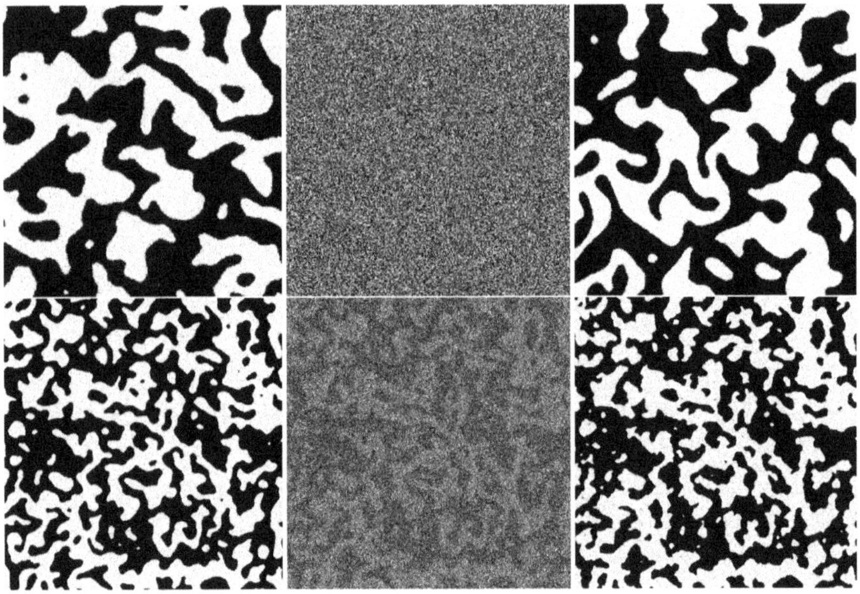

Figure 13. The sequence of outputs along the MCMC simulation before and after the system is disturbed. The first row (when β is set to zero) shows that the system evolved to a different stable configuration after the perturbation. The second row (when β is set to β^*) indicates that the system was able to recover a significant part from its previous stable configuration.

would actually see is that $\vec{F}_A^B = \vec{F}_B^A$. As a consequence, to decrease/increase the system's temperature would be like moving towards the future/past. In fact, the basic notion of time in that system would be compromised, since time would be a perfectly reversible process (just similar to a spatial dimension, in which we can move in both directions). In other words, we would not distinguish whether the system is moving forward or moving backwards in time.

8. Conclusions

The definition of what is information in a complex system is a fundamental concept in the study of many problems. In this paper, we discussed the roles of two important statistical measures in isotropic pairwise Markov random fields composed of Gaussian variables: Shannon entropy and Fisher information. By using the pseudo-likelihood function of the GMRF model, we derived analytical expressions for these measures. The definition of a Fisher curve as a geometric representation for the study and analysis of complex systems allowed us to reveal the intrinsic non-linear relation between these information-theoretic measures and gain insights about the behavior of such systems. Computational experiments demonstrate the effectiveness of the proposed tools in decoding information from the underlying spatial dependence structure of a Gaussian-Markov random field. Typical informative patterns in a complex systems are located in the boundaries of the clusters. One of the main conclusions of this scientific investigation concerns the notion of time in a complex system. The obtained results suggest that the relationship between Fisher information and entropy determines whether the system is moving forward or backward in time. Apparently, in the natural orientation (when the system is evolving forward in time), when β is growing, that is, the temperature of the system is reducing, an increase in Fisher information leads to an increase in the system's entropy, and when β is reducing, that is the temperature of the system is growing, a decrease in the system's entropy leads to a decrease in the Fisher information. In future works we expect to completely characterize the metric tensor that represents the geometric structure of the isotropic pairwise GMRF model by specifying all the elements of the Fisher information matrix. Future investigations should also include the definition and analysis of the proposed tools in other Markov random field models, such as the Ising and Potts pairwise interaction models. Besides, a topic of interest concerns the investigation of minimum and maximum information paths in graphs to explore intrinsic similarity measures between objects belonging to a common surface or manifold in $\Re^n$. We believe this study could bring benefits to some pattern recognition and data analysis computational applications.

Acknowledgments: The author would like to thank CNPQ(Brazilian Council for Research and Development) for the financial support through research grant number 475054/2011-3.

Conflicts of Interest: Conflict of Interest

The authors declare no conflict of interest.

References

1. Shannon, C.; Weaver, W. *The Mathematical Theory of Communication*; University of Illinois Press: Urbana, Chicago, IL & London, USA, 1949.
2. Rényi, A. On measures of information and entropy. In Proceedings of the 4th Berkeley Symposium on Mathematics, Statistics and Probability, Berkeley, CA, USA, 20 June–30 July 1960; University of California Press: Berkeley, CA, USA, 1961. pp. 547–561
3. Tsallis, C. Possible generalization of Boltzmann-Gibbs statistics. *J. Stat. Phys.* **1988**, *52*, 479–487.
4. Bashkirov, A. Rényi entropy as a statistical entropy for complex systems. *Theor. Math. Phys.* **2006**, *149*, 1559–1573.
5. Jaynes, E. Information theory and statistical mechanics. *Phys. Rev.* **1957**, *106*, 620–630.
6. Grad, H. The many faces of entropy. *Comm. Pure. Appl. Math.* **1961**, *14*, 323–254.
7. Adler, R.; Konheim, A.; McAndrew, A. Topological entropy. *Trans. Am. Math. Soc.* **1965**, *114*, 309–319.
8. Goodwyn, L. Comparing topological entropy with measure-theoretic entropy. *Am. J. Math.* **1972**, *94*, 366–388.
9. Samuelson, P.A. Maximum principles in analytical economics. *Am. Econ. Rev.* **1972**, *62*, 249–262.
10. Costa, M. Writing on dirty paper. *IEEE T. Inform. Theory* **1983**, *29*, 439–441.
11. Dembo, A.; Cover, T.; Thomas, J. Information theoretic inequalities. *IEEE T. Inform. Theory* **1991**, *37*, 1501–1518.
12. Cover, T.; Zhang, Z. On the maximum entropy of the sum of two dependent random variables. *IEEE T. Inform. Theory* **1994**, *40*, 1244–1246.
13. Frieden, B.R. *Science from Fisher Information: A Unification*; Cambridge University Press: Cambridge, London, 2004.
14. Frieden, B.R.; Gatenby, R.A. *Exploratory Data Analysis Using Fisher Information*; Springer: London, UK, 2006.
15. Lehmann, E.L. *Theory of Point Estimation*; Wiley: New York, NY, USA, 1983.
16. Bickel, P.J. *Mathematical Statistics*; Holden Day: New York, NY, USA, 1991.
17. Casella, G.; Berger, R.L. *Statistical Inference*, 2nd ed.; Duxbury: New York, NY, USA, 2002.
18. Amari, S. Nagaoka, H. *Methods of information geometry (Translations of mathematical monographs vol. 191)*; AMS Bookstore: Tokyo, Japan, 2000.
19. Kass, R.E. The geometry of asymptotic inference. *Stat. Sci.* **1989**, *4*, 188–234.
20. Anandkumar, A.; Tong, L.; Swami, A. Detection of Gauss-Markov random fields with nearest-neighbor dependency. *IEEE T. Inform. Theory* **2009**, *55*, 816–827.
21. Gómez-Villegas, M.A.; Main, P.; Susi, R. The effect of block parameter perturbations in Gaussian Bayesian networks: Sensitivity and robustness. *Inform. Sci.* **2013**, *222*, 439–458.
22. Moura, J.; Balram, N. Recursive structure of noncausal Gauss-Markov random fields. *IEEE T. Inform. Theory* **1992**, *38*, 334–354.
23. Moura, J.; Goswami, S. Gauss-Markov random fields (GMrf) with continuous indices. *IEEE Trans. Inform. Theory* **1997**, *43*, 1560–1573.

24. Besag, J. Spatial interaction and the statistical analysis of lattice systems. *J. Roy. Stat. Soc. B Stat. Meth.* **1974**, *36*, 192–236.
25. Besag, J. Statistical analysis of non-lattice data. *The Statistician* **1975**, *24*, 179–195.
26. Hammersley, J.; Clifford, P. (University of California, Berkeley, Oxford and Bristol). Markov Field on Finite Graphs and Lattices. Unpublished work, 1971.
27. Efron, B.F.; Hinkley, D.V. Assessing the accuracy of the ml estimator: Observed versus expected fisher information. *Biometrika* **1978**, *65*, 457–487.
28. Isserlis, L. On a formula for the product-moment coefficient of any order of a normal frequency distribution in any number of variables. *Biometrika* **1918**, *12*, 134–139.
29. Jensen, J.; Künsh, H. On asymptotic normality of pseudo likelihood estimates for pairwise interaction processes. *Ann. Inst. Stat. Math.* **1994**, *46*, 475–486.
30. Winkler, G. *Image Analysis, Random Fields and Markov Chain Monte Carlo Methods: A Mathematical Introduction*; Springer-Verlag New York, Inc.: Secaucus, NJ, USA, 2006.
31. Liang, G.; Yu, B. Maximum pseudo likelihood estimation in network tomography. *IEEE T. Signal Proces.* **2003**, *51*, 2043–2053.

entropy

MDPI

Article

Network Decomposition and Complexity Measures: An Information Geometrical Approach

Masatoshi Funabashi

Sony Computer Science Laboratories, inc. Takanawa muse bldg. 3F, 3-14-13, Higashi Gotanda, Shinagawa-ku, Tokyo 141-0022, Japan; E-Mail: masa_funabashi@csl.sony.co.jp; Tel.: +81-3-5448-4380; Fax: +81-3-5448-4273

Received: 28 March 2014; in revised form: 24 June 2014 / Accepted: 14 July 2014 / Published: 23 July 2014

Abstract: We consider the graph representation of the stochastic model with n binary variables, and develop an information theoretical framework to measure the degree of statistical association existing between subsystems as well as the ones represented by each edge of the graph representation. Besides, we consider the novel measures of complexity with respect to the system decompositionability, by introducing the geometric product of Kullback–Leibler (KL-) divergence. The novel complexity measures satisfy the boundary condition of vanishing at the limit of completely random and ordered state, and also with the existence of independent subsystem of any size. Such complexity measures based on the geometric means are relevant to the heterogeneity of dependencies between subsystems, and the amount of information propagation shared entirely in the system.

Keywords: information geometry; complexity measure; complex network; system decompositionability; geometric mean

1. Introduction

Complex systems sciences emphasize on the importance of non-linear interactions that can not be easily approximated linearly. In other word, the degrees of non-linear interactions are the source of complexity. The classical reductionism approach generally decomposes a system into its components with linear interactions, and tries to evaluate whether the whole property of the system can still be reproduced. If this decomposition of a system destroys too much information to reproduce the system's whole property, the plausibility of such reductionism is lost. Inversely, if we can evaluate how much information is ignored by the decomposition, we can assume how much complexity of the whole system is lost. This gives us a way to measure the complexity of a system with respect to the system decomposition.

In stochastic systems described as a set of joint distributions, the interaction can basically be expressed as the statistical association between the variables. The simplest reductionism approach is to separate the whole system into some subsets of variables, and assume the independence between them. If such decomposition does not affect the system's property, the isolated subsystem is independent from the rest. On the other hand, if the decomposition loses too much information, then the subsystem is inside of a larger subsystem with strong internal dependencies and can not be easily separated.

The stochastic models have often been represented with the use of graph representation, and treated with the name of complex network [1–3]. Generally, the nodes represent the variables and the weights on the edges are the statistical association between them. However, if we consider the information contained in the different orders of dependencies among variables, the graph with a single kind of edges is not sufficient to express the whole information of the system [4]. An edge of a graph with n nodes contains the information of statistical association up to the n-th order dependencies among n variables. If we try to decompose the system independently by cutting these information, we have to consider what it means to cut the edge of the graph from the information theoretical point of view.

Indeed, analysis on the degree of dependencies existing between variables derived many definition of complexity in stochastic model [5], which have been mostly studied with information theoretical perspective. Beginning with seminal works of Lempel and Ziv (e.g., [6]), computation-oriented definition of complexity takes deterministic formalization and measures the necessary information to reproduce a given symbolic sequence exactly, which is classified with the name of *algorithmic complexity* [7–9].

On the other hand, statistical approach to complexity, namely *statistical complexity*, assumes some stochastic model as theoretical basis, and refers to the structure of information source on it in measure-theoretic way [10–12].

One of the most classical statistical complexities is the mutual information between two stochastic variables, and its generalized form to measure dependence between n variables is proposed (e.g., [13]) and explored in relevance to statistical models and theories by several authors [14–16].

We should also recall that complexity is not necessary conditioned only by information theory, but rather motivated from the organization of living system such as brain activity. The TSE complexity shows further extension of generalized mutual information into biological context, where complexity exists as the heterogeneity between different system hierarchies [17]. These statistical complexities are all based on the boundary condition of vanishing at the limit of completely random and ordered state [18].

The complexity measure is usually the projection from system's variables to one-dimensional quantity, which is composed to express the degree of characteristic that we define to be important in what means *"complexity"*. Since the complexity measure is always a many-to-one association, it has both aspects of compressing information to classify the system from simple to complex, and losing resolution of the system's phase space. If the system has n variables, we generally need n independent complexity measures to completely characterize the system with real-value resolution. The problematics of defining a complexity measure is situated on the edge of balancing the information compression on system's complexity with theoretical support, and the resolution of the system identification to be maintained high enough to avoid trivial classification. The latter criterion increases its importance as the system size becomes larger. The better complexity measure is therefore a set of indices, with as less number as possible, which characterizes major features related to the complexity of the system. In this sense, the ensemble of complexity measures is also analogous to the feature space of support vector machine. A non-trivial set of complexity measures need to be complementary to each other in parameter space for the possible best discrimination of different systems.

In this paper, we first consider the stochastic system with binary variables and theoretically develop a way to measure the information between subsystems, which is consistent to the information represented by the edges of the graph representation.

Next, we particularly focus on the generalized mutual information as a start point of the argument, and further consider to incorporate network heterogeneity into novel measures of complexity with respect to the system's decompositionability. This approach will be revealed to be complementary to TSE complexity as the difference between arithmetic and geometric means of information.

2. System Decomposition

Let us consider the stochastic system with n binary variables $\mathbf{x} = (x_1, \cdots, x_n)$ where $x_i \in \{0, 1\}$ $(1 \le i \le n)$. We denote the joint distribution of $\mathbf{x}$ by $p(\mathbf{x})$. We define the decomposition $p^{dec}(\mathbf{x})$ of $p(\mathbf{x})$ into two subsystems $\mathbf{y^1} = (x^1_1, \cdots, x^1_{n_1})$ and $\mathbf{y^2} = (x^2_1, \cdots, x^2_{n_2})$ $(n_1 + n_2 = n$, $\mathbf{y^1} \cup \mathbf{y^2} = \mathbf{x}$, $\mathbf{y^1} \cap \mathbf{y^2} = \phi)$ as follows:

$$p^{dec}(\mathbf{x}) = p(\mathbf{y^1})p(\mathbf{y^2}), \tag{1}$$

where $p(\mathbf{y^1})$ and $p(\mathbf{y^2})$ are the joint distributions of $\mathbf{y^1}$ and $\mathbf{y^2}$, respectively. For simplicity, hereafter we denote the system decomposition using the smallest subscript of variables in each subsystem. For example, in case $n = 4$, $\mathbf{y^1} = (x_1, x_3)$ and $\mathbf{y^2} = (x_2, x_4)$, we describe the decomposed system $p^{dec}(\mathbf{x})$

as < 1212 >. The system decomposition means to cut all statistical association between the two subsystems, which is expressed as setting the independent relation between them.

We will further consider the Equation (1) in terms of the graph representation. We define the undirected graph $\Gamma := (V, E)$ of the system $p(\mathbf{x})$, whose vertices $V = \{x_1, \cdots, x_n\}$ and edges $E = V \times V$ represent the variables and the statistical association, respectively. To express the system, we set the value of each vertex as the value of the corresponding variable, and the weight of each edge as the degree of dependency between the connected variables.

There is however a problem considering the representation with a single kind of edge. The statistical association among variables is not only between two variables, but can be independently defined among plural variables up to the n-th order. Therefore, the exact definition of the weight of the edges remains unclear. To clarify these problematics, we consider the hierarchical marginal distributions $\jmath$ as another coordinates of the system $p(\mathbf{x})$ as follows:

$$\jmath = (\jmath^{\mathbf{1}}; \jmath^{\mathbf{2}}; \cdots; \jmath^{\mathbf{n}}), \tag{2}$$

where

$$\begin{aligned} \jmath^{\mathbf{1}} &= (\eta_1, \cdots, \eta_i, \cdots, \eta_n), (1 < i < n), \\ \jmath^{\mathbf{2}} &= (\eta_{1,2}, \cdots, \eta_{i,j}, \cdots, \eta_{n-1,n}), (1 < i < j < n), \\ &\vdots \\ \jmath^{\mathbf{n}} &= \eta_{1,2,\cdots,n}, \end{aligned} \tag{3}$$

and

$$\begin{aligned} \eta_1 &= \sum_{i_2,\cdots,i_n \in \{0,1\}} p(1, i_2, \cdots, i_n), \\ &\vdots \\ \eta_n &= \sum_{i_1,\cdots,i_{n-1} \in \{0,1\}} p(i_1, \cdots, i_{n-1}, 1), \\ \eta_{1,2} &= \sum_{i_3,\cdots,i_n \in \{0,1\}} p(1, 1, i_3, \cdots, i_n), \\ &\vdots \\ \eta_{n-1,n} &= \sum_{i_1,\cdots,i_{n-2} \in \{0,1\}} p(i_1, \cdots, i_{n-2}, 1, 1), \\ &\vdots \\ \eta_{1,2,\cdots,n} &= p(1, 1, \cdots, 1). \end{aligned} \tag{4}$$

Since the definition of $\jmath$ is a linear transformation of $p(\mathbf{x})$, both coordinates have the degrees of freedom $\sum_{k=1}^{n} {}_nC_k$.

The subcoordinates $\jmath^{\mathbf{1}}$ are simply the set of marginal distributions of each variable. The subcoordinates $\jmath^{\mathbf{k}}$ $(1 < k \leq n)$ include the statistical association among k variables, that can not be expressed with the coordinates less than the k-th order. This means that the different statistical associations exist independently in each order among the corresponding sets of the variables. The statistical association represented by the weight of a graph edge $\{x_i, x_j\}$ is therefore the superposition of the different dependencies defined on every subset of $\mathbf{x}$ including x_i and x_j.

To measure the degree of statistical association in each order, the information geometry established the following setting [19]. We first define another coordinates ` $= ($`$^{\mathbf{1}};$`$^{\mathbf{2}}; \cdots;$`$^{\mathbf{n}})$ that are the dual coordinates of $\jmath$ with respect to the Legendre transformation of the exponential family's potential function $\psi(`)$ to its conjugate potential $\phi(\jmath)$ as follows:

$$\begin{aligned} `^{\mathbf{1}} &= (\theta_1, \cdots, \theta_n), \\ `^{\mathbf{2}} &= (\theta_{1,2}, \cdots, \theta_{n-1,n}), \\ &\vdots \\ `^{\mathbf{n}} &= \theta_{1,2,\cdots,n}, \end{aligned} \tag{5}$$

where

$$\begin{aligned} \psi(`) &= \log \frac{1}{p(0, \cdots, 0)}, \\ \phi(\jmath) &= \sum_i \theta_i \eta_i + \sum_{i<j} \theta_{i,j} \eta_{i,j} + \cdots + \theta_{1,2,\cdots,n} \eta_{1,2,\cdots,n} - \psi(`), \\ \theta_i &= \frac{\partial \phi(\jmath)}{\partial \eta_i}, (1 \le i \le n), \\ \theta_{i,j} &= \frac{\partial \phi(\jmath)}{\partial \eta_{i,j}}, (1 \le i < j \le n), \\ &\vdots \\ \theta_{1,2,\cdots,n} &= \frac{\partial \phi(\jmath)}{\partial \eta_{1,2,\cdots,n}}. \end{aligned} \tag{6}$$

Note that $\jmath$ can be inversely derived from $`$, following Legendre transformation between $\phi(\jmath)$ and $\psi(`)$:

$$\begin{aligned} \eta_i &= \frac{\partial \psi(`)}{\partial \theta_i}, (1 \le i \le n), \\ \eta_{i,j} &= \frac{\partial \psi(`)}{\partial \theta_{i,j}}, (1 \le i < j \le n), \\ &\vdots \\ \eta_{1,2,\cdots,n} &= \frac{\partial \psi(`)}{\partial \theta_{1,2,\cdots,n}}. \end{aligned} \tag{7}$$

Using the coordinates $`$, the system is described in the form of the exponential family as follows:

$$p(\mathbf{x}) = \sum_i \theta_i x_i + \sum_{i<j} \theta_{i,j} x_i x_j + \cdots + \theta_{1,2,\cdots,n} x_1 x_2 \cdots x_n - \psi(`). \tag{8}$$

The information geometry revealed that the exponential family of probability distribution forms a manifold with a dual-flat structure. More precisely, the coordinates $`$ form a flat manifold with respect to the Fisher information matrix as the Riemannian metric, and $\alpha$-connection with $\alpha = 1$. Dually to $`$, the coordinates $\jmath$ are flat with respect to the same metric but α-connection with $\alpha = -1$. It is known that $`$ and $\jmath$ are orthogonal to each other with respect to the Fisher information matrix. This structure give us a way to decompose the degree of statistical association among variables into separated elements of arbitrary orders. We define the so-called k-cut mixture coordinates $\imath^{\mathbf{k}}$ as follows [14].

$$\imath^{\mathbf{k}} = (\jmath^{\mathbf{k}-}; `^{\mathbf{k}+}), \tag{9}$$

$$\jmath^{\mathbf{k}-} = (\jmath^{\mathbf{1}}, \cdots, \jmath^{\mathbf{k}}), \tag{10}$$

$$`^{\mathbf{k}+} = (`^{\mathbf{k+1}}, \cdots, `^{\mathbf{n}}). \tag{11}$$

We also define the k-cut mixture coordinates $\boldsymbol{\imath}_0^{\mathbf{k}} = (\boldsymbol{\jmath}^{\mathbf{k}-}; 0, \cdots, 0)$ with no dependency above the k-th order. We denote the system specified with $\boldsymbol{\imath}^{\mathbf{k}}$ and $\boldsymbol{\imath}_0^{\mathbf{k}}$ as $p(\mathbf{x}, \boldsymbol{\imath}^{\mathbf{k}})$ and $p(\mathbf{x}, \boldsymbol{\imath}_0^{\mathbf{k}})$, respectively.

Then the degree of the statistical association more than the k-th order in the system can be measured by the Kullback-Leibler (KL-) divergence $D[p(\mathbf{x}, \boldsymbol{\imath}) : p(\mathbf{x}, \boldsymbol{\imath}_0^{\mathbf{k}})]$.

$$2N \cdot D[p(\mathbf{x}, \boldsymbol{\imath}) : p(\mathbf{x}, \boldsymbol{\imath}_0^{\mathbf{k}})] \sim \chi^2 \left(\sum_{i=k+1}^{n} {}_nC_i \right), \tag{12}$$

where $D[\cdot : \cdot]$ is the KL-divergence from the first system to the second one.

Here, the decomposition is performed according to the orders of statistical association, which does not spatially distinguish the vertices. If we define the weight of an edge $\{x_i, x_j\}$ with the KL-divergence, the above k-cut coordinates $\boldsymbol{\imath}^{\mathbf{k}}$ are not appropriate to measure the information represented in each edge. We need to set another mixture coordinates so that to separate only the existing information between x_i and x_j regardless of its order.

Let us return to the definition of the system decomposition and consider on the dual-flat coordinates ` and $\boldsymbol{\jmath}$.

Proposition 1. *The independence between the two decomposed systems* $\mathbf{y^1} = (x_1^1, \cdots, x_{n_1}^1)$ *and* $\mathbf{y^2} = (x_1^2, \cdots, x_{n_2}^2)$ *can be expressed on the new coordinates* $\boldsymbol{\jmath}^{\mathbf{dec}}$ *as follows:*

$$\begin{aligned}
\eta_i^{dec} &= \eta_i, (1 \le i \le n), \\
\eta_{i,j}^{dec} &= \begin{cases} \eta_{i,j}, (1 \le i < j \le n), & \textit{if } \{x_i, x_j\} \subseteq \mathbf{y^1} \textit{ or } \subseteq \mathbf{y^2} \\ \eta_i \eta_j, (1 \le i < j \le n), & \textit{else} \end{cases}, \\
\eta_{i,j,k}^{dec} &= \begin{cases} \eta_{i,j,k}, (1 \le i < j < k \le n), & \textit{if } \{x_i, x_j, x_k\} \subseteq \mathbf{y^1} \textit{ or } \subseteq \mathbf{y^2} \\ \eta_{i,j}\eta_k, (1 \le i < j < k \le n), & \textit{else if } \{x_i, x_j\} \subseteq \mathbf{y^1} \textit{ or } \subseteq \mathbf{y^2} \\ \eta_i \eta_{j,k}, (1 \le i < j < k \le n), & \textit{else if } \{x_j, x_k\} \subseteq \mathbf{y^1} \textit{ or } \subseteq \mathbf{y^2} \\ \eta_j \eta_{i,k}, (1 \le i < j < k \le n), & \textit{else (if } \{x_i, x_k\} \subseteq \mathbf{y^1} \textit{ or } \subseteq \mathbf{y^2}) \end{cases}, \\
&\vdots \\
\eta_{1,2,\cdots,n}^{dec} &= \eta_{s[i,k_1,\cdots,k_{n_1-1}]} \eta_{s[j,l_1,\cdots,l_{n_2-1}]}, (x_i \in \mathbf{y^1}, x_j \in \mathbf{y^2}),
\end{aligned} \tag{13}$$

where $s[\cdots]$ *is the ascending sort of the internal sequence.*

Then the corresponding dual coordinates `$^{\mathbf{dec}}$ *take* 0 *elements as follows:*

$$\begin{aligned}
\theta_{i,j}^{dec} &= 0, \quad (1 \le i < j << n), \quad \textit{if } \{x_i, x_j\} \cap \mathbf{y^1} \ne \phi \textit{ and } \{x_i, x_j\} \cap \mathbf{y^2} \ne \phi \\
\theta_{i,j,k}^{dec} &= 0, \quad (1 \le i < j < k \le n), \quad \textit{if } \{x_i, x_j, x_k\} \cap \mathbf{y^1} \ne \phi \textit{ and } \{x_i, x_j, x_k\} \cap \mathbf{y^2} \ne \phi \\
&\vdots \\
\theta_{1,2,\cdots,n}^{dec} &= 0.
\end{aligned} \tag{14}$$

Proof. For simplicity, we show the cases of $n = 2$ and $n = 3$ for the first node separation.

For $n = 2$, the above defined $\jmath^{\mathbf{dec}}$ for the system decomposition $< 12 >$ give its dual coordinates $\grave{}^{\mathbf{dec}}$ as follows:

$$
\begin{aligned}
\theta_1^{dec} &= \log \frac{\eta_1^{dec} - \eta_{1,2}^{dec}}{1 - \eta_1^{dec} - \eta_2^{dec} + \eta_{1,2}^{dec}} &&= \log \frac{\eta_1}{1 - \eta_1}, \\
\theta_2^{dec} &= \log \frac{\eta_2^{dec} - \eta_{1,2}^{dec}}{1 - \eta_1^{dec} - \eta_2^{dec} + \eta_{1,2}^{dec}} &&= \log \frac{\eta_2}{1 - \eta_2}, \\
\theta_{1,2}^{dec} &= \log \frac{\eta_{1,2}^{dec}(1 - \eta_1^{dec} - \eta_2^{dec} + \eta_{1,2}^{dec})}{(\eta_1^{dec} - \eta_{1,2}^{dec})(\eta_2^{dec} - \eta_{1,2}^{dec})} &&= 0,
\end{aligned} \tag{15}
$$

which means the first and second node is independent.

For $n = 3$, the above defined $\jmath^{\mathbf{dec}}$ for the system decomposition $< 122 >$ give its dual coordinates $\grave{}^{\mathbf{dec}}$ as follows:

$$
\begin{aligned}
\theta_1^{dec} &= \log \frac{\eta_1^{dec} - \eta_{1,2}^{dec} - \eta_{1,3}^{dec} + \eta_{1,2,3}^{dec}}{1 - \eta_1^{dec} - \eta_2^{dec} - \eta_3^{dec} + \eta_{1,2}^{dec} + \eta_{1,3}^{dec} + \eta_{2,3}^{dec} - \eta_{1,2,3}^{dec}} = \log \frac{\eta_1}{1 - \eta_1}, \\
\theta_2^{dec} &= \log \frac{\eta_2^{dec} - \eta_{1,2}^{dec} - \eta_{1,3}^{dec} + \eta_{1,2,3}^{dec}}{1 - \eta_1^{dec} - \eta_2^{dec} - \eta_3^{dec} + \eta_{1,2}^{dec} + \eta_{1,3}^{dec} + \eta_{2,3}^{dec} - \eta_{1,2,3}^{dec}} = \log \frac{\eta_2 - \eta_{2,3}}{1 - \eta_2 - \eta_3 + \eta_{2,3}}, \\
\theta_3^{dec} &= \log \frac{\eta_3^{dec} - \eta_{1,3}^{dec} - \eta_{2,3}^{dec} + \eta_{1,2,3}^{dec}}{1 - \eta_1^{dec} - \eta_2^{dec} - \eta_3^{dec} + \eta_{1,2}^{dec} + \eta_{1,3}^{dec} + \eta_{2,3}^{dec} - \eta_{1,2,3}^{dec}} = \log \frac{\eta_3 - \eta_{2,3}}{1 - \eta_2 - \eta_3 + \eta_{2,3}},
\end{aligned} \tag{16}
$$

$$
\begin{aligned}
\theta_{1,2}^{dec} &= \log \frac{(\eta_{1,2}^{dec} - \eta_{1,2,3}^{dec})(1 - \eta_1^{dec} - \eta_2^{dec} - \eta_3^{dec} + \eta_{1,2}^{dec} + \eta_{1,3}^{dec} + \eta_{2,3}^{dec} - \eta_{1,2,3}^{dec})}{(\eta_1^{dec} - \eta_{1,2}^{dec} - \eta_{1,3}^{dec} + \eta_{1,2,3}^{dec})(\eta_2^{dec} - \eta_{1,2}^{dec} - \eta_{2,3}^{dec} + \eta_{1,2,3}^{dec})} \\
&= 0, \\
\theta_{1,3}^{dec} &= \log \frac{(\eta_{1,3}^{dec} - \eta_{1,2,3}^{dec})(1 - \eta_1^{dec} - \eta_2^{dec} - \eta_3^{dec} + \eta_{1,2}^{dec} + \eta_{1,3}^{dec} + \eta_{2,3}^{dec} - \eta_{1,2,3}^{dec})}{(\eta_1^{dec} - \eta_{1,2}^{dec} - \eta_{1,3}^{dec} + \eta_{1,2,3}^{dec})(\eta_3^{dec} - \eta_{1,3}^{dec} - \eta_{2,3}^{dec} + \eta_{1,2,3}^{dec})} \\
&= 0, \\
\theta_{2,3}^{dec} &= \log \frac{(\eta_{2,3}^{dec} - \eta_{1,2,3}^{dec})(1 - \eta_1^{dec} - \eta_2^{dec} - \eta_3^{dec} + \eta_{1,2}^{dec} + \eta_{1,3}^{dec} + \eta_{2,3}^{dec} - \eta_{1,2,3}^{dec})}{(\eta_2^{dec} - \eta_{1,2}^{dec} - \eta_{2,3}^{dec} + \eta_{1,2,3}^{dec})(\eta_3^{dec} - \eta_{1,3}^{dec} - \eta_{2,3}^{dec} + \eta_{1,2,3}^{dec})} \\
&= \log \frac{\eta_{2,3}(1 - \eta_2 - \eta_3 + \eta_{2,3})}{(\eta_2 - \eta_{2,3})(\eta_3 - \eta_{2,3})},
\end{aligned} \tag{17}
$$

$$
\begin{aligned}
\theta_{1,2,3}^{dec} &= \log \left[\frac{\eta_{1,2,3}^{dec}}{(\eta_{1,2}^{dec} - \eta_{1,2,3}^{dec})(\eta_{1,3}^{dec} - \eta_{1,2,3}^{dec})(\eta_{2,3}^{dec} - \eta_{1,2,3}^{dec})} \right. \\
&\quad \left. \times \frac{(\eta_1^{dec} - \eta_{1,2}^{dec} - \eta_{1,3}^{dec} + \eta_{1,2,3}^{dec})(\eta_2^{dec} - \eta_{1,2}^{dec} - \eta_{2,3}^{dec} + \eta_{1,2,3}^{dec})(\eta_3^{dec} - \eta_{1,3}^{dec} - \eta_{2,3}^{dec} + \eta_{1,2,3}^{dec})}{(1 - \eta_1^{dec} - \eta_2^{dec} - \eta_3^{dec} + \eta_{1,2}^{dec} + \eta_{1,3}^{dec} + \eta_{2,3}^{dec} - \eta_{1,2,3}^{dec})} \right] \\
&= 0,
\end{aligned} \tag{18}
$$

which means the first node is independent from the other nodes.

The generalization is possible with the use of recurrence formula between system size n and $n + 1$, according to the symmetry of the model and Legendre transformation between $\jmath^{\mathbf{dec}}$ and $\grave{}^{\mathbf{dec}}$ coordinates.

Numerical proof can be obtained by computing directly 0 elements of $\grave{}^{\mathbf{dec}}$ from $\jmath^{\mathbf{dec}}$. □

The definition of $\jmath^{\mathbf{dec}}$ means to decompose the hierarchical marginal distributions $\jmath$ into the products of the subsystems' marginal distributions, in case the subscripts traverse the two subsystems. Therefore, only the statistical associations between two subsystems are set to be independent, while the internal dependencies of each subsystem remain unchanged. This is analytically equivalent to compose another mixture coordinates ¸, namely the $< \cdots >$-cut coordinates, with proper description of the system decomposition with $< \cdots >$. The ¸ consists of the $\jmath$ coordinates with the subscripts that do not traverse between the decomposed subsystems, and the ` coordinates whose subscripts traverse between them.

For simplicity, we only describe here the case $n = 4$ and the decomposition $< 1133 >$ (the set of the first, second, and the third, fourth nodes each form a subsystem). The system $p(\mathbf{x})$ is expressed with the $< 1133 >$-cut coordinates ¸ as

$$
\begin{aligned}
\tilde{\varsigma}_1 &= \eta_1, \\
&\vdots \\
\tilde{\varsigma}_4 &= \eta_4, \\
\tilde{\varsigma}_{1,2} &= \eta_{1,2}, \\
\tilde{\varsigma}_{1,3} &= \theta_{1,3}, \\
\tilde{\varsigma}_{1,4} &= \theta_{1,4}, \\
\tilde{\varsigma}_{2,3} &= \theta_{2,3}, \\
\tilde{\varsigma}_{2,4} &= \theta_{2,4}, \\
\tilde{\varsigma}_{3,4} &= \eta_{3,4}, \\
\tilde{\varsigma}_{1,2,3} &= \theta_{1,2,3}, \\
&\vdots \\
\tilde{\varsigma}_{2,3,4} &= \theta_{2,3,4}, \\
\tilde{\varsigma}_{1,2,3,4} &= \theta_{1,2,3,4}.
\end{aligned}
\tag{19}
$$

The decomposed system with no statistical association between two subsystems have the following coordinates ¸$^{\mathbf{dec}}$, which is, in any decomposition, equivalent to set all θ in ¸ as 0:

$$
\begin{aligned}
\tilde{\varsigma}_1^{dec} &= \eta_1, \\
&\vdots \\
\tilde{\varsigma}_4^{dec} &= \eta_4, \\
\tilde{\varsigma}_{1,2}^{dec} &= \eta_{1,2}, \\
\tilde{\varsigma}_{1,3}^{dec} &= 0, \\
\tilde{\varsigma}_{1,4}^{dec} &= 0, \\
\tilde{\varsigma}_{2,3}^{dec} &= 0, \\
\tilde{\varsigma}_{2,4}^{dec} &= 0, \\
\tilde{\varsigma}_{3,4}^{dec} &= \eta_{3,4}, \\
\tilde{\varsigma}_{1,2,3}^{dec} &= 0, \\
&\vdots \\
\tilde{\varsigma}_{2,3,4}^{dec} &= 0, \\
\tilde{\varsigma}_{1,2,3,4}^{dec} &= 0.
\end{aligned}
\tag{20}
$$

This is analytically equivalent to the definition of the decomposition (13)–(14) in case of $< 1133 >$. Therefore, the KL-divergence $D[p(\mathbf{x}, ¸) : p(\mathbf{x}, ¸^{\mathbf{dec}})]$ measures the information lost by the system decomposition. The following asymptotic agreement to χ^2 test also holds.

Proposition 2.

$$2N \cdot D[p(\mathbf{x}, ¸) : p(\mathbf{x}, ¸^{\mathbf{dec}})] \sim \chi^2(\sharp_\theta(¸)), \tag{21}$$

where $\sharp_\theta(¸)$ *is the number of* ` *coordinates appearing in the* ¸ *coordinates.*

3. Edge Cutting

We further expand the concept of system decomposition to eventually quantify the total amount of information expressed by an edge of the graph. Let us consider to cut an edge $\{x_i, x_j\}$ $(1 \le i < j \le n)$ of the graph with n vertices. Hereafter we call this operation as the edge cutting $i - j$. In the same way as the system decomposition, the edge cutting corresponds to modify the ȷ coordinates to produce $ȷ^{\mathbf{ec}}$ coordinates as follows:

$$\begin{aligned} \eta^{ec}_{i,j} &= \eta_i \eta_j, \\ \eta^{ec}_{s[i,j,k_1]} &= \eta_{s[i,k_1]} \eta_{s[j,k_1]}, \\ \eta^{ec}_{s[i,j,k_1,k_2]} &= \eta_{s[i,k_1,k_2]} \eta_{s[j,k_1,k_2]}, \\ &\vdots \\ \eta^{ec}_{s[i,j,k_1,\cdots,k_{n-2}]} &= \eta_{s[i,k_1,\cdots,k_{n-2}]} \eta_{s[j,k_1,\cdots,k_{n-2}]}, \\ (\{i,j,k_1,\cdots,k_{n-2}\} &= \{1,\cdots,n\}), \end{aligned} \tag{22}$$

and the rest of $ȷ^{\mathbf{ec}}$ remains the same as those of ȷ.

The formation of $ȷ^{ec}$ from ȷ consists of replacing the k-th order elements ($k \ge 3$) of ȷ including both i and j in its subscripts, with the product of the $k - 1$-th order ȷ in maximum subgraphs ($k - 1$ vertices) each including i or j. This means that all orders of statistical association including the variables x_i and x_j are set to be independent only between them. Other relations that do not include simultaneously x_i and x_j remain unchanged.

Certain combinations of edge cuttings coincide with system decompositions. For example, in case $n = 4$, the edge cuttings $1 - 2$, $1 - 3$, and $1 - 4$ are equivalent to the system decomposition $< 1222 >$.

We define the $i - j$-cut mixture coordinates ¸ for orthogonal decomposition of the statistical association represented by the edge $\{x_i, x_j\}$. Although actual calculation can be performed only with ȷ coordinates, this generalization is necessary to have a geometrical definition of the orthogonality. For simplicity, we only describe the ¸ in the case of $n = 4$:

$$
\begin{aligned}
\zeta_1 &= \eta_1, \\
&\vdots \\
\zeta_4 &= \eta_4, \\
\zeta_{1,2} &= \theta_{1,2}, \\
\zeta_{1,3} &= \eta_{1,3}, \\
\zeta_{1,4} &= \eta_{1,4}, \\
\zeta_{2,3} &= \eta_{2,3}, \\
\zeta_{2,4} &= \eta_{2,4}, \\
\zeta_{3,4} &= \eta_{3,4}, \\
\zeta_{1,2,3} &= \theta_{1,2,3}, \\
\zeta_{1,2,4} &= \theta_{1,2,4}, \\
\zeta_{1,3,4} &= \eta_{1,3,4}, \\
\zeta_{2,3,4} &= \eta_{2,3,4}, \\
\zeta_{1,2,3,4} &= \theta_{1,2,3,4},
\end{aligned} \tag{23}
$$

where orthogonality between the elements of $\boldsymbol{\theta}$ and $\boldsymbol{\eta}$ holds with respect to the Fisher information matrix.

Calculating the dual coordinates $\boldsymbol{\eta}^{\mathbf{ec}}$ of $\boldsymbol{\theta}^{\mathbf{ec}}$, we can define the coordinates $\boldsymbol{\zeta}^{\mathbf{ec}}$ of the system after the edge cutting $1-2$ as follows:

$$
\begin{aligned}
\zeta_1^{ec} &= \eta_1, \\
&\vdots \\
\zeta_4^{ec} &= \eta_4, \\
\zeta_{1,2}^{ec} &= \theta_{1,2}^{ec}, \\
\zeta_{1,3}^{ec} &= \eta_{1,3}, \\
\zeta_{1,4}^{ec} &= \eta_{1,4}, \\
\zeta_{2,3}^{ec} &= \eta_{2,3}, \\
\zeta_{2,4}^{ec} &= \eta_{2,4}, \\
\zeta_{3,4}^{ec} &= \eta_{3,4}, \\
\zeta_{1,2,3}^{ec} &= \theta_{1,2,3}^{ec}, \\
\zeta_{1,2,4}^{ec} &= \theta_{1,2,4}^{ec}, \\
\zeta_{1,3,4}^{ec} &= \eta_{1,3,4}, \\
\zeta_{2,3,4}^{ec} &= \eta_{2,3,4}, \\
\zeta_{1,2,3,4}^{ec} &= \theta_{1,2,3,4}^{ec}.
\end{aligned}
$$

Note that the edge cutting can not be defined simply by setting the corresponding elements of $\boldsymbol{\eta}^{\mathbf{ec}}$ as 0.

Then the KL-divergence $D[p(\mathbf{x},\boldsymbol{\zeta}) : p(\mathbf{x},\boldsymbol{\zeta}^{\mathbf{ec}})]$ represent the total amount of information represented by the edge $1-2$.

The following asymptotic agreement to χ^2 test also holds:

Proposition 3.

$$2N \cdot D[p(\mathbf{x}, _{,}) : p(\mathbf{x}, _{,}^{\mathbf{ec}})] \sim \chi^2(1 + \sum_{k=1}^{n-2} {}_{n-2}C_k). \tag{24}$$

We call this χ^2 value or the KL-divergence itself as edge information of edge $1 - 2$.

4. Generalized Mutual Information as Complexity with Respect to the Total System Decomposition

In previous sections, we have introduced a measure of complexity in terms of system decomposition, by measuring the KL-divergence between a given system and its independently decomposed subsystems. We consider here the total system decomposition, and measure the informational distance I between the system and the totally decomposed system where each element are independent.

$$I := \sum_{i=1}^{n} H(x_i) - H(x_1, \cdots, x_n), \tag{25}$$

where

$$H(\mathbf{x}) := -\sum_{\mathbf{x}} p(\mathbf{x}) \log(\mathbf{x}). \tag{26}$$

This quantity is the generalization of mutual information, and is named in various ways such as generalized mutual information, integration, complexity, multi-information, *etc.* according to different authors. For simplicity, we call the I as "*multi-information* taking after [15]. This quantity can be interpreted as a measure of complexity that sums up the order-wise statistical association existing in each subset of components with information geometrical formalization [14]

For simplicity, we denote the multi-information I of n-dimensional stochastic binary variables as follows, using the notation of the system decomposition:

$$I = D[< 111 \cdots 1 >:< 123 \cdots n >]. \tag{27}$$

5. Rectangle-Bias Complexity

The multi-information contains some degrees of freedom in case $n > 2$. That is, we can define a set of distributions $\{p(\mathbf{x}) | I = const.\}$ with different parameters but the same I value. This fact can be clearly explained with the use of information geometry. From the Pythagorean relation, we obtain the followings in case of $n = 3$:

$$\begin{aligned} D[< 111 >:< 113 >] + D[< 113 >:< 123 >] &= D[< 111 >:< 123 >], \\ D[< 111 >:< 121 >] + D[< 121 >:< 123 >] &= D[< 111 >:< 123 >], \\ D[< 111 >:< 122 >] + D[< 122 >:< 123 >] &= D[< 111 >:< 123 >]. \end{aligned} \tag{28}$$

Using these relations, we can schematically represent the decomposed systems on a circle diagram with diameter $\sqrt{I}$. This representation is based on the analogous algebra between Pythagorean relation of KL-divergence, and that of Euclidian geometry where the circumferential angle of a semi-circular arc is always $\frac{\pi}{2}$.

Figure1 represents two different cases with the same I value in case $n = 3$. The distance between two systems in the same diagram corresponds to the square root value of KL-divergence between them. Clearly the left and right figures represent different dependencies between nodes, although they both have the same I value. Such geometrical variation is possible by the abundance of degree of freedom in dual coordinates compared to the given constraint. There exist 7 degrees of freedom in η or θ coordinates for $n = 3$, while the only constraint is the invariance of I value, which only reduce 1 degree of freedom. The remaining 6 degrees of freedom can then be deployed to produce geometrical variation in the circle diagram. As for considering system decomposition, the left figure is difficult to obtain decomposed systems without losing much information. While in the right figure there exists relatively easy decomposition $< 122 >$, which loses less information than any decomposition in the left figure. We call such degree of facility of decomposition with respect to the losing information as *system decompositionability*. In this sense, the left system is more complex although the 2 systems both have the same I value. Especially, in case $D[< 111 >:< 122 >] = D[< 111 >:< 113 >] = D[< 111 >:< 121 >]$, the system does not have any easiest way of decomposition, and any isolation of a node loses significant amount of information.

To further incorporate such geometrical structure reflecting system decompositionability into a measure of complexity, we consider a mathematical way to distinguish between these two figures. Although the total sum of KL-divergence along the sequence of system decomposition is always identical to I by Pythagorean relation, their product can vary according to the geometrical composition in the circle diagram. This is analogous to the isoperimetric inequality of rectangle, where regular tetragon gives the maximum dimensions amongst constant perimeter rectangles.

We propose provisionary a new measure of complexity as follows, namely *rectangle-bias complexity* C_r:

$$C_r = \frac{1}{|SD| - 2} \sum_{<\cdots>\in SD} D[< 11\cdots 1 >:< \cdots >] \cdot D[< \cdots >:< 12\cdots n >], \tag{29}$$

where SD is the set of possible system decomposition in n binary variables, and $|SD|$ is the element number of SD. For example, $SD = \{< 111 >, < 122 >, < 121 >, < 113 >, < 123 >\}$ and $|SD| = 5$ for $n = 3$. This measure distinguishes between the two systems in Figure 1, and gives larger value for the left figure. It also gives maximum value in case $D[< 111 >:< 122 >] = D[< 111 >:< 113 >] = D[< 111 >:< 121 >]$. We propose provisionary a new measure of complexity as follows, namely *rectangle-bias complexity* C_r:

$$C_r = \frac{1}{|SD| - 2} \sum_{<\cdots>\in SD} D[< 11\cdots 1 >:< \cdots >] \cdot D[< \cdots >:< 12\cdots n >], \tag{30}$$

where SD is the set of possible system decomposition in n binary variables, and $|SD|$ is the element number of SD. For example, $SD = \{< 111 >, < 122 >, < 121 >, < 113 >, < 123 >\}$ and $|SD| = 5$ for $n = 3$. This measure distinguishes between the two systems in Figure 1, and gives larger value for the left figure. It also gives maximum value in case $D[< 111 >:< 122 >] = D[< 111 >:< 113 >] = D[< 111 >:< 121 >]$.

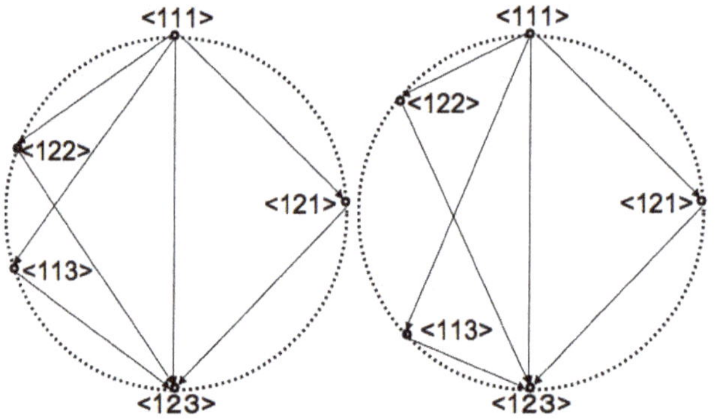

Figure 1. *Circle diagrams of system decomposition in 3-node network.* Both systems have the same value of multi-information I that is expressed as the identical diameter length of the circles. 2 variations are shown, where the left system is more complex (C_r high) in a sense any system decomposition requires to lose more information than the easiest one ($< 122 >$) in the right figure (C_r low).

6. Complementarity between Complexities Defined with Arithmetic and Geometric Means

We evaluate the possibility and the limit of rectangle-bias complexity C_r comparing with other proposed measures of complexity.

The Interests in measuring network heterogeneity have been developed toward the incorporation of multi-scale characteristics into complexity measures. The TSE complexity is motivated from the structure of the functional differentiation of brain activity, which measures the difference of neural integration between all sizes of subsystems and the whole system [17]. Biologically motivated TSE complexity is also investigated from theoretical point of view, to further attribute desirable property as an universal complexity measure independent of system size [20]. The hierarchical structure of the exponential family in information geometry also leads to the order-wise description of statistical association, which can be regarded as a multi-scale complexity measure [14]. The relation between the order-wise dependencies and the TSE complexity is theoretically investigated to establish the order-wise component correspondence between them [15].

These indices of network heterogeneity, however, all depend on the arithmetic mean of the component-wise information theoretical measure. We show that these arithmetic means still miss to measure certain modularity based on the statistical independence between subsystems.

Figure 2 present the simplified cases where complexity measures with arithmetic means fail to distinguish. We consider the two systems with different heterogeneity but identical multi-information I. Here, the multi-information can not reflect the network heterogeneity. The TSE complexity and its information geometrical correspondence in [15] has a sensitivity to measure the network heterogeneity, but since the arithmetic mean is taken over all subsystems, they do not distinguish the component-wise break of symmetry between different scales. The renormalized TSE complexity with respect to the multi-information I still has the same insensitivity. Even by incorporating the information of each subsystem scale, the arithmetic mean can balance out between the scale-wise variations, and a large range of the heterogeneity in different scale can realize the same value of these complexities. For the application in neuroscience, the assumption of a model with simple parametric heterogeneity and the comparison of TSE complexity between different I values alleviate this limitation [17].

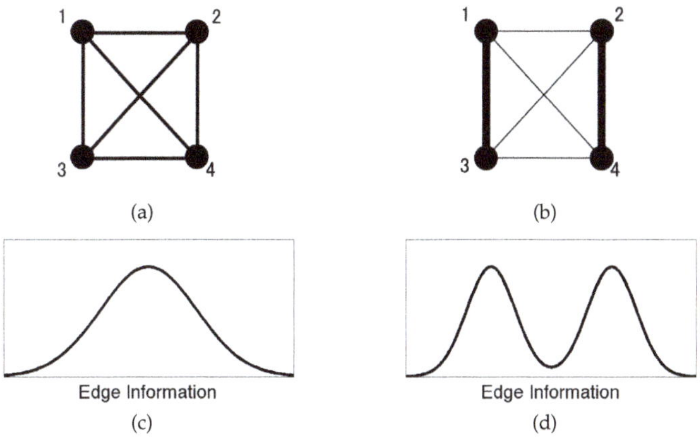

Figure 2. *Schematic examples of stochastic systems with identical multi-information I where complexity measures with arithmetic mean fail to distinguish.* (**a**): Example 1 of stochastic system with homogeneous mean of edge information and symmetric fluctuation of its heterogeneity; (**b**): Example 2 of heterogeneous stochastic system with bimodal edge information distribution and identical multi-information I and complexity based on arithmetic mean as example 1; (**c**): schematic representation of the distribution of statistical association (edge information) in upper network; (**d**): schematic representation of the distribution of statistical association (edge information) in upper network.

In contrast to complexities with arithmetic mean, the rectangle-bias complexity C_r is related to the geometrical mean. The C_r can distinguish the two systems in Figure 2, giving relatively high C_r value to the left system and low value to the right one.

This does not mean , however, that the C_r has a finer resolution than other complexity measures. The constant conditions of complexity measures are the constraints on $\sum_{k=1}^{n} {}_nC_k$ degrees of freedom in model parameter space, which define different geometrical composition of corresponding submanifolds. We basically need $\sum_{k=1}^{n} {}_nC_k$ independent measures to assure the real-value resolution of network feature characterization. Complexities with arithmetic and geometric means are just giving complementary information on network heterogeneity, or different constant-complexity submanifolds structure in statistical manifold as depicted in Figure 3. Therefore, it is also possible to construct a class of systems that has identical I and C_r values but different TSE complexity. Complexity measures should be utilized in combination, with respect to the non-linear separation capacity of network features of interest.

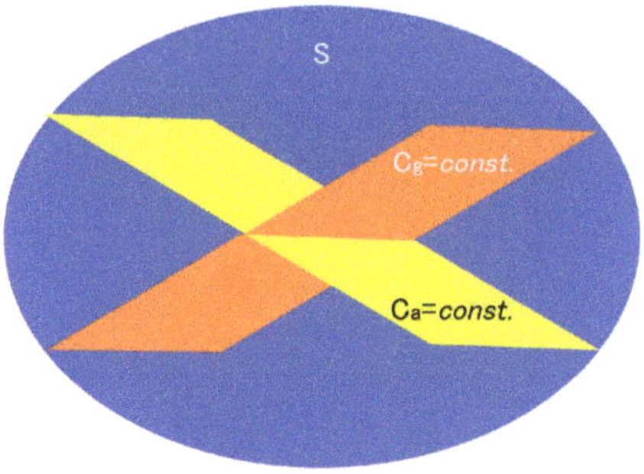

Figure 3. *Schematic representation of complementarity between complexity measures based on arithmetic mean (C_a) and geometric mean (C_g) of informational distance.* An example of the $n-1$ dimensional constant-complexity submanifolds with respect to $C_a = const.$ and $C_g = const.$ conditions are depicted with yellow and orange surface, respectively. The dimension of the whole statistical manifold S is the parameter number n.

7. Cuboid-Bias Complexity with Respect to System Decompositionability

We consider the expansion of C_r into general system size n. The $n \geq 4$ situation is different from $n = 3$ and less in the existence of a hierarchical structure between system decompositions.

Figure 4 shows the hierarchy of the system decompositions in case $n = 4$. Such hierarchical structure between system decompositions is not homogeneous with respect to the subsystems number, and depends on the isomorphic types of decomposed systems. This fact produces certain difference of meaning in complexity between each KL-divergences when considering the system decompositionability.

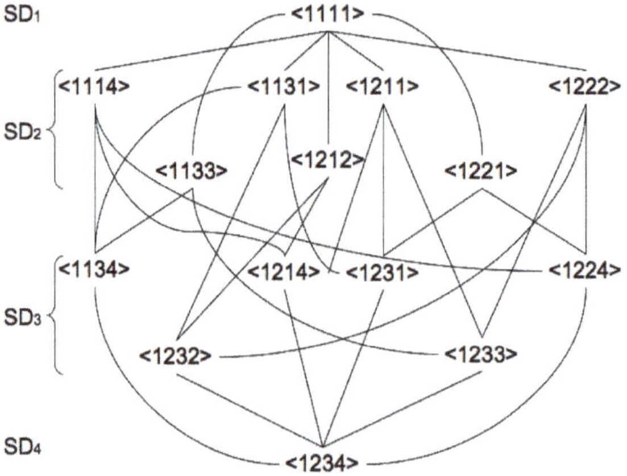

Figure 4. *Hierarchy of system decomposition for 4 nodes network ($n = 4$).* Possible sequences of $Seq = \{SD_1(i_s) \to SD_2(i_s) \to SD_3(i_s) \to SD_4(i_s) | 1 \leq i_s \leq |Seq| = 18\}$ are connected with the lines.

A simple example in 4 nodes network is shown in Figure 5.

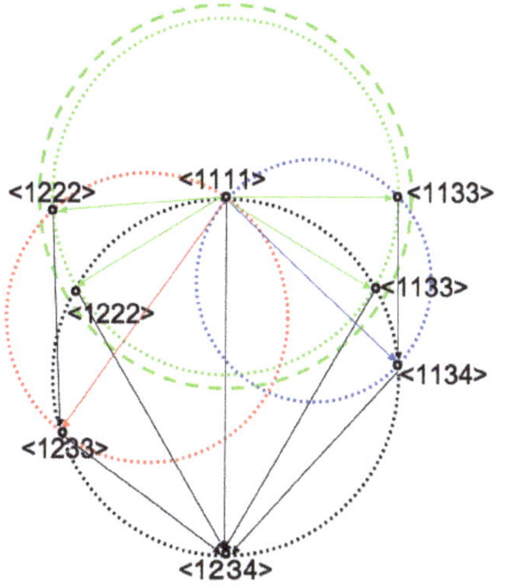

Figure 5. *Hierarchical effect of sequential system decomposition on cuboid volume and rectangle surface on circle graph.* We consider to increase the diameter of the green circle from dotted to dashed one without changing those of the red and blue circles, which gives different effect on the change of $D[< 1222 >:< 1233 >]$ and $D[< 1133 >:< 1134 >]$ according to the hierarchical structure of the decomposition sequences.

We consider the modification of 2 KL-divergences in the figure, $D[< 1111 >:< 1222 >]$ and $D[< 1111 >:< 1133 >]$ from the diameter of green dotted circle to the dashed one.

The joint distribution $P(x_1, x_2, x_3, x_4)$ of a discrete distribution with 4 binary variables (x_1, x_2, x_3, x_4) $(x_1, x_2, x_3, x_4 \in \{0, 1\})$ have $2^4 - 1 = 15$ parameters, which define the dual-flat coordinates of statistical manifold in information geometry.

On the other hand, the possible system decompositions exist as the followings in $n = 4$:

$$\begin{aligned} SD \ := \ & \{< 1111 >, < 1114 >, < 1131 >, < 1211 >, < 1222 >, \\ & < 1133 >, < 1212 >, < 1221 >, < 1134 >, < 1214 >, \\ & < 1231 >, < 1224 >, < 1232 >, < 1233 >, < 1234 >\}. \end{aligned} \tag{31}$$

Since the number of possible system decompositions is 15, and each is associated with the modification of different sets of $P(x_1, x_2, x_3, x_4)$ parameters, the system decompositions and KL-divergences between them can be defined independently. This also holds even under the constant condition of I value or other complexity measures except the ones imposing dependency between system decompositions.

This means that we can independently modify the diameter of green dotted circle in Figure 5, without changing the diameters of the red and blue circles, which define the system decompositions $< 1233 >$ and $< 1134 >$ in the sub-hierarchy of $< 1222 >$ and $< 1133 >$, respectively. Other KL-divergences can also be maintained as given constant values for the same reason.

The rectangle-biased complexity C_r increases its value with such modification, but does not reflect the heterogeneity of KL-divergences according to the hierarchy of system decompositions. If we consider the system decompositionability as the mean facility to decompose the given system into its finest components with respect to the "all" possible system decompositions, such hierarchical difference also has a meaning in the definition of complexity.

The effect of modifying the diameter of the green dotted circle is different between the decomposition sequences $< 1111 > \to < 1222 > \to < 1233 > \to < 1234 >$ and $< 1111 > \to <$

1133 >→< 1134 >→< 1234 >. The decrease of the KL-divergence $D[< 1222 >:< 1233 >]$ is less than $D[< 1133 >:< 1134 >]$ since the diameter of the red dotted circle is larger than the blue one in Figure 5. This means that the effect of changing the same amount of KL-divergences in $D[< 1111 >:< 1222 >]$ and $D[< 1111 >:< 1133 >]$ produces larger effect on the sequence < 1111 >→< 1133 >→< 1134 >→< 1234 > than < 1111 >→< 1222 >→< 1233 >→< 1234 >, if compared at the sequence level. The rectangle-biased complexity C_r does not reflect such characteristics since it does not distinguish between the hierarchical structure between the diameters of the green, red and blue dotted circles.

To incorporate such hierarchical effect in a complexity measure with geometric mean, we have the natural expansion of the rectangle-biased complexity C_r as the *cuboid-bias complexity* C_c, which is defined as follows:

$$C_c := \frac{1}{|Seq|} \sum_{i_s=1}^{|Seq|} \prod_{i=1}^{n-1} D[SD_i(i_s) : SD_{i+1}(i_s)], \tag{32}$$

where Seq represents the possible sequences of hierarchical system decompositions as follows:

$$Seq = \{SD_1(i_s) \to SD_2(i_s) \to \cdots SD_i(i_s) \cdots \to SD_n(i_s) | 1 \leq i_s \leq |Seq|\}. \tag{33}$$

The elements $SD_i(i_s)$ of Seq corresponds to the system decomposition, which is aligned according to the hierarchy with the following algorithmic procedure (based on [15]):

(1) Initialization: Set the initial sets of system decomposition of all sequences in Seq as the whole system $SD_1(i_s) :=< 111 \cdots 1 > (1 \leq i_s \leq |Seq|)$.
(2) Step $i \to i+1$: If the system decomposition is the total system decomposition ($SD_i(i_s) :=< 123 \cdots n >$), then stop. Otherwise, choose a non-decomposed subsystem $SS_i(i_s)$ of the system decomposition $SD_i(i_s)$, and further divide it into two independent subsystems $SS_i^1(i_s)$ and $SS_i^2(i_s)$ different for each i_s. $SD_{i+1}(i_s)$ is then defined as a system decomposition of total system that further separates independently subsystems $SS_i^1(i_s)$ and $SS_i^2(i_s)$, in addition to the previous decomposition $SD_i(i_s)$.
(3) Go to the next step $i+1 \to i+2$.

The value of $|Seq|$ corresponds to the number of different sequences generated by this algorithm. For example, $|Seq| = 3$ and $|Seq| = 18$ holds for $n = 3$ and $n = 4$, respectively. The general analytical form $|Seq|_n$ of $|Seq|$ with system size n is obtained as the following recurrence formula:

$$|Seq|_n = \sum_{i=1}^{\lfloor \frac{n}{2} \rfloor} {}_nC_i |Seq|_{n-i} |Seq|_i, \tag{34}$$

where $\lfloor \cdot \rfloor$ is a floor function and with formal definition of $|Seq|_1 := 1$.

The products of KL-divergences according to the hierarchical sequences of system decompositions in Equation (32) is related to the volume of $n-1$-dimensional cuboids in the circle diagram. An example in case of $n = 4$ is presented in Figure 5, where two cuboids with 3 orthogonal edges of the different decomposition sequences < 1111 >→< 1222 >→< 1233 >→< 1234 > and < 1111 >→< 1133 >→< 1134 >→< 1234 > are depicted, whose cuboid volumes are

$$\sqrt{D[< 1111 >:< 1222 >]D[< 1222 >:< 1233 >]D[< 1233 >:< 1234 >]}, \tag{35}$$

and

$$\sqrt{D[< 1111 >:< 1133 >]D[< 1133 >:< 1134 >]D[< 1134 >:< 1234 >]}, \tag{36}$$

respectively.

In the same way as C_r, we took in the definition of C_c the arithmetic average of cuboid volumes so that to renormalize the combinatorial increase of the decomposition paths ($|Seq|$) according to the system size n.

Note that on the other hand we did not renormalize the rectangle-bias complexity C_r and the cuboid-bias complexity C_c by taking the exact geometrical mean of each product of KL-divergences such as $\sqrt[n-1]{\prod_{i=1}^{n-1} D[SD_i(i_s) : SD_{i+1}(i_s)]}$. This is for further accessibility to theoretical analysis such as variational method (see "Further Consideration" section), and does not change qualitative behavior of C_r and C_c since the power root is a monotonically increasing function. This treatment can be interpreted as taking the $(n-1)$-th power of the geometric means for the hierarchical sequences of KL-divergences.

A more comprehensive example on the utility of the cuboid-bias complexity C_c with respect to the rectangle-biased one C_r is shown in Figure 6. We consider the 6 nodes networks ($n = 6$) with the same I and C_r values but different heterogeneity. The system in the top left figure has a circularly connected structure with medium intensity, while that of the top right figure has strongly connected 3 subsystems. These systems have qualitatively five different ways of system decomposition that are the basic generators of all hierarchical sequences $Seq = \{SD_1(i_s) \rightarrow \cdots \rightarrow SD_5(i_s) | 1 \leq i_s \leq |Seq|\}$ for these networks. The five basial system decompositions are shown with the number ①, ②, ②$'$, ③ and ④ in top figures.

The circle diagrams of these systems are depicted in the middle figures. To suppose the same constant value of C_r in both systems, the following condition is satisfied in the middle right figure: $D[< 111111 >:$ ②$]$ $<$ $D[< 111111 >:$ ①in Middle Left figure$]$ $<$ $D[< 111111 >:$ ①$]$ $<$ $D[< 111111 >:$ ②in Middle Left figure$]$ $<$ $D[< 111111 >:$ ③$]$ $<$ $D[< 111111 >:$ ④$]$. Furthermore, the total surface of right triangles sharing the circle diameter as hypotenuse in the middle left and the middle right figures are conditioned to be identical, therefore the rectangle-bias complexity C_r fails to distinguish.

On the other hand, under the same condition, the cuboid-bias complexity C_c distinguishes between these two systems and gives higher value to the left one. The volume of 5-dimensional cuboids of the decomposition sequence $< 111111 > \xrightarrow{\text{①②②}'\text{③④}} < 123456 >$ are schematically shown in the bottom figures, maintaining the quantitative difference between KL-divergences. Since the multi-information I is identical between the two systems, so is the values of KL-divergence $D[< 111111 >:< 123456 >]$, which is the sum of the KL-divergences along the sequence $< 111111 > \xrightarrow{\text{①②②}'\text{③④}} < 123456 >$ from the Pythagorean theorem. This means that the inequality between the cuboid volumes can be represented as the isoperimetric inequality of high-dimensional cuboid. As a consequence, the left system has quantitatively higher value of C_c than the right one. The cuboid-bias complexity C_c is also sensitive to such heterogeneity.

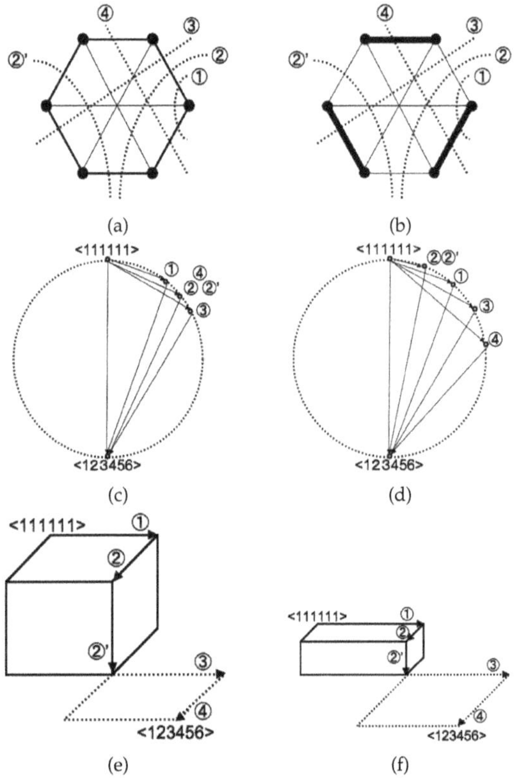

Figure 6. *Meaning of taking geometric mean over the sequence of system decomposition in cuboid-bias complexity* C_c. (**a**): Example of 6-node network with circularly connected structure with medium intensity. Edge width is proportional to edge information; (**b**): Example of 6-node network with strongly connected 3 subsystems. Edge width is proportional to edge information. The multi-information I of the two systems in Top figures are conditioned to be identical; The dotted lines schematically represent possible system decompositions. (**c**,**d**): Circle diagrams of each system decomposition in upper networks; The total surface of right triangles sharing the circle diameter as hypotenuse in (c) and (d) are conditioned to be identical, therefore the rectangle-bias complexity C_r fails to distinguish. (**e**,**f**): 5-dimensional cuboids of upper networks (Figure 6a,b) whose edges are the root of KL-divergences for the strain of system decomposition $< 111111 > \xrightarrow{①②②'③④} < 123456 >$. Only the first 3-dimensional part is shown with solid line, and the remaining 2-dimensional part is represented with dotted line. The volume of cuboid in (e) is larger than the one in (f), according to the isoperimetric inequality of high-dimensional cuboid. The total squared length of each side is identical between two cuboids, which represents multi-information $I = D[< 111111 >:< 123456 >]$.

8. Regularized Cuboid-Bias Complexity with Respect to Generalized Mutual Information

We further consider the geometrical composition of system decompositions in the circle diagram and insist the necessity of renormalizing the cuboid-bias complexity C_c with the multi-information I, which gives another measure of complexity namely "*regularized cuboid-bias complexity* C_c^R."

We consider the situation in actual data where the multi-information I varies. Figure 7 shows the $n = 3$ cases where the C_c fails to distinguish. Both the blue and red systems are supposed to have the same C_c value by adjusting the red system to have relatively smaller values of KL-divergences

$D[<111>:<122>]$ and $D[<113>:<123>]$ than the blue one. Such conditioning is possible since the KL-divergences are independent parameters with each other.

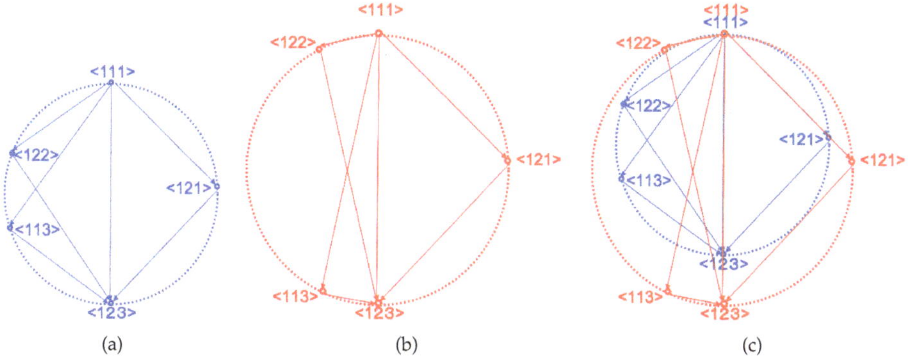

Figure 7. *Examples of the 3-node systems with identical cuboid-bias complexity C_c but different multi-information I on circle graph.* (**a**): System with smaller I but larger C_c^R; (**b**): System with larger I but smaller C_c^R; (**c**): Superposition of the above two systems. The regularized cuboid-bias complexity C_c^R distinguishes between the blue and red systems.

Although the C_c value is identical, the two systems have different geometrical composition of system decompositions in the circle diagram. The red system has relatively easier way of decomposition $<111>\rightarrow<122>$ if renormalized with the total system decomposition $<111>\rightarrow<123>$. This relative decompositionability with respect to the renormalization with the multi-information I can be clearly understood by superimposing the circle diagram of the two systems and comparing the angles between each and total decomposition paths (bottom figure). The red system has larger angle between the decomposition paths $<111>\rightarrow<122>$ and $<111>\rightarrow<123>$ than any others in the blue system, which represents the relative facility of the decomposition under renormalization with I. In this term, the paths $<111>\rightarrow<121>$ in the red and blue system do not change its relative facility, and the paths $<111>\rightarrow<113>$ are easier in the blue system.

To express the system decompositionability based on these geometrical compositions in a comprehensive manner, we define the *regularized cuboid-bias complexity* C_c^R as follows:

$$\begin{aligned} C_c^R &:= \frac{1}{|Seq|}\sum_{i_s=1}^{|Seq|}\prod_{i=1}^{n-1}\frac{D[SD_i(i_s):SD_{i+1}(i_s)]}{D[<11\cdots1>:<12\cdots n>]} \\ &:= \frac{C_c}{D[<11\cdots1>:<12\cdots n>]^{n-1}} \\ &:= \frac{C_c}{I^{n-1}}. \end{aligned} \tag{37}$$

The red system then has quantitatively smaller C_c^R value than the blue system in Figure 7.

9. Modular Complexity with Respect to the Easiest System Decomposition Path

We have considered so far the system decompositionability with respect to the all possible decomposition sequences. This was also a way to avoid the local fluctuation of the network heterogeneity to be reflected in some specific decomposition paths. On the other hand, the easiest decomposition is particularly important when considering the modularity of the system. If there exists hierarchical structure of modularity in different scales with different coherence of the system, the KL-divergence and the sequence of the easiest decomposition gives much information.

Figure 8 schematically shows a typical example where there exist two levels of modularity. Such structure with different scales of statistical coherence appears as functional segregation in neural systems [17], and is expected to be observed widely in complex systems.

The hierarchical topology of the easiest decomposition path reflects these structures. For example, in the system of Figure 8, the decompositions between $< 1\ 1\ \cdots\ 1 >$ and $< 1\,1\,1\,1\,5\,5\,5\,5\,9\,9\,9\,9\,13\,13\,13\,13 >$ are easier than those inside of the 4-node subsystems. The values of KL-divergence also reflect the hierarchy, giving relatively low values for the decomposition between the 4-node subsystems, and high values inside of them. By examining the shortest decomposition path and associated KL-divergences in possible *Seq*, one can project the hierarchical structure of the modularity existing in the system.

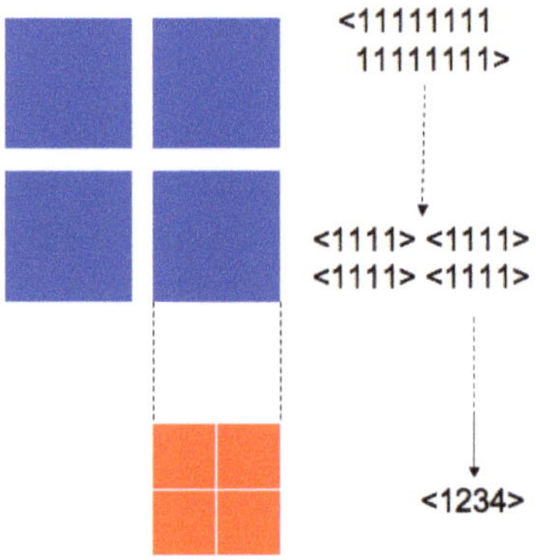

Figure 8. *Example of* 16*-node system* $< 11 \cdots 1 >$ *that has different levels of modularity.* The four 4-node subsystems $< 1111 >$ (blue blocks) are loosely connected and easy to be decomposed, while inside each component (red blocks) is tightly connected. The degree of connection represents statistical dependency or edge information between subsystems. Such hierarchical structure can be detected by observing the decomposition path of the modular complexity C_m.

For this reason, we define the *modular complexity* C_m as follows, which is the shortest path component of the cuboid-bias complexity C_c:

$$C_m := \prod_{i=1}^{n-1} D[SD_i(i_{min}) : SD_{i+1}(i_{min})], \tag{38}$$

where the index i_{min} of the sequence $SD_1(i_{min}) \to SD_2(i_{min}) \to \cdots \to SD_n(i_{min})$ is chosen as follows:

$$i_{min} = \{i_1\} \cap \{i_2\} \cap \cdots \cap \{i_{n-1}\}, \tag{39}$$

where

$$\begin{aligned} \{i_1\} &= \operatorname*{argmin}_{i_s} \{D[SD_1(i_s) : SD_2(i_s)] | 1 \le i_s \le |Seq|\}, \\ \{i_2\} &= \operatorname*{argmin}_{i_1} \{D[SD_2(i_1) : SD_3(i_1)] | i_1 \in \{i_1\}\}, \\ &\vdots \\ \{i_{n-1}\} &= \operatorname*{argmin}_{i_{n-2}} \{D[SD_{n-1}(i_{n-2}) : SD_n(i_{n-2})] | i_{n-1} \in \{i_{n-1}\}\}, \end{aligned} \tag{40}$$

which gives eventually

$$i_{min} = i_{n-1}. \tag{41}$$

This means that beginning from the undecomposed state $< 11 \cdots 1 >$, we continue to choose the shortest decomposition path in the next hierarchy of system decomposition. The minimization of the path length is guaranteed by the sequential minimization since the geometric mean of isometric path division is bounded below by its minimum component. i_{min} is unique if the system is completely heterogenous (*i.e.*, $D[SD_1(i_k) : SD_2(i_k)] \neq D[SD_1(i_l) : SD_2(i_l)], 1 \leq i_k < i_l \leq |Seq|$), otherwise plural decomposition paths that give the same C_m value are possible according to the homogeneity of the system. Besides its value, the modular complexity C_m should be utilized with the sequence information of the shortest decomposition path to evaluate the modularity structure of a system.

The cases where C_m are identical but C_c are different can be composed by varying the system decompositions other than in the shortest path $SD_1(i_{min}) \rightarrow SD_2(i_{min}) \rightarrow \cdots \rightarrow SD_n(i_{min})$ without modifying the index $i_m in$. There exist also inverse examples with identical C_c and different C_m, due to the complementarity between C_m and C_c.

We finally define the *regularized modular complexity* C_m^R as follows, for the same reason as defining C_c^R from C_c;

$$\begin{aligned} C_m^R &:= \prod_{i=1}^{n-1} \frac{D[SD_i(i_{min}) : SD_{i+1}(i_{min})]}{D[< 11 \cdots 1 >:< 12 \cdots n >]} \\ &:= \frac{C_m}{D[< 11 \cdots 1 >:< 12 \cdots n >]^{n-1}} \\ &:= \frac{C_m}{I^{n-1}}. \end{aligned} \tag{42}$$

Proposition 4. *The cuboid-bias complexities C_c and C_c^R are bounded by the modular complexities C_m and C_m^R respectively:*

$$C_c \leq C_m, \tag{43}$$
$$C_c^R \leq C_m^R. \tag{44}$$

And they coincide at the maximum values under the given multi-information I:

$$\max\{C_m | I = const.\} = \max\{C_c | I = const.\}, \tag{45}$$
$$\max\{C_m^R\} = \max\{C_c^R\}. \tag{46}$$

These relations (43)–(46) are numerically shown in the "Numerical Comparison" section.

The superiority of the modular complexities is due to the hierarchical dependency of KL-divergence value in decomposition paths. In the shortest decomposition path defining modular complexities, the easier system decomposition relatively increase its value since they incorporate more number of edge cutting. Since we eventually cut all edges to obtain $< 12 \cdots n >$ at the end of the decomposition sequence, collecting the edges with relatively weak edge information and cutting them together augment the value of the product of KL-divergences. The modular complexities are then the maximum value components among the possible decomposition paths calculated in cuboid-bias complexities:

$$C_m = \max\left\{ \prod_{i=1}^{n-1} D[SD_i(i_s) : SD_{i+1}(i_s)] \,\middle|\, 1 \leq i_s \leq |Seq| \right\}, \tag{47}$$

$$C_m^R = \max\left\{ \prod_{i=1}^{n-1} \frac{D[SD_i(i_s) : SD_{i+1}(i_s)]}{D[< 11 \cdots 1 >:< 12 \cdots n >]^{n-1}} \,\middle|\, 1 \leq i_s \leq |Seq| \right\}. \tag{48}$$

The difference between the cuboid-bias complexities and the modular complexities is an index of the geometrical variation of decomposed systems in the circle graph, which reflects the fluctuation of the sequence-wise system decompositionability. If the variation of the system decompositionability for each system decomposition is large, accordingly the modular complexities tend to give higher values than the cuboid-bias complexities.

10. Numerical Comparison

We numerically investigate the complementarity between the proposed complexities, C_c, C_c^R, C_m, and C_m^R. Since the minimum node number giving non-trivial meaning to these measures is $n = 4$, the corresponding dimension of parameter space is $\sum_{k=1}^{n} {}_nC_k = 15$. The constant-complexity submanifolds are therefore difficult to visualize due to the high dimensionality. For simplicity, we focus on the 2-dimensional subspace of this parameter space whose first axis ranging from random to maximum dependencies of the system, and the second one representing the system decompositionability of $< 1133 >$.

For this purpose, we introduce the following parameters α and β $(0 \leq \alpha, \beta \leq 1)$ in the η-coordinates of the discrete distribution with 4-dimensional binary stochastic variable:

$$
\begin{aligned}
\eta_1 &= \eta_0, \\
\eta_2 &= \eta_0, \\
\eta_3 &= \eta_0, \\
\eta_4 &= \eta_0, \\
\eta_{1,2} &= \eta_1\eta_2 + \alpha(\eta_0 - \epsilon - \eta_1\eta_2), \\
\eta_{3,4} &= \eta_3\eta_4 + \alpha(\eta_0 - \epsilon - \eta_3\eta_4), \\
\eta_{1,3} &= \eta_1\eta_3 + \alpha\beta(\eta_0 - \epsilon - \eta_1\eta_3), \\
\eta_{1,4} &= \eta_1\eta_4 + \alpha\beta(\eta_0 - \epsilon - \eta_1\eta_4), \\
\eta_{2,3} &= \eta_2\eta_3 + \alpha\beta(\eta_0 - \epsilon - \eta_2\eta_3), \\
\eta_{2,4} &= \eta_2\eta_4 + \alpha\beta(\eta_0 - \epsilon - \eta_2\eta_4), \\
\eta_{1,2,3} &= \eta_{1,2}\eta_3 + \alpha\beta(\eta_0 - 2\epsilon - \eta_{1,2}\eta_3), \\
\eta_{1,2,4} &= \eta_{1,2}\eta_4 + \alpha\beta(\eta_0 - 2\epsilon - \eta_{1,2}\eta_4), \\
\eta_{1,3,4} &= \eta_1\eta_{3,4} + \alpha\beta(\eta_0 - 2\epsilon - \eta_1\eta_{3,4}), \\
\eta_{2,3,4} &= \eta_2\eta_{3,4} + \alpha\beta(\eta_0 - 2\epsilon - \eta_2\eta_{3,4}), \\
\eta_{1,2,3,4} &= \eta_{1,2}\eta_{3,4} + \alpha\beta(\eta_0 - 3\epsilon - \eta_{1,2}\eta_{3,4}).
\end{aligned}
\tag{49}
$$

Where α represents the degree of statistical association from random $(\alpha = 0)$ to maximum $(\alpha = 1)$, and β control the system decompositionability of $< 1133 >$. If $\beta = 1$, the system has the maximum KL-divergence $D[< 1111 >:< 1133 >]$ under the constraint of α parameter, and $\beta = 0$ gives $D[< 1111 >:< 1133 >] = 0$.

ϵ is the minimum value of the joint distribution of 4-dimensional variable, which is defined to be more than 0 to avoid singularity in the dual-flat coordinates of statistical manifold. $\epsilon = 1.0 \times 10^{-10}$ and $\eta_0 = 0.5$ was chosen for the calculation.

The system with maximum statistical association under given η_0 corresponds to the $\alpha = \beta = 1$ condition in given parameters, whose ȷ-coordinates become as follows:

$$
\begin{aligned}
\eta_1 &= \eta_0, \\
&\vdots \\
\eta_4 &= \eta_0, \\
\eta_{1,2} &= \eta_0 - \epsilon, \\
&\vdots \\
\eta_{3,4} &= \eta_0 - \epsilon, \\
\eta_{1,2,3} &= \eta_0 - 2\epsilon, \\
&\vdots \\
\eta_{2,3,4} &= \eta_0 - 2\epsilon, \\
\eta_{1,2,3,4} &= \eta_0 - 3\epsilon, .
\end{aligned} \tag{50}
$$

On the other hand, the totally decomposed system corresponds to the $\alpha = 0$ condition, and the ȷ-coordinates are:

$$
\begin{aligned}
\eta_1 &= \eta_0, \\
&\vdots \\
\eta_4 &= \eta_0, \\
\eta_{1,2} &= \eta_0\eta_0, \\
&\vdots \\
\eta_{3,4} &= \eta_0\eta_0, \\
\eta_{1,2,3} &= \eta_0\eta_0\eta_0, \\
&\vdots \\
\eta_{2,3,4} &= \eta_0\eta_0\eta_0, \\
\eta_{1,2,3,4} &= \eta_0\eta_0\eta_0\eta_0.
\end{aligned} \tag{51}
$$

Note that the completely deterministic case $\eta_0 = 1.0$ and $\alpha = \beta = 1$ gives $I = 0$.

The intuitive meaning of these parameters α and β are also schematically depicted in Figure 9 bottom right.

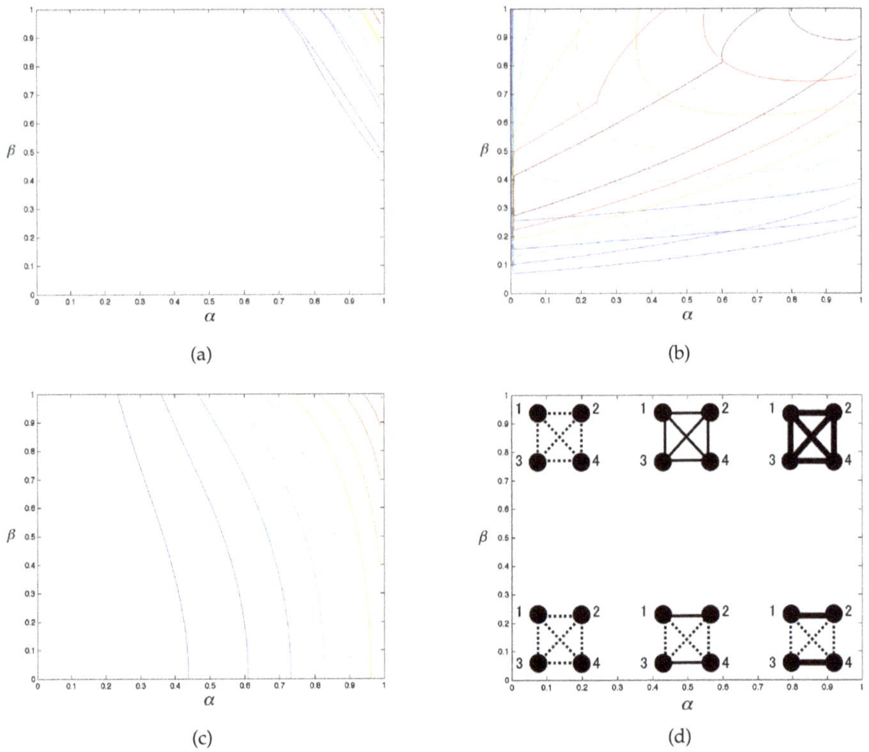

Figure 9. *Contour plot of the complexity landscape of I, C_c, C_m, C_c^R, and C_m^R on α-β plane.* (**a**): Contour plot superposition of C_c and C_m. (**b**): Contour plot superposition of C_c^R and C_m^R. (**c**): Contour plot of I. The color of contour plots corresponds to the color gradient of 3D plots in Figure 10; (**d**): Schematic representation of the system in different regions of α-β plane. Edge width represents the degree of edge information, and independence is depicted with dotted line.

Figure 10 shows the landscape of the proposed complexities on the α-β plane. Their contour plots are depicted in Figure 9. The proposed complexities each differs from others in almost everywhere points on α-β plane except at the intersection lines. Therefore, these measures serve as the independent features of the system, each has its specific meaning with respect to the system decompositionability. The α-β plane shows a section of the actual structure of the complementarity expressed in Figure 3 between the proposed complexity measures.

The relations between the cuboid-bias complexities and modular complexities in Equations (43)–(46) are also numerically confirmed. The modular complexities are superior than the corresponding cuboid-bias complexities, and coincide at the parameter $\alpha = \beta = 1$ giving maximum values and dependencies in this parameterization.

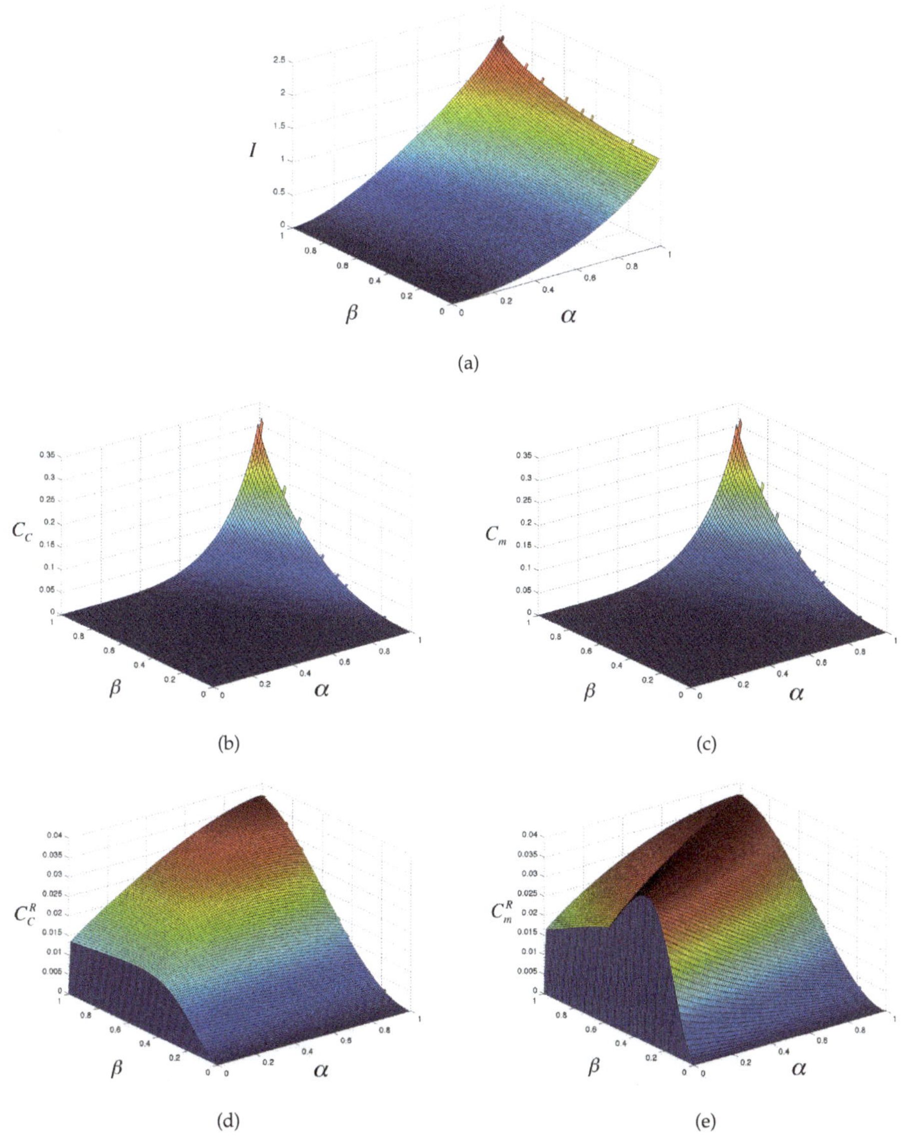

Figure 10. *Landscape of complexities I, C_c, C_m, C_c^R, and C_m^R on α-β plane.* (**a**): Multi-information I; (**b**): Cuboid-bias complexity C_c. (**c**): Modular complexity C_m;(**d**): Regularized cuboid-bias complexity C_c^R; (**e**): Regularized modular complexity C_m^R. All complexity measures show the complementarity intersecting with each other, satisfying the boundary conditions vanishing at $\alpha = 0$ and $\beta = 0$ except the multi-information I. Note that regularized complexities C_c^R and C_m^R show singularity of convergence at $\alpha \to 0$ due to the regularization of infinitesimal value.

In general case without the parameterization with α, β and η_0, the boundary conditions of C_c, C_c^R, C_m and C_m^R include that of the multi-information I, which vanish at the completely random or ordered state. This is common to other complexity measures such as the LMC complexity, and fit to the basic intuition on the concept of complexity situated equivalently far from the completely predictable and disordered states [21,22].

The proposed complexities further incorporate boundary conditions that vanish with the existence of a completely independent subsystem of any size. This means that the C_c, C_c^R, C_m and C_m^R of a system become 0 if we add another independent variable. This property does not reflect the intuition of complexity defined by the arithmetic average of statistical measures. The proposed complexity can better find its meaning in comparison to other complexity measures such as the multi-information I, and by interactively changing the system scale to avoid trivial results with small independent subsystem. For example, the proposed complexities could be utilized as the information criteria for the model selection problems, especially with an approximative modular structure based on the statistical independency of data between subsystems. We insist that the complementarity principle between plural complexity measures of different foundation is the key to understand the complexity in a comprehensive manner.

To characterize the property of C_c, C_c^R, C_m and C_m^R in relation to the diverse composition of each system decomposition, it is useful to consider the geometry of their contour structure, as compared in Figure 9. The contour can be formalized as $C_c, C_c^R, C_m, C_m^R = const.$ for each complexity measure, and $D[< 11\cdots 1 >: SD_i(i_s)] = const.$ $(1 \leq i \leq n-1, 1 \leq i_s \leq |Seq|)$ for each system decomposition. For that purpose, analysis with algebraic geometry can be considered as a prominent tool. Algebraic geometry investigates the geometrical property of polynomial equations [23]. The complexities C_c, C_c^R, C_m and C_m^R can be interpreted as polynomial functions by taking each system decomposition as novel coordinates, therefore directly accessible to algebraic geometry. However, if we want to investigate the contour of the complexities on the $\mathbf{p}$ parameter space, logarithmic function appears as the definition of KL-divergence, which is a transcendental function and outreach the analytical requirement of algebraic geometry. To introduce compatibility between the $\mathbf{p}$ parameter space of information geometry and algebraic geometry, it suffices to describe the model by replacing the logarithmic functions as another n variables such as $\mathbf{q} = \log \mathbf{p}$, and reconsider the intersection between the result from algebraic geometry on the coordinates $(\mathbf{p}, \mathbf{q})$ and $\mathbf{q} = \log \mathbf{p}$ condition. The contour of C_c, C_c^R, C_m and C_m^R is also important to seek for the utility of these measures as a potential to interpret the dynamics of statistical association as geodesics.

11. Further Consideration

11.1. Pythagorean Relations in System Decomposition and Edge Cutting

We further look back at the system decomposition and edge cutting in terms of the Pythagorean relation between KL-divergences, which is based on the orthogonality between ` and ȷ coordinates.

In system decomposition, the distribution of decomposed system is analytically obtained from the product of subsystems' η coordinates, which is equivalent to set all θ^{dec} parameters as 0 in mixture coordinate ξ^{dec}. From the consistency of θ^{dec} parameters in ξ^{dec} being 0 in all system decompositions, we have the Pythagorean relation according to the inclusion relation of system decomposition. For example, the following holds:

$$
\begin{aligned}
D[< 1111 >:< 1234 >] \;\; = \;\; & D[< 1111 >:< 1222 >] \\
+ \;\; & D[< 1222 >:< 1233 >] \\
+ \;\; & D[< 1233 >:< 1234 >].
\end{aligned}
\tag{52}
$$

The proof is in the same way as k-cut coordinates isolating k-tuple statistical association between variables [14].

On the other hand, the edge cutting previously defined using the product of remaining maximum cliques' η coordinates does not coincides with the $\theta^{ec} = 0$ condition in mixture coordinates ξ^{ec}. We have defined the η^{ec} values of edge cutting based only on the orthogonal relation between η and θ

coordinates, by generalizing the rule of system decomposition in η^{ec} coordinates, and did not consider the Pythagorean relation between different edge cuttings.

It is then possible to define another way of edge cutting using $\theta^{ec} = 0$ condition in ζ^{ec}. Indeed, in k-cut mixture coordinates, $\theta^{k+} = 0$ condition is derived from the independent condition of the variables in all orders, and k-tuple statistical association is measured by reestablishing the η parameters for the statistical association up to $k-1$-tuple order. In the same way, we can set $\theta^{dec} = 0$ condition for ζ^{dec} of a system decomposition, and reestablish edges with respect to the η parameters, except the one in focus for edge cutting.

As a simple example, consider the system decomposition $< 1222 >$ and edge cutting $1-2$ in 4-node graph. We have the mixture coordinate ζ^{dec} for the system decomposition as follows:

$$\begin{aligned}
\zeta^{dec}_{1,2} &= \theta^{dec}_{1,2} = 0,\\
\zeta^{dec}_{1,3} &= \theta^{dec}_{1,3} = 0,\\
\zeta^{dec}_{1,4} &= \theta^{dec}_{1,4} = 0,\\
\zeta^{dec}_{1,2,3} &= \theta^{dec}_{1,2,3} = 0,\\
\zeta^{dec}_{1,3,4} &= \theta^{dec}_{1,3,4} = 0,\\
\zeta^{dec}_{1,2,3,4} &= \theta^{dec}_{1,2,3,4} = 0,
\end{aligned} \tag{53}$$

where all the rest of ζ^{dec} coordinates is equivalent to that of η coordinates.

We then consider the new way of edge cutting $1-2$ by recovering the statistical association in edges $1-3$ and $1-4$ from system decomposition $< 1222 >$, orthogonally to that of edge $1-2$. The new mixture coordinate ζ^{EC} changes to the following:

$$\begin{aligned}
\zeta^{EC}_{1,2} = \theta^{EC}_{1,2} &= 0,\\
\zeta^{EC}_{1,3} &= \eta_{1,3},\\
\zeta^{EC}_{1,4} &= \eta_{1,4},\\
\zeta^{EC}_{1,2,3} = \theta^{EC}_{1,2,3} &= 0,\\
\zeta^{EC}_{1,3,4} &= \eta_{1,3,4},\\
\zeta^{EC}_{1,2,3,4} = \theta^{EC}_{1,2,3,4} &= 0,
\end{aligned} \tag{54}$$

and the rest is equivalent to that of η coordinates.

This new ζ^{EC} is also compatible with k-cut coordinates formalization for its simple $\theta^{EC} = 0$ conditions. To obtain ζ^{EC} for arbitrary edge cutting $i-j$, one should take θ^{EC} containing i and j in its subscript, set them to 0, and combine with η coordinates for the rest of the subscript. For plural edge cuttings $i-j, \cdots, k-l$ $(1 \leq i,j,k,l \leq n)$, it suffices to take θ^{EC} containing i and j, ... , k and l in its subscript respectively, then set them to 0.

We finally obtain the Pythagorean relation between edge cuttings. Denoting the general edge cutting(s) coordinates as $\zeta^{i-j,\cdots,k-l}$, the following holds for the example of system decomposition $< 1222 >$:

$$\begin{aligned}
D[< 1111 >:< 1222 >] &= D[< 1111 >: p(\zeta^{1-2})]\\
&+ D[p(\zeta^{1-2}) : p(\zeta^{1-2,1-3})]\\
&+ D[p(\zeta^{1-2,1-3}) : p(\zeta^{1-2,1-3,1-4})].
\end{aligned} \tag{55}$$

Despite the consistency with the dual structure between θ and η, we do not generally have analytical solution to determine η^{EC} values from $\theta^{EC} = 0$ conditions. We should call for some numerical algorithm to solve $\theta^{EC} = 0$ conditions with respect to η^{EC} values, which are in general high-degree simultaneous polynomials. Furthermore, numerical convergence of the solution has to be very strict, since tiny deviation from the conditions can become non-negligible by passing fractional function and logarithmic function of θ coordinates.

On the other hand, the previously defined edge cutting with ζ^{ec} using the product between subgraphs' η coordinates is analytically simple and does not need to consider the other edges' recovery from system decomposition or independence hypothesis. We then chose the previous way of edge cutting for both calculability and clarity of the concept.

There have been many attempts to approximate complex network by low-dimensional system with the use of statistical physics and network theory. As a contemporary example, moment-closure approximation provides a various way to abstract essential dynamics e.g., in discrete adaptive network [24]. Although the approximation takes several theoretical assumptions such as random graph approximation, it is difficult to quantitatively reproduce the dynamics even in some simplest model. This is partly due to homogeneous treatment of statistics such as truncation into pair-wise order. The edge cutting can offer a complementary view on the evaluation of moment-closure approximations. Using orthogonal decomposition between edge information, one can evaluate which part of network link and which order of statistics contain essential information, which does not necessary conform to top-down theoretical treatment.

11.2. Complexity of the Systems with Continuous Phase Space

We have developed the concept of system decompositionability based on discrete binary variables. One can also apply the same principle to continuous variable.

For an ergodic map $G : X \to X$ in continuous space X, KS entropy $h(\mu, G)$ is defined as the maximum of entropy rate with respect to all possible system decomposition A, when the invariant measure μ exists:

$$h(\mu, G) = \sup_A h(\mu, G, A). \tag{56}$$

where A is the disjoint decomposition of X that consists of non-trivial sets a_i, whose total number is $n(A)$, defined as

$$X = \bigcup_{i=1}^{n(A)} a_i, \tag{57}$$

$$a_i \cap a_j = \phi,\ i \neq j,\ 1 \leq i, j \leq n(A), \tag{58}$$

meaning the natural expansion of system decomposition into continuous space.

The entropy rate $h(\mu, G, A)$ in Equation (56) is defined as

$$h(\mu, G, A) = \lim_{n \to \infty} \frac{1}{n} H(\mu, A \vee G^{-1}(A) \vee \cdots \vee G^{-n+1}(A)), \tag{59}$$

according to the entropy $H(\mu, A)$ based on the decomposition $A = \{a_i\}$

$$H(\mu, A) = -\sum_{i=1}^{n(A)} \mu(a_i) \ln \mu(a_i), \tag{60}$$

and the product $C = A \vee B$ as

$$\begin{aligned} C &= A \vee B \\ &= \{c_i = a_j \cap b_k | 1 \leq j \leq n(A), 1 \leq k \leq n(B)\}. \end{aligned} \tag{61}$$

In a more general case, topological entropy $h_T(G)$ is defined simply with the number of decomposed subsystem elements by preimages as follows, without requiring ergodicity, therefore neither the existence of invariant measure μ:

$$h_T(G) = \sup_A \lim_{n \to \infty} \frac{1}{n} \ln n(A \vee G^{-1}(A) \vee \cdots \vee G^{-n+1}(A)). \tag{62}$$

Topological entropy takes the maximum value of the possible preimage divisions, in order to measure the complexity in terms of the mixing degree of the orbits. For example, if the KS entropy is positive as $h(\mu, G) > 0$, the dynamics of G on an invariant set of invariant measure μ is chaotic for almost everywhere initial conditions. As for the positive topological entropy $h_T(G) > 0$, the dynamics of G contain chaotic orbits, but not necessary as attractive chaotic invariant set, since $h_T(G) \geq h(\mu, G)$ and the KS entropy can be negative.

Although these definitions are useful to characterize the existence of chaotic dynamics, the system decompositionability is another property representing different aspect of the system complexity. It is rather the matter of the existence of independent dynamics components, or the degree of orbit localization between arbitrary system decompositions. We propose the following "*geometric topological entropy*" $h_g(G)$ applying the same principle of taking geometric product between all hierarchical structure of the system decomposition A.

$$h_g(G) := \prod_{\sigma(A)>0} \lim_{n \to \infty} \frac{1}{n} \ln n(A \vee G^{-1}(A) \vee \cdots \vee G^{-n+1}(A)), \tag{63}$$

where $\sigma(A) > 0$ means to take all components of A having positive Lebesgue measure on X.

This gives 0 if the preimage of certain $a_i \in A$ is a_i itself, meaning there exist a subsystem a_i whose range is invariant under G, closed by itself. The system X can be completely divided into a_i and the rest. This corresponds to the existence of an independent subsystem in cuboid-bias and modular complexities. In case such independent components do not exist, it still reflects the degree of orbit localization for all possible system decompositions in multiplicative manner. The condition $\sigma(A) > 0$ is to avoid trivial case such as the existence of unstable limit cycle, whose Lebesgue measure is 0.

Typical example giving $h_g(G) = 0$ is the function having independent ergodic components, such as the Chirikov-Taylor map with appropriate parameter [25].

12. Conclusions and Discussion

We have theoretically developed a framework to measure the degree of statistical association existing between subsystems as well as the ones represented by each edge of the graph representation. We then reconsidered the problem of how to define complexity measures in terms of the construction of non-linear feature space. We defined new type of complexity based on the geometrical product of KL-divergence representing the degree of system decompositionability. Different complexity measures as well as newly proposed ones are compared on a complementarity basis on statistical manifold.

Application of presented theory can encompass a large field of complex systems and data science, such as social network, genetic expression network, neural activities, ecological database, and any kind of complex networks with binary co-occurrence matrix data e.g., [26–29], databases: [30–34]. Continuous variables are also accessible by appropriate discretization of information source with e.g., entropy maximization principle.

In contrast to arithmetic mean of information over the whole system, geometric mean has not been investigated sufficiently in the analysis of complex network. However in different fields, theoretical ecology has already pointed out the importance of geometric mean when considering the long-term fitness of a species population in a randomly varying environment [35,36]. Long-term fitness refers to the ecological complexity of its survival strategy under large stochastic fluctuation. Here, we can find useful analogy between the growth rate of a population in ecology and the spatio-temporal propagation rate of information between subsystems in general. If we take an arbitrary subsystem and consider the amount of information it can exchange with all other subsystems, the proposed complexity measures with geometric mean reflect the minimum amount with amongst all possible other subsystems, which can not be distinguished with arithmetic mean. The propagation rate of a population in ecology and the information transmission in complex network hold mathematically analogous structure. In population ecology, the variance of growth rate is crucial to evaluate the long-term survival of the population. Even if the arithmetic mean of growth rate is high, large variance will lead to low geometric mean even with a small amount of exceptionally small fitness situation, which ecologically means extinction of an entire species. In stochastic network, the variance of system decompositionability is essential to evaluate the amount of information shared between subsystems, or information persistence in the entire network. Even the multi-information I is high, large heterogeneity of edge information can lead to informational isolation of certain subsystem, which means extinction of its information. If such subsystem is situated on the transmission pathway, information cannot propagate across these nodes. Therefore, the proposed complexity measures C_C, C_C^R, C_m and C_m^R generally reflect the minimum amount of information propagation rate spread entirely on the system without exception of isolated division.

Some recent studies on adaptive network focus on the evolution of network topology in response to node activity, such as game-theoretic evolution of strategies [37], opinion dynamics on an evolving network [38], epidemic spreading on an adaptive network [39], *etc.* Analysis of coevolution network between variables and interactions can capture important dynamical feature of complex systems. In contrast to topological network analysis, the newly proposed complexity measures can complement its statistical dynamics analysis. In addition to the topological change of network model, (e.g., linking dynamics of game theory, opinion community network structure, contact network of epidemics transmission), one can evaluate the emerged statistical association between the variables that does not necessary coincide with the network topology. Interesting feature of non-linear dynamics is the unexpected correlation between distant variables, which is quantified as Tsallis entropy [40]. The complementary relation between concrete interaction and resulting statistical association can provide a twofold methodology to characterize the coevolutionary dynamics of adaptive network. Such strategy can promote integrated science from laboratory experiments to open-field *in natura* situation, where actual multi-scale problematics remain to be solved [41].

Arithmetic and geometric means can be integrated in a mutual formula called generalized mean [42]. Therefore, the proposed complexity measures with geometric mean of KL-divergence is an expansion of preexisting complexity measures with mixture coordinates. Table 1 summarizes the generalization of complexity measure in this article. Based on the k-cut coordinates ı, the weighted sum of KL-divergence representing k-tuple order of statistical association derived complexity measures with (weighted) arithmetic mean such as multi-information I and TSE complexity. On the other hand, we showed that subsystem-wise correlation can also be isolated with the use of mixture coordinates, namely $< \cdots >$-cut coordinates ¸. To quantify the heterogeneity of system decompositionability, we generally took a weighted geometric mean of KL-divergence in C_C, C_C^R, C_m and C_m^R. Here, the shortest path selection of C_m and C_m^R, and regularization of C_C^R and C_m^R with respect to multi-information I can be interpreted as the weight function of geometric mean. This perspective brings a definition of a generalized class of complexity measures based on the mixture coordinates and generalized mean of KL-divergence. Information discrepancy can also be generalized from KL-divergence to Bregman divergence, providing access to the concept of multiple centroids in large stochastic data analysis such as image processing [43]. The blank columns of the Table 1 imply the possibility of other complexity measures in this class. For example, the weighted geometric mean of KL-divergence defined between k-cut coordinates is expected to yield complexity measures that are sensitive to the heterogeneity of correlation orders. The weighted arithmetic mean of KL-divergence defined between $< \cdots >$-cut coordinates should be sensitive to the mean decompositionability of arbitrary subsystem. Since these measures take analytically different form on mixture coordinates and/or mean functions, their derivatives do not coincide, which give independent information of the system on the complementary basis on statistical manifold, as long as the number of complexity measures are inferior to the freedom degree of the system.

Table 1. Classification of complexity measures with KL-divergence on mixture coordinates.

		Generalized Mean of KL-Divergence	
		Arithmetic Mean	**Geometric Mean**
Mixture Coordinates	k-cut	TSE complexity, I	
	$< \cdots >$-cut		C_C, C_C^R, C_m, C_m^R

Acknowledgments: This study was partially supported by CNRS, the long term study abroad support program of the university of Tokyo, and the French government (Promotion Simone de Beauvoir).

Conflicts of Interest: Conflicts of Interest

The author declares no conflict of interest.

References

1. Boccalettia, S.; Latorab, V.; Morenod, Y.; Chavezf, M.; Hwang, D.U. Complex Networks: Structure and Dynamics. *Phys. Rep.* **2006**, *424*, 175–308.
2. Strogatz, S.H. Exploring Complex Networks. *Nature* **2001**, *410*, 268–276.
3. Wasserman, S.; Faust, K. *Social Network Analysis*; Cambridge University Press: Cambridge, UK, 1994.
4. Funabashi, M.; Cointet, J.P.; Chavalarias, D. Complex Network. In *Studies in Computational Intelligence*; Springer: Berlin/Heidelberg, Germany, 2009; Volume 207, pp. 161–172.
5. Badii, R.; Politi, A. *Complexity: Hierarchical Structures and Scaling in Physics*; Cambridge University Press: Cambridge, UK, 2008.
6. Lempel, A.; Ziv, J. On the Complexity of Finite Sequences. *IEEE Trans. Inf. Theory* **1976**, *22*, 75–81.
7. Li, M.; Vitanyi, P. Texts in Computer Science. In *An Introduction to Kolmogorov Complexity and Its Applications*, 2nd ed.; Springer: Berlin/Heidelberg, Germany, 1997.
8. Cover, T.M.; Thomas, J.A. *Elements of Information Theory*; Wiley: New York, NY, USA, 2006.
9. Bennett, C. On the Nature and Origin of Complexity in Discrete, Homogeneous, Locally-Interacting Systems. *Found. Phys.* **1986**, *16*, 585–592.
10. Grassberger, P. Toward a Quantitative Theory of Self-Generated Complexity. *Int. J. Theor. Phys.* **1986**, *25*, 907–938.
11. Crutchfield, J.P.; Feldman, D.P. Regularities Unseen, Randomness Observed: The Entropy Convergence Hierarchy. *Chaos* **2003**, *15*, 25–54.
12. Crutchfield, J.P. Inferring Statistical Complexity. *Phys. Rev. Lett.* **1989**, *63*, 105–108.
13. Prichard, D.; Theiler, J. Generalized Redundancies for Time Series Analysis. *Physica D* **1995**, *84*, 476–493.
14. Amari, S. Information Geometry on Hierarchy of Probability Distributions. *IEEE Trans. Inf. Theory* **2001**, *47*, 1701–1711.
15. Ay, N.; Olbrich, E.; Bertschinger, N.; Jost, J. *A Unifying Framework for Complexity Measures of Finite Systems*; Report 06-08-028; Santa Fe Institute: Santa Fe, NM, USA, 2006.
16. MacKay, R.S. Nonlinearity in Complexity Science. *Nonlinearity* **2008**, *21*, T273–T281.
17. Tononi, G.; Sporns, O.; Edelman, M. A Measure for Brain Complexity: Relating Functional Segregation and Integration in the Nervous System. *Proc. Natl. Acad. Sci. USA* **1994**, *91*, 5033.
18. Feldman, D.P.; Crutchfield, J.P. Measures of statistical complexity: Why? *Phys. Lett. A* **1998**, *238*, 244–252.
19. Nakahara, H.; Amari, S. Information-Geometric Measure for Neural Spikes. *Neural Comput.* **2002**, *14*, 2269–2316.
20. Olbrich, E.; Bertschinger, N.; Ay, N.; Jost, J. How Should Complexity Scale with System Size? *Eur. Phys. J. B* **2008**, *63*, 407–415.
21. Feldman, D.P.; Crutchfield, J.P. Measures of Statistical Complexity: Why? *Phys. Lett. A* **1998**, *238*, 244–252.
22. Lopez-Ruiz, R.; Mancini, H.; Calbet, X. A Statistical Measure of Complexity. *Phys. Lett. A* **1995**, *209*, 321–326.

23. Hodge, W.; Pedoe, D. *Methods of Algebraic Geometry*; Cambridge Mathematical Library, Cambridge University Press: Cambridge, UK, 1994; Volume 1–3.
24. Demirel, G.; Vazquez, F.; Bohme, G.; Gross, T. Moment-closure Approximations for Discrete Adaptive Networks. *Physica D* **2014**, *267*, 68–80.
25. Fraser, G., Ed. *The New Physics for the Twenty-First Century*; Cambridge University Press: Cambridge, UK, 2006; p. 335.
26. Scott, J. *Social Network Analysis: A Handbook*; SAGE Publications Ltd.: London, UK, 2000.
27. Geier, F.; Timmer, J.; Fleck, C. Reconstructing Gene-Regulatory Networks from Time Series, Knock-Out Data, and Prior Knowledge. *BMC Syst. Biol.* **2007**, *1*, doi:10.1186/1752-0509-1-11.
28. Brown, E.N.; Kass, R.E.; Mitra, P.P. Multiple Neural Spike Train Data Analysis: State-of-the-Art and Future Challenges. *Nat. Neurosci.* **2004**, *7*, 456–461.
29. Yee, T.W. The Analysis of Binary Data in Quantitative Plant Ecology. Ph.D. Thesis, The University of Auckland, New Zealand, 1993.
30. Stanford Large Network Dataset Collection. Available online: http://snap.stanford.edu/data/ (accessed on 19 July 2014).
31. BioGRID. Available online: http://thebiogrid.org/ (accessed on 19 July 2014).
32. Neuroscience Information Framework. Available online: http://www.neuinfo.org/ (accessed on 19 July 2014).
33. Global Biodiversity Information Facility. Available online: http://www.gbif.org/ (accessed on 19 July 2014).
34. UCI Network Data Repository. Available online: http://networkdata.ics.uci.edu/index.php (accessed on 19 July 2014).
35. Lewontin, R.C.; Cohen, D. On Population Growth in a Randomly Varying Environment. *Proc. Natl. Acad. Sci. USA* **1969**, *62*, 1056–1060.
36. Yoshimura, J.; Clark, C.W. Individual Adaptations in Stochastic Environments. *Evol. Ecol.* **1969**, *5*, 173–192.
37. Wu, B.; Zhou, D.; Wang, L. Evolutionary Dynamics on Stochastic Evolving Networks for Multiple-Strategy Games. *Phys. Rev. E* **2011**, *84*, 046111.
38. Fu, F.; Wang, L. Coevolutionary Dynamics of Opinions and Networks: From Diversity to Uniformity. *Phys. Rev. E* **2008**, *78*, 016104.
39. Gross, T.; D'Lima, C.J.D.; Blasius, B. Epidemic Dynamics on an Adaptive Network. *Phys. Rev. Lett.* **2006**, *96*, 208701.
40. Tsallis, C. Possible Generalization of Boltzmann-Gibbs Statistics. *J. Stat. Phys.* **1988**, *52*, 479–487.
41. Quintana-Murci, L.; Alcais, A.; Abel, L.; Casanova, J.L. Immunology in natura: Clinical, Epidemiological and Evolutionary Genetics of Infectious Diseases. *Nat. Immunol.* **2007**, *8*, 1165–1171.
42. Hardy, G.; Littlewood, J.; Polya, G. *Inequalities*; Cambridge University Press: Cambridge, UK, 1967; Chapter 3.
43. Nielsen, F.; Nock, R. Sided and symmetrized Bregman centroids. *IEEE Trans. Inf. Theory* **2009**, *55*, 2882–2904.

entropy

MDPI

Article

The Entropy-Based Quantum Metric

Roger Balian

Institut de Physique Théorique, CEA/Saclay, F-91191 Gif-sur-Yvette Cedex, France;
E-Mail: roger@balian.fr

Received: 15 May 2014; in revised form: 25 June 2014 / Accepted: 11 July 2014 /
Published: 15 July 2014

Abstract: The von Neumann entropy $S(\hat{D})$ generates in the space of quantum density matrices $\hat{D}$ the Riemannian metric $\mathrm{d}s^2 = -\mathrm{d}^2 S(\hat{D})$, which is physically founded and which characterises the amount of quantum information lost by mixing $\hat{D}$ and $\hat{D} + \mathrm{d}\hat{D}$. A rich geometric structure is thereby implemented in quantum mechanics. It includes a canonical mapping between the spaces of states and of observables, which involves the Legendre transform of $S(\hat{D})$. The Kubo scalar product is recovered within the space of observables. Applications are given to equilibrium and non equilibrium quantum statistical mechanics. There the formalism is specialised to the relevant space of observables and to the associated reduced states issued from the maximum entropy criterion, which result from the exact states through an orthogonal projection. Von Neumann's entropy specialises into a relevant entropy. Comparison is made with other metrics. The Riemannian properties of the metric $\mathrm{d}s^2 = -\mathrm{d}^2 S(\hat{D})$ are derived. The curvature arises from the non-Abelian nature of quantum mechanics; its general expression and its explicit form for q-bits are given, as well as geodesics.

Keywords: quantum entropy; metric; q-bit; information; geometry; geodesics; relevant entropy

---

1. A Physical Metric for Quantum States

Quantum physical quantities pertaining to a given system, termed as "observables" $\hat{O}$, behave as non-commutative random variables and are elements of a C*-algebra. We will consider below systems for which these observables can be represented by n-dimensional Hermitean matrices in a finite-dimensional Hilbert space $\mathcal{H}$. In quantum (statistical) mechanics, the "state" of such a system encompasses the expectation values of all its observables [1]. It is represented by a density matrix $\hat{D}$, which plays the rôle of a probability distribution, and from which one can derive the expectation value of $\hat{O}$ in the form

$$< \hat{O} >= \mathrm{Tr}\, \hat{D}\hat{O} = (\hat{D}; \hat{O})\,. \tag{1}$$

Density matrices should be Hermitean ($< \hat{O} >$ is real for $\hat{O} = \hat{O}^{\dagger}$), normalised (the expectation value of the unit observable is $\mathrm{Tr}\, \hat{D} = 1$) and non-negative (variances $< \hat{O}^2 > - < \hat{O} >^2$ are non-negative). They depend on $n^2 - 1$ real parameters. If we keep aside the multiplicative structure of the set of operators and focus on their linear vector space structure, Equation (1) appears as a linear mapping of the space of observables onto real numbers. We can therefore regard the observables and the density operators $\hat{D}$ as elements of two dual vector spaces, and expectation values (1) appear as scalar products.

It is of interest to define a metric in the space of states. For instance, the distance between an exact state $\hat{D}$ and an approximation $\hat{D}_{\mathrm{app}}$ would then characterise the quality of this approximation. However, all physical quantities come out in the form (1) which lies astride the two dual spaces of observables and states. In order to build a metric having physical relevance, we need to rely on another meaningful quantity which pertains only to the space of states.

We note at this point that quantum states are probabilistic objects that gather information about the considered system. Then, the amount of missing information is measured by von Neumann's entropy

$$S(\hat{D}) \equiv -\operatorname{Tr} \hat{D} \ln \hat{D} \,. \tag{2}$$

Introduced in the context of quantum measurements, this quantity is identified with the thermodynamic entropy when $\hat{D}$ is an equilibrium state. In non-equilibrium statistical mechanics, it encompasses, in the form of "relevant entropy" (see Section 5 below), various entropies defined through the maximum entropy criterion. It is also introduced in quantum computation. Alternative entropies have been introduced in the literature, but they do not present all the distinctive and natural features of von Neumann's entropy, such as additivity and concavity.

As $S(\hat{D})$ is a concave function, and as it is the sole physically meaningful quantity apart from expectation values, it is natural to rely on it for our purpose. We thus define [2] the distance $\mathrm{d}s$ between two neighbouring density matrices $\hat{D}$ and $\hat{D} + \mathrm{d}\hat{D}$ as the square root of

$$\mathrm{d}s^2 = -\mathrm{d}^2 S(\hat{D}) = \operatorname{Tr} \mathrm{d}\hat{D} \mathrm{d} \ln \hat{D} \,. \tag{3}$$

This Riemannian metric is of the Hessian form since the metric tensor is generated by taking second derivatives of the function $S(\hat{D})$ with respect to the $n^2 - 1$ coordinates of $\hat{D}$. We may take for such coordinates the real and imaginary parts of the matrix elements, or equivalently (Section 6) some linear transform of these (keeping aside the norm $\operatorname{Tr} \hat{D} = 1$).

2. Interpretation in the Context of Quantum Information

The simplest example, related to quantum information theory, is that of a q-bit (two-level system or spin $\frac{1}{2}$) for which $n = 2$. Its states, represented by 2×2 Hermitean normalised density matrices $\hat{D}$, can conveniently be parameterised, on the basis of Pauli matrices, by the components $r_\mu = D_{12} + D_{21}$, $i(D_{12} - D_{21})$, $D_{11} - D_{22}$ ($\mu = 1, 2, 3$) of a 3-dimensional vector $\mathbf{r}$ lying within the unit Poincaré–Bloch sphere ($r \leq 1$). From the corresponding entropy

$$S = \frac{1+r}{2} \ln \frac{2}{1+r} + \frac{1-r}{2} \ln \frac{2}{1-r} \,, \tag{4}$$

we derive the metric

$$\mathrm{d}s^2 = \frac{1}{1-r^2} \left(\frac{\mathbf{r} \cdot \mathrm{d}\mathbf{r}}{r} \right)^2 + \frac{1}{2r} \ln \frac{1+r}{1-r} \left\| \frac{\mathbf{r} \times \mathrm{d}\mathbf{r}}{r} \right\|^2 \,, \tag{5}$$

which is a natural Riemannian metric for q-bits, or more generally for positive 2×2 matrices. The metric tensor characterizing (5) diverges in the vicinity of pure states $r = 1$, due to the singularity of the entropy (2) for vanishing eigenvalues of $\hat{D}$. However, the distance between two arbitrary (even pure) states $\hat{D}'$ and $\hat{D}''$ measured along a geodesic is always finite. We shall see (Equation (29)) that for $n = 2$ the geodesic distance s between two neighbouring pure states $\hat{D}'$ and $\hat{D}''$, represented by unit vectors $\mathbf{r}'$ and $\mathbf{r}''$ making a small angle $\delta\varphi \sim |\mathbf{r}' - \mathbf{r}''|$, behaves as $\delta s^2 \sim \delta\varphi^2 \ln(4\sqrt{\pi}/\delta\varphi)$. The singularity of the metric tensor manifests itself through this logarithmic factor.

Identifying von Neumann's entropy to a measure of missing information, we can give a simple interpretation to the distance between two states. Indeed, the concavity of entropy expresses that some information is lost when two statistical ensembles described by different density operators merge. By mixing two equal size populations described by the neighbouring distributions $\hat{D}' = \hat{D} + \frac{1}{2}\delta\hat{D}$ and $\hat{D}'' = \hat{D} - \frac{1}{2}\delta\hat{D}$ separated by a distance δs, we lose an amount of information given by

$$\mathbf{\Delta S} \equiv \mathbf{S}\left(\hat{\mathbf{D}}\right) - \frac{\mathbf{S}(\hat{\mathbf{D}}') + \mathbf{S}(\hat{\mathbf{D}}'')}{\mathbf{2}} \sim \frac{\mathbf{ffis}^{\mathbf{2}}}{\mathbf{8}} \,, \tag{6}$$

and thereby directly related to the distance δs defined by (3). The proof of this equivalence relies on the expansion of the entropies $S(\hat{D}')$ and $S(\hat{D}'')$ around $\hat{D}$, and is valid when $\mathrm{Tr}\,\delta\hat{D}^2$ is negligible compared to the smallest eigenvalue of $\hat{D}$. If $\hat{D}'$ and $\hat{D}''$ are distant, the quantity $8\Delta S$ cannot be regarded as the square of a distance that would be generated by a local metric. The equivalence (6) for neighbouring states shows that $\mathrm{d}s^2$ is the metric that is the best suited to measure losses of information my mixing.

The singularity of δs^2 at the edge of the positivity domain of $\hat{D}$ may suggest that the result (6) holds only within this domain. In fact, this equivalence remains nearly valid even in the limit of pure states because ΔS itself involves a similar singularity. Indeed, if the states $\hat{D}' = |\psi' >< \psi'|$ and $\hat{D}'' = |\psi'' >< \psi''|$ are pure and close to each other, the loss of information $\mathbf{\Delta S}$ behaves as $8\mathbf{\Delta S} \sim \mathbf{ffi'^2}\ln(\mathbf{4/ffi'})$ where $\delta\varphi^2 \sim 2\,\mathrm{Tr}\,\delta D^2$. This result should be compared to various geodesic distances between pure quantum states, which behave as $\delta s^2 \sim \delta\varphi^2 \ln(4\sqrt{\pi}/\delta\varphi$ for the present metric, and as $\delta s^2_{\mathrm{BH}} = 4\delta s^2_{\mathrm{FS}} \sim \delta\varphi^2 \sim \mathrm{Tr}(\hat{D}' - \hat{D}'')^2$ for the Bures – Helstrom and the quantum Fubini – Study metrics, respectively (see Section 7; these behaviours hold not only for $n = 2$ but for arbitrary n since only the space spanned by $|\psi' >$ and $|\psi'' >$ is involved). Thus, among these metrics, only $\mathrm{d}s^2 = -\mathrm{d}^2 S$ can be interpreted in terms of information loss, whether the states $\hat{D}'$ and $\hat{D}''$ are pure or mixed.

At the other extreme, around the most disordered state $\hat{D} = \hat{I}/n$, in the region $\| n\hat{D} - \hat{I} \| \ll 1$, the metric becomes Euclidean since $\mathrm{d}s^2 = \mathrm{Tr}\,\mathrm{d}\hat{D}\mathrm{d}\ln\hat{D} \sim n\,\mathrm{Tr}(\mathrm{d}\hat{D})^2$ (for $n = 2$, $\mathrm{d}s^2 = \mathrm{d}r^2$). For a given shift $\mathrm{d}\hat{D}$, the qualitative change of a state $\hat{D}$, as measured by the distance $\mathrm{d}s$, gets larger and larger as the state $\hat{D}$ becomes purer and purer, that is, when the information contents of $\hat{D}$ increases.

3. Geometry of Quantum Statistical Mechanics

A rich geometric structure is generated for both states and observables by von Neumann's entropy through introduction of the metric $\mathrm{d}s^2 = -\mathrm{d}^2 S$. Now, this metric (3) supplements the algebraic structure of the set of observables and the above duality between the vector spaces of states and of observables, with scalar product (1). Accordingly, we can define naturally within the space of states scalar products, geodesics, angles, curvatures.

We can also regard the coordinates of $\mathrm{d}\hat{D}$ and $\mathrm{d}\ln\hat{D}$ as covariant and contravariant components of the same infinitesimal vector (Section 6). To this aim, let us introduce the mapping

$$\hat{D} \equiv \frac{e^{\hat{X}}}{\mathrm{Tr}\,e^{\hat{X}}} \tag{7}$$

between $\hat{D}$ in the space of states and $\hat{X}$ in the space of observables. The operator $\hat{X}$ appears as a parameterisation of $\hat{D}$. (The normalisation of $\hat{D}$ entails that $\hat{X}$, defined within an arbitrary additive constant operator $X_0\,\hat{I}$, also depends on $n^2 - 1$ independent real parameters.) The metric (3) can then be re-expressed in terms of $\hat{X}$ in the form

$$\mathrm{d}s^2 = \mathrm{Tr}\,\mathrm{d}\hat{D}\mathrm{d}\hat{X} = \mathrm{Tr}\int_0^1 \mathrm{d}\xi \hat{D} e^{-\xi\hat{X}} \mathrm{d}\hat{X} e^{\xi\hat{X}} \mathrm{d}\hat{X} - (\mathrm{Tr}\,\hat{D}\mathrm{d}\hat{X})^2 = \mathrm{d}^2 \ln \mathrm{Tr}\,e^{\hat{X}} = \mathrm{d}^2 F\,, \tag{8}$$

where we introduced the function

$$F(\hat{X}) \equiv \ln \mathrm{Tr}\,e^{\hat{X}} \tag{9}$$

of the observable $\hat{X}$(The addition of $X_0\hat{I}$ to $\hat{X}$ results in the addition of the irrelevant constant X_0 to F). This mapping provides us with a natural metric in the space of observables, from which we recover the scalar product between $\mathrm{d}\hat{X}_1$ and $\mathrm{d}\hat{X}_2$ in the form of a Kubo correlation in the state $\hat{D}$. The metric (8) has been quoted in the literature under the names of Bogoliubov–Kubo–Mori.

4. Covariance and Legendre Transformation

We can recover the above geometric mapping (7) between $\hat{D}$ and $\hat{X}$, or between the covariant and contravariant coordinates of $\mathrm{d}\hat{D}$, as the outcome of a Legendre transformation, by considering the function $F(\hat{X})$. Taking its differential $\mathrm{d}F = \mathrm{Tr}\, e^{\hat{X}}\mathrm{d}\hat{X} / \mathrm{Tr}\, e^{\hat{X}}$, we identify the partial derivatives of $F(\hat{X})$ with the coordinates of the state $\hat{D} = e^{\hat{X}} / \mathrm{Tr}\, e^{\hat{X}}$, so that $\hat{D}$ appears as conjugate to $\hat{X}$ in the sense of Legendre transformations. Expressing then $\hat{X}$ as function of $\hat{D}$ and inserting into $F - \mathrm{Tr}\, \hat{D}\hat{X}$, we recognise that the Legendre transform of $F(\hat{X})$ is von Neumann's entropy $F - \mathrm{Tr}\, \hat{D}\hat{X} = S(\hat{D}) = -\mathrm{Tr}\, \hat{D} \ln \hat{D}$. The conjugation between $\hat{D}$ and $\hat{X}$ is embedded in the equations

$$\mathrm{d}F = \mathrm{Tr}\, \hat{D}\mathrm{d}\hat{X}\,; \quad \mathrm{d}S = -\,\mathrm{Tr}\, \hat{X}\mathrm{d}\hat{D}\,. \tag{10}$$

Legendre transformations are currently used in equilibrium thermodynamics. Let us show that they come out in this context directly as a special case of the present general formalism. The entropy of thermodynamics is a function of the extensive variables, energy, volume, particle numbers, etc. Let us focus for illustration on the energy U, keeping the other extensive variables fixed. The thermodynamic entropy $S(U)$, a function of the single variable U, generates the inverse temperature as $\beta = \partial S / \partial U$. Its Legendre transform is the Massieu potential $F(\beta) = S - \beta U$. In order to compare these properties with the present formalism, we recall how thermodynamics comes out in the framework of statistical mechanics. The thermodynamic entropy $S(U)$ is identified with the von Neumann entropy (2) of the Boltzmann–Gibbs canonical equilibrium state $\hat{D}$, and the internal energy with $U = \mathrm{Tr}\, \hat{D}\hat{H}$. In the relation (7), the operator $\hat{X}$ reads $\hat{X} = -\beta\hat{H}$ (within an irrelevant additive constant). By letting U or β vary, we select within the spaces of states and of observables a one-dimensional subset. In these restricted subsets, $\hat{D}$ is parameterised by the single coordinate U, and the corresponding $\hat{X}$ by the coordinate $-\beta$.
By specialising the general relations (10) to these subsets, we recover the thermodynamic relations $\mathrm{d}F = -U\mathrm{d}\beta$ and $\mathrm{d}S = \beta\mathrm{d}U$. We also recover, by restricting the metric (3) or (8) to these subsets, the current thermodynamic metric $\mathrm{d}s^2 = -(\partial^2 S / \partial U^2)\mathrm{d}U^2 = -\mathrm{d}U\mathrm{d}\beta$.

More generally, we can consider the Boltzmann–Gibbs states of equilibrium statistical mechanics as the points of a manifold embedded in the full space of states. The thermodynamic extensive variables, which parameterise these states, are the expectation values of the conserved macroscopic observables, that is, they are a subset of the expectation values (1) which parameterise arbitrary density operators. Then the standard geometric structure of thermodynamics simply results from the restriction of the general metric (3) to this manifold of Boltzmann–Gibbs states. The commutation of the conserved observables simplifies the reduced thermodynamic metric, which presents the same features as a Fisher metric (see Section 6).

5. Relevant Entropy and Geometry of the Projection Method

The above ideas also extend to non-equilibrium quantum statistical mechanics [2–4]. When introducing the metric (3), we indicated that it may be used to estimate the quality of an approximation. Let us illustrate this point with the Nakajima–Zwanzig–Mori–Robertson projection method, best introduced through maximum entropy. Consider some set $\{\hat{A}_k\}$ of "relevant observables", whose time-dependent expectation values $a_k \equiv < \hat{A}_k > = \mathrm{Tr}\, \hat{D}\hat{A}_k$ we wish to follow, discarding all other variables. The exact state $\hat{D}$ encodes the variables $\{a_k\}$ that we are interested in, but also the expectation values (1) of the other observables that we wish to eliminate. This elimination is performed by associating at each time with $\hat{D}$ a "reduced state" $\hat{D}_\mathrm{R}$ which is equivalent to $\hat{D}$ as regards the set $a_k = \mathrm{Tr}\, \hat{D}_\mathrm{R}\hat{A}_k$, but which provides no more information than the values$\{a_k\}$. The former condition provides the constraints $< \hat{A}_k > = a_k$, and the latter condition is implemented by means of the maximum entropy criterion: One expresses that, within the set of density matrices compatible with these constraints, $\hat{D}_\mathrm{R}$ is the one which maximises von Neumann's entropy (2), that is, which contains solely the information about the relevant variables a_k. The least biased state $\hat{D}_\mathrm{R}$ thus defined has the form $\hat{D}_\mathrm{R} = e^{\hat{X}_\mathrm{R}} / \mathrm{Tr}\, e^{\hat{X}_\mathrm{R}}$, where $\hat{X}_\mathrm{R} \equiv \sum_k \lambda_k \hat{A}_k$ involves the time-dependent Lagrange multipliers λ_k, which are related to the set a_k through $\mathrm{Tr}\, \hat{D}_\mathrm{R}\hat{A}_k = a_k$.

The von Neumann entropy $S(\hat{D}_{\rm R}) \equiv S_{\rm R}\{a_k\}$ of this reduced state $\hat{D}_{\rm R}$ is called the "relevant entropy" associated with the considered relevant observables $\hat{A}_k$. It measures the amount of missing information, when only the values $\{a_k\}$ of the relevant variables are given. During its evolution, $\hat{D}$ keeps track of the initial information about all the variables $< \hat{O} >$ and its entropy $S(\hat{D})$ remains constant in time. It is therefore smaller than the relevant entropy $S(\hat{D}_{\rm R})$ which accounts for the loss of information about the irrelevant variables. Depending on the choice of relevant observables $\{\hat{A}_k\}$, the corresponding relevant entropies $S_{\rm R}\{a_k\}$ encompass various current entropies, such as the non-equilibrium thermodynamic entropy or Boltzmann's H-entropy.

The same structure as the one introduced above for the full spaces of observables and states is recovered in this context. Here, for arbitrary values of the parameters λ_k, the exponents $\hat{X}_{\rm R} = \sum_k \lambda_k \hat{A}_k$ constitute a subspace of the full vector space of observables, and the parameters $\{\lambda_k\}$ appear as the coordinates of $\hat{X}_{\rm R}$ on the basis $\{\hat{A}_k\}$. The corresponding states $\hat{D}_{\rm R}$, parameterised by the set $\{a_k\}$, constitute a subset of the space of states, the manifold $\mathcal{R}$ of "reduced states"(Note that this manifold is not a hyperplane, contrary to the space of relevant observables; it is embedded in the full vector space of states, but does not constitute a subspace). By regarding $S_{\rm R}\{a_k\}$ as a function of the coordinates $\{a_k\}$, we can define a metric $\mathrm{d}s^2 = -\mathrm{d}^2 S_{\rm R}\{a_k\}$ on the manifold $\mathcal{R}$, which is the restriction of the metric (3). Its alternative expression $\mathrm{d}s^2 = \sum_k \mathrm{d}a_k \mathrm{d}\lambda_k = \mathrm{d}^2 F_{\rm R}\{\lambda_k\}$, where $F_{\rm R}\{\lambda_k\} \equiv \ln \mathrm{Tr} \exp \sum_k \lambda_k \hat{A}_k$, is a restriction of (8). The correspondence between the two parameterisations $\{a_k\}$ and $\{\lambda_k\}$ is again implemented by the Legendre transformation which relates $S_{\rm R}\{a_k\}$ and $F_{\rm R}\{\lambda_k\}$.

The projection method relies on the mapping $\hat{D} \mapsto \hat{D}_{\rm R}$ which associates $\hat{D}_{\rm R}$ to $\hat{D}$. It consists in replacing the Liouville–von Neumann equation of motion for $\hat{D}$ by the corresponding dynamical equation for $\hat{D}_{\rm R}$ on the manifold $\mathcal{R}$, or equivalently for the coordinates $\{a_k\}$ or for the coordinates $\{\lambda_k\}$, a programme that is in practice achieved through some approximations. This mapping is obviously a projection in the sense that $\hat{D} \mapsto \hat{D}_{\rm R} \mapsto \hat{D}_{\rm R}$, but moreover the introduction of the metric (3) shows that the vector $\hat{D} - \hat{D}_{\rm R}$ in the space of states is perpendicular to the manifold $\mathcal{R}$ at the point $\hat{D}_{\rm R}$. This property is readily shown by writing, in this metric, the scalar product $\mathrm{Tr}\, \mathrm{d}\hat{D}\, \mathrm{d}\hat{X}'$ of the vector $\mathrm{d}\hat{D} = \hat{D} - \hat{D}_{\rm R}$ by an arbitrary vector $\mathrm{d}\hat{D}'$ in the tangent plane of $\mathcal{R}$. The latter is conjugate to any combination $\mathrm{d}\hat{X}'$ of observables $\hat{A}_k$, and this scalar product vanishes because $\mathrm{Tr}\, \hat{D}\hat{A}_k = \mathrm{Tr}\, \hat{D}_{\rm R}\hat{A}_k$. Thus the mapping $\hat{D} \mapsto \hat{D}_{\rm R}$ appears as an orthogonal projection, so that the relevant state $\hat{D}_{\rm R}$ associated with $\hat{D}$ may be regarded as its best possible approximation on the manifold $\mathcal{R}$.

6. Properties of the Metric

The metric tensor can be evaluated explicitly in a basis where the matrix $\hat{D}$ is diagonal. Denoting by D_i its eigenvalues and by $\mathrm{d}D_{ij}$ the matrix elements of its variations, we obtain from (3)

$$\mathrm{d}s^2 = \mathrm{Tr} \int_0^\infty \mathrm{d}\xi \left(\frac{\mathrm{d}\hat{D}}{\hat{D}+\xi} \right)^2 = \sum_{ij} \frac{\ln D_i - \ln D_j}{D_i - D_j} \mathrm{d}D_{ij} \mathrm{d}D_{ji} \, . \tag{11}$$

(For $D_i = D_j$,whether or not $i = j$, the ratio is defined as $1/D_i$ by continuity.) In the same basis, the form (8) of the metric reads

$$\mathrm{d}s^2 = \frac{1}{Z} \sum_{ij} \frac{e^{X_i} - e^{X_j}}{X_i - X_j} \mathrm{d}X_{ij} \mathrm{d}X_{ji} - \left(\frac{\sum_i e^{X_i} \mathrm{d}X_{ii}}{Z} \right)^2 , \tag{12}$$

with $Z = \sum_i e^{X_i}$(For $X_i = X_j$, the ratio is e^{X_i}). The singularity of the metric (11) in the vicinity of vanishing eigenvalues of $\hat{D}$, in particular near pure states (end of Section 2), is not apparent in the representation (12) of this metric, because the mapping from $\hat{D}$ to $\hat{X}$ sends the eignevalue X_i to $-\infty$ when D_i tends to zero.

Let us compare the expression (11) with the corresponding classical metric, which is obtained by starting from Shannon's entropy instead of von Neumann's entropy. For discrete probabilities p_i, we have then $S\{p_i\} = -\sum_i p_i \ln p_i$ and hence the same definition $\mathrm{d}s^2 = -\mathrm{d}^2 S\{p_i\}$ as above of an entropy-based metric yields $\mathrm{d}s^2 = \sum_i \mathrm{d}p_i^2/p_i$, which is identified with the Fisher information metric. The present metric thus appears as the extension to quantum statistical mechanics of the Fisher metric when the latter is interpreted in terms of entropy. In fact, the terms of (11) which involve the diagonal elements $i = j$ of the variations $\mathrm{d}\hat{D}$ reduce to $\mathrm{d}D_{ii}^2/D_i$. This result was expected since density matrices behave as probability distributions if both $\hat{D}$ and $\mathrm{d}\hat{D}$ are diagonal.

Let us more generally consider in (11), instead of solely diagonal variations $\mathrm{d}D_{ii}$, variations $\mathrm{d}D_{ij}$ with indices i and j such that $|D_i - D_j| \ll D_i + D_j$. The expansion of D_i and D_j around $\frac{1}{2}(D_i + D_j)$ in the corresponding ratios of (11) yields $(\ln D_i - \ln D_j)/(D_i - D_j) \sim 2/(D_i + D_j)$. The considered terms of (11) are therefore the same as in the Bures–Helstrom metric

$$\mathrm{d}s_{\mathrm{BH}}^2 = \sum_{ij} \frac{2}{D_i + D_j} \mathrm{d}D_{ij} \mathrm{d}D_{ji} , \tag{13}$$

introduced long ago as an extension to matrices of the Fisher metric [5]. We thus recover this Bures–Helstrom metric as an approximation of the present entropy-based metric $\mathrm{d}s^2 = -\mathrm{d}^2 S(\hat{D})$. For $n = 2$, $\mathrm{d}s_{\mathrm{BH}}^2$ is obtained from the expression (5) of $\mathrm{d}s^2$ by omitting the factor $\tanh^{-1} r/r$ entering the second term.

In order to express the properties of the Riemannian metric (3) in a general form, which will exhibit the tensor structure, we use a Liouville representation. There, the observables $\hat{O} = O_\mu \hat{\Omega}^\mu$, regarded as elements of a vector space, are represented by their coordinates O_μ on a complete basis $\hat{\Omega}^\mu$ of n^2 observables. The space of states is spanned by the dual basis $\hat{\Sigma}_\mu$, such that $\mathrm{Tr}\, \hat{\Omega}^\nu \hat{\Sigma}_\mu = \delta_\mu^\nu$, and the states $\hat{D} = D^\mu \hat{\Sigma}_\mu$ are represented by their coordinates D_μ. Thus, the expectation value (1) is the scalar product $D^\mu O_\mu$. In the matrix representation which appears as a special case, μ denotes a pair of indices i, j, $\hat{\Omega}^\mu$ stands for $|\, j >< i\,|$, $\hat{\Sigma}_\mu$ for $|\, i >< j\,|$, O_μ denotes the matrix element O_{ji} and D^μ the element D_{ij}. For the q-bit ($n = 2$) considered in Section 2, we have chosen the Pauli operators $\hat{\sigma}^\mu$ as basis $\hat{\Omega}^\mu$ for observables, and $\frac{1}{2}\hat{\sigma}_\mu$ as dual basis $\hat{\Sigma}_\mu$ for states, so that the coordinates $D^\mu = \mathrm{Tr}\, \hat{D}\hat{\Omega}^\mu$ of $\hat{D} = \frac{1}{2}(\hat{I} + r^\mu \hat{\sigma}_\mu)$ are the components r^μ of the vector $\mathbf{r}$ (The unit operator $\hat{I}$ is kept aside since $\hat{D}$ is normalised and since constants added to $\hat{X}$ are irrelevant). The function $F\{X\} = \ln \mathrm{Tr}\, e^{\hat{X}}$ of the coordinates X_μ of the observable $\hat{X}$, and the von Neumann entropy $S\{D\}$ as function of the coordinates D^μ of the state $\hat{D}$, are related by the Legendre transformation $F = S + D^\mu X_\mu$, and the relations (10) are expressed by $D^\mu = \partial F/\partial X_\mu$, $X_\mu = -\partial S/\partial D^\mu$. The metric tensor is given by

$$g^{\mu\nu} = \frac{\partial^2 F}{\partial X_\mu \partial X_\nu}, \quad g_{\mu\nu} = -\frac{\partial^2 S}{\partial D^\mu \partial D^\nu} , \tag{14}$$

and the correspondence issued from (7) between covariant and contravariant infinitesimal variations of $\hat{X}$ and $\hat{D}$ is implemented as $\mathrm{d}D^\mu = g^{\mu\nu}\mathrm{d}X_\nu$, $\mathrm{d}X_\mu = g_{\mu\nu}\mathrm{d}D^\nu$.

These expressions exhibit the Hessian nature of the metric. This property simplifies the expression of the Christoffel symbol, which reduces to

$$\Gamma_{\mu\nu\rho} = -\frac{1}{2} \frac{\partial^3 S}{\partial D^\mu \partial D^\nu \partial D^\rho} , \tag{15}$$

and which provides a parametric representation $\hat{D}(t)$ of the geodesics in the space of states through

$$\frac{\mathrm{d}^2 D^\mu}{\mathrm{d}t^2} + g^{\mu\sigma}\Gamma_{\sigma\nu\rho}\frac{\mathrm{d}D^\nu}{\mathrm{d}t}\frac{\mathrm{d}D^\rho}{\mathrm{d}t} = 0 . \tag{16}$$

Then, the Riemann curvature tensor comes out as

$$R_{\mu\rho\ \nu\sigma} = g^{\xi\zeta}(\Gamma_{\mu\sigma\xi}\Gamma_{\nu\rho\zeta} - \Gamma_{\mu\nu\xi}\Gamma_{\rho\sigma\zeta}) , \tag{17}$$

the Ricci tensor and the scalar curvature as

$$R_{\mu\nu} = g^{\rho\sigma} R_{\mu\rho\,\nu\sigma}, \quad R = g^{\mu\nu} R_{\mu\nu}\,, \tag{18}$$

We have noted that the classical equivalent of the entropy-based metric $\mathrm{d}s^2 = -\mathrm{d}^2 S$ is the Fisher metric $\sum_i \mathrm{d}p_i^2/p_i$, which as regards the curvature is equivalent to a Euclidean metric. While the space of classical probabilities is thus flat, the above equations show that the space of quantum states is curved. This curvature arises from the non-commutation of the observables, it vanishes for the completely disordered state $\hat{D} = \hat{I}/n$. Curvature can thus be used as a measure of the degree of classicality of a state.

7. Geometry of the Space of q-Bits

In the illustrative example of a q-bit, the operator $\hat{X} = \chi_\mu \hat{\sigma}^\mu$ associated with $\hat{D}$ is parameterised by the 3 components of the vector χ_μ ($\mu = 1, 2, 3$), related to $\mathbf{r}$ by $\chi = \tanh^{-1} r$ and $\chi_\mu/\chi = r^\mu/r$. The metric tensor given by (5) is expressed as

$$\begin{aligned} g_{\mu\nu} &= K r_\mu r_\nu + \frac{\chi}{r}\delta_{\mu\nu}\,, \quad K \equiv \frac{1}{r}\frac{\mathrm{d}}{\mathrm{d}r}\frac{\chi}{r} = \frac{1}{r^2}\left(\frac{1}{1-r^2} - \frac{\chi}{r}\right), \\ g^{\mu\nu} &= (1-r^2)p^{\mu\nu} + \frac{r}{\chi}q^{\mu\nu}\,. \end{aligned} \tag{19}$$

(We have defined $r_\mu = r^\mu$, $\delta_{\mu\nu} = \delta^\mu_{\ \nu} = \delta^{\mu\nu}$ so as to introduce the projectors $r^\mu r^\nu/r^2 \equiv p^{\mu\nu} \equiv \delta^{\mu\nu} - q^{\mu\nu}$ in the Euclidean 3-dimensional space, and thus to simplify the subsequent calculations.) In polar coordinates $\mathbf{r} = (r, \theta, \varphi)$, the infinitesimal distance takes the form

$$\mathrm{d}s^2 = \mathrm{d}r\mathrm{d}\chi + r\chi(\mathrm{d}\theta^2 + \sin^2\theta\mathrm{d}\varphi^2)\,. \tag{20}$$

We determine from (15) and (19) the explicit form

$$\Gamma_{\mu\nu\rho} = \frac{K}{2}\left(r_\mu\delta_{\nu\rho} + r_\nu\delta_{\mu\rho} + r_\rho\delta_{\mu\nu}\right) + \frac{1}{2r}\frac{\mathrm{d}K}{\mathrm{d}r} r_\mu r_\nu r_\rho \tag{21}$$

of the Christoffel symbol. By raising its first index with $g^{\mu\nu}$ and using polar coordinates, we obtain from (16) the equations of geodesics for $n = 2$. Within the Poincaré–Bloch sphere the geodesics are deduced by rotations from a one-parameter family of curves which lie in the $\theta = \frac{1}{2}\pi$, $|\varphi| \leq \frac{1}{2}\pi$ half-plane and which are symmetric with respect to the $\varphi = 0$ axis. This family is characterized by the equations (where $\chi = \tanh^{-1} r$):

$$\frac{\mathrm{d}^2 r}{\mathrm{d}t^2} + \frac{r}{1-r^2}\left(\frac{\mathrm{d}r}{\mathrm{d}t}\right)^2 - \frac{r}{2}\left[1 + \frac{\chi}{r}\left(1-r^2\right)\right]\left(\frac{\mathrm{d}\varphi}{\mathrm{d}t}\right)^2 = 0\,, \tag{22}$$

$$\frac{\mathrm{d}^2\varphi}{\mathrm{d}t^2} + \frac{1}{r}\frac{\mathrm{d}r}{\mathrm{d}t}\frac{\mathrm{d}\varphi}{\mathrm{d}t} + \frac{1}{\chi}\frac{\mathrm{d}\chi}{\mathrm{d}t}\frac{\mathrm{d}\varphi}{\mathrm{d}t} = 0\,, \tag{23}$$

and the boundary conditions at $t = 0$:

$$r(0) = a\,, \quad \varphi(0) = 0\,, \quad \frac{\mathrm{d}r(0)}{\mathrm{d}t} = 0\,, \quad \frac{\mathrm{d}\varphi(0)}{\mathrm{d}t} = \frac{1}{k}\,, \quad k^2 = a\tanh^{-1} a\,. \tag{24}$$

Equation (23) provides, using the boundary conditions (24):

$$\frac{\mathrm{d}\varphi}{\mathrm{d}t} = \frac{k}{r\chi}\,. \tag{25}$$

Insertion of (25) into (22) gives rise to an equation for $r(t)$, which can be integrated by regarding t as a function of $\zeta = \arcsin r$. One obtains:

$$\left(\frac{\mathrm{d}r}{\mathrm{d}t}\right)^2 = \left(1 - r^2\right)\left(1 - \frac{k^2}{r\chi}\right) . \tag{26}$$

The scale of t has been fixed by relating to $r(0)$ the boundary condition (24) for $\mathrm{d}\varphi(0)/\mathrm{d}t$, a choice which ensures that $\mathrm{d}s^2 = \mathrm{d}r\mathrm{d}\chi + r\chi\mathrm{d}\varphi^2 = \mathrm{d}t^2$, and hence that the parameter t measures the distance along geodesics.

For $k = 0$, we obtain $r = |\sin t|$, $\varphi = \pm\pi/2$. Thus, the longest geodesics are the diameters of the Poincaré–Bloch sphere. We find the value π for their "length", that is, for the geodesic distance between two orthogonal pure states. At the other extreme, when the middle point $r = a$, $\varphi = 0$ of a geodesic lies close to the surface $r = 1$ of the sphere, the asymptotic form of the equation (26) is solved as

$$t = \pm 2k\sqrt{\pi}\mathrm{e}^{-k^2}\operatorname{erf}\xi\,, \quad \xi = \sqrt{\frac{1}{2}\ln\frac{1-a}{1-r}}\,, \quad k^2 = \frac{1}{2}\ln\frac{2}{1-a} \tag{27}$$

(by taking ξ as variable instead of r). The determination of the explicit equations of such short geodesic curves is achieved by integrating (25) into

$$\varphi = \frac{t}{k} = \pm 2\sqrt{\pi}\mathrm{e}^{-k^2}\operatorname{erf}\xi\,. \tag{28}$$

From (27) and (28) we can determine the geodesic distance between two neighbouring pure states $\hat{D}' = |\psi'\rangle\langle\psi'|$ and $\hat{D}'' = |\psi''\rangle\langle\psi''|$ represented by the points $r_{\max} = 1$, $\varphi_{\max} = \pm\frac{1}{2}\delta\varphi$ with $\delta\varphi$ small. At these two points, we have $\xi \to \infty$, $\operatorname{erf}\xi = 1$, and this determines k in terms of $\frac{1}{2}\delta\varphi$ through (28). The length of the geodesic that joins them, given by (27), is:

$$\delta s^2 = \delta\varphi^2 \ln\frac{4\sqrt{\pi}}{\delta\varphi}\,, \quad \delta\varphi = \arccos|\langle\psi' \mid \psi''\rangle|\,. \tag{29}$$

Thus, in spite of its singularity for $r = 1$, the present 3-dimensional metric (5) in the space r, θ, φ defines distances between pure states represented by points on the surface $r = 1$ of the Poincaré–Bloch sphere. However, It should be noted that the presence of the logarithmic factor in (29) forbids such distances to be generated by a 2-dimensional metric in the space θ, φ. In fact, the distance (29) is measured along a geodesic that penetrates the sphere $r = 1$, because no geodesic is tangent to the surface of this sphere nor lies on its surface.

In contrast, all geodesics produced by the Bures–Helstrom metric are tangent to the surface of the sphere, or are its great circles. They are given by Equations (25) and (26), where χ is replaced by r and k by a; the solution of these equations provides the ellipses

$$r\cos\varphi = a\cos t\,, \quad r\sin\varphi = \sin t\,. \tag{30}$$

Here as above, the largest distance π is reached for orthogonal pure states represented by opposite points on the sphere, but now a peculiarity occurs. Whereas the metric $\mathrm{d}s^2 = -\mathrm{d}^2S$ produces a single geodesic, the diameter joining these two points (with "length" π), the Bures metric produces a double infinity of geodesics, the half-ellipses (30) having as long axis this diameter, and having all the same "length" π. Other pairs of pure states are joined by geodesics which are arcs of great circles, and their Bures distance $\delta s_{\mathrm{BH}} = \delta\varphi$ is identified with the ordinary length of the arc. Here for $n = 2$ as in the general case, the 3-dimensional Bures–Helstrom metric admits a restriction to pure states generated by a 2-dimensional metric, which is identified with the quantum Fubini–Study metric, itself defined only for pure states by $s_{\mathrm{FS}} = \arccos|\langle\psi' \mid \psi''\rangle| = \frac{1}{2}s_{\mathrm{BH}}$.

Returning to the metric $ds^2 = d^2S$, the Riemann curvature is obtained from (17) as

$$R^{\mu}{}_{\rho\ \nu\sigma} = \frac{K}{4}\left[(r^2 + \frac{r}{\chi} - 1)(q^{\mu}{}_{\sigma}q_{\nu\rho} - q^{\mu}{}_{\nu}q_{\rho\sigma}) + (r^2 - \frac{r}{\chi} + 1)(p^{\mu}{}_{\sigma}q_{\nu\rho} - p^{\mu}{}_{\nu}q_{\rho\sigma}) \right. \tag{31}$$

$$\left. + \frac{r}{\chi}\frac{1}{1-r^2}(r^2 - \frac{r}{\chi} + 1)(q^{\mu}{}_{\sigma}p_{\nu\rho} - q^{\mu}{}_{\nu}p_{\rho\sigma})\right] .$$

Contracting with $g^{\rho\sigma}$ the indices of (30) as in (18), we finally derive the Ricci curvature

$$R^{\mu}{}_{\nu} = -\frac{Kr}{2\chi}\left(r^2\delta^{\mu}{}_{\nu} + \frac{\chi - r}{\chi}p^{\mu}{}_{\nu}\right) , \tag{32}$$

and the scalar curvature

$$R = -\frac{Kr}{2\chi}\left(3r^2 + \frac{\chi - r}{\chi}\right) . \tag{33}$$

Both are negative in the whole Poincaré sphere. In the limit $r \to 0$, the curvature R vanishes as $R \sim -\frac{10}{9}r^2$, as expected from the general argument of Section 2: a weakly polarised spin behaves classically. At the other extreme $r \to 1$, R behaves as $R \sim -2\left[(1-r) \mid \ln(1-r) \mid\right]^{-1}$; it diverges, again as expected: pure states have the largest quantum nature.

The metric $ds^2 = -d^2S$, introduced above in the context of quantum mechanics for mixed states (and their pure limit) and information theory, might more generally be useful to characterise distances in spaces of positive matrices.

Conflicts of Interest: Conflicts of Interest

The author declares no conflict of interest.

References

1. Thirring, W. Quantum Mechanics of Large Systems. In *A Course of Mathematical Physics*; Volume 4; Springler-Verlag: New York, NY, USA, 1983.
2. Balian, R.; Alhassid, Y.; Reinhardt, H. Dissipation in many-body systems: A geometric approach based on information theory. *Phys. Rep.* **1986**, *131*, 1–146.
3. Balian, R. Incomplete descriptions and relevant entropies. *Am. J. Phys.* **1999**, *67*, 1078–1090.
4. Balian, R. Information in statistical physics. *Stud. Hist. Philos. Mod. Phys.* **2005**, *36*, 323–353.
5. Bures, D. An extension of Kakutani's theorem. *Trans. Am. Math. Soc.* **1969**,*135*, 199–212.

entropy

MDPI

Article

Extending the Extreme Physical Information to Universal Cognitive Models via a Confident Information First Principle

Xiaozhao Zhao [1], Yuexian Hou [1,2,*], Dawei Song [1,3] and Wenjie Li [2]

[1] School of Computer Science and Technology, Tianjin University, Tianjin 300072, China; E-Mails: 0.25eye@gmail.com (X.Z.); dawei.song2010@gmail.com (D.S.)

[2] Department of Computing, The Hong Kong Polytechnic University, Hung Hom, Kowloon, Hong Kong, China; E-Mail: cswjli@comp.polyu.edu.hk

[3] Department of Computing and Communications, The Open University, Milton Keynes MK76AA, UK

* E-Mail: yxhou@tju.edu.cn; Tel.: +86-022-27406538.

Received: 25 March 2014; in revised form: 6 June 2014 / Accepted: 20 June 2014 / Published: 1 July 2014

Abstract: The principle of extreme physical information (EPI) can be used to derive many known laws and distributions in theoretical physics by extremizing the physical information loss K, *i.e.*, the difference between the observed Fisher information I and the intrinsic information bound J of the physical phenomenon being measured. However, for complex cognitive systems of high dimensionality (e.g., human language processing and image recognition), the information bound J could be excessively larger than I ($J \gg I$), due to insufficient observation, which would lead to serious over-fitting problems in the derivation of cognitive models. Moreover, there is a lack of an established exact invariance principle that gives rise to the bound information in universal cognitive systems. This limits the direct application of EPI. To narrow down the gap between I and J, in this paper, we propose a confident-information-first (CIF) principle to lower the information bound J by preserving confident parameters and ruling out unreliable or noisy parameters in the probability density function being measured. The confidence of each parameter can be assessed by its contribution to the expected Fisher information distance between the physical phenomenon and its observations. In addition, given a specific parametric representation, this contribution can often be directly assessed by the Fisher information, which establishes a connection with the inverse variance of any unbiased estimate for the parameter via the Cramér–Rao bound. We then consider the dimensionality reduction in the parameter spaces of binary multivariate distributions. We show that the single-layer Boltzmann machine without hidden units (SBM) can be derived using the CIF principle. An illustrative experiment is conducted to show how the CIF principle improves the density estimation performance.

Keywords: information geometry; Boltzmann machine; Fisher information; parametric reduction

1. Introduction

Information has been found to play an increasingly important role in physics. As stated in Wheeler [1]: "All things physical are information-theoretic in origin and this is a participatory universe...Observer participancy gives rise to information; and information gives rise to physics". Following this viewpoint, Frieden [2] unifies the derivation of physical laws in major fields of physics, from the Dirac equation to the Maxwell-Boltzmann velocity dispersion law, using the extreme physical information principle (EPI). More specifically, a variety of equations and distributions can be derived by extremizing the physical information loss K, *i.e.*, the difference between the observed Fisher information I and the intrinsic information bound J of the physical phenomenon being measured.

The first quantity, I, measures the amount of information as a finite scalar implied by the data with some suitable measure [2]. It is formally defined as the trace of the Fisher information matrix [3]. In addition to I, the second quantity, the information bound J, is an invariant that characterizes the information that is intrinsic to the physical phenomenon [2]. During the measurement procedure, there may be some loss of information, which entails $I = \kappa J$, where $\kappa \leq 1$ is called the efficiency coefficient of the EPI process in transferring the Fisher information from the phenomenon (specified by J) to the output (specified by I). For closed physical systems, in particular, any solution for I attains some fraction of J between 1/2 (for classical physics) and one (for quantum physics) [4].

However, it is usually not the case in cognitive science. For complex cognitive systems (e.g., human language processing and image recognition), the target probability density function (pdf) being measured is often of high dimensionality (e.g., thousands of words in a human language vocabulary and millions of pixels in an observed image). Thus, it is infeasible for us to obtain a sufficient collection of observations, leading to excessive information loss between the observer and nature. Moreover, there is a lack of an established exact invariance principle that gives rise to the bound information in universal cognitive systems. This limits the direct application of EPI in cognitive systems.

In terms of statistics and machine learning, the excessive information loss between the observer and nature will lead to serious over-fitting problems, since the insufficient observations may not provide necessary information to reasonably identify the model and support the estimation of the target pdf in complex cognitive systems. Actually, a similar problem is also recognized in statistics and machine learning, known as the model selection problem [5]. In general, we would require a complex model with a high-dimensional parameter space to sufficiently depict the original high-dimensional observations. However, over-fitting usually occurs when the model is excessively complex with respect to the given observations. To avoid over-fitting, we would need to adjust the complexity of the models to the available amount of observations and, equivalently, to adjust the information bound J corresponding to the observed information I.

In order to derive feasible computational models for cognitive phenomenon, we propose a confident-information-first (CIF) principle in addition to EPI to narrow down the gap between I and J (thus, a reasonable efficiency coefficient κ is implied), as illustrated in Figure 1. However, we do not intend to actually derive the distribution laws by solving the differential equations of the extremization of the new information loss K'. Instead, we assume that the target distribution belongs to some general multivariate binary distribution family and focus on the problem of seeking a proper information bound with respect to the constraint of the parametric number and the given observations.

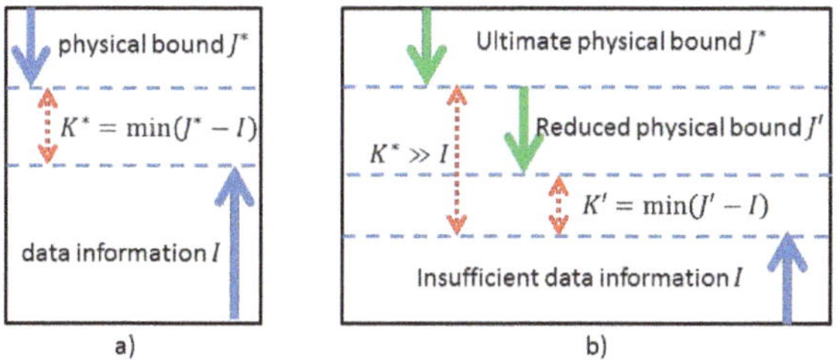

Figure 1. (**a**) The paradigm of the extreme physical information principle (EPI) to derive physical laws by the extremization of the information loss K^* ($K^* = J/2$ for classical physics and $K^* = 0$ for quantum physics); (**b**) the paradigm of confident-information-first (CIF) to derive computational models by reducing the information loss K' using a new physical bound J'.

The key to the CIF approach is how to systematically reduce the physical information bound for high-dimensional complex systems. As stated in Frieden [2], the information bound J is a functional form that depends upon the physical parameters of the system. The information is contained in the variations of the observations (often imperfect, due to insufficient sampling, noise and intrinsic limitations of the "observer"), and can be further quantified using the Fisher information of system parameters (or coordinates) [3] from the estimation theory. Therefore, the physical information bound J of a complex system can be reduced by transforming it to a simpler system using some parametric reduction approach. Assuming there exists an ideal parametric model S that is general enough to represent all system phenomena (which gives the ultimate information bound in Figure 1), our goal is to adopt a parametric reduction procedure to derive a lower-dimensional sub-model M (which gives the reduced information bound in Figure 1) for a given dataset (usually insufficient or perturbed by noises) by reducing the number of free parameters in S.

Formally speaking, let $q(\xi)$ be the ideal distribution with parameters ξ that describes the physical system and $q(\xi + \Delta\xi)$ be the observations of the system with some small fluctuation $\Delta\xi$ in parameters. In [6], the averaged information distance $I(\Delta\xi)$ between the distribution and its observations, the so-called shift information, is used as a disorder measure of the fluctuated observations to reinterpret the EPI principle. More specifically, in the framework of information geometry, this information distance could also be assessed using the Fisher information distance induced by the Fisher–Rao metric, which can be decomposed into the variation in the direction of each system parameter [7]. In principle, it is possible to divide system parameters into two categories, *i.e.*, the parameters with notable variations and the parameters with negligible variations, according to their contributions to the whole information distance. Additionally, the parameters with notable contributions are considered to be confident, since they are important for reliably distinguishing the ideal distribution from its observation distributions. On the other hand, the parameters with negligible contributions can be considered to be unreliable or noisy. Then, the CIF principle can be stated as the parameter selection criterion that maximally preserves the Fisher information distance in an expected sense with respect to the constraint of the parametric number and the given observations (if available), when projecting distributions from the parameter space of S into that of the reduced sub-model M. We call it the distance-based CIF. As a result, we could manipulate the information bound of the underlying system by preserving the information of confident parameters and ruling out noisy parameters.

In this paper, the CIF principle is analyzed in the multivariate binary distribution family in the mixed-coordinate system [8]. It turns out that, in this problematic configuration, the confidence of a parameter can be directly evaluated by its Fisher information, which also establishes a connection with the inverse variance of any unbiased estimate for the parameter via the Cramér–Rao bound [3]. Hence, the CIF principle can also be interpreted as the parameter selection procedure that keeps the parameters with reliable estimates and rules out unreliable or noisy parameters. This CIF is called the information-based CIF. Note that the definition of confidence in distance-based CIF depends on both Fisher information and the scale of fluctuation, and the confidence in the information-based CIF (*i.e.*, Fisher information) can be seen as a special case of confidence measure with respect to certain coordinate systems. This simplification allows us to further apply the CIF principle to improve existing learning algorithms for the Boltzmann machine.

The paper is organized as follows. In Section 2, we introduce the parametric formulation for the general multivariate binary distributions in terms of information geometry (IG) framework [7]. Then, Section 3 describes the implementation details of the CIF principle. We also give a geometric interpretation of CIF by showing that it can maximally preserve the expected information distance (in Section 3.2.1), as well as the analysis on the scale of the information distance in each individual system parameter (in Section 3.2.2). In Section 4, we demonstrate that a widely used cognitive model, *i.e.*, the Boltzmann machine, can be derived using the CIF principle. Additionally, an illustrative experiment is conducted to show how the CIF principle can be utilized to improve the density estimation performance of the Boltzmann machine in Section 5.

2. The Multivariate Binary Distributions

Similar to EPI, the derivation of CIF depends on the analysis of the physical information bound, where the choice of system parameters, also called "Fisher coordinates" in Frieden [2], is crucial. Based on information geometry (IG) [7], we introduce some choices of parameterizations for binary multivariate distributions (denoted as statistical manifold S) with a given number of variables n, *i.e.*, the open simplex of all probability distributions over binary vector $x \in \{0,1\}^n$.

2.1. Notations for Manifold S

In IG, a family of probability distributions is considered as a differentiable manifold with certain parametric coordinate systems. In the case of binary multivariate distributions, four basic coordinate systems are often used: p-coordinates, η-coordinates, θ-coordinates and mixed-coordinates [7,9]. Mixed-coordinates is of vital importance for our analysis.

For the p-coordinates $[p]$ with n binary variables, the probability distribution over 2^n states of x can be completely specified by any $2^n - 1$ positive numbers indicating the probability of the corresponding exclusive states on n binary variables. For example, the p-coordinates of $n = 2$ variables could be $[p] = (p_{01}, p_{10}, p_{11})$. Note that IG requires all probability terms to be positive [7].

For simplicity, we use the capital letters $I, J, \ldots$ to index the coordinate parameters of probabilistic distribution. To distinguish the notation of Fisher information (conventionally used in literature, e.g., data information I and information bound J in Section 1) from the coordinate indexes, we make explicit explanations when necessary from now on. An index I can be regarded as a subset of $\{1,2,\ldots,n\}$. Additionally, p_I stands for the probability that all variables indicated by I equal to one and the complemented variables are zero. For example, if $I = \{1,2,4\}$ and $n = 4$, then $p_I = p_{1101} = Prob(x_1 = 1, x_2 = 1, x_3 = 0, x_4 = 1)$. Note that the null set can also be a legal index of the p-coordinates, which indicates the probability that all variables are zero, denoted as $p_{0\ldots0}$.

Another coordinate system often used in IG is η-coordinates, which is defined by:

$$\eta_I = E[X_I] = Prob\{\prod_{i\in I} x_i = 1\} \tag{1}$$

where the value of X_I is given by $\prod_{i\in I} x_i$ and the expectation is taken with respect to the probability distribution over x. Grouping the coordinates by their orders, the η-coordinate system is denoted as $[\eta] = (\eta_i^1, \eta_{ij}^2, \ldots, \eta_{1,2\ldots n}^n)$, where the superscript indicates the order number of the corresponding parameter. For example, η_{ij}^2 denotes the set of all η parameters with the order number two.

The θ-coordinates (natural coordinates) are defined by:

$$\log p(x) = \sum_{I\subseteq\{1,2,\ldots,n\}, I\neq NullSet} \theta^I X_I - \psi(\theta) \tag{2}$$

where $\psi(\theta) = \log(\sum_x exp\{\sum_I \theta^I X_I(x)\})$ is the cumulant generating function and its value equals to $-\log Prob\{x_i = 0, \forall i \in \{1,2,...,n\}\}$. The θ-coordinate is denoted as $[\theta] = (\theta_1^i, \theta_2^{ij}, \ldots, \theta_n^{1,\ldots,n})$, where the subscript indicates the order number of the corresponding parameter. Note that the order indices locate at different positions in $[\eta]$ and $[\theta]$ following the convention in Amari *et al.* [8].

The relation between coordinate systems $[\eta]$ and $[\theta]$ is bijective. More formally, they are connected by the Legendre transformation:

$$\theta^I = \frac{\partial\phi(\eta)}{\partial\eta_I}, \eta_I = \frac{\partial\psi(\theta)}{\partial\theta^I} \tag{3}$$

where $\psi(\theta)$ is given in Equation (2) and $\phi(\eta) = \sum_x p(x;\eta)\log p(x;\eta)$ is the negative of entropy. It can be shown that $\psi(\theta)$ and $\phi(\eta)$ meet the following identity [7]:

$$\psi(\theta) + \phi(\eta) - \sum \theta^I \eta_I = 0 \tag{4}$$

Next, we introduce mixed-coordinates, which is important for our derivation of CIF. In general, the manifold S of probability distributions could be represented by the l-mixed-coordinates [8]:

$$[\zeta]_l = (\eta_i^1, \eta_{ij}^2, \ldots, \eta_{i,j,\ldots,k}^l, \theta_{l+1}^{i,j,\ldots,k}, \ldots, \theta_n^{1,\ldots,n}) \tag{5}$$

where the first part consists of η-coordinates with order less or equal to l (denoted by $[\eta^{l-}]$) and the second part consists of θ-coordinates with order greater than l (denoted by $[\theta_{l+}]$), $l \in \{1, ..., n-1\}$.

2.2. Fisher Information Matrix for Parametric Coordinates

For a general coordinate system $[\xi]$, the i-th row and j-th column element of the Fisher information matrix for $[\xi]$ (denoted by G_ξ) is defined as the covariance of the scores of $[\xi_i]$ and $[\xi_j]$ [3], *i.e.*,

$$g_{ij} = E[\frac{\partial \log p(x;\xi)}{\partial \xi_i} \cdot \frac{\partial \log p(x;\xi)}{\partial \xi_j}]$$

under the regularity condition for the pdf that the partial derivatives exist. The Fisher information measures the amount of information in the data that a statistic carries about the unknown parameters [10]. The Fisher information matrix is of vital importance to our analysis, because the inverse of Fisher information matrix gives an asymptotically tight lower bound to the covariance matrix of any unbiased estimate for the considered parameters [3]. Another important concept related to our analysis is the orthogonality defined by Fisher information. Two coordinate parameters ξ_i and ξ_j are called orthogonal if and only if their Fisher information vanishes, *i.e.*, $g_{ij} = 0$, meaning that their influences on the log likelihood function are uncorrelated.

The Fisher information for $[\theta]$ can be rewritten as $g_{IJ} = \frac{\partial^2 \psi(\theta)}{\partial \theta^I \partial \theta^J}$, and for $[\eta]$, it is $g^{IJ} = \frac{\partial^2 \phi(\eta)}{\partial \eta_I \partial \eta_J}$ [7]. Let $G_\theta = (g_{IJ})$ and $G_\eta = (g^{IJ})$ be the Fisher information matrices for $[\theta]$ and $[\eta]$, respectively. It can be shown that G_θ and G_η are mutually inverse matrices, *i.e.*, $\sum_J g^{IJ} g_{JK} = \delta_K^I$, where $\delta_K^I = 1$ if $I = K$ and zero otherwise [7]. In order to generally compute G_θ and G_η, we develop the following Propositions 1 and 2. Note that Proposition 1 is a generalization of Theorem 2 in Amari *et al.* [8].

Proposition 1. *The Fisher information between two parameters θ^I and θ^J in $[\theta]$, is given by:*

$$g_{IJ}(\theta) = \eta_{I \cup J} - \eta_I \eta_J \tag{6}$$

Proof. in Appendix A. □

Proposition 2. *The Fisher information between two parameters η_I and η_J in $[\eta]$, is given by:*

$$g^{IJ}(\eta) = \sum_{K \subseteq I \cap J} (-1)^{|I-K|+|J-K|} \cdot \frac{1}{p_K} \tag{7}$$

where $|\cdot|$ denotes the cardinality operator.

Proof. in Appendix B. □

Based on the Fisher information matrices G_η and G_θ, we can calculate the Fisher information matrix G_ζ for the l-mixed-coordinate system $[\zeta]_l$, as follows:

Proposition 3. *The Fisher information matrix G_ζ of the l-mixed-coordinates $[\zeta]_l$ is given by:*

$$G_\zeta = \begin{pmatrix} A & 0 \\ 0 & B \end{pmatrix} \tag{8}$$

where $A = ((G_\eta^{-1})_{I_\eta})^{-1}$, $B = ((G_\theta^{-1})_{J_\theta})^{-1}$, G_η *and* G_θ *are the Fisher information matrices of* $[\eta]$ *and* $[\theta]$, *respectively,* I_η *is the index set of the parameters shared by* $[\eta]$ *and* $[\zeta]_l$, *i.e.,* $\{\eta_i^1, ..., \eta_{i,j,...,k}^l\}$, *and* J_θ *is the index set of the parameters shared by* $[\theta]$ *and* $[\zeta]_l$, *i.e.,* $\{\theta_{l+1}^{i,j,...,k}, \ldots, \theta_n^{1,...,n}\}$.

Proof. in Appendix C. □

3. The General CIF Principle

In this section, we propose the CIF principle to reduce the physical information bound for high-dimensionality systems. Given a target distribution $q(x) \in S$, we consider the problem of realizing it by a lower-dimensionality submanifold. This is defined as the problem of parametric reduction for multivariate binary distributions. The family of multivariate binary distributions has been proven to be useful when we deal with discrete data in a variety of applications in statistical machine learning and artificial intelligence, such as the Boltzmann machine in neural networks [11,12] and the Rasch model in human sciences [13,14].

Intuitively, if we can construct a coordinate system so that the confidences of its parameters entail a natural hierarchy, in which high confident parameters are significantly distinguished from and orthogonal to lowly confident ones, then we can conveniently implement CIF by keeping the high confident parameters unchanged and setting the lowly confident parameters to neutral values. Therefore, the choice of coordinates (or parametric representations) in CIF is crucial to its usage. This strategy is infeasible in terms of p-coordinates, η-coordinates or θ-coordinates, since the orthogonality condition cannot hold in these coordinate systems. In this section, we will show that the l-mixed-coordinates $[\zeta]_l$ meets the requirement of CIF.

In principle, the confidence of parameters should be assessed according to their contributions to the expected information distance between the ideal distribution and its fluctuated observations. This is called the distance-based CIF (see Section 1). For some coordinated systems, e.g., the mixed-coordinate system $[\zeta]_l$, the confidence of a parameter can also be directly evaluated by its Fisher information. This is called the information-based CIF (see Section 1). The information-based CIF (*i.e.*, Fisher information) can be seen as an approximation to distance-based CIF, since it neglects the influence of parameter scaling to the expected information distance. However, considering the standard mixed-coordinates $[\zeta]_l$ for the manifold of multivariate binary distributions, it turns out that both distance-based CIF and information-based CIF entail the same submanifold M (refer to Section 3.2 for detailed reasons).

For the purpose of legibility, we will start with the information-based CIF, where the parameter's confidence is simply measured using its Fisher information. After that, we show that the information-based CIF leads to an optimal submanifold M, which is also optimal in terms of the more rigorous distance-based CIF.

3.1. The Information-Based CIF Principle

In this section, we will show that the l-mixed-coordinates $[\zeta]_l$ meet the requirement of the information-based CIF. According to Proposition 3 and the following Proposition 4, the confidences of coordinate parameters (measured by Fisher information) in $[\zeta]_l$ entail a natural hierarchy: the first part of high confident parameters $[\eta^{l-}]$ are separated from the second part of low confident parameters $[\theta_{l+}]$. Additionally, those low confident parameters $[\theta_{l+}]$ have the neutral value of zero.

Proposition 4. *The diagonal elements of A are lower bounded by one, and those of B are upper bounded by one.*

Proof. in Appendix D. □

Moreover, the parameters in $[\eta^{l-}]$ are orthogonal to the ones in $[\theta_{l+}]$, indicating that we could estimate these two parts independently [9]. Hence, we can implement the information-based CIF for parametric reduction in $[\zeta]_l$ by replacing low confident parameters with neutral value zero and reconstructing the resulting distribution. It turns out that the submanifold of S tailored by information-based CIF becomes $[\zeta]_{l_t} = (\eta_i^1, ..., \eta_{ij...k}^l, 0, \ldots, 0)$. We call $[\zeta]_{l_t}$ the l-tailored-mixed-coordinates.

To grasp an intuitive picture for the CIF strategy and its significance w.r.t mixed-coordinates, let us consider an example with $[p] = (p_{001} = 0.15, p_{010} = 0.1, p_{011} = 0.05, p_{100} = 0.2, p_{101} = 0.1, p_{110} = 0.05, p_{111} = 0.3)$. Then, the confidences for coordinates in $[\eta]$, $[\theta]$ and $[\zeta]_2$ are given by the diagonal elements of the corresponding Fisher information matrices. Applying the two-tailored CIF in mixed-coordinates, the loss ratio of Fisher information is 0.001%, and the ratio of the Fisher information of the tailored parameter (θ_3^{123}) to the remaining η parameter with the smallest Fisher information is 0.06%. On the other hand, the above two ratios become 7.58% and 94.45% (in η-coordinates) or 12.94% and 92.31% (in θ-coordinates), respectively. We can see that $[\zeta]_2$ gives us a much better way to tell apart confident parameters from noisy ones.

3.2. The Distance-Based CIF: A Geometric Point-of-View

In the previous section, the information-based CIF entails a submanifold of S determined by the l-tailored-mixed-coordinates $[\zeta]_{l_t}$. A more rigorous definition for the confidence of coordinates is the distance-based confidence used in the distance-based CIF, which relies on both of the coordinate's Fisher information and its fluctuation scaling. In this section, we will show that the the submanifold M determined by $[\zeta]_{l_t}$ is also an optimal submanifold M in terms of the distance-based CIF. Note that, for other coordinate systems (e.g., arbitrarily rescaling coordinates), the information-based CIF may not entail the same submanifold as the distance-based CIF.

Let $q(x)$, with coordinate ζ_q, denote the exact solution to the physical phenomenon being measured. Additionally, the act of observation would cause small random perturbations to $q(x)$, leading to some observation $q'(x)$ with coordinate $\zeta_q + \Delta\zeta_q$. When two distributions $q(x)$ and $q'(x)$ are close, the divergence between $q(x)$ and $q'(x)$ on manifold S could be assessed by the Fisher information distance: $D(q, q') = (\Delta\zeta_q \cdot G_\zeta \cdot \Delta\zeta_q)^{1/2}$, where G_ζ is the Fisher information matrix and the perturbation $\Delta\zeta_q$ is small. The Fisher information distance between two close distributions $q(x)$ and $q'(x)$ on manifold S is the Riemannian distance under the Fisher–Rao metric, which is shown to be the square root of the twice of the Kullback–Leibler divergence from $q(x)$ to $q'(x)$ [8]. Note that we adopt the Fisher information distance as the distance measure between two close distributions, since it is shown to be the unique metric meeting a set of natural axioms for the distribution metrics [7,15,16], e.g., the invariant property with respect to reparametrizations and the monotonicity with respect to the random maps on variables.

Let M be a smooth k-dimensionality submanifold in S ($k < 2^n - 1$). Given the point $q(x) \in S$, the projection [8] of $q(x)$ on M is the point $p(x)$ that belongs to M and is closest to $q(x)$ with respect to the Kullback–Leibler divergence (K-L divergence) [17] from the distribution $q(x)$ to $p(x)$. On the submanifold M, the projections of $q(x)$ and $q'(x)$ are $p(x)$ and $p'(x)$, with coordinates ζ_p and $\zeta_p + \Delta\zeta_p$, respectively, shown in Figure 2.

Let the preserved Fisher information distance be $D(p, p')$ after projecting on M. In order to retain the information contained in observations, we need the ratio $\frac{D(p,p')}{D(q,q')}$ to be as large as possible in the expected sense, with respect to the given dimensionality k of M. The next two sections will illustrate that CIF leads to an optimal submanifold M based on different assumptions on the perturbations $\Delta\zeta_q$.

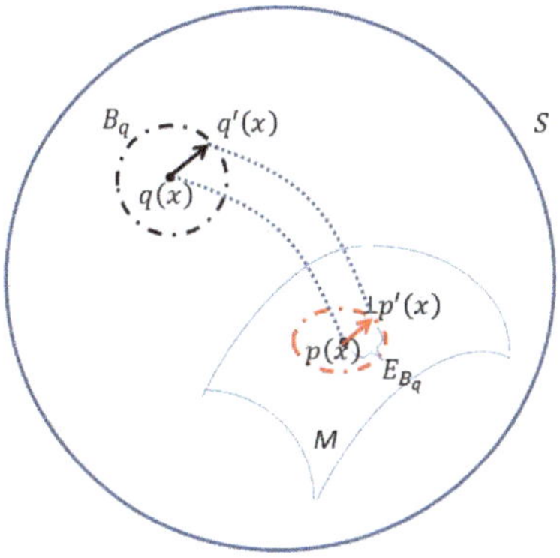

Figure 2. By projecting a point $q(x)$ on S to a submanifold M, the l-tailored mixed-coordinates $[\zeta]_{l_t}$ gives a desirable M that maximally preserves the expected Fisher information distance when projecting a ε-neighborhood centered at $q(x)$ onto M.

3.2.1. Perturbations in Uniform Neighborhood

Let B_q be a ε-sphere surface centered at $q(x)$ on manifold S, *i.e.*, $B_q = \{q' \in S|||KL(q, q') = \varepsilon\}$, where $KL(\cdot, \cdot)$ denotes the K-L divergence and ε is small. Additionally, $q'(x)$ is a neighbor of $q(x)$ uniformly sampled on B_q, as illustrated in Figure 2. Recall that, for a small ε, the K-L divergence can be approximated by half of the squared Fisher information distance. Thus, in the parameterization of $[\zeta]_l$, B_q is indeed the surface of a hyper-ellipsoid (centered at $q(x)$) determined by G_ζ. The following proposition shows that the general CIF would lead to an optimal submanifold M that maximally preserves the expected information distance, where the expectation is taken upon the uniform neighborhood, B_q.

Proposition 5. *Consider the manifold S in l-mixed-coordinates $[\zeta]_l$. Let k be the number of free parameters in the l-tailored-mixed-coordinates $[\zeta]_{l_t}$. Then, among all k-dimensional submanifolds of S, the submanifold determined by $[\zeta]_{l_t}$ can maximally preserve the expected information distance induced by the Fisher–Rao metric.*

Proof. in Appendix E. □

3.2.2. Perturbations in Typical Distributions

To facilitate our analysis, we make a basic assumption on the underlying distributions $q(x)$ that at least $(2^n - 2^{n/2})$ p-coordinates are of the scale ϵ, where ϵ is a sufficiently small value. Thus, residual p-coordinates (at most $2^{n/2}$) are all significantly larger than zero (of scale $\Theta(1/2^{(n/2)})$), and their sum approximates one. Note that these assumptions are common situations in real-world data collections [18], since the frequent (or meaningful) patterns are only a small fraction of all of the system states.

Next, we introduce a small perturbation Δp to the p-coordinates $[p]$ for the ideal distribution $q(x)$. The scale of each fluctuation Δp_I is assumed to be proportional to the standard variation of corresponding p-coordinate p_I by some small coefficients (upper bounded by a constant a), which can be approximated by the inverse of the square root of its Fisher information via the Cramér–Rao bound. It turns out that we can assume the perturbation Δp_I to be $a\sqrt{p_I}$.

In this section, we adopt the l-mixed-coordinates $[\zeta]_l = (\eta^{l-}; \theta_{l+})$, where $l = 2$ is used in the following analysis. Let $\Delta\zeta_q = (\Delta\eta^{2-}; \Delta\theta_{2+})$ be the incremental of mixed-coordinates after the perturbation. The squared Fisher information distance $D^2(p, p') = \Delta\zeta_q \cdot G_\zeta \cdot \Delta\zeta_q$ could be decomposed into the direction of each coordinate in $[\zeta]_l$. We will clarify that, under typical cases, the scale of the

Fisher information distance in each coordinate of θ_{l+} (reduced by CIF) is asymptotically negligible, compared to that in each coordinate of η^{l-} (preserved by CIF).

The scale of squared Fisher information distance in the direction of η_I is proportional to $\Delta\eta_I \cdot (G_\zeta)_{I,I} \cdot \Delta\eta_I$, where $(G_\zeta)_{I,I}$ is the Fisher information of η_I in terms of the mixed-coordinates $[\zeta]_2$. From Equation (1), for any I of order one (or two), η_I is the sum of 2^{n-1} (or 2^{n-2}) p-coordinates, and the scale is $\Theta(1)$. Hence, the incremental $\Delta\eta^{2-}$ is proportional to $\Theta(1)$, denoted as $a \cdot \Theta(1)$. It is difficult to give an explicit expression of $(G_\zeta)_{I,I}$ analytically. However, the Fisher information $(G_\zeta)_{I,I}$ of η_I is bounded by the (I, I)-th element of the inverse covariance matrix [19], which is exactly $1/g^{I,I}(\theta) = \frac{1}{\eta_I - \eta_I^2}$ (see Proposition 3). Hence, the scale of $(G_\zeta)_{I,I}$ is also $\Theta(1)$. It turns out that the scale of squared Fisher information distance in the direction of η_I is $a^2 \cdot \Theta(1)$.

Similarly, for the part θ_{2+}, the scale of squared Fisher information distance in the direction of θ^J is proportional to $\Delta\theta^J \cdot (G_\zeta)_{J,J} \cdot \Delta\theta^J$, where $(G_\zeta)_{J,J}$ is the Fisher information of θ^J in terms of the mixed-coordinates $[\zeta]_2$. The scale of θ^J is maximally $f(k)|log(\sqrt{\epsilon})|$ based on Equation (2), where k is the order of θ^J and $f(k)$ is the number of p-coordinates of scale $\Theta(1/2^{(n/2)})$ that are involved in the calculation of θ^J. Since we assume that $f(k) \leq 2^{(n/2)}$, the maximum scale of θ^J is $2^{(n/2)}|log(\sqrt{\epsilon})|$. Thus, the incremental $\Delta\theta^J$ is of a scale bounded by $a \cdot 2^{(n/2)}|log(\sqrt{\epsilon})|$. Similar to our previous deviation, the Fisher information $(G_\zeta)_{J,J}$ of θ^J is bounded by the (J, J)-th element of the inverse covariance matrix, which is exactly $1/g_{J,J}(\eta)$ (see Proposition 3). Hence, the scale of $(G_\zeta)_{J,J}$ is $(2^k - f(k))^{-1}\epsilon$. In summary, the scale of squared Fisher information distance in the direction of θ^J is bounded by the scale of $a^2 \cdot \Theta(2^n \epsilon \frac{|log(\sqrt{\epsilon})|^2}{2^k - f(k)})$. Since ϵ is a sufficiently small value and a is constant, the scale of squared Fisher information distance in the direction of θ^J is asymptotically zero.

In summary, in terms of modeling the fluctuated observations of typical cognitive systems, the original Fisher information distance between the physical phenomenon ($q(x)$) and observations ($q'(x)$) is systematically reduced using CIF by projecting them on an optimal submanifold M. Based on our above analysis, the scale of Fisher information distance in the directions of $[\eta^{l-}]$ preserved by CIF is significantly larger than that of the directions $[\theta_{l+}]$ reduced by CIF.

4. Derivation of Boltzmann Machine by CIF

In the previous section, the CIF principle is uncovered in the $[\zeta]_l$ coordinates. Now, we consider an implementation of CIF when l equals to two, which gives rise to the single-layer Boltzmann machine without hidden units (SBM).

4.1. Notations for SBM

The energy function for SBM is given by:

$$E_{SBM}(x;\xi) = -\frac{1}{2}x^T U x - b^T x \tag{9}$$

where $\xi = \{U, b\}$ are the parameters and the diagonals of U are set to zero. The Boltzmann distribution over x is $p(x;\xi) = \frac{1}{Z}exp\{-E_{SBM}(x;\xi)\}$, where Z is a normalization factor. Actually, the parametrization for SBM could be naturally expressed by the coordinate systems in IG (e.g., $[\theta] = (\theta_1^i = b_i, \theta_2^{ij} = U_{ij}, \theta_3^{ijk} = 0, ..., \theta_n^{1,2,...,n} = 0)$).

4.2. The Derivation of SBM using CIF

Given any underlying probability distribution $q(x)$ on the general manifold S over $\{x\}$, the logarithm of $q(x)$ can be represented by a linear decomposition of θ-coordinates, as shown in Equation (2). Since it is impractical to recognize all coordinates for the target distribution, we would like to only approximate part of them and end up with a k-dimensional submanifold M of S, where k ($\ll 2^n - 1$) is the number of free parameters. Here, we set k to be the same dimensionality as SBM, *i.e.*, $k = \frac{n(n+1)}{2}$, so that all candidate submanifolds are comparable to the submanifold endowed by

SBM (denoted as M_{sbm}). Next, the rationale underlying the design of M_{sbm} can be illustrated using the general CIF.

Let the two-mixed-coordinates of $q(x)$ on S be $[\zeta]_2 = (\eta_i^1, \eta_{ij}^2, \theta_3^{i,j,k}, \ldots, \theta_n^{1,\ldots,n})$. Applying the general CIF on $[\zeta]_2$, our parametric reduction rule is to preserve the high confident part parameters $[\eta^{2-}]$ and replace low confident parameters $[\theta_{2+}]$ by a fixed neutral value of zero. Thus, we derive the two-tailored-mixed-coordinates: $[\zeta]_{2_t} = (\eta_i^1, \eta_{ij}^2, 0, \ldots, 0)$, as the optimal approximation of $q(x)$ by the k-dimensional submanifolds. On the other hand, given the two-mixed-coordinates of $q(x)$, the projection $p(x) \in M_{sbm}$ of $q(x)$ is proven to be $[\zeta]_p = (\eta_i^1, \eta_{ij}^2, 0, \ldots, 0)$ [8]. Thus, SBM defines a probabilistic parameter space that is derived from CIF.

4.3. The Learning Algorithms for SBM

Let $q(x)$ be the underlying probability distribution from which samples $D = \{d_1, d_2, \ldots, d_N\}$ are generated independently. Then, our goal is to train an SBM (with stationary probability $p(x)$) based on D that realizes $q(x)$ as faithfully as possible. Here, we briefly introduce two typical learning algorithms for SBM: maximum-likelihood and contrastive divergence [11,20,21].

Maximum-likelihood (ML) learning realizes a gradient ascent of log-likelihood of D:

$$\Delta U_{ij} = \varepsilon \frac{\partial l(\xi; D)}{\partial U_{ij}} = \varepsilon(E_q[x_i x_j] - E_p[x_i x_j]) \tag{10}$$

where ε is the learning rate and $l(\xi; D) = \frac{1}{N}\sum_{n=1}^{N} \log(d_n; \xi)$. $E_q[\cdot]$ and $E_p[\cdot]$ are expectations over $q(x)$ and $p(x)$, respectively. Actually, $E_q[x_i x_j]$ and $E_p[x_i x_j]$ are the coordinates η_{ij}^2 of $q(x)$ and $p(x)$, respectively. $E_q[x_i x_j]$ could be unbiasedly estimated from the sample. Markov chain Monte Carlo [22] is often used to approximate $E_p[x_i x_j]$ with an average over samples from $p(x)$.

Contrastive divergence (CD) learning realizes the gradient descent of a different objective function to avoid the difficulty of computing the log-likelihood gradient, shown as follows:

$$\Delta U_{ij} = -\varepsilon \frac{\partial(KL(q_0||p) - KL(p_m||p))}{\partial U_{ij}} = \varepsilon(E_{q_0}[x_i x_j] - E_{p_m}[x_i x_j]) \tag{11}$$

where q_0 is the sample distribution, p_m is the distribution by starting the Markov chain with the data and running m steps and $KL(\cdot||\cdot)$ denotes the K-L divergence. Taking samples in D as initial states, we could generate a set of samples for $p_m(x)$. Those samples can be used to estimate $E_{p_m}[x_i x_j]$.

From the perspective of IG, we can see that ML/CD learning is to update parameters in SBM, so that its corresponding coordinates $[\eta^{2-}]$ are getting closer to the data (along with the decreasing gradient). This is consistent with our theoretical analysis in Section 3 and Section 4.2 that SBM uses the most confident information (*i.e.*, $[\eta^{2-}]$) for approximating an arbitrary distribution in an expected sense.

5. Experimental Study: Incorporate Data into CIF

In the information-based CIF, the actual values of the data were not used to explicitly effect the output PDF (e.g., the derivation of SBM in Section 4). The data constrains the state of knowledge about the unknown pdf. In order to force the estimate of our probabilistic model to obey the data, we need to further reduce the difference between data information and physical information bound. How can this be done?

In this section, the CIF principle will also be used to modify existing SBM training algorithm (*i.e.*, CD-1) by incorporating data information. Given a particular dataset, the CIF can be used to further recognize less-confident parameters in SBM and to reduce them properly. Our solution here is to apply CIF to take effect on the learning trajectory with respect to specific samples and, hence, further confine the parameter space to the region indicated by the most confident information contained in the samples.

5.1. A Sample-Specific CIF-Based CD Learning for SBM

The main modification of our CIF-based CD algorithm (CD-CIF for short) is that we generate the samples for $p_m(x)$ based on those parameters with confident information, where the confident information carried by certain parameter is inherited from the sample and could be assessed using its Fisher information computed in terms of the sample.

For CD-1 (*i.e.*, m=1), the firing probability for the i-th neuron after a one-step transition from the initial state $x^{(0)} = \{x_1^{(0)}, x_2^{(0)}, \ldots, x_n^{(0)}\})$ is:

$$p(x_i^{(1)} = 1|x^{(0)}) = \frac{1}{1 + exp\{-\sum_{j\neq i} U_{ij}x_j^{(0)} - b_i\}} \tag{12}$$

For CD-CIF, the firing probability for the i-th neuron in Equation (12) is modified as follows:

$$p(x_i^{(1)}=1|x^{(0)})=\frac{1}{1+exp\{-\sum_{(j\neq i)\&(F(U_{ij})>\tau)} U_{ij}x_j^{(0)}-b_i\}} \tag{13}$$

where τ is a pre-selected threshold, $F(U_{ij}) = E_{q_0}[x_ix_j] - E_{q_0}[x_ix_j]^2$ is the Fisher information of U_{ij} (see Equation (6)) and the expectations are estimated from the given sample D. We can see that those weights whose Fisher information are less than τ are considered to be unreliable w.r.t D. In practice, we could setup τ by the ratio r to specify the proportion of the total Fisher information T_{FI} of all parameters that we would like to remain, i.e., $\sum_{U_{ij}>\tau, i<j} F(U_{ij}) = r * T_{FI}$.

In summary, CD-CIF is realized in two phases. In the first phase, we initially "guess" whether certain parameter could be faithfully estimated based on the finite sample. In the second phase, we approximate the gradient using the CD scheme, except for when the CIF-based firing function in Equation (13) is used.

5.2. Experimental Results

In this section, we empirically investigate our justifications for the CIF principle, especially how the sample-specific CIF-based CD learning (see Section 5) works in the context of density estimation.

Experimental Setup and Evaluation Metric: We utilize the random distribution uniformly generated from the open probability simplex over 10 variables as underlying distributions, whose samples size N may vary. Three learning algorithms are investigated: ML, CD-1 and our CD-CIF. K-L divergence is used to evaluate the goodness-of-fit of the SBM's trained by various algorithms. For sample size N, we run 100 instances (20 (randomly generated distributions) $\times$ 5 (randomly running)) and report the averaged K-L divergences. Note that we focus on the case that the variable number is relatively small ($n = 10$) in order to analytically evaluate the K-L divergence and give a detailed study on algorithms. Changing the number of variables only offers a trivial influence for the experimental results, since we obtained qualitatively similar observations on various variable numbers (not reported here).

Automatically Adjusting r for Different Sample Sizes: The Fisher information is additive for i.i.d. sampling. When sample sizes change, it is natural to require that the total amount of Fisher information contained in all tailored parameters is steady. Hence, we have $\alpha = (1-r)N$, where α indicates the amount of Fisher information and becomes a constant when the learning model and the underlying distribution family are given. It turns out that we can first identify α using the optimal r w.r.t several distributions generated from the underlying distribution family and then determine the optimal r's for various sample sizes using $r = 1 - \alpha/N$. In our experiments, we set $\alpha = 35$.

Density Estimation Performance: The averaged K-L divergences between SBMs (learned by ML, CD-1 and CD-CIF with the r automatically determined) and the underlying distribution are shown in Figure 3a. In the case of relatively small samples ($N \leq 500$) in Figure 3a, our CD-CIF method shows significant improvements over ML (from 10.3% to 16.0%) and CD-1 (from 11.0% to 21.0%). This is because we could not expect to have reliable identifications for all model parameters from insufficient samples, and hence, CD-CIF gains its advantages by using parameters that could be confidently estimated. This result is consistent with our previous theoretical insight that Fisher information gives a reasonable guidance for parametric reduction via the confidence criterion. As the sample size increases ($N \geq 600$), CD-CIF, ML and CD-1 tend to have similar performances, since, with relatively large samples, most model parameters can be reasonably estimated, hence the effect of parameter reduction using CIF gradually becomes marginal. In Figure 3b and Figure 3c, we show how sample size affects the interval of r. For $N = 100$, CD-CIF achieves significantly better performances for a wide range of r. While, for $N = 1,200$, CD-CIF can only marginally outperform baselines for a narrow range of r.

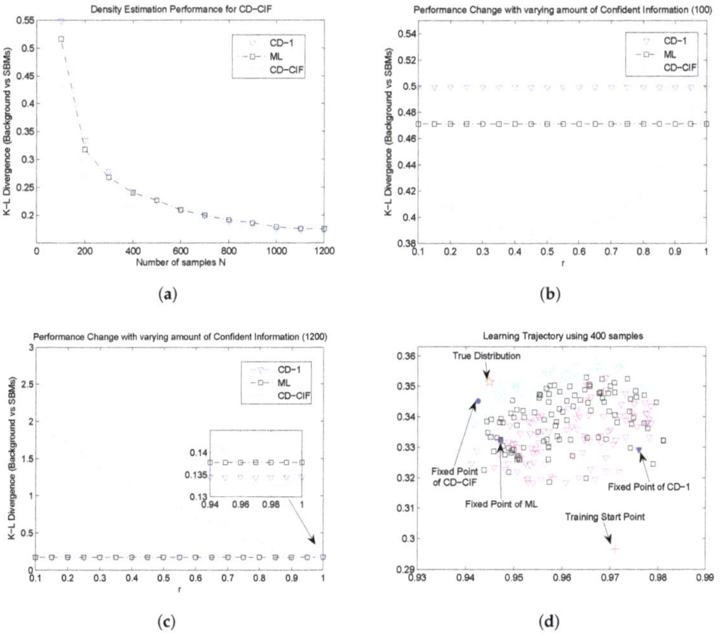

Figure 3. (a): the performance of CD-CIF on different sample sizes; (b) and (c): The performances of CD-CIF with various values of r on two typical sample sizes, *i.e.*, 100 and 1200; (d) illustrates one learning trajectory of the last 100 steps for ML (squares), CD-1 (triangles) and CD-CIF (circles).

Effects on Learning Trajectory: We use the 2D visualizing technology SNE [20] to investigate learning trajectories and dynamical behaviors of three comparative algorithms. We start three methods with the same parameter initialization. Then, each intermediate state is represented by a 55-dimensional vector formed by its current parameter values. From Figure 3d, we can see that: (1) In the final 100 steps, the three methods seem to end up staying in different regions of the parameter space, and CD-CIF confines the parameter in a relatively thinner region compared to ML and CD-1; (2) The true distribution is usually located on the side of CD-CIF, indicating its potential for converging to the optimal solution. Note that the above claims are based on general observations, and Figure 3d is shown as an illustration. Hence, we may conclude that CD-CIF regularizes the learning trajectories in a desired region of the parameter space using the sample-specific CIF.

6. Conclusions

Different from the traditional EPI, the CIF principle proposed in this paper aims at finding a way to derive computational models for universal cognitive systems by a dimensionality reduction approach in parameter spaces: specifically, by preserving the confident parameters and reducing the less confident parameters. In principle, the confidence of parameters should be assessed according to their contributions to the expected information distance between the ideal distribution and its fluctuated observations. This is called the distance-based CIF. For some coordinated systems, e.g., the mixed-coordinate system $[\zeta]_l$, the confidence of a parameter can also be directly evaluated by its Fisher information, which establishes a connection with the inverse variance of any unbiased estimate for the parameter via the Cramér–Rao bound. This is called the information-based CIF. The criterion of information-based CIF (*i.e.*, Fisher information) can be seen as an approximation to distance-based CIF, since it neglects the influence of parameter scaling to the expected information distance. However, considering the standard mixed-coordinates $[\zeta]_l$ for the manifold of multivariate binary distributions, it turns out that both distance-based CIF and information-based CIF entail the same optimal submanifold M.

The CIF provides a strategy for the derivation of probabilistic models. The SBM is a specific example in this regard. It has been theoretically shown that the SBM can achieve a reliable representation in parameter spaces by using the CIF principle.

The CIF principle can also be used to modify existing SBM training algorithms by incorporating data information, such as CD-CIF. One interesting result shown in our experiments is that: although CD-CIF is a biased algorithm, it could significantly outperform ML when the sample is insufficient. This suggests that CIF gives us a reasonable criterion for utilizing confident information from the underlying data, while ML lacks a mechanism to do so.

In the future, we will further develop the formal justification of CIF w.r.t various contexts (e.g., distribution families or models).

Acknowledgments: We would like to thank the anonymous reviewers for their valuable comments. We also thank Mengjiao Xie and Shuai Mi for their helpful discussions. This work is partially supported by the Chinese National Program on Key Basic Research Project (973 Program, Grant No. 2013CB329304 and 2014CB744604), the Natural Science Foundation of China (Grant Nos. 61272265, 61070044, 61272291, 61111130190 and 61105072).

Appendix

Appendix A Proof of Proposition 1

Proof. By definition, we have:

$$g_{IJ} = \frac{\partial^2 \psi(\theta)}{\partial \theta^I \partial \theta^J}$$

where $\psi(\theta)$ is defined by Equation (4). Hence, we have:

$$g_{IJ} = \frac{\partial^2 (\sum_I \theta^I \eta_I - \phi(\eta))}{\partial \theta^I \partial \theta^J} = \frac{\partial \eta_I}{\partial \theta^J}$$

By differentiating η_I, defined by Equation (1), with respect to θ^J, we have:

$$\begin{aligned} g_{IJ} &= \frac{\partial \eta_I}{\partial \theta^J} = \frac{\partial \sum_x X_I(x)(exp\{\sum_I \theta^I X_I(x) - \psi(\theta)\})}{\partial \theta^J} \\ &= \sum_x X_I(x)[X_J(x) - \eta_J]p(x;\theta) = \eta_{I \cup J} - \eta_I \eta_J \end{aligned}$$

This completes the proof. □

Appendix B Proof of Proposition 2

Proof. By definition, we have:

$$g^{IJ} = \frac{\partial^2 \phi(\eta)}{\partial \eta_I \partial \eta_J}$$

where $\phi(\eta)$ is defined by Equation (4). Hence, we have:

$$g^{IJ} = \frac{\partial^2(\sum_J \theta^J \eta_J - \psi(\theta))}{\partial \eta_I \partial \eta_J} = \frac{\partial \theta^I}{\partial \eta_J}$$

Based on Equations (2) and (1), the θ^I and p_K could be calculated by solving a linear equation of $[p]$ and $[\eta]$, respectively. Hence, we have:

$$\theta^I = \sum_{K \subseteq I} (-1)^{|I-K|} log(p_K); \quad p_K = \sum_{K \subseteq J} (-1)^{|J-K|} \eta_J$$

Therefore, the partial derivation of θ^I with respect to η_J is:

$$g^{IJ} = \frac{\partial \theta^I}{\partial \eta_J} = \sum_K \frac{\partial \theta^I}{\partial p_K} \cdot \frac{\partial p_K}{\partial \eta_J} = \sum_{K \subseteq I \cap J} (-1)^{|I-K|+|J-K|} \cdot \frac{1}{p_K}$$

This completes the proof. □

Appendix C Proof of Proposition 3

Proof. The Fisher information matrix of $[\zeta]$ could be partitioned into four parts: $G_\zeta = \begin{pmatrix} A & C \\ D & B \end{pmatrix}$. It can be verified that in the mixed coordinate, the θ-coordinate of order k is orthogonal to any η-coordinate less than k-order, implying the corresponding element of the Fisher information matrix is zero ($C = D = 0$) [23]. Hence, G_ζ is a block diagonal matrix.

According to the Cramér–Rao bound [3], a parameter (or a pair of parameters) has a unique asymptotically tight lower bound of the variance (or covariance) of the unbiased estimate, which is given by the corresponding element of the inverse of the Fisher information matrix involving this parameter (or this pair of parameters). Recall that I_η is the index set of the parameters shared by $[\eta]$ and $[\zeta]_l$ and that J_θ is the index set of the parameters shared by $[\theta]$ and $[\zeta]_l$; we have $(G_\zeta^{-1})_{I_\zeta} = (G_\eta^{-1})_{I_\eta}$ and $(G_\zeta^{-1})_{J_\zeta} = (G_\theta^{-1})_{J_\theta}$, *i.e.*, $G_\zeta^{-1} = \begin{pmatrix} (G_\eta^{-1})_{I_\eta} & 0 \\ 0 & (G_\theta^{-1})_{J_\theta} \end{pmatrix}$. Since G_ζ is a block tridiagonal matrix, the proposition follows. □

Appendix D Proof of Proposition 4

Proof. Assume the Fisher information matrix of $[\theta]$ to be: $G_\theta = \begin{pmatrix} U & X \\ X^T & V \end{pmatrix}$, which is partitioned based on I_η and J_θ. Based on Proposition 3, we have $A = U^{-1}$. Obviously, the diagonal elements of U are all smaller than one. According to the succeeding Lemma A1, we can see that the diagonal elements of A (*i.e.*, U^{-1}) are greater than one.

Next, we need to show that the diagonal elements of B are smaller than 1. Using the Schur complement of G_θ, the bottom-right block of G_θ^{-1}, *i.e.*, $(G_\theta^{-1})_{J_\theta}$, equals to $(V - X^T U^{-1} X)^{-1}$. Thus, the diagonal elements of B: $B_{jj} = (V - X^T U^{-1} X)_{jj} < V_{jj} < 1$. Hence, we complete the proof. □

Lemma A1. *With a $l \times l$ positive definite matrix H, if $H_{ii} < 1$, then $(H^{-1})_{ii} > 1, \forall i \in \{1, 2, \dots, l\}$.*

Proof. Since H is positive definite, it is a Gramian matrix of l linearly independent vectors $v_1, v_2, \dots, v_l$, *i.e.*, $H_{ij} = \langle v_i, v_j \rangle$ ($\langle \cdot, \cdot \rangle$ denotes the inner product). Similarly, H^{-1} is the Gramian matrix of l linearly independent vectors $w_1, w_2, \dots, w_l$ and $(H^{-1})_{ij} = \langle w_i, w_j \rangle$. It is easy to verify that $\langle w_i, v_i \rangle = 1, \forall i \in \{1, 2, \dots, l\}$. If $H_{ii} < 1$, we can see that the norm $\|v_i\| = \sqrt{H_{ii}} < 1$. Since $\|w_i\| \times \|v_i\| \geq \langle w_i, v_i \rangle = 1$, we have $\|w_i\| > 1$. Hence, $(H^{-1})_{ii} = \langle w_i, w_i \rangle = \|w_i\|^2 > 1$. □

Appendix E Proof of Proposition 5

Proof. Let B_q be a ε-ball surface centered at $q(x)$ on manifold S, *i.e.*, $B_q = \{q' \in S|||KL(q,q') = \varepsilon\}$, where $KL(\cdot,\cdot)$ denotes the Kullback–Leibler divergence and ε is small. ζ_q is the coordinates of $q(x)$. Let $q(x)+dq$ be a neighbor of $q(x)$ uniformly sampled on B_q and $\zeta_{q(x)+dq}$ be its corresponding coordinates. For a small ε, we can calculate the expected information distance between $q(x)$ and $q(x)+dq$ as follows:

$$E_{B_q} = \int [(\zeta_{q(x)+dq} - \zeta_q)^T G_\zeta (\zeta_{q(x)+dq} - \zeta_q)]^{\frac{1}{2}} dB_q \tag{A1}$$

where G_ζ is the Fisher information matrix at $q(x)$.

Since Fisher information matrix G_ζ is both positive definite and symmetric, there exists a singular value decomposition $G_\zeta = U^T \Lambda U$ where U is an orthogonal matrix and Λ is a diagonal matrix with diagonal entries equal to the eigenvalues of G_ζ (all ≥ 0).

Applying the singular value decomposition into Equation (A1), the distance becomes:

$$E_{B_q} = \int [(\zeta_{q(x)+dq} - \zeta_q)^T U^T \Lambda U (\zeta_{q(x)+dq} - \zeta_q)]^{\frac{1}{2}} dB_q \tag{A2}$$

Note that U is an orthogonal matrix, and the transformation $U(\zeta_{q(x)+dq} - \zeta_q)$ is a norm-preserving rotation.

Now, we need to show that among all tailored k-dimensional submanifolds of S, $[\zeta]_{l_t}$ is the one that preserves maximum information distance. Assume $I_T = \{i_1, i_2, \ldots, i_k\}$ is the index of k coordinates that we choose to form the tailored submanifold T in the mixed-coordinates $[\zeta]$. According to the fundamental analytical properties of the surface of the hyper-ellipsoid and the orthogonality of the mixed-coordinates, there exists a strict positive monotonicity between the expected information distance E_{B_q} for T and the sum of eigenvalues of the sub-matrix $(G_\zeta)_{I_T}$, where the sum equals to the trace of $(G_\zeta)_{I_T}$. That is, the greater the trace of $(G_\zeta)_{I_T}$, the greater the expected information distance E_{B_q} for T.

Next, we show that the sub-matrix of G_ζ specified by $[\zeta]_{l_t}$ gives a maximum trace. Based on Proposition 4, the elements on the main diagonal of the sub-matrix A are lower bounded by one and those of B upper bounded by one. Therefore, $[\zeta]_{l_t}$ gives the maximum trace among all sub-matrices of G_ζ. This completes the proof. □

Author Contributions: Author Contributions

Theoretical study and proof: Yuexian Hou and Xiaozhao Zhao. Conceived and designed the experiments: Xiaozhao Zhao, Yuexian Hou, Dawei Song and Wenjie Li. Performed the experiments: Xiaozhao Zhao. Analyzed the data: Xiaozhao Zhao, Yuexian Hou. Wrote the manuscript: Xiaozhao Zhao, Dawei Song, Wenjie Li and Yuexian Hou. All authors have read and approved the final manuscript.

Conflicts of Interest: Conflicts of Interest

The authors declare no conflict of interest.

References

1. Wheeler, J.A. *Time Today*; Cambridge University Press: Cambridge, UK, 1994; pp. 1–29.
2. Frieden, B.R. *Science from Fisher Information: A Unification*; Cambridge University Press: Cambridge, UK, 2004.
3. Rao, C.R. Information and the accuracy attainable in the estimation of statistical parameters. *Bull. Calcutta Math. Soc.* **1945**, *37*, 81–91.
4. Frieden, B.R.; Gatenby, R.A. Principle of maximum Fisher information from Hardy's axioms applied to statistical systems. *Phys. Rev. E* **2013**, *88*, 042144.
5. Burnham, K.P.; Anderson, D.R. *Model Selection and Multimodel Inference: A Practical Information—Theoretic Approach*; Springer: Berlin/Heidelberg, Germany, 2002.
6. Vstovsky, G.V. Interpretation of the extreme physical information principle in terms of shift information. *Phys. Rev. E* **1995**, *51*, 975–979.

7. Amari, S.; Nagaoka, H. *Methods of Information Geometry*; Translations of Mathematical Monographs; Oxford University Press: Oxford, UK, 1993.
8. Amari, S.; Kurata, K.; Nagaoka, H. Information geometry of Boltzmann machines. *IEEE Trans. Neural Netw.* **1992**, *3*, 260–271.
9. Hou, Y.; Zhao, X.; Song, D.; Li, W. Mining pure high-order word associations via information geometry for information retrieval. *ACM Trans. Inf. Syst.* **2013**, *31*, 12:1–12:32.
10. Kass, R.E. The geometry of asymptotic inference. *Stat. Sci.* **1989**, *4*, 188–219.
11. Ackley, D.H.; Hinton, G.E.; Sejnowski, T.J. A learning algorithm for Boltzmann machines. *Cogn. Sci.* **1985**, *9*, 147–169.
12. Hinton, G.E.; Salakhutdinov, R.R. Reducing the Dimensionality of Data with Neural Networks. *Science* **2006**, *313*, 504–507.
13. Rasch, G. *Probabilistic Models for Some Intelligence and Attainment Tests*; Danish Institute for Educational Research: Copenhagen, Denmark, 1960.
14. Bond, T.; Fox, C. *Applying the Rasch Model: Fundamental Measurement in the Human Sciences*; Psychology Press: London, UK, 2013.
15. Gibilisco, P. *Algebraic and Geometric Methods in Statistics*; Cambridge University Press: Cambridge, UK, 2010.
16. Čencov, N.N. *Statistical Decision Rules and Optimal Inference*; American Mathematical Society: Washington, D.C., USA, 1982.
17. Kullback, S.; Leibler, R.A. On Information and Sufficiency. *Ann. Math. Stat.* **1951**, *22*, 79–86.
18. Buhlmann, P.; van de Geer, S. *Statistics for High-Dimensional Data: Methods, Theory And Applications*; Springer: Berlin/Heidelberg, Germany, 2011.
19. Bobrovsky, B.; Mayer-Wolf, E.; Zakai, M. Some classes of global Cramér-Rao bounds. *Ann. Stat.* **1987**, *15*, 1421–1438.
20. Hinton, G.E. Training products of experts by minimizing contrastive divergence. *Neural Comput.* **2002**, *14*, 1771–1800.
21. Carreira-Perpinan, M.A.; Hinton, G.E. On contrastive divergence learning. In Proceedings of the International Workshop on Artificial Intelligence and Statistics, Key West, FL, USA, 6–8 January 2005; pp. 33–40.
22. Gilks, W.R.; Richardson, S.; Spiegelhalter, D. Introducing markov chain monte carlo. In *Markov Chain Monte Carlo in Practice*; Chapman and Hall/CRC: London, UK, 1996; pp. 1–19.
23. Nakahara, H.; Amari, S. Information geometric measure for neural spikes. *Neural Comput.* **2002**, *14*, 2269–2316.

MDPI
St. Alban-Anlage 66
4052 Basel
Switzerland
Tel. +41 61 683 77 34
Fax +41 61 302 89 18
www.mdpi.com

Entropy Editorial Office
E-mail: entropy@mdpi.com
www.mdpi.com/journal/entropy

www.ingramcontent.com/pod-product-compliance
Lightning Source LLC
Chambersburg PA
CBHW051711210326
41597CB00032B/5437
* 9 7 8 3 0 3 8 9 7 6 3 2 5 *